Nanosciences and Nanotechnology

Jean-Michel Lourtioz · Marcel Lahmani
Claire Dupas-Haeberlin · Patrice Hesto
Editors

Nanosciences and Nanotechnology

Evolution or Revolution?

EUROPEAN MATERIALS RESEARCH SOCIETY

Springer

Editors
Jean-Michel Lourtioz
Institut d'Electronique Fondamentale
Université Paris-Sud
Orsay
France

Marcel Lahmani
Club NanoMicroTechnologie
Villebon-sur-Yvette
France

Claire Dupas-Haeberlin
École Normale Supérieure de Cachan
Paris
France

Patrice Hesto
Institut d'Electronique Fondamentale
Université Paris-Sud
Orsay
France

Translation by Stephen N. Lyle

ISBN 978-3-319-19359-5 ISBN 978-3-319-19360-1 (eBook)
DOI 10.1007/978-3-319-19360-1

Library of Congress Control Number: 2015945598

Springer Cham Heidelberg New York Dordrecht London

Translation from the French language edition: *Nanosciences et Nanotechnologies: Évolution ou révolution?* by Stephen N. Lyle, © Belin, France 2014. All rights reserved

Printed on acid-free paper

Springer International Publishing AG Switzerland is part of Springer Science+Business Media (www.springer.com)

Foreword to the French Edition

Nanotechnology can produce objects with hitherto unimagined architectures and it is unique, not only for its spinoffs in every area of science and technology, but also for its economic and social consequences.

It is interesting to observe that specialists have some difficulty agreeing on a precise definition of nanoscience and nanotechnology. Physicists, chemists, and biologists often refer to different size ranges, but they all agree that the nanoworld is characterised by a radical change in the physical, chemical, or biological properties of the objects in question. Indeed, nanotechnological development is not just the business of miniaturising matter. It can be used to produce novel and often unexpected features, such as the appearance of remarkable optical properties, the creation of molecular motors, and the possibility of navigating within human cells. It should thus come as no surprise that nanoscience and nanotechnology find many often cross-disciplinary applications in areas as varied as electronics and information systems, energy, optics, materials, and nanomedicine. So nanoscience is neither physics, nor chemistry, nor biology, but all these disciplines at the same time, as they converge toward the nano scale. Furthermore, in the nanoworld, simple ethical and philosophical questions can sometimes be difficult to answer. For example, a protein can be synthesised by a nanomachine consisting of an mRNA associated with a ribosome. On this scale, are we already dealing with the living or can we treat this using a mechanistic description of elementary molecules?

It would be hard to say at the present time whether the 'nano' field will simply continue as it is today, whether it will gradually fade from view, or whether it will tend toward different concepts and applications. Breakthrough technologies are unpredictable by their very definition. But I am quite convinced that international competition will be determined by a country's ability to promote cross-disciplinary interactions in the truest sense of the term, of the kind that occur in the cafeteria when research staff and students of different fields are housed on the same site and can exchange ideas about new concepts and methods. It is sad to say that such points of exchange and interdisciplinarity have not yet been put to the fore in France.

The main goal of this book is thus to present an integrated and cross-disciplinary description of nanoscience and nanotechnology. It aims at a very broad readership from undergraduates to researchers who would like to obtain a clear overview of the ideas and applications of nanoscience and nanotechnology in an area where they may not have specialist knowledge. But since all studies of future prospects suggest that nanotechnology will lead to economic development and the creation of wealth, the book also takes into account the executives and decision-makers across a range of industrial sectors. In fact, the different chapters can be read independently, while the deliberately selective bibliographies focus on the main advances in the given field and can be used as a starting point for further investigation.

The very general introduction to the book deals with the nanometric scale, describing the nanoworld and its applications in the fields of nanomaterials, nanoelectronics, and health. The basic principles involved in making nano-objects are also discussed, viz., the bottom–up and top–down approaches. The remainder of the book is divided into four parts.

The first of these takes a detailed look at nanophysics, nanoelectronics, and nanophotonics. It shows that quantum effects and surface phenomena play a crucial role in the nanoworld and lead to remarkable properties. Boxed explanations provide a comprehensible overview of the physical principles underlying the radically different properties observed in the nanoworld, describing among other things the tunnel effect, which makes short work of potential barriers on the nanoscale, and the notions of spin and magnetic domains. The author of Chap. 2 also discusses the way nanophysics may develop tomorrow in his opinion, with ever greater involvement of quantum effects. It remains an open question whether it will evolve into picophysics, i.e., physics on the picometric scale, which corresponds to the world of individual atoms, manipulating them tomorrow as we build up structures today from individual molecules.

The second part focuses on the ways nanochemistry can contribute to the discovery and design of new (nano)materials. As the author of Chap. 5 so clearly explains, chemistry is already a nanoscience, since it manipulates and synthesises nanometric entities, from atoms to molecules, polymers, and colloids with varying degrees of complexity. Novel carbon-based nanomaterials such as fullerenes, graphene, carbon nanotubes, and nanodiamonds are also discussed. A large part of the chapter is devoted to the role of molecular and supramolecular chemistry in the field of nanomaterials, as exemplified by the so-called click chemistry', which aims to functionalise nano-objects or self-assemblies resulting from interactions between molecular entities.

The third part devoted to nanobiotechnology concentrates mainly on nanomedicine. This discipline has become a field of investigation in its own right, as attested by the attribution of a chair at the *Collège de France* in 2009–2010. As explained in the ensuing chapters, the application of nanotechnology to medicine includes therapy through targeted drug delivery, in vitro and in vivo medical diagnostics, and combinations of the two (theranostics), but also regenerative medicine for the reconstruction of organs and tissues. All these areas are discussed, as are the consequent ethical and regulatory questions. The preclinical development of a nanomedicine is also illustrated by a concrete example, namely the use of squalenisation to design anti-cancer and anti-infective nanodrugs. Apart from nanomedicine, two further chapters deal with matters of toxicology. One treats the impact of nanotechnology on human health, and the other environmental risks.

The last part, entitled *Nanotechnology and Society*, provides a detailed overview of industrial and research institutions currently engaged in this field, both in France and in the rest of the world. It also describes the various courses available in France and across Europe, and shows how the cross-disciplinary nature of nanoscience and nanotechnology makes it possible to teach science at every level of education without necessarily having to make sacrifices to the fundamentals of the traditional disciplines. The field of nanotechnology is undoubtedly a subject of tough industrial competition in the worldwide pursuit of patents and funding. As demonstrated in the last chapter of this part, such intense competition requires strict international regulation, taking into account the potential risks inherent in nanotechnology.

The present work illustrates the wealth and diversity of knowledge generated by nanoscience and nanotechnology, and it will help readers to form their own opinion in answer to the question raised by the underlying theme of this book: will nanoscience and nanotechnology turn out to be an evolution or a revolution?

July 2013

Patrick Couvreur
Université Paris-Sud and Institut
Universitaire de France
Chatenay-Malabry

Preface to the French Edition

The series of books on the nanosciences originally instigated by Marcel Lahmani and Philippe Houdy already comprises four multi-author books with a largely didactic objective, each tackling one of the main scientific disciplines, and aimed at readers who have already reached graduate level. The first volume, entitled *Nanotechnologies et nanophysique*, was published in 2004 and reedited twice in 2006 and 2009. This was followed in 2006 by a second volume, entitled *Nanomatériaux et nanochimie*, reedited in 2012. Then came *Nanobiotechnologies et nanobiologie* in 2007 and *Nanotoxicologie et nanoéthique* in 2010.[1]

The original French versions of the series were nominated four times for the Roberval Prize and the series as a whole was awarded the *Trophée Roberval*, the highest award attributed by this institution. Naturally, all the credit must go to the researchers and engineers who wrote the contributions, along with Belin for the French publication and Springer for the English version, but not forgetting perhaps the stubborn determination of those who initiated the project.

The present work differs significantly from its predecessors. Indeed, the main motivation here is to give a didactic overview of the key areas of science that relate in some way to nanotechnology, and this for a broad non-specialist public. The book thus stands out by the diversity of themes covered, each illustrated by a broad range of examples, leading to numerous applications of immediate interest to the world of industry. It refers to all the scientific disciplines whose development is in some way affected by nanoscientific endeavour, from physics in the form of electronics and photonics, through chemistry and materials science to biology and medicine, not forgetting the inherent toxicological issues. In the last part of the book, there is a guide to, or rather an inventory of, all the current actors on the French scene, including research institutes, educational establishments, and industrial sites. This is intended to provide students, researchers, and engineers with

---

[1]Concerning the English edition, the first volume, entitled *Nanotechnologies and Nanophysics*, published in 2007, was followed by a second, entitled *Nanomaterials and Nanochemistry*, the same year. Then came *Nanobiotechnology and Nanobiology* in 2010 and *Nanotoxicology and Nanoethics* in 2011.

a clearer picture of all the nano activities going on in our country. The book ends with a review of possible social implications of nanotechnology and current regulation. A glossary of some 200 terms has been compiled to provide the reader with the terminology needed to get a good grasp of the subject matter in each chapter.

The second aim was to update the themes in each of the disciplines covered in the previous volumes and thereby bring out the cross-disciplinary nature of nanoscience and nanotechnology and the high level of interaction they generate between research and industrial applications. Particular stress has been put on recent progress in research at the interfaces between physics and chemistry, physics and biology, chemistry and medicine, and between physics and chemistry and the life sciences.

To illustrate the links between research and industry, we have included with the main chapters short articles describing the prospects for nanotechnology in industrial sectors which will in certain cases come as something of a surprise to the general public.

And last but not least, our third aim has been to answer the following question: will nanoscience and nanotechnology lead to a scientific and industrial revolution, as was announced some fifteen years ago?

The reader will soon realise the enormous amount of research that has been carried out in recent years, and will no doubt understand the need to pursue this exploration of the nanoworld for some time to come, knowing that chance will surely surprise us with further extraordinary discoveries as only she knows how!

Acknowledgements

First and foremost, we thank all the authors of the different chapters for their commitment to this undertaking and the quality of their contributions. The choice of authors was made with a view to updating the topics discussed in the previous volumes of the series, including the latest work by several researchers, teachers, and professionals responsible for presenting recent nanotechnological applications. There is also an exhaustive inventory of the relevant resources available in France, together with the latest regulations currently in force.

The main task of the editorial committee has been to dialogue with the authors and edit the resulting manuscripts with an eye for didactic content and the aim of producing a coherent whole. The entire project was supervised by Jean-Michel Lourtioz.

We would particularly like to thank Jean Dutour for the many illustrations, Jean-François Pône for his invaluable assistance at various stages in the preparation of the book, Stephen N. Lyle for his excellent translation of the French edition, EMRS and Springer for their financial suppport to the English edition and Cécile Foullon, who provided the interface with the publisher Belin, for the huge task of formatting and finalising the submitted texts to make them accessible to as broad a readership as possible.

We also extend our warmest thanks to Patrick Couvreur who kindly accepted to write the foreword to our book.

This project was backed both morally and financially by UniverSud Paris and Labex NanoSaclay, to whom we express our gratitude. From 2007 to 2014, the association UniverSud Paris has brought together three universities and three *Grandes Ecoles*,[2] and in 2015 it has been incorporated into a still larger association, of international importance, the University of Paris-Saclay. The Labex NanoSaclay will house all the nano research teams of this future university.

September 2013

Jean-Michel Lourtioz
Marcel Lahmani
Claire Dupas-Haeberlin
Patrice Hesto

[2] *Université Paris-Sud, Université de Versailles–St-Quentin-en-Yvelines, Université d'Évry–Val-d'Essonne, École Centrale de Paris, École supérieure d'électricité*, and *École normale supérieure de Cachan.*

Contents

1 Introduction 1
Louis Laurent
1.1 There's Plenty of Room at the Bottom 1
1.2 A Short History 2
1.3 The Nanoworld 4
1.3.1 Matter on the Scale of a Few Atoms 4
1.3.2 Electricity 6
1.4 How Can We Make Such Tiny Objects? 7
1.4.1 Self-Assembly 9
1.4.2 Nanofabrication 10
1.4.3 Mimicking Nature 11
1.5 What's It All For? 11
1.5.1 Nanomaterials 12
1.5.2 Nanoelectronics 12
1.5.3 Biology, Nanomedicine, and Health 13
1.6 The Debate 13
1.7 Prospects 17

Part I Nanophysics, Nanoelectronics, and Nanophotonics

2 Overview of the Field 23
Jean-Michel Lourtioz
2.1 Is There a Road Map for Nanophysics? The Example of Nanoelectronics 24
2.2 Higher Performance and More Versatile Nanotechnologies . . . 28
2.2.1 Two Complementary Approaches 28
2.2.2 General Features of Fabrication Processes 29
2.2.3 High-Resolution Lithography 30
2.2.4 Etching 31
2.2.5 Atomic Scale Deposition and Surface Treatments . . . 32

2.2.6 Nanoimprinting: Cheap Nanometers per Square Meter 34
2.2.7 Nano 'Lego' and New Methods for 3D Assembly . . . 36
2.3 Gaining in Length: From Clusters and Quantum Dots to Nanotubes, Nanowires, and Nanoribbons 37
2.3.1 Fabrication of Nano-Objects 38
2.3.2 Physical Properties of Nano-Objects 44
2.3.3 Lycurgus Cup 46
2.4 Viewing and Manipulating Nanostructures: A Diversity of Microscopes 48
2.4.1 Very High Resolution Transmission Electron Microscopy 49
2.4.2 Scanning Tunnelling Microscopy: Imaging and Manipulation 50
2.4.3 Atomic Force Microscopy 52
2.4.4 Scanning Near-Field Optical Microscopy: Observation and Manipulation 53
2.4.5 A Few Remarks 56
2.5 Tomorrow's Nanoelectronics. The Challenge of Zero Consumption 56
2.5.1 MOS Transistors and Current Technology 57
2.5.2 Spin Electronics and Nanomagnetism 61
2.5.3 Quantum Components 67
2.5.4 Components Involving Few Electrons 68
2.5.5 Molecular Electronics 69
2.5.6 Memristors and Neurally-Inspired Architectures 70
2.5.7 Conclusion 71
2.6 Is Nanophotonics an Exact Parallel of Nanoelectronics? 72
2.6.1 Optical Micro- and Nanosources, Lasers, and Controlled Photon Sources 73
2.6.2 Photonic Crystals and Metamaterials 76
2.6.3 Plasmonics for Photovoltaics or Biosensing 78
2.6.4 Telecommunications and Optical Interconnects: Links Between Photonics and Electronics 80
2.6.5 Photonics at Extreme Wavelengths: Ultraviolet and Terahertz 81
2.7 Nanophysics: Genuinely Cross-Disciplinary Physics 83
2.7.1 Nanowires in Electromechanical Microsystems 83
2.7.2 Nanobiotechnology: Convergence Between the Physical Sciences and the Life Sciences 85
2.8 Nanophysics: What Future? 88
References 90

3 Applications I. From Nanometers to Megawatts: Photovoltaic Applications 95
Bruno Ghyselen

4 Applications II: The Secret Revolution of Magnetic Sensors 103
Claude Fermon
4.1 Spin Valves and Magnetic Sensors 104
4.2 A First Application: Angle Sensors 105
4.3 An Essential Application for the Automobile 105
4.4 Detecting Very Weak Magnetic Fields 106
4.5 Measuring Currents 107
4.6 Biomagnetism and Ultralow Field MRI 107
4.7 Prospects 109
Reference 109

Part II Nanomaterials and Nanochemistry

5 Overview of the Field 113
Christine Vautrin-Ul
5.1 Introduction: Nanochemistry 113
5.2 Carbon and Its Nanostates 116
5.2.1 Carbon-Based Nanomaterials: Zero-Dimensional Structures or Nanoparticles 117
5.2.2 Carbon-Based Nanomaterials. One-Dimensional Structures: Carbon Nanotubes 123
5.2.3 Carbon-Based Nanomaterials. Two-Dimensional Structures: Graphene 126
5.2.4 Carbon-Based Nanomaterials. Three-Dimensional Structures 132
5.3 Nanochemistry in the Service of Nanomaterials: From Molecular to Supramolecular Chemistry 137
5.3.1 Click Chemistry: Molecular Methods 138
5.3.2 Self-Assembly and Supramolecular Chemistry 141
5.4 Functionalisation of Materials 148
5.4.1 The Many Roads to Functionalisation by Chemical Methods in Gas Phase or in Solution 149
5.4.2 Example of Chemical Functionalisation: Self-Assembled Monolayers 150
5.4.3 Electrochemical Functionalisation: Electrografting by Reduction of Diazonium Salts 151

5.5 Different Approaches to Chemical Synthesis of Nanomaterials . . . 153
5.5.1 Synthesis of Metal Nanoparticles: Colloidal Solutions of Transition Metals . . . 154
5.5.2 Molecular Imprinting for Nanostructured Materials . . . 162
5.5.3 Multifunctional Hybrid Nanomaterials Produced by Soft Chemistry . . . 163
5.6 Conclusion . . . 171
References . . . 172

6 Applications I: Nanolubricants . . . 175
Fabrice Dassenoy
6.1 Why Nano? . . . 176
6.2 Tribological Properties of Nanoparticles . . . 177
6.3 Mechanism of Lubrication by IF-MeS_2 Nanoparticles . . . 178
6.4 Conclusion . . . 181
References . . . 181

7 Applications II: Nanocrystallites in Cement: What Future After Two Thousand Years? . . . 183
Abdi Feylessoufi
7.1 A Material Full of Surprises . . . 183
7.2 The Roman's Were Here . . . 184
7.3 Late Rediscovery and a Burst of Development . . . 185
7.4 Ultra-High Performance Concretes . . . 186
7.5 Toward Nanometric Sizes . . . 187
7.6 More Rigid Nanocrystallites . . . 188
7.7 Progress in Nanocrystallite Thermodynamics . . . 188
7.8 Better Resistance to Wear and Tear . . . 188
7.9 The Beginning of the Twenty-First Century: A Clear Future . . . 189
7.10 The Scientific Challenge . . . 192
7.11 The Industrial Challenge . . . 194
7.12 The Many Remaining Obstacles . . . 195
References . . . 197

8 Applications III. Nano-TiO_2 . . . 199
François Tardif
8.1 Applications of 250–350 nm TiO_2 Particles: From Paints to Foods . . . 200
8.2 Titanium Dioxide Nanoparticles (<100 nm): From Anti-UV Films to Elimination of Pollutants . . . 200
8.2.1 Anti-UV Films . . . 201
8.2.2 Application to Photocatalysis . . . 202
8.2.3 Application to Photovoltaics . . . 202

8.3 Toxicity of Nano-TiO_2 204
References 205

Part III Nanobioscience, Nanomedicine, and Nanotoxicology

9 Nanobiosciences: New Ideas and Tools for Investigating and Manipulating Living Systems 209
Maxime Dahan
9.1 Nano-Objects as Functional Probes for Nanoscale Exploration of Living Systems 210
9.2 Tracking Single Biomolecules in the Cell 211
9.3 Manipulating Living Systems on the Molecular and Cellular Scales 214
9.4 Nano-Objects for Diagnosis and Therapy: The Prospect of Nanomedicine 216
9.5 Conclusion and Prospects 217
References 218

10 Nanomedicine and Nanotechnology for Medicine 219
Patrick Boisseau
10.1 Nanomedicine or Nanotechnology for Medicine? 221
10.2 The Ideal Length Scale in Medicine 222
10.3 The Nanoscale: A Medical Eldorado? 224
10.4 Medical Diagnostics 226
10.4.1 Medical Laboratories 226
10.4.2 Medical Imaging Revisited 229
10.4.3 Smaller and Smaller Implants and Sensors 231
10.4.4 Personalised Medicine 233
10.5 Therapy 234
10.5.1 Drug Nanodroplets and Nanocarriers 235
10.5.2 Drug Pumps 237
10.5.3 Physical Destruction of Tumours 238
10.6 Combining Diagnosis and Therapy: Theranostics 240
10.7 Regenerative Medicine 242
10.7.1 Stem Cells and Cell Therapy 243
10.7.2 Biomaterials 244
10.8 Is Nanotechnology Essential to Medicine? 246
10.9 Nanomedicine and Ethical Questions 247
10.10 Regulating Nanodrugs 248
10.11 A Newly Emerging Industry 249
10.12 The Situation in France Today 250
10.13 Nanomedicine: Evolution or Revolution? 251
References 252

11 Squalenoylation: A Novel Technology for Anticancer and Antibiotic Drugs with Enhanced Activity 253
Patrick Couvreur
11.1 Origin of the Concept of Squalenoylation 255
11.2 Coupling Squalene with Drug Molecules 256
11.3 Characteristics and Morphology of the Main Squalene Derivatives 257
11.4 Applications to Cancer Treatment 257
11.4.1 Nanoparticles of Gemcitabine Coupled with Squalene 257
11.4.2 Nanoparticles of Doxorubicin Coupled with Squalene 263
11.4.3 Nanoparticles of SiRNA Coupled with Squalene (SQsiRNA) 264
11.5 Application to Treatment of Infectious Diseases 264
11.5.1 Antiretroviral Nucleoside Analogues 264
11.5.2 Intracellular Antibiotherapy 265
11.6 Application to the Treatment of Neurological Disorders 266
11.7 Magnet Guidance, Imaging, and Theranostics 267
11.8 From Squalenoylation to Terpenoylation 269
11.9 Conclusion 271
References 271

12 Health Impacts of Nanomaterials 273
Francelyne Marano and Rina Guadagnini
12.1 Comparing the Health Risks of Ultrafine Atmospheric Particles and Nanoparticles 276
12.2 Nanoparticle Exposure Routes and Potential Health Effects 277
12.2.1 Exposure by the Respiratory Route 277
12.2.2 Exposure by the Cutaneous Route 282
12.2.3 Exposure by the Oral Route 283
12.3 Effects of Nanoparticles on the Cardiovascular System 284
12.4 Effects of Nanoparticles on the Nervous System 284
12.5 Conclusion 285
References 286

13 Environmental Risks of Nanotechnology: A New Challenge? 287
Jean-Yves Bottero
13.1 From Naturally Occurring Nanoparticles to Manufactured Nanoparticles 288
13.1.1 Naturally Occurring Nanoparticles 288
13.1.2 Nanoparticle By-Products of Human and Industrial Activities 289
13.1.3 Manufactured Nanoparticles 289

13.2 The Importance of Contact with Water 291
13.2.1 Do Nanoparticles Solubilise According to the Same Laws as Microparticles? 291
13.2.2 Transition from Hydrophobic to Hydrophilic Properties 292
13.2.3 A Possible Trojan Horse Effect 292
13.2.4 Aggregation Properties Associated with Reactivity at the Water or Solvent Interface 293
13.3 Nanoparticle Transport in Porous Media 297
13.4 Main Nanoparticle Toxicity Mechanisms: Experiments on Living Organisms 300
13.5 Assessing the Risks of Nanomaterials: Toward Predictive Models? 303
13.5.1 Difficulties Inherent in Risk Assessment 303
13.5.2 Experimental Approach to Risk Assessment Integrating Collateral Damage 305
13.5.3 Risk Forecasting Models for Nanomaterials 306
13.6 Concluding Remarks 309
References 309

Part IV Nanotechnology and Society

14 Research in Nanoscience and Nanotechnology: The French Research System 315
Patrice Hesto and Jean-Michel Lourtioz
14.1 Upstream Research in France 315
14.2 Research and Development in France 318
14.2.1 Organising the Nano Community 318
14.2.2 French Programs and Financing 319
14.3 International Programs 320
14.4 Conclusion 324

15 Training in Nanoscience and Nanotechnology 325
Sylvie Retailleau, Claire Dupas-Haeberlin, Jean-Michel Lourtioz and Arnaud Bournel
15.1 Opportunities for Cross-Disciplinary Training 325
15.2 BMD and the Role of Nano Teaching in France and Worldwide 328
15.3 Professional Opportunites After Nano Training 334
15.4 Outreach: Nanoscience at School 334
15.5 Outreach: From Pupils to the General Public 336
15.6 Prospects 337
References 338

16 Nanotechnology and Industry . . . 339
Patrice Hesto, Jean-Michel Lourtioz, Claire Dupas-Haeberlin, Marcel Lahmani and Thomas Dubouchet
16.1 The French Industrial Landscape . . . 340
16.1.1 Multisector Applications . . . 340
16.1.2 Some Key Industrial Sectors . . . 343
16.1.3 Industrial Actors in France . . . 345
16.1.4 Zoom on Start-Up Companies . . . 346
16.1.5 But What About the Risks? . . . 347
16.2 The Industrial Situation Worldwide . . . 348
16.2.1 A Growth Market . . . 348
16.2.2 State Involvement . . . 349
16.2.3 French Provisions for the Development of Nanotechnology . . . 352
16.3 By Way of Conclusion . . . 354
References . . . 356

17 Societal Approach to Nanoscience and Nanotechnology: When Technology Reflects and Shapes Society . . . 357
Françoise Roure
17.1 Making the Nanoworld Accessible, Informing, and Regulating: An Ethical, Legal, and Societal Imperative . . . 359
17.1.1 Informing . . . 360
17.1.2 Regulating . . . 364
17.2 The Scope of Social Disruption Induced by Nanotechnology . . . 367
17.2.1 Manipulating the Genome . . . 367
17.2.2 Toward the New Green Deal? . . . 371
17.2.3 Developing Responsible Innovation Upstream of Supply . . . 373
17.2.4 Potential Risks and Gains of Nanotechnology Through Biogenetics . . . 373
17.3 The Systemic Nature of Risks Induced by Nanotechnology . . . 378
17.3.1 The Effects . . . 378
17.3.2 Consequences of the Systemic Nature of Risk on the Assessment of Changes Induced by Nanotechnology . . . 380
17.4 Legal Problems Raised by Nanotechnology . . . 383
17.5 Ethical Questions . . . 384
17.5.1 Need for an International Normative Framework Applicable to Nanotechnology . . . 385
17.5.2 Labelling: Helping the Consumer to Choose in a Complex and Uncertain Situation . . . 386

17.6 The Key Role of Databases and a Better Understanding of Technology 387
17.7 Religion, French Transhumanism, and the Question of Ends 392
17.7.1 Hybrid Immersive Systems, the C Factor of NBIC Convergence, and Freedom of Thought . . . 393
17.7.2 Societal Questions Arising from Potential Uses of Nanoscale Technologies 396
17.8 The Fragile Beginnings of World Governance in the Field of Nanotechnology 397
17.8.1 The Long and Yet Unlikely Path Toward Diplomatic Action 397
17.8.2 When Different Interests Can Coexist on Harmonised Prerequisites 398
17.8.3 What Has Been Achieved by Intergovernmental Relations? 398
17.8.4 A More Mature and More Assertive European Impetus 400
References 403

Glossary 405

Index 417

Contributors

Patrick Boisseau Commissariat à l'Energie Atomique et aux Energies Alternatives (CEA), Campus MINATEC, Grenoble, France

Jean-Yves Bottero Centre Européen de recherche et d'enseignements des géosciences de l'environnement (CEREGE) UMR 7330, CNRS/Université d'Aix-Marseille, Aix-en-Provence, France

Arnaud Bournel Institut d'Electronique Fondamentale, Université Paris-Sud, Orsay, France

Patrick Couvreur Institut Galien, UMR CNRS 8612, Université Paris-Sud, Châtenay-Malabry, France

Maxime Dahan Laboratoire de Physico-chimie, Institut Curie, CNRS UMR 168, Paris, France

Fabrice Dassenoy Laboratoire de Tribologie et Dynamique des Systèmes, École Centrale de Lyon, Écully, France

Thomas Dubouchet Nanothinking, Paris, France

Claire Dupas-Haeberlin ENS Cachan, Cachan, France

Claude Fermon CEA Saclay, Gif-sur-Yvette, France

Abdi Feylessoufi Centre de Recherche sur la Matière Cohésive, Université d'Evry, Evry, France

Bruno Ghyselen SOITEC, Parc Technologique des Fontaines, Bernin, France

Rina Guadagnini Laboratoire des Réponses Moléculaires et Cellulaires aux Xénobiotiques (RMCX), Unité de Biologie Fonctionnelle et Adaptative (BFA), CNRS EAC 4413 case 7073 Bâtiment Buffon, Université Paris Diderot-Paris 7, Paris, France

Patrice Hesto Paris, France

Marcel Lahmani Club NanoMicroTechnologie, Villebon-sur-Yvette, France

Louis Laurent ANSES, Maisons-Alfort, France

Jean-Michel Lourtioz Institut d'Electronique Fondamentale Bât 220, Université Paris-Sud, Orsay, France

Francelyne Marano Laboratoire des Réponses Moléculaires et Cellulaires aux Xénobiotiques (RMCX), Unité de Biologie Fonctionnelle et Adaptative (BFA), CNRS EAC 4413 case 7073 Bâtiment Buffon, Université Paris Diderot-Paris 7, Paris, France

Sylvie Retailleau Institut d'Electronique Fondamentale, Université Paris-Sud, Orsay, France

Françoise Roure Ministère de l'Economie, de l'Industrie et du Numérique, Paris, France

François Tardif CEA Liten, Grenoble, France

Christine Vautrin-Ul Centre de Recherche Sur la Matière divisée, Université d'Orléans, Orléans, France

Chapter 1
Introduction

Louis Laurent

Abstract During the twentieth century, progress in physics, chemistry, and biology brought a detailed understanding of the structure and properties of both living and inert matter on the nanometric scale, i.e., lengths of the order of one billionth of a metre. By the end of the 1950s, instruments were being developed to observe, manipulate, and assemble matter and devices on this same length scale. It was thus clear that the convergence of all this knowledge would open the way to spectacular applications, and yet it was not until the 1990s that nanotechnology truly came into its own. The most visible application was without doubt nanoelectronics, today present in an increasing number of products which are in the process of changing our lives. But nanotechnology is being put to work in many other sectors such as materials, sensors, energy, and medical applications. Thousands of products contain nanosized ingredients and, given the scope of these developments, concern has been expressed, particularly about the possible toxicity of nanoparticles and inadequate control of industrial applications. Since its inception, nanotechnology has been closely associated with the notion of economic growth, but its maturity will also depend on an understanding of the associated risks and its contribution to the crucial future questions of sustainable development.

1.1 There's Plenty of Room at the Bottom

Nanoscience and nanotechnology get their definition primarily from a scale of length, viz., the nanometer, which is a billionth of a meter. To see what this means, consider a hair. This will grow about a centimeter every month. Carrying out a simple calculation, we discover that this hair will grow at a rate of about four nm per second. But why does the nanometer play such an important role? To answer this, we must turn to the very constituents of matter.

The matter around us is made up either of atoms, which can be represented as tiny spheres with diameters of the order of a few tenths of a nanometer, or of molecules,

L. Laurent (✉)
ANSES, 14 Rue Pierre Et Marie Curie, 94701 Maisons-Alfort, France
e-mail: louis.laurent@anses.fr

J.-M. Lourtioz et al. (eds.), *Nanosciences and Nanotechnology*,
DOI 10.1007/978-3-319-19360-1_1

which are strongly bound assemblies of atoms measuring a few nanometers across. In other words, the smallest amount of a substance we can speak about does indeed concern this length scale.[1] Atoms are so small that, in our everyday lives, we are unlikely ever to feel as though we are made up of such things. Even what we consider to be microscopic comprises a huge number of atoms. For example, a bacterium in the form of a rod 3000 nm long contains a hundred billion atoms, enough to build up the complex machinery that allows this bacterium to live. Simpler than the bacteria, viruses measure a hundred or so nanometers and still contain tens of millions of atoms. Likewise, sometimes highly complex microscopic systems can be made by industry, such as microprocessors, with features now measuring only a few tens of nanometers. This was the 'nanoworld' that Richard Feynman, Nobel Prize for Physics, was referring to in the title of a talk he gave on nanoscience back in 1959: *There's plenty of room at the bottom.*

1.2 A Short History

The idea that the world is made of atoms is not a recent one. It is attributed to the Greek philosopher Leucippus and his disciple Democritus (circa 400 BC). However, it was not until the nineteenth century that progress in the sciences turned this into a serious hypothesis and then a reality, with implicit reference to the nanometric length scale. Between the end of the nineteenth century and the middle of the twentieth, techniques were developed for observing matter on this scale. For example, in 1931, two German engineers Ernst Ruska and Max Knoll invented the electron microscope. The underlying idea can be understood by comparison with the optical microscope, except that light is replaced by electrons which probe matter in much finer detail. After ten years or so, the electron microscope could already achieve a resolution of 10 nm, and further progress was made right through the twentieth century. A resolution of 0.1 nm, the size of a single atom, was reached in the 1990s. Figure 1.1 shows an example of such an image.

It was also in the 1980s that techniques were improved for laboratory manipulation of matter on the nanometric scale. A symbolic step was the invention of the scanning tunnelling microscope (STM) by two researchers at the IBM research center in Zurich, Heinrich Rohrer and Gerd Binnig. If the electron microscope is likened to a kind of ultra-powerful eye, the scanning tunnelling microscope could be described as a sort of finger that investigates matter by prodding it. A fine tip is displaced across the surface of the matter, so delicately that it can 'feel' the very atoms themselves on the surface. In the original version, invented in 1981, this sensing was carried out using electricity. As we shall see later on, each atom on the surface 'attracts' electricity. Then, from 1986, another version of the microscope appeared, measuring

[1] However, it should not be thought that there is nothing smaller than an atom. The atom is itself a complex object, made up of elementary particles much smaller than the nanometer, but this matter is no longer of the kind we see around us.

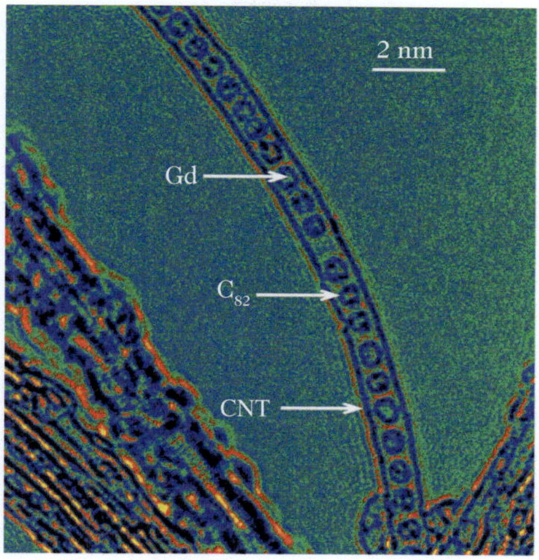

Fig. 1.1 Bright-field electron microscope image of Russian nanodolls showing tube-shaped structures made of carbon atoms, known as carbon nanotubes (CNT). Inside these nanotubes are small spheres called fullerenes, each made from 82 carbon atoms (C_{82}). Each fullerene itself encloses a gadolinium (Gd) atom. Courtesy of A. Gloter, Solid State Physics, Orsay, CNRS France–AIST Japan

the attraction exerted by each atom on the tip. This was the atomic force microscope (AFM). These much cheaper devices soon became commonplace in research establishments. Even more spectacularly, the tip of an atomic force microscope can actually displace a single atom. A symbolic event here was the feat achieved by Don Eigler who, in 1989, deposited 35 xenon atoms on a nickel surface to write the IBM logo. (Don Eigler worked for IBM.) This symbolic image is easily found on the Internet even today.

In parallel with these technical achievements came a growing understanding of the behaviour of matter on the microscopic scale. A good example is provided by the science of colloids. These are tiny pieces of matter with nanometric dimensions suspended in air or water. They are present throughout nature, e.g., milk is a colloid suspension. They were first identified in the nineteenth century, and were observed at the beginning of the twentieth century using the ultramicroscope invented by the chemist Richard Zsigmondy. However, it was only later in the twentieth century that colloid science came into its own, with industrial applications to materials, foods, cosmetics, and pharmaceuticals.

Another example is provided by molecular biology, which seeks to explain the way living organisms operate through the interactions between molecules taking place in cells. This science appeared in the 1930s, and over the next fifty years or so gave rise to a completely new outlook on cell dynamics and the advent of biotechnology.

A last example, but no doubt the most spectacular of all, is the development of microelectronics. The extraordinary expansion in this field at the end of the twentieth century brought the pursuit of ever smaller devices into the limelight. The first integrated circuit was invented in 1958 by Jack Kilby of Texas Instruments. This device integrates several components including transistors on a wafer of semiconducting materials and is ideally disposed to miniaturisation, provided that one has machines capable of making and positioning tiny components with sufficient accuracy.

As time went by, these different areas of understanding came together to form a field in its own right, the field of nanoscience (or nanotechnology when we speak of applications), taking up Richard Feynman's far-sighted challenge. The main lines of research were many and varied and the applications even more diverse. The unifying feature came above all from the length scale at which the different phenomena were taking place. The word 'nanotechnology' itself is attributed to professor Norio Taniguchi of Tokyo university who, in 1974, used this term to describe techniques for fabricating semiconductors. However, it was not until the end of the 1990s that the new field reached maturity. Various nanotechnological instruments, including the famous scanning tunnelling microscope, were by then available commercially and becoming standard equipment in research centers. At the same time, the microelectronics industry was well on the way down the road to miniaturisation and had already reached the scale of 200 nm, with the symbolic landmark of 100 nm in view. The prefix 'nano' was becoming common currency in the world of research and governments the world over were launching their first nanoscience and nanotechnology programmes, the former when concerned with improving understanding and the latter when they had applications in mind.

1.3 The Nanoworld

So what caused this sudden interest? The point is that, largely because it is being observed at these very small length scales, matter has very different properties, and while many of these properties can be deduced from laws of physics that have been around for over a century, it is only recently that observation and fabrication techniques have actually been able to demonstrate their existence, let alone build devices that put them to use.

1.3.1 Matter on the Scale of a Few Atoms

Researchers routinely manipulate small clusters of atoms, deposit ultrathin layers, and shape devices to an accuracy of 10 nm. The granular nature of matter on the atomic scale then becomes fully manifest. Solids and liquids look like more or less regular assemblies of atoms or molecules which hold onto one another via forces ensuring overall cohesion. The atoms will sit a few tenths of a nanometer apart. A cube of side 10 nm will contain roughly 25,000 atoms, although the exact number will depend on the material. Another important point is this: the smaller the objects, the greater the proportion of atoms that will lie close to the boundary of the nano-object. For example, 20 % of the atoms in the above 10 nm cube will be located at its surface. If we ourselves were microscopic and observed the world, two properties in particular would look strange.

The first of these is thermal agitation. On this scale, each atom, molecule, or cluster will be in permanent vibratory motion. When Scottish botanist Robert Brown observed this motion in pollen grains in 1827, it gave him the impression that the grains were alive. In a solid, an atom will often simply oscillate vigorously, otherwise held in place by its neighbour, although it may sometimes pass between them. In a liquid, it will move around and regularly change neighbours. But in a gas, it will hurtle headlong like a cannonball until it comes up against another atom or the wall of the container. And the hotter the matter, the more agitated the atoms become. At room temperature, this agitation corresponds to speeds of a few hundred meters a second. Sometimes we must cool an object under investigation in order to reduce this agitation.

Another disconcerting phenomenon is cohesion. Two small fragments of matter, whatever kind of matter it may be, will strongly attract one another like two magnets. These forces are basically of electrical origin. Atoms are made up of a positively charged nucleus surrounded by negatively charged electrons in such a way that the object as a whole is electrically neutral. However, these charges reorganise themselves all the time, and a (sophisticated) calculation reveals that the attraction between opposite charges wins through. This phenomenon is often exploited by researchers to build nano-objects. When separate components, molecules, or atoms are brought together, they will stick to one another, often forming very regular patterns. This force has a range of only a few nanometers and so would be quite imperceptible on the scale of our own world. But it is this force that ensures that solids maintain their cohesion or that water forms droplets, unless they are heated up so much that the atoms are agitated enough to escape completely.

Forces

Let us take a closer look at the forces exerted between two objects in the nanoworld. On this scale, it is more convenient to use the nanonewton as unit of force. This is one billionth of a newton. This force is roughly equal to the weight of a sphere of radius 3 hundredths of a millimeter filled with water and containing 3.4 million billion water molecules.

Strong Bonds When two atoms join up in a chemical reaction by sharing their electrons, remaining firmly and permanently attached to one another at a separation of only a fraction of a nanometer, we say that they form a covalent bond. On the scale of the nanoworld, the force in a covalent bond is truly colossal. Indeed, to separate the two atoms forming a hydrogen molecule, one must exert a traction force of the order of a nanonewton, i.e., as we have seen, equal to the weight of a million billion hydrogen atoms. Such a force is hard to imagine on our own scale. If the two hydrogen atoms corresponded to two peas each weighing one gram, the force between them would be equal to the weight of 100 million Airbus 380 aircraft. This explains why hydrogen and many other molecules are so stable, making it very difficult to remove their atoms.

Weak Bonds When they are placed a distance of the order of a nanometer apart, i.e., several atomic diameters, atoms and molecules do not join together. When they come close enough, however, they attract under the influence of electric force, any gravitational attraction being completely negligible in comparison. These forces, known as van der Waals[2] forces, are of the order of a thousandth of a nanonewton, which is a hundred to thousand times weaker than covalent binding forces. They are nevertheless very strong. If the atoms were two peas placed a centimeter apart, they would attract with a force corresponding to 100,000 Airbus 380 aircraft. This is why, on the length scale of the nanoworld, everything is 'sticky'. However, there is another phenomenon that prevents matter from clumping together, and this is thermal agitation. When it gets hot enough, atoms that are stuck together begin to dissociate due to repeated impacts from neighbouring atoms. Room temperature often suffices for this dissociation. In other case, more heat must be supplied.

Sticky Nano-Objects Attractive forces play a major role in the dynamics of nano-objects. To begin with, they ensure cohesion, unless it gets too hot. However, they also cause nano-objects to stick easily to anything passing within their range. So when a cluster of carbon atoms of radius about 50 nm, which will contain around 60 million atoms, comes close to a container wall, each atom in the cluster will be attracted to that wall. It will thus stick to it, attracted by a force of several nanonewtons, i.e., 200 million times its own weight.

1.3.2 Electricity

When electrons move through matter, we observe an electric current on our scale. The 10 A that trigger a circuit breaker correspond to 60 billion billion electrons going by every second. Although each electron is not really a little grain of electricity carrying its charge, this is a convenient way on our scale to describe the current passing through a metal.

We have been making nanoscale electrical devices for several years now. The basic building blocks can be nanocavities a few nanometers across which store electrons, nanowires which carry a current, or films a few atoms in thickness. Some of these devices are found only in the laboratory, while others are mass-produced by the microelectronics industry. There is also the prospect of 'ultimate' electronic devices in which electrons are manipulated one by one. On the nanoscale, quantum mechanics becomes an essential tool for describing the behaviour of the electric current. In other words, while it may be satisfactory to treat the electron as a little grain of electricity when it is viewed from afar, the situation is quite different when we zoom in closer.

[2]Johannes Diderik van der Waals was awarded the Nobel Prize for Physics in 1910 for his work on the equation of state of gases and liquids. His research on the continuity of fluid states, and in particular gases and liquids, led him to discover short range cohesive forces.

An electron then looks fuzzy, spreading over several nanometers. Here we have to speak of its wave function, and it looks as though several possible electrons coexist, each being associated with its own position and speed. The phenomena observed have no equivalent in our own world.

The best known of these phenomena is the tunnel effect. In our macroscopic world, if we cut a copper wire carrying a current, this current will no longer pass through it. But this is not quite true on the nanometric scale. If the wire is discontinued over a very tiny distance of the order of a fraction of a nanometer, the electrons arriving at one side of the break already have one 'foot' on the other side. These electrons can then cross the insulating gap. A current passes as though by magic across the gap, although its strength falls off exponentially with the leap it must make. This phenomenon is exploited in scanning tunnelling microscopes (STM). A fine tip is displaced just above a conducting surface, at a distance of a few tenths of a nanometer. A current thus passes with a strength of a few billionths of an ampere, depending sensitively on the tip–surface separation. By displacing the tip and measuring the current at each of its positions, we can thus reconstruct an image of the conductor surface, and we can in a certain sense even 'see' the individual surface atoms. In other situations, the tunnel effect can be something of a nuisance, especially in the manufacture of nanodevices. Indeed, it becomes impossible to use insulators below a certain thickness, simply because a current can then cross them. This is why microscopic transistors in processors or memories are prone to leakage currents which increase the electricity consumption.

Another relevant effect is quantisation. When an electron is trapped in a tiny region of space, it can only exist in a certain number of configurations. There is an analogy with organ pipes. Only sounds of a specific frequency can be emitted from a pipe of given height. This phenomenon, first observed for electrons in atoms, explains why each atom will only emit its own specific spectrum of light. However, the same phenomenon can now be reproduced by enclosing electrons in tiny boxes known as quantum dots. Like atoms, these have specific optical properties.

1.4 How Can We Make Such Tiny Objects?

A nanometer is a truly small length, much shorter than anything that could be perceived by our senses. So how could we possibly make objects on such a scale? Paradoxically, it is rather straightforward, since matter assembles itself spontaneously in nanometric form. For example, if we place a droplet of petrol of a few cubic millimeters on a bowl full of water, it will spread out to form a film a few hundred nm thick. The resulting iridescence is due to the play of light bouncing back and forth from one face to the other of this thin layer of petrol. The surface of a soap bubble may be of similar thickness and the same kind of iridescence is observed.

Even more surprising, in our everyday lives, we produce large amounts of nano-objects without ever appealing to nanotechnology. The main source of nanoparticles is the combustion of all kinds of materials, such as wood, cigarettes, petrol, and so on, all of which generate vast numbers of nanoparticles.

The Nanop Project

This project, which was financed by the *Agence française de sécurité sanitaire de l'environnement et du travail* (AFSSET), was led by researchers from different institutes.[3] Experimental work was carried out in the *maison automatisée pour des recherches innovantes sur l'air* (MARIA), which is an experimental house run by the *Centre scientifique et technique du bâtiment* in Champs-sur-Marne (in the Seine-et-Marne, France). It contains all the standard household goods that a family would use on an everyday basis (see Fig. 1.2). Various devices measure the pollution generated by everyday activities and in each room of the house.

This experimental house was used to analyse nanoparticle emissions produced during daily life, including their concentration, chemical composition, persistence in the air, and propagation throughout the house. This project showed that many such activities do indeed produce these particles, including in particular petrol heating systems and cooking. Peaks of concentration can reach a million nanoparticles per cubic centimeter. These particles propagate extremely quickly from one room to another.

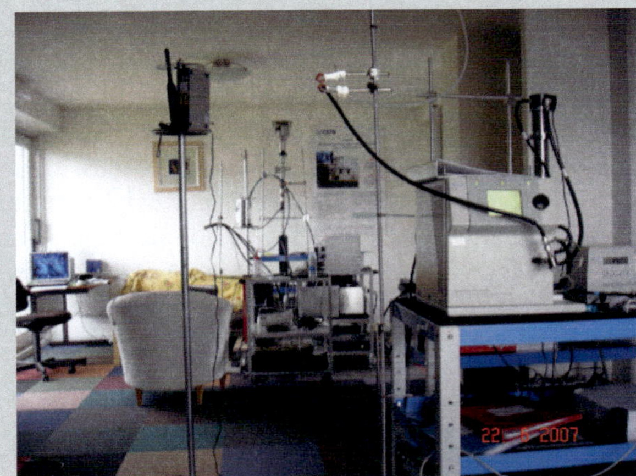

Fig. 1.2 Analysis of particles emitted in the living room. Courtesy of Corinne Mandin, *Centre scientifique et technique du bâtiment*

[3]These include the *Institut national de l'environnement industriel et des risques* (INERIS), the *Centre scientifique et technique du bâtiment* (CSTB), the *Institut de recherches sur la catalyse et l'environnement de Lyon* (IRCELYON), the *Laboratoire d'étude des particules inhalées de la ville de Paris* (LEPI), the *École des hautes études en santé publique* (EHESP), and the *université Paris-Est*.

In industry, many processes have been developed to produce nano-objects in a controlled way. These fall into three main groups: self-assembly, nanofabrication, and mimicking nature.

1.4.1 Self-Assembly

This first kind of process is useful for making nanoparticles and nanomaterials or treating surfaces. It exploits the tendency of atoms to stick together to form aggregates or even fibres. In scientific jargon, researchers call this the bottom–up approach, because we start out with atoms, at the bottom of the scale, and build structures up from there.

Sometimes when molecules are put in close proximity they assemble spontaneously into aggregates which may be nanometric. This tendency to self-assemble is exploited to fabricate nanoparticles in a great many industrial processes. One can simply heat the right kind of molecule or trigger a chemical reaction so that they break and the resulting pieces will tend to stick to one another. Nanoparticles or fibres produced in this way are then incorporated into a material whose properties one would like to modify, thereby producing a nanomaterial. Another process commonly used in industry is the production of thin films, with thicknesses measured in nanometers. Rather than have the atoms stick to one another, one arranges for them to stick to a surface.

Chemistry is also used to modify the surface of nano-objects such as nanoparticles or nanostructures machined on a surface. Molecules are grafted on to give them specific properties: therapeutic properties for a drug, properties of molecular recognition for diagnosis, and properties of chemical affinity for incorporation into a polymer matrix. The application of this kind of chemistry to nano-objects is referred to as nanochemistry.

Nanotrees in the Alps

These are in fact silicon nanostructures which grow on their own in the presence of a vapour containing silicon. The atoms arrange themselves spontaneously into structures with a trunk and 'nanobranches', giving the whole thing the appearance of a fir tree (see Fig. 1.3). These structures are particularly interesting for potential applications in the field of energy storage and energy conversion in photovoltaic cells.

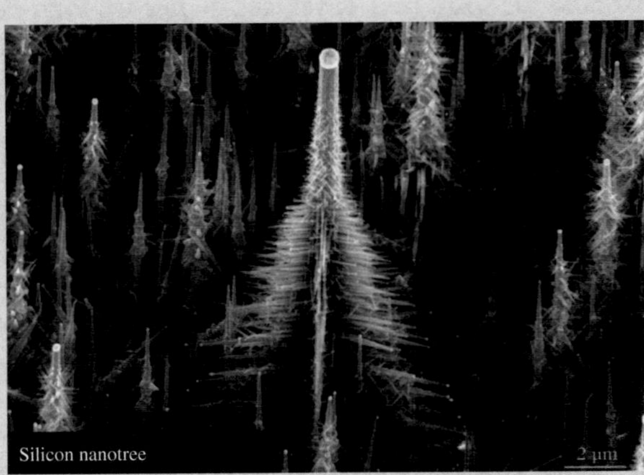

Fig. 1.3 Silicon nanotree. Image courtesy of Fabrice Oehler, CEA/INAC/SPMMM/ SiNAPS

1.4.2 *Nanofabrication*

This second kind of process encompasses a whole range of techniques for producing devices rather than materials. In this case, the elements are machined to an accuracy specified in nanometers, and above all they are positioned with similar accuracy. This is known as the top–down approach, because one starts out at the top length scale with the material to be machined and ends up with patterns and motifs of almost atomic precision.

Microelectronics began to develop in the 1960 s and with it, so did lithography. The material to be machined, usually silicon, is coated with a light-sensitive resist. The image of some very fine motif that one would like to engrave is then projected onto this resin, thereby transferring the motif to the resin. The silicon is subsequently etched at the places where it is no longer protected by the resist, in the case where the latter is qualified as 'positive'. However, it is also possible, depending on the resist, to modify the silicon at places that are not protected, depositing metals or implanting atoms and hence changing its electrical properties. The cycle of operations is then repeated above the first film to develop, complete, or modify the pattern. A microprocessor is thus built up gradually, layer by layer.

In fifty years or so, we have reached an astonishing level of control in these processes, engraving surfaces with lines measuring a few tens of nanometers. While these processes were inspired by technology that largely predates the nano era, the machines designed to achieve such technical feats have become considerably more sophisticated. They require ultra clean conditions since the tiniest grain of dust would look like a huge rock in comparison. They ensure remarkable positioning accuracy in such a way that successive layers are precisely superposed one above the other. They exploit new developments in optics, using ultraviolet light with ingenious expedients for getting beyond the classical limits of optical resolution. As a result, factories can now cost billions of euros to set up and run.

1.4.3 Mimicking Nature

From the standpoint of a biologist, living beings are highly complex machines in which most of the components operate on the nanoscale. They make admirable use of the properties of the nanoworld, and in particular the binding forces between molecules and thermal agitation. These components, which are often proteins housed in cells, carry out a range of tasks from the production of the molecules necessary to life to the destruction of waste products, transport of nutrients, information handling, energy transfer, communications, and so on.

Will we be able to do things as well as this one day? In his book *Engines of Creation: The Coming Era of Nanotechnology*, which appeared in 1986, Eric Drexler predicted that mankind would build 'assemblers', nanorobots that would assemble matter atom by atom.[4] So humans could eventually do as well as nature herself, or even better. These ideas remain in the domain of science fiction. We are still a long way from being able to compete with nature when it comes to nanofabrication. However, research has since taken a different road. Our understanding of living beings is being refined all the time, and the tools for reading and writing the genetic code are now available to an increasing number of scientists. Even in the 1970s, we were able to insert extra genes into cells to produce drugs, for example. The term used here is 'synthetic biology'. This research aims to use the same methods as nature to fabricate macromolecules such as drugs or biofuels, but also nano-assemblies such as the components of a vaccine.

1.5 What's It All For?

A key difference between nanoscience and other fields of research is that it is cross-disciplinary, underlying extremely varied applications, as the reader will soon realise in the following chapters. Nanotechnology has already taken its place in our everyday lives. This is the topic investigated by the Woodrow Wilson Institute in its project on emergent nanotechnologies. The institute regularly publishes inventories of products appearing on the Internet which involve some mention of nanotechnology.[5] They note a steady increase in the number of such products: 54 in 2005, then 1317 in 2010, and probably 2000 in 2013.

[4] K.E. Drexler: Engines of Creation: The Coming Era of Nanotechnology, Doubleday, New York (1986).

[5] http://www.nanotechproject.org/inventories/.

However, nanotechnology should not be considered as an industrial sector like the automobile industry, for example. This is hardly surprising, because the word ‘nanotechnology’ derives from a length scale, not a type of product. So there are nanoscale ingredients in almost every area of application, either through the presence of microelectronic devices or, more often, through the incorporation of nano-objects in existing products. These ingredients are sometimes mere details, representing only a few percent of the value of the product. Without claiming to be exhaustive, nanotechnological applications can be divided into three main categories: nanomaterials, nanoelectronics, and nanobiology (including nanodrugs).

1.5.1 Nanomaterials

This was the oldest field of applications if we take into account certain materials going back several centuries, whose properties derive from inclusions or motifs of nanometric dimensions. Consider for example steels, or glasses coloured by metal precipitates, but also materials made from living materials like wood, bone, leather, or paper. Having said this, it should be remembered that these were in a certain sense unintended nanotechnologies. The difference today is that industry knowingly appeals to nanometric ingredients, incorporating them into materials or depositing them on their surface. These nanomaterials, some of which have utterly novel properties, can be used in an extremely wide range of applications. It would be difficult to make a complete list, so disparate would it appear, but one can nevertheless cite some examples. An important one is the improvement of material properties such as strength, permeability, and hardness, e.g., strengthening of plastics or concretes by incorporating nanoparticles or nanofibres. Another example is the development of materials with large surface area for absorbents or filters, antipollution surfaces, and anti-bacterial surfaces. The latter application in particular has seen considerable development over the past few years. The idea is to fix silver nanoparticles to a given surface to kill bacteria. More generally, the most commonly used nanoparticles are silver, carbon, silica (nanosands), and titanium dioxide, which absorbs ultraviolet sunlight and destroys atmospheric pollutants.

1.5.2 Nanoelectronics

Since the 1960s, the electronics industry has been following its road map with steady miniaturisation of circuits and components. In the last 50 years, engineers have achieved extraordinary technical feats. The smallest detail on circuits mass-produced in 1960 was of the order of 55,000 nm, but it went down to 1,000 nm in 1990, then 32 nm by 2010. Smaller transistors are faster and more of them can be placed per unit area. This has led to a considerable increase in computational power per euro of components bought. Regarding products for the general public, computational powers are now measured in billions of operations per second. There have been a great many applications, such as image and video processing, automatic translation, voice recognition, and others. Today one finds calculators in many everyday products, from domestic appliances to cars, not forgetting smartphones. Another striking example of what miniaturisation can achieve is provided by data storage on computer hard disks. In the 1990s, the capacity of the hard disk on a PC was of the order of 100 megabytes. In just 20 years, this capacity has been increased by a factor of 10,000, so that today each point memorised is written on a spot measuring about 15 nm across, i.e., the area occupied by a few thousand atoms. In other words, a week's worth of video can be stored on one square centimeter.

1.5.3 Biology, Nanomedicine, and Health

This area has seen less development than the last two, but the prospects are very promising. Technology inspired by microelectronics is already commonly implemented in analytical methods involving just a few molecules. Another field with much promise is the development of nanodrugs. Ideally, drugs are molecules that affect the way a living thing works, but if possible, acting only on the part that requires treatment. For example, an ideal antibiotic should in principle block some process that is crucial to a bacterium, but without interacting with the cells of the patient. Likewise, for a product used in chemotherapy, the idea is that it should be toxic for cancer cells, but not for healthy ones. Unfortunately, in reality, treatments are often more invasive than one would like.

Nanotechnology opens up new prospects here. The drug molecule is replaced by a more complex system, a drug nanovector, which will carry the molecule precisely to the sick tissue. The molecule is encapsulated to protect it during transport to the target region, and it is provided with antennas to fix it exclusively onto the sick part. At the same time, this makes it possible to use more aggressive molecules. Clinical trials, especially for the treatment of certain tumours, are already under way.

1.6 The Debate

Nanoscience and nanotechnology comprise a ragbag of subjects whose unifying feature is the nanometer. Such a situation is not unique in the history of science. One may think of the science of complex systems which deals with phenomena as diverse as ecosystems, the brain, the economy, or the climate, the common denominator being complexity. However, nanotechnology is particular in the sense that it leads to applications in almost every field of life.

Another key factor is that nanoscience reached maturity at the same time as the so-called knowledge economy, at the beginning of the 2000s. This new development was part of what has been called the Lisbon strategy. The idea is that nations should build upon innovation to reconcile growth, sustainable development, and social cohesion. Among all the areas of science able to contribute to such goals, nanotechnology and information technology stand out. These two areas have seen considerable development in Europe, the United States, and above all in Asia. One of the symbolic events that has contributed to making nanotechnology a flagship activity was President Clinton's launch in 2001 of a national initiative, a wide-ranging programme that would soon exceed the billion dollar mark, followed closely by many similar undertakings on every continent.

The rapid growth of nanotechnology at the beginning of the 2000s was accompanied by often exaggerated claims from the various stakeholders. It was frequently described as a new revolution that would raise thousands of billions of dollars and profoundly change the structure of society. In the United States in particular, there were claims that nanotechnology could play a role in cyber-humanity, immortality, and artificial life forms, while various science fiction writers described futuristic scenarios in which nanotechnology would become a commonplace feature of our lives, or else would cause major disasters. In the latter category, one of the best known novels is *Prey*, published in 2002 by Michael Crichton. Combined with the extensive and extensively publicised research programmes being set up just about everywhere, such excessive claims soon led to a raging controversy over the nanotechnologies. For each extravagant promise of benefits, there was a matching fear of risk: new synthetic life forms would suddenly appear and start taking over, there would be changes to human nature, populations would be controlled by implanted chips, and new weapons would be developed. Since then, over the past ten years, such wild claims have faded into the background and the debate over risks has refocused on more concrete issues, which are not in fact specific to nanotechnology, even though it sometimes has to be taken into account in particular ways.

Toxicity

Millions of tons of nanoparticles of all different kinds are already produced industrially and new species are periodically introduced onto the market. For the main part, nanoparticles are integrated into nanomaterials, but they can nevertheless escape when the material gets worn through use or ages in a waste disposal site. These nanoparticles are also present in factories where nanomaterials are made and may affect the personnel and the environment. One must therefore inquire as to their toxicity. Indeed, there is no particular reason to assume that new kinds of particles recently introduced onto the market would present no risk to humans, or more generally, to ecosystems. The case of asbestos fibres[6] is there to remind us that this is not just a theoretical matter. So what are the risks with nanofibres?

[6]These generally have diameters of the order of a 1000 nm, but some are 10 times thinner.

We are still a long way from understanding all the mechanisms involved in the toxicity of chemical products where this question is already relevant, but the situation is even more worrying for nanoparticles, if only because we have so much less experience with them. It has been clearly demonstrated that nanoparticles can easily enter the body, for example by inhalation, move around inside us, and sometimes damage cells via various mechanisms. Appropriate toxicity tests still need to be developed, and for each species of nanoparticle we will need to specify a level of exposure below which there is judged to be no risk.

One thing that causes concern among the general public is the large scale arrival of commercial products containing nanoparticles, without specific regulation, or even any obligation to inform the consumer of their presence. In France, an important step was taken toward traceability of nanotechnological products in the law of 12 July 2010 known as Grenelle 2. This requires all persons fabricating, importing, or distributing substances liable to release nanoparticles to declare them to the relevant authority. On 30 June 2013, 3400 such declarations were made by more than 930 declarants. At the same time, studies of the risks involved in nanoparticles are making some progress, but it will probably take some time before we are able to build up a firm understanding of these risks.

Nanoparticle Surface Areas

One property that is often cited when discussing the risks relating to nanoparticles is the high ratio of surface area to volume. Since‘ all interactions occur via the surface, this vastly increases their capacity to interact with the environment and the cells in our bodies. To illustrate this, consider a cube of matter of side one centimeter (see Fig. 1.4). This has a surface area of 6 cm^2. If we now cut this cube up into smaller cubes of side one mm, we obtain 1,000 such cubes, and taken together, these have a surface area of 60 cm^2. If we go on to cut these up into cubes of side 10 nm, we end up with a billion billion cubes and a total surface area of 600 m^2. And all this from one cubic centimeter of matter, a volume easily held on a spoon. Quite generally, this is known as the specific surface area of a nanopowder, a quantity measured in square meters per gram. It is not unusual to find commercial powders with specific surface areas of a few hundred square meters per gram.

Fig. 1.4 Increase in the ratio of surface area to volume in the transition from a homogeneous solid to a set of nanoparticles

The main characteristic of a material reduced to a highly divided powder is thus its enormous surface area compared with its volume or its weight. And it is often precisely this enormous surface area that motivates potential users of nanoparticles. For example, when we wish to eliminate a pollutant from a certain environment by introducing particles to neutralise it, we expect the process to increase in efficiency as the particle size is reduced, since this will increase the surface area for exchange to take place.

Who decides?

Scientists push forward our knowledge and understanding, and at some point a new idea for an application can spring up. A long process is then set in motion to make this application possible, starting with technical trials, then industrialisation, and finally the arrival of new products on the market. This raises the following question: To what extent did citizen or consumer decide to realise this application? The question becomes more pressing still if the negative aspects of these new products, such as suggested or proven risks, effects on the environment, the establishment of monopolies, and so on, tend to outweigh apparent benefits like efficiency, reduced cost of a service, sustainable development, and so on. In some cases, the only possibility for any kind of influence left to the consumer, viz., the decision not to purchase, is effectively eliminated because nothing on the label indicates the presence of the ingredient that causes concern. Nanotechnological products fall perfectly into this category, with the explosion of new products over the past five years, despite the urgent need to address the question of risk. And the benefits are not always there. So while nanomedicine brings much hope and new products resulting from the miniaturisation of transistors and memories have led to a boom, other nanoproducts are clearly futile, with no obvious benefits.

Protection of Privacy

The development of information technology means that more and more details of our various activities are stored in memory, whether this be commercial or administrative data or just personal information to be found on the social networks, including now in the form of video. As early as the 1960s, there was already concern over the abusive exploitation of such data by an Orwellian state, or indeed by private individuals. But at that time there was no question of nanotechnology, whereas today, this innovation looks as though it can only complicate the situation, since more data can be intercepted, processed, and stored, thereby considerably aggravating the problem.

1.7 Prospects

Recall the title of this book: *Nanoscience and Nanotechnology: Evolution or Revolution?* What is the current situation? What has been happening over the last ten or fifteen years? Perceptions vary:

- For most scientists, nanoscience and nanotechnology are just the continuation of a long series of discoveries made through the twentieth century, along with the development of measurement and manipulation techniques on the atomic scale which allow us to observe and identify the novel properties of matter on this scale.
- For some of those who administrate science, but also for a whole range of commentators, the appreciation has been rather different, referring to an industrial revolution in which it was essential to invest in order to remain competitive.
- For those in industry, the exploitation of nanotechnologies generated by research has led to the creation of new products or improved existing products in some way, depending on the sector.
- For the citizen or the consumer, the advent of these technologies, mainly at the beginning of the 2000s, was rather sudden, without due consideration of their added value or risks. Within a few years, nanotechnology has thus become a subject of social dialogue and sometimes controversy.

So is this more like an evolution than a revolution? The actual situation is undoubtedly more complex. This would be to forget that, generally speaking, there is no direct route from a scientific discovery to a breakthrough product capable of transforming society, since this kind of product usually incorporates the results of many such discoveries. Nanoscience and nanotechnology are no different in this respect. Having said this, however, taking stock of the way things are going these days, it must be agreed that, through the miniaturisation of components they make possible, nanoscience and nanotechnology, associated with progress in other domains, have already made a major contribution to the revolution in information technology and communications. The cross-disciplinary dynamics of nanoscientific research will be illustrated in many other ways throughout the book, associated with new scientific horizons in the fields of energy, the environment, health, and medicine.

We may thus attempt to extrapolate the prospects from current research, and reflect upon what use can be made of them. Many such prospects will be further discussed in the following chapters. To this end, we shall consider three important examples.

Information Processing

The miniaturisation of components can already be exploited to make devices capable of processing and storing very large amounts of information. For the moment, this is without doubt the most tangible manifestation of the impact of nanotechnology on our everyday lives. Recent developments such as Internet, smartphones, social networks, and geopositioning, have had a profound impact on society and will probably continue to do so in the future. The literature is full of forthcoming innovations involving devices made possible by miniaturisation which may soon escape the confines of science fiction. As an example, we could before long find ourselves with smartphones that record videos of our every waking hour, enabling us to retrieve scenes on demand, or we could interact with machines simulating our conscience, and omnipresent sensors making our lives entirely 'transparent' and accessible on the social networks. Unless of course the world evolves in some other direction and we prefer to anticipate the increasing scarcity of energy and raw materials, realising sustainable information technologies and ensuring that personal data acquired in this way should remain private.

Novel Properties of Nanomaterials for Energy, the Environment, and Therapeutics

Nanomaterials are called upon whenever there is a need to improve materials or bestow them with new mechanical, optical, or electrical properties. However, over the past two decades, society's outlook has changed considerably. It has become increasingly important to save energy and raw materials. Another growing concern is risk. In the past, various promising substances used in materials have turned out to be dangerous, to the point where some have even been forbidden. The classic example is asbestos, but there have been others. We have become wary of mass-produced molecules and nanoparticles which are incorporated in materials and later turn up in the environment, the air in our houses, and our bodies.

On the other hand, nanomaterials are also a source of hope, because they may provide ways of avoiding certain molecules that have become undesirable, saving energy by improving insulation and reducing the weight of transported goods, and improving the efficiency of systems such as batteries or solar cells. The future of nanomaterials will clearly depend on the benefits they can procure and progress in understanding the toxicity and environmental properties of nanoparticles. For this reason some institutes have opted for a new orientation by setting up what they call green nanotechnologies. These aim to produce less toxic materials, or ones that make more efficient use of energy.

Moreover, one should not focus only on materials. Active molecules could be replaced by nano-objects incorporating not only these molecules but also other components, such as a capsule or a securement system. This idea, already used to make nanodrugs, could be adapted to many other substances as a way of improving efficiency or reducing toxicity.

Active Systems: Biomedical Applications

In the overlap of the two previous topics, there may be progress with active materials. We now know how to produce very small sensors and calculators, as well as actuators (a kind of motor) almost on the molecular scale. There is nothing to prevent us from integrating these into a material capable of exchanging with its environment, or even acting on its environment. These are likely to be relatively expensive materials, but could be adapted to highly specific applications.

One can get a better grasp of this idea by drawing a parallel with computer technology. From the centralised computers of the 1960s, we made the transition to personal computers in the 1980s, then to embedded systems, i.e., very small computers integrated into other devices in the 2000s, and the trend toward increasing integration of computer systems into materials themselves will doubtless continue in the mid-term. In the same way, factories were equipped with a centralised machine at the beginning of the twentieth century, but individual motors were widespread by the 1950s with the development of the automobile, and are now commonly integrated into dozens of everyday consumer products. In the end, we can easily imagine a more far-reaching integration of motor functions in materials, just as in biological tissues. In short, just what would be needed to make the sorting hat in the Harry Potter series.

In the biomedical field, an application of choice must certainly be tissue regeneration. Intelligent systems could be made to control the growth of new healthy cells by providing the molecular scaffolding and at the same time delivering the chemical signals needed for the growth of the tissues themselves. Naturally, very different applications can be envisaged here.

What we have just summarised so briefly represents only a tiny cross-section of the possibilities. The aim is to put nanoscience and nanotechnology in perspective, but as always, predictions are hazardous in such a field and it is important to keep an open mind.

Part I
Nanophysics, Nanoelectronics, and Nanophotonics

Chapter 2
Overview of the Field

Jean-Michel Lourtioz

Abstract The nanometre is not by a long way the smallest length scale explored today by physicists, but it is at this scale that the properties of matter change radically with, among other things, the manifestation of quantum effects which can be exploited in electronics and photonics. The development of nanoscience and nanotechnology has thus been driven forward by the quest to better understand the physical properties of matter on the nanoscale and thereby to miniaturise electronic and photonic components. The first parts of this chapter show how we can today produce nano-objects by the top–down approach, how we can visualise, manipulate, and assemble them, and how some of their astonishing properties can be illustrated by quantum dots, nanotubes, nanowires, and a material like graphene. The following parts deal with nanoelectronics and nanophotonics. We also discover the current limits on the miniaturisation of the key component in electronics, the MOS transistor. These limitations have inspired scientists to investigate a broad range of new components and circuits, like those proposed by the new field of spintronics. When we consider nanophotonics, we discover a field of research with the same breadth as nanoelectronics, illustrated by recent progress in the development of micro- and nanosources, the ideas of photonic crystals and metamaterials, and applications to photovoltaic conversion and optical interconnects. Finally, through the consideration of microsystems and nanobiotechnology, the last part of the chapter shows how research at the interfaces between nanophysics, chemistry, biology, and medicine will guarantee discovery, innovation, and significant industrial applications, some of which will be presented in subsequent chapters.

J.-M. Lourtioz (✉)
Institut d'Electronique Fondamentale,
Université Paris-Sud, Bât 220, 91405 Orsay Cedex, France
e-mail: jean-michel.lourtioz@u-psud.fr

J.-M. Lourtioz et al. (eds.), *Nanosciences and Nanotechnology*,
DOI 10.1007/978-3-319-19360-1_2

2.1 Is There a Road Map for Nanophysics? The Example of Nanoelectronics

As long as there have been physicists, they have been drawn toward the two infinities: the infinitely small and the infinitely big. In this exploration, the nanometric scale, taken to extend from the nanometer, just above the dimension of atoms, to a few hundred nanometers, would merely be an intermediate scale of no particular interest if it were not for the fact that it corresponds to a size where objects or devices exhibit novel physical properties, not to mention important potential applications in electronics and optics. Indeed, quantum effects such as the discretisation of energy levels or the dual personality of the electron as both wave and particle begin to manifest themselves once one gets down to the nanoscale. At this point, macroscopic physics begins to fail because the dimensions of an entity become a crucial factor in explaining its electronic, optical, thermal, mechanical, and other properties. Still more generally, the relative importance of different physical effects are altered when one reaches the diminished size of nano-objects. For example, since the ratio of surface area to volume goes as the reciprocal of the linear dimension of the object, surface phenomena grow in importance, leading to a considerable increase in exchanges with the surrounding medium.

The motivation for nanophysics and nanoscience in general derives from the desire to understand the physics of nano-objects and to discover their unusual properties. The development of new measurement tools by physicists in the 1980s, e.g., scanning tunneling microscopy (STM), atomic force microscopy (AFM), and scanning near-field optical microscopy (SNOM), combined with significant progress in fabrication methods, e.g., high resolution lithography and etching, self-assembled growth of nano-objects, have given a considerable boost to research and innovation in these fields over the last two decades.

Light can now be manipulated over length scales much shorter than the wavelength and it can be made to 'surf' over the surface of metals by virtue of their free electrons. Most semiconductor components for fibre optic telecommunications and light-emitting diodes now commonly found in lighting systems make use of electrons in quantum well or quantum dot nanostructures, with greatly increased performance in terms of efficiency, current, threshold, and modulation frequency.

With the miniaturisation of transistors, it is now commonplace to use the quantum tunnel effect to write logical information in the semiconductor memories of digital cameras, cell phones, printers, USB keys, laptop computers, and sound recording systems such as digital music players (see below). The development of magnetic nanostructures has led to new kinds of non-volatile random access memory, where information can be stored indefinitely without the need for cooling. Spin electronics has also opened up new prospects for very low power circuits, where the electron spin will be used to complement its charge (see Sect. 2.5). While not exhaustive, the above examples show how quickly fundamental notions are turned into applications. Clearly, the race toward miniaturisation in electronics has also played a key role in this rapid transition.

The Tunnel Effect Makes Short Work of Nanoscale Potential Barriers

A solid is made up of atoms, themselves comprising a positively charged nucleus and negatively charged electrons. In conducting materials, some of the electrons are only weakly bound to the nucleus and can move around under the effects of imposed stresses, such as thermal agitation (heat), an electric field, or a concentration gradient. These are the conduction electrons.

When two conducting materials are brought close together, in the classical picture, i.e., if we treat the electron as a particle, a certain minimum energy must be supplied to the electron to get it to go from one material to the other. This energy is usually of the order of a few electronvolts, depending on the electrodes and the intervening medium. We say that there is a potential barrier between the two conductors that the electrons must overcome in order to go from one to the other (see Fig. 2.1).

However, quantum mechanics tells us that we may also associate a wave with the electron. If we then send an electron toward the potential barrier, a large part of the wave will be reflected, but another part will be transmitted through the barrier. This is the quantum tunnel effect. The transmitted part, which depends on the height and thickness of the barrier, determines the probability that the electron will get through.

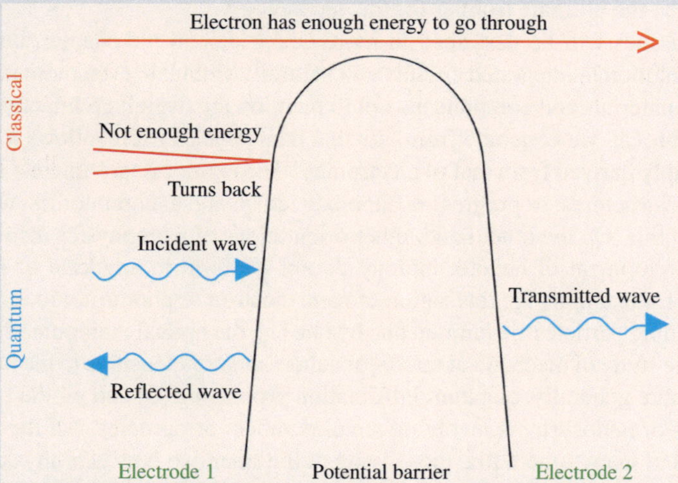

Fig. 2.1 Potential barrier and tunneling

For this phenomenon to reach any significant level, the barrier must not be too high (1–2 eV) and it must not be too thick (a few nanometers). If many electrons are injected toward a potential barrier, there will be a quantum tunnel current, and this will increase with the number of electrons in the emitting electrode and with the space available in the collecting electrode. In more scientific terms, it will be greater for a higher density of filled electron states in the emitting electrode and for a higher density of empty electron states in the collecting electrode. It will also increase in proportion to the transmission coefficient.

This quantum tunnel effect is the underlying principle exploited in scanning tunneling microscopy (see Sect. 2.4.2). It is also used to activate memory bits in electronic circuits (charging and discharging insulated gates).

Since the first integrated circuit in the 1970s, the development of silicon-based microelectronics has followed the celebrated law formulated by Gordon Moore, according to which the number of transistors incorporated in a chip is multiplied by four every three years (see below). Generations of integrated circuits have thus gone by, and each time the integration density has doubled, thereby reducing the characteristic size of the basic electronic component, the MOS (metal oxide semiconductor) transistor, by a factor of 0.7. By the beginning of the 2000s, the dimensions finally fell below the 100 nm mark. However, Moore's law itself predicts a major technological breakthrough, because the scale reduction of the MOS transistor below 10 nm would appear to be the ultimate frontier for this technology.

These issues will be described in more detail later in the chapter, but the predicted revolution in integrated circuits will naturally stimulate even more research on new nanomaterials and components, not to mention the overall architecture of these devices. But can we conclude from this that nanophysics itself follows a road map that is simply derived from that of electronics? The extraordinary interest in carbon-based nanostructures, or progress in nanomagnetism and nanophotonics, would tend to confirm this. On the other hand, other orientations of nanophysics inspired today by the development of nanotechnology do not conform to any kind of road map. Consider, for example, the realisation of nano electron or photon sources, where the corresponding particles are emitted one by one [1], the optical manipulation of nano-objects, the study of mechanical micro- or nanoresonators sensitive to the photon flux [2], and more generally, quantum information processing [3] and all the fundamental studies of nanosystems involving small numbers of particles. All these lines of research and many others like those lying at the interface between physics and the life sciences may also contribute in their turn to the discoveries and innovations of tomorrow.

From Micro- to Nanoelectronics. A Deterministic Evolution

The evolution of microelectronics has certainly been the fastest of all the industrial developments. The transistor was invented in 1947 by John Bardeen, Walter Brattain, and William Shockley, winning them the Nobel Prize in 1956. The first integrated circuit, with planar technology, was invented in 1958 by Jack Kilby, who was awarded the Nobel Prize in 2000. It was produced industrially by Fairchild in 1959 (patent Robert Noyce) and contained one transistor. Today, the highest performance microprocessors contain more than two billion transistors!

This evolution was predicted in 1965 by Gordon Moore, well before the fabrication of the first microprocessor by INTEL[1] in 1971. This was based on purely economic considerations. In particular, the more transistors can be manufactured at the same time, the less they will cost per transistor, while on the other hand the efficiency will be reduced.[2] One can then evaluate the best compromise between the number of transistors and the efficiency to obtain a minimum price per component. It was thus predicted that the number of transistors per chip would be doubled every year. In reality, this was somewhat optimistic, with a progression closer to a factor of 1.5 every year since 1959 (see Fig. 2.2). This prediction,

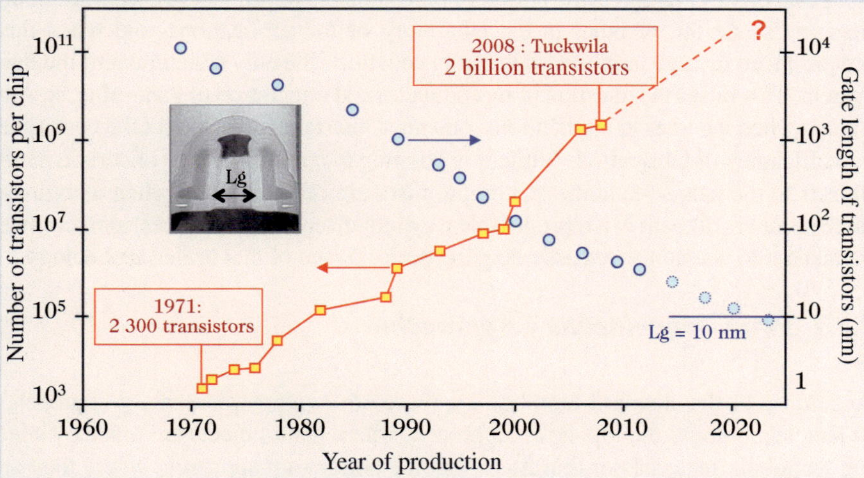

Fig. 2.2 *Red* Evolution of the number of transistors per chip, deduced from figures reported by various manufacturers. *Blue* Gate length of MOS transistors as a function of year of production. After 2011, figures are predictions by the International Technology Roadmap for Semiconductors (ITRS). *Inset* Cross-section of a nanoscale MOS transistor showing the position of the gate

[1]Company founded by Robert Noyce and Gordon Moore.

[2]If we produce a billion objects each with a 99.9999999 % efficiency, the total efficiency will be of the order of 35 %.

known as Moore's law, a kind of road map for semiconductors, is in fact readjusted every year by the main producers of integrated circuits, who produce the International Technology Roadmap for Semiconductors (ITRS). The figure shows certain fluctuations about a steady average development.

Many attempts have been made to produce road maps in other areas than silicon-based electronics or data storage with varying degrees of success. Regarding nanotechnologies, we may certainly expect the present plethora of new ideas to converge toward several general consumer products that will benefit from mass production, bearing in mind that the microprocessor market represents more than 40 billion dollars a year.

2.2 Higher Performance and More Versatile Nanotechnologies

When we talk of making objects to the nearest millionth of a millimeter, the first issue that comes to mind is the notion of precision. However, nano-objects are rarely made as a one-off, whether in the laboratory or for applications, and when they are produced in large numbers, they rarely constitute the only constituent of the final system. This raises the question of reproduction and integration of nano-objects. And finally, when it comes to applications, one must also take into account the cost, time, and efficiency of fabrication. While it is possible to control the size of certain nano-objects to the nearest nanometer, compromises are often required when integrating large numbers of them in a reproducible way into macroscopic devices, and this must be taken into account when assessing the performance of this or that technology.

2.2.1 Two Complementary Approaches

According to the standard terminology, there are two complementary approaches to nanotechnology: the top–down approach, where nano-objects are made by starting from bulk materials or thin films, and the bottom–up approach, where they are obtained directly by assembling atoms or molecules (see Fig. 2.3). As one can imagine, it is easier to produce very small objects by starting out with atoms or molecules, but then the size dispersion of these objects and irregular spatial distribution become major problems when building devices. On the other hand, it is more difficult to obtain objects of nanometric size from bulk materials, but when a technique has been perfected, collective fabrication is possible with high efficiency.

In studies of single nano-objects, physicists use both approaches, although they sometimes appeal to chemists, who have greater know-how when it comes to supramolecular construction. When studying more elaborate systems or devices

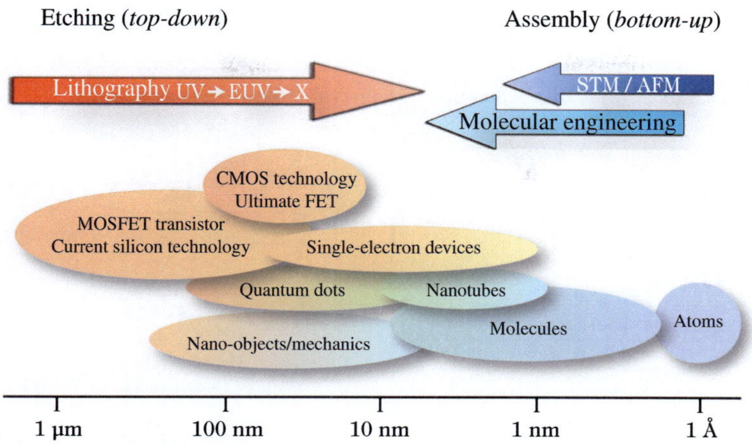

Fig. 2.3 Complementary approaches to nanotechnology: the top–down approach, where nano-objects are made by starting out with bulk materials or thin films, and the bottom–up approach, where they are obtained directly by assembling atoms or molecules

involving nano-objects, the situation is different and the top–down approach is mainly used. Nanoelectronics is still the most prominent example of this approach (see the discussion above). So despite the apparent conservatism, not only has nanotechnology gained in precision since the 2000s, but it has considerably diversified, with applications to an ever increasing range of materials. In the 1980s, only about a dozen materials were used in microelectronics devices, while today more than 60 elements of Mendeleev's periodic table are employed in the construction of electronic circuits!

2.2.2 General Features of Fabrication Processes

Figure 2.4 summarises the main processes involved in nanopatterning a thin film. We consider the example of planar technologies, in which all components are constructed on the same surface of a substrate which serves as support. Lithographic processes are used to specify the dimensions of motifs.

In order to etch a motif on a thin film that has been previously deposited (Fig. 2.4b), its surface is first protected by a photosensitive resist. This resist is then exposed to UV radiation through a mask, whereby the mask patterns are reproduced in the resin after development. The remaining resist serves as a protection. Unprotected regions of the material are then etched to the required depth. Chemical species can also be implanted, in particular for doping semiconductors. The nanopatterned material is revealed when the resist is removed. In addition to these processes, which can be repeated several times if necessary, thin film deposition (see Fig. 2.4a) and surface treatments are used to generate complex systems like integrated circuits.

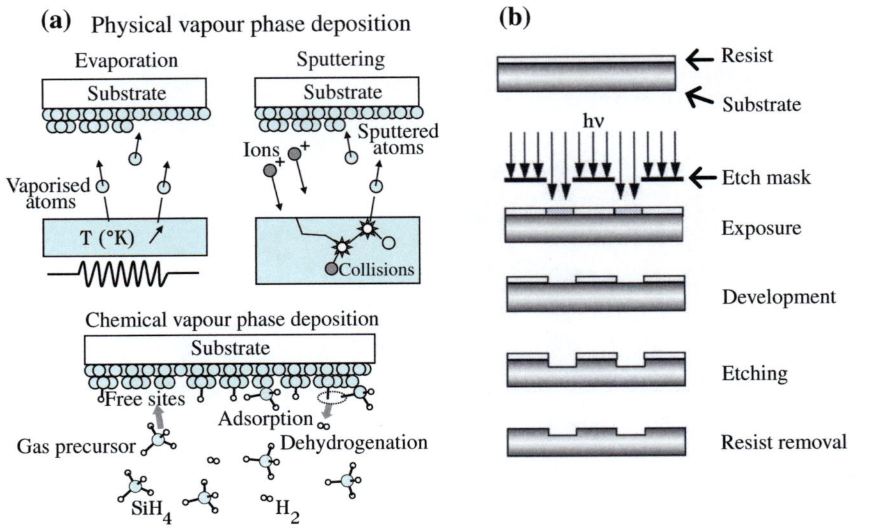

Fig. 2.4 **a** Depositing a thin film of material by physical or chemical vapour phase deposition. **b** Nanopatterning by lithography and etching of a thin film of material or substrate. Courtesy of Daniel Bouchier, *Institut d'électronique fondamentale*, Orsay

Nanotechnology is well exemplified by the fabrication of microprocessors produced on silicon wafers of diameter 300 mm. The finest motifs are of order 30 nm in size.[3] The highest performance integrated circuits require about 20 deposition–photolithography–etching cycles and can take several months. Billions of components are processed at the same time (several tens of circuits, each involving around a billion transistors).

2.2.3 High-Resolution Lithography

The UV radiation currently used for lithography has a wavelength of 193 nm, and it is only through extremely sophisticated tactics that we are able to produce motifs with dimensions of 50 nm. For future generations of integrated circuits, the aim is to use extreme UV radiation in the wavelength range 10–15 nm [4].

Focused electron or ion beams can also be used to write motifs directly onto the resist. As for exposure to UV radiation, the impact of the electrons or ions leads to a local alteration of the polymer resist which is then revealed by development. The wavelengths of the beams that scan across the resists are very short. For example, the wavelength of an electron beam of energy 100 keV is 0.005 nm, while that of

[3]By 2020, sizes of around 10 nm are expected and wafers will have diameters of 450 mm.

an He ion beam of energy 30 keV is 0.0001 nm [5], thus allowing an accuracy of nanometric order in the motifs at the present time.

In contrast to UV exposure, electron or ion beam lithography cannot be used to process whole wafers and the motifs must be transferred one by one. Having said this, while electron beam lithography may be at a disadvantage when it comes to producing integrated circuits in a semiconductor production unit, it has nevertheless become the standard fabrication technique for the primary masks used in photolithography.

2.2.4 Etching

Etching processes used for nanotechnology must cater for two main difficulties: the large number of different materials to be processed and the nanometric size of the motifs. Schematically, methods fall into two main categories: selective etches which affect a given material without acting on the others, and ion beam etches which, by ion bombardment, rip off atoms of the target layer but make little distinction between materials, including the mask.

Selective etches are usually chemical, but not necessarily in the aqueous phase. The most widely used are plasma etches which allow lateral excavation among other things. Ion beam etches are techniques which, depending on the kind of ions used, may have either a physical or a chemical effect. In the first case, the ions do not react with the target and their only effect is to tear off atoms close to the surface of the bombarded material. This type of etching is not selective. In the second case, the ions react with the surface atoms to form volatile species, thus constituting a selective etch process. This method known as reactive ion etching (RIE) was originally developed in the 1980s. In practice, it requires a specific study for each new material. With plasma etching, it has become one of the most widely used etching techniques today.

Ion beam etching gives the best resolution, with the highest performance obtained by using a highly focused ion beam. This is the technique known as focused ion beam etching (FIB). With sufficiently heavy and energetic ions, nanopatterning of the target material is achieved directly under the impact of the ions without the need to deposit a resist or any other intermediate process. Direct etching is accurate to within a few nanometers, as can be seen from the two examples in Fig. 2.5. The etching of nanopores of diameter a few nanometers through fine insulating membranes has potential applications in the analysis of single biomolecules [6].

Despite its high resolution, the FIB technique is limited by the fact that a single beam is used to etch, so that the motifs are produced one by one. For this reason, the process is currently reserved for laboratory studies. However, RIE can process whole silicon wafers of diameter 300 mm, or even 450 mm, in a single bombardment, with great uniformity and high reproducibility. This is the etching process used industrially to make circuits.

More generally, depending on the target object, one or other of the two techniques, plasma or ion beam etching, will be chosen for the task at hand. For example, to make a suspended element, lateral etching is required and plasma etching wins the day.

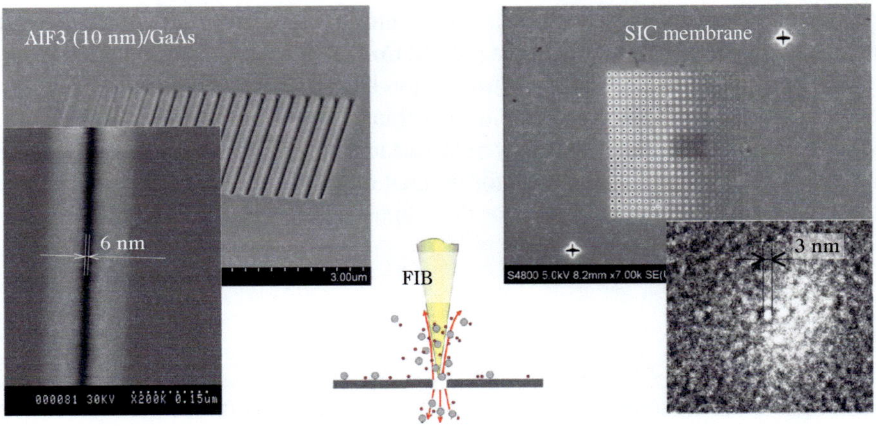

Fig. 2.5 Nanoetching by focused ion beam (FIB). Images produced by experiments at the photonics and nanostructure laboratory (Marcoussis). *Center* Schematic of etching process. *Left* Grating made from nanometric grooves (~6 nm) etched in gallium arsenide using an aluminium fluoride (AlF_3) etch mask, and enlarged view of a single groove (*inset*). *Right* Array of nanopores (~3 nm) through a silicon carbide membrane of thickness 20 nm. Enlarged view of a single pore (*inset*). Courtesy of Jacques Gierak, *Laboratoire de photonique et nanostructures*, Marcoussis

On the other hand, if a highly vertical etch is needed to produce trenches with high aspect ratio, ion beam etching will be preferred.

2.2.5 Atomic Scale Deposition and Surface Treatments

Deposition and growth techniques for materials and surface treatments are as well placed as lithography and etching when it comes to precision. Many examples could be cited thanks to the large number of specific techniques and materials. In the following, we shall restrict discussion to two examples that correspond well to recent developments in nanotechnology: the deposition of a thin film that perfectly respects substrate contours and a surface treatment designed to remove asperities.

Since the end of the 1970s, with the advent of molecular beam epitaxy, we have become accustomed to the deposition of atomic monolayers. For example, quantum wells and semiconductor superlattices for optoelectronics, or magnetic films for giant magnetoresistance (see Sect. 2.5). In this case, these are planar layers deposited on a substrate whose surface roughness has previously been minimised in some way. Now in many other cases, it may be desirable rather to deposit a thin film of material that perfectly conforms to surface features such as hollows and humps on a patterned substrate. This is precisely what is achieved by the aptly named atomic layer deposition (ALD). First suggested by Russian scientists in the 1960s [7], the technique was revived in the 2000s when it was applied to the deposition of thin films of oxides and metal compounds on silicon.

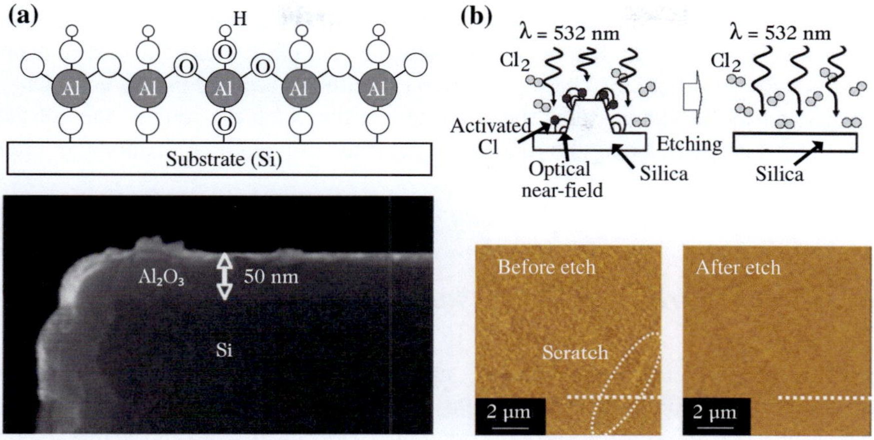

Fig. 2.6 **a** Atomic layer deposition (ALD) applied to growth of an aluminium oxide film on silicon. *Upper* Schematic of an aluminium oxide (Al_2O_3) monolayer terminated by hydroxyl (OH^-) ions. *Lower* Deposition of a film of thickness 50 nm on a silicon step, realised at the *Institut d'électronique, de microélectronique et de nanotechnologie* (Lille). **b** Reducing the rugosity of a silica surface by chlorine etching under a UV laser beam. Taken from [8], courtesy of Takashi Yatsui and with the kind permission of Springer Science and Business Media. *Upper* Schematic view. *Lower* Comparison between the state of the surface before and after etching. The *dotted ellipse* indicates a scratch observed before etching and which disappears after the treatment. *Straight dotted lines* indicate the place on the sample where the rugosity was measured. This is close to the angstrom after etching

Consider the example of a thin film of aluminium oxide on silicon, as shown in Fig. 2.6a. The monolayer is deposited in a small number of operations that are repeated until the required film thickness is attained. The surface is first hydroxylated and then exposed to a chemical precursor in the gaseous phase which brings aluminium atoms to the surface. By the very nature of the chemical reaction, there can only be a single aluminium atom for each initial hydroxyl group. This is what makes ALD a self-limiting process. The reaction products and excess precursor are then removed. The new surface is exposed to water vapour, which oxidises the aluminium. The deposition chamber is purged once again and the final result is observed to be the formation of an aluminium oxide monolayer terminated by hydroxyl ions. The initial surface conditions are thus recovered and the process can be repeated. Thin oxide films are thereby deposited in such a way as to conform perfectly to the surface of a nanopatterned substrate.

In contrast to the last example, it is also extremely useful at times to be able to remove holes and bumps on the surface of a substrate in order to be able to deposit exceptionally flat films. The electronics industry commonly uses mechanochemical polishing to plane surfaces after deposition of rough metal or silica films. This method has also been successfully extended to circuits incorporating carbon nanotubes. In the case of silica surfaces, a more recent method uses UV laser-assisted chemical etching to achieve record flatness (see Fig. 2.6b) [8]. The surface is exposed to chlorine in the presence of a UV laser beam. The photon energy is chosen to be less than the dissociation energy of the chlorine. It can then be shown that the chlorine can only dissociate at points of high substrate rugosity, i.e., at points where the optical near-field is stronger due to the sharp point effect of either a protrusion or a hole. Once dissociated to the atomic state, the chlorine can then etch the rough zones of the substrate, thereby reducing the rugosity to the scale of a tenth of a nanometer after treatment.

2.2.6 Nanoimprinting: Cheap Nanometers per Square Meter

The lithography, etch, deposition, or surface treatment processes we have just described are all expensive. For example, the cost of an exposure machine for electronics on a 300 mm substrate is of the order of 10 M€. Naturally, such an investment will be profitable in the long term if used for mass production. However, it represents a significant outlay that could usefully be reduced. Ultraviolet nanoimprint lithography (UV-NIL) may well be able to meet the challenge of providing cheap nanolithography in the years to come. The basic idea is to make a mould that can reproduce the nanoscale motifs of a template by mechanical pressure on resist-coated substrates, and this as often as one would like.

Nanoimprinting was first suggested in 1995 using a silicon mould [9]. However, the high temperatures and pressures required to make the imprint resist malleable enough to obtain an accurate imprint limited this process to a small number of applications. A major step forward was taken by replacing the silicon mould by one made of quartz, which is UV transparent, and using a liquid photoresist that is polymerised by UV light. Then from around 2002–2003, UV nanoimprint lithography was finally recognised as a key technology, able to replicate nanoscale motifs over large substrate areas at room temperature and under moderate pressures [10]. Since then, UV-NIL has continued to progress steadily in terms of resolution, quality, and throughput. In order to further reduce production costs, a new variant known as soft nanoimprinting looks particularly promising. Here the mould is made out of a flexible polymer. Apart from the low cost, this flexibility means that the mould can conform to the shape of substrates that are themselves flexible or curved.

Figure 2.7a shows how extremely dense nanoscale motifs, in this case an array of holes of diameter 20 nm with period 60 nm on a 25 mm^2 silicon surface, can now be reproduced using this technique [11]. The use of flexible moulds has also led to the recent development of a new method known as roll-to-roll nanoimprinting, with impressive performance in terms of printed surface and throughput [12]. This method can be applied equally well to rigid and flexible substrates. Figure 2.7b shows an optical grating on glass. A motif with period 700 nm was etched over an area of $10 \times 25\,cm^2$ in less than 15 s. Welcome to the nano printing press!

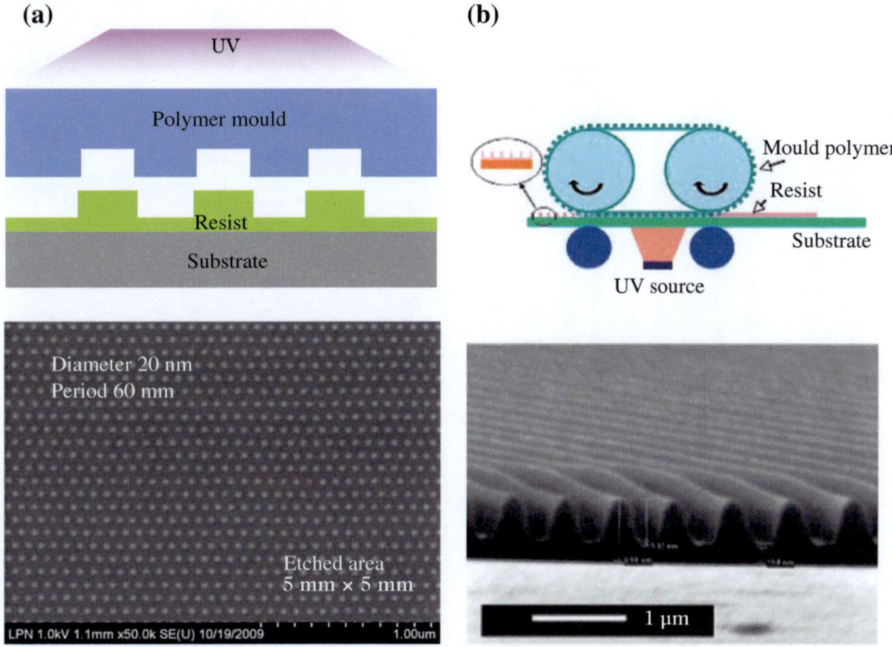

Fig. 2.7 **a** Soft nanoimprint techniques using UV. *Upper* Schematic view. *Lower* Replication of nanoscale motifs (20 nm) with period 60 nm on a 25 mm^2 silicon surface. Courtesy of Anne-Marie Haghiri, *Laboratoire de photonique et de nanostructures*, Marcoussis. **b** Roll-to-roll nanoimprinting. *Upper* Schematic view. *Lower* Optical grating on a $10 \times 25\,cm^2$ glass surface using this process. Taken from [12], © American Physical Society, 2012

2.2.7 Nano 'Lego' and New Methods for 3D Assembly

The development of micro- and nanosystems with ever more diverse features (see Sect. 2.7) also gives rise to new methods for assembling materials and micro- and nanostructures. Traditionally, electronics has used planar technologies where assembly consists in repeating certain basic processes, such as deposition of thin films, lithography, etching, and surface treatments, until the required functional device is obtained. The introduction of new materials and new structures in a given sector, e.g., the silicon industry, may make assembly still more complex whenever the relevant technologies are not compatible with the technology of the sector itself. The alternative is to make these materials and structures separately on other substrates, then transfer them to the target substrate.

The process known as wafer bonding exploits the van der Waals adhesive forces which come into play whenever two ultra-flat pieces of material are set in contact (see the introductory chapter to this book). Bonding can be permanent or only temporary. Once it has been achieved, the intermediate transfer substrate is either thinned or eliminated, or sometimes debonded for reuse. This technique thus diversifies the nano 'lego' available to electronics and photonics for 3D assembly. Figure 2.8a shows the example of a telecommunications microlaser (see Sects. 2.6.1 and 2.6.2) in which the active layers have been fabricated on an indium phosphide (InP) substrate and then transferred to a silicon substrate. In the final stage, the microlaser is etched after transfer [13].

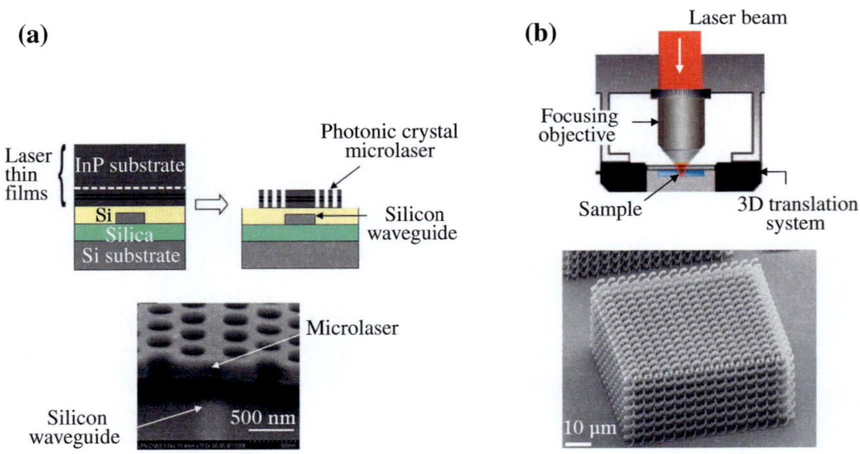

Fig. 2.8 **a** Telecommunications microlaser made by wafer bonding on silicon. Adapted from [13]. *Upper* Schematic view. The thin films of the laser are fabricated on InP, then transferred by wafer bonding onto a silicon-on-insulator (SOI) substrate. *Lower* Scanning electron microscope image of the resulting 3D structure. **b** Laser lithography. *Upper* Schematic view. *Lower* Chiral photonic crystal made using this process in a polymer resist. Adapted from [14], © Wiley, 2009

Another innovation in 3D fabrication methods is laser lithography, which is currently being applied to an increasing number of fields, from photonics to microfluidics and the life sciences in general. The method involves writing submicrometric motifs into a resist, not only at its surface, but also in the bulk of the material using a focused laser beam that scans the sample. The wavelength of the laser is chosen so that the resist is transparent to it. The resist conformation is thus modified only over a small region of the focal spot, where the optical intensity is high enough to produce harmonic conversion effects, doubling the frequency of the laser light. It is the UV light generated in this way that modifies the resist. The focal spot of the laser can therefore be located somewhere in the bulk of the resist, whence motifs can be realised in three dimensions. Owing to the very limited region in which the resist is polymerised, the resolution of the motifs can be comfortably submicrometric, being around a hundred nanometers at the present time. Figure 2.8b shows the example of a chiral photonic crystal made using this process. It comprises a stack of eight dielectric layers over a thickness of around 20 μm [14]. Another application of this technique uses the block of nanostructured resist as a 3D mould for fabricating other materials which are deposited in the hollowed regions. Once the filling material has been deposited, the resist is then removed by dissolving it.

2.3 Gaining in Length: From Clusters and Quantum Dots to Nanotubes, Nanowires, and Nanoribbons

For the general public, the prefix 'nano' conjures up an image of a nanoparticle as something rather like a grain of dust, but much smaller and possibly toxic, which may insidiously find its way into our living space, or even into our bodies. In physics, this familiar and somewhat simplistic vision is replaced by more scientific terms such as clusters, islands, and quantum dots, with shapes that often look at first glance like spheres or slightly flattened ellipsoids. The physical properties of such nano-objects, and in particular their electronic confinement properties, are roughly equivalent in the three space directions. One then speaks of 3D electron confinement. For this reason semiconductor quantum dots are often treated as artificial atoms.

Carbon chemistry completely changed all this at the beginning of the 1990s by introducing new geometries into the world of nanophysics, with the discovery of the fullerenes[4] and even more so with the discovery of carbon nanotubes. While the fullerenes C_{60} are an exact nanoscale replica of a football (see Fig. 2.9b), carbon nanotubes are long, thin objects, with lengths up to a thousand times greater than their diameters (see Fig. 2.9c). During the 2000s, physicists turned their attention to the fabrication and study of nano-tubes and nanowires, since these can be used to make an interface between the macroscopic and nanoscopic worlds and so in a sense are complementary to closed systems like quantum dots and clusters. Indeed,

[4]This discovery was rewarded by the Nobel Prize for Chemistry in 1996, awarded to Harold Kroto, Robert Curl, and Richard Smalley.

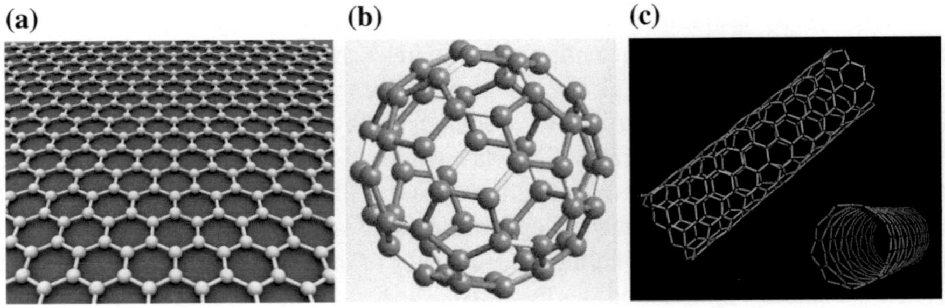

Fig. 2.9 Atomic structures of graphene, fullerene C_{60}, and a single-walled carbon nanotube. © NCCR Nanoscale Science Institute of Physics, University of Basel. Bonds between neighbouring atoms are shown as lines

we shall see that some growth techniques lend themselves directly to incorporation of quantum wells or quantum dots in semiconductor nanowires.

In the following, we shall focus on the two main types of nano-objects just mentioned. We shall discuss both fabrication and physical properties. The applications of these nanostructures to electronics and optics will be treated in Sects. 2.5 and 2.6, respectively. A special mention will be made of graphene (see Fig. 2.9a). The discovery of its remarkable properties led to the attribution of the Nobel Prize for Physics in 2010.[5] However, we shall only briefly mention the molecular assemblies that complete the full panoply of nano-objects exploited by physicists, since their synthesis falls clearly into the domain of the chemist. We shall end with an example of nanotechnology that came well before its time, viz., the Lycurgus cup at the British Museum.

2.3.1 *Fabrication of Nano-Objects*

Quantum Dots in Colloidal Solution

One of the two main methods for making semiconductor quantum dots uses the chemical route provided by colloidal suspensions. Research into the fabrication of semiconductor nanocrystals comprising only a few thousand atoms began at the Ioffe Institute in Saint Petersburg at the beginning of the 1980s and led to the first clear results at the beginning of the 1990s. Today, this has culminated in the synthesis of extremely uniform cadmium selenide nanocrystals (CdSe) whose surface can be treated for specific applications such as cell imaging [15]. They are now produced industrially and have been commercialised since the end of 2002. At the present time the method known as organometallic precursor decomposition is the most effective. In the case of CdSe, for example, the cadmium and selenium precursors are injected

[5] Awarded to the physicists Andre Geim and Konstantin Novoselov, of Russian origin but working at the University of Manchester.

into an organic solvent heated to around 300 °C. A precipitate of CdSe then forms. These core particles can subsequently be coated with a thick enough shell of zinc sulfide (ZnS) or cadmium sulfide (CdS) to obtain a stable luminescence level and avoid undesirable scintillation effects (see Fig. 2.10). By surface treatment, chemical ligands can be attached to bind the quantum dot to a biomolecule, and this can also be made 'visible' through luminescence. Molecular marking by quantum dots has now become a widely used technique for biological analysis and imaging, and a boon for the biophysicist (see Chap. 9).

Quantum Dots by Epitaxy

The other important method for fabricating semiconductor quantum dots is epitaxy in Stranski–Krastanov growth mode [16]. This mode occurs when a thin semiconductor film is grown on a substrate that is itself a semiconductor but has a different crystal structure (or in the technical jargon, has a different unit cell, i.e., different lattice parameters). The notion of critical thickness comes in here. For example, the lattice mismatch between indium arsenide (InAs) and gallium arsenide (GaAs) is 6.7 %, which corresponds to a critical thickness of slightly less than two monatomic layers. This mismatch leads to stresses that tend to dislocate a thicker epitaxial layer. Specific growth conditions are thus required, in particular regarding temperature and supply of matter, to prevent the layer from dislocating. It then deforms and this gives rise to 3D islands which will are the future quantum dots (see Fig. 2.11).

This growth mode can be used to integrate quantum dots at high densities in optoelectronic devices. At the wavelengths of optical telecommunications, quantum dot lasers now perform extremely well in optical pulsed mode [17]. It turns out that epitaxially fabricated quantum dots are equally well suited to the realisation of miniature lasers with very low threshold current or light sources, where photons are emitted one by one with very high output (see Sect. 2.6). These have great potential for future quantum information systems.

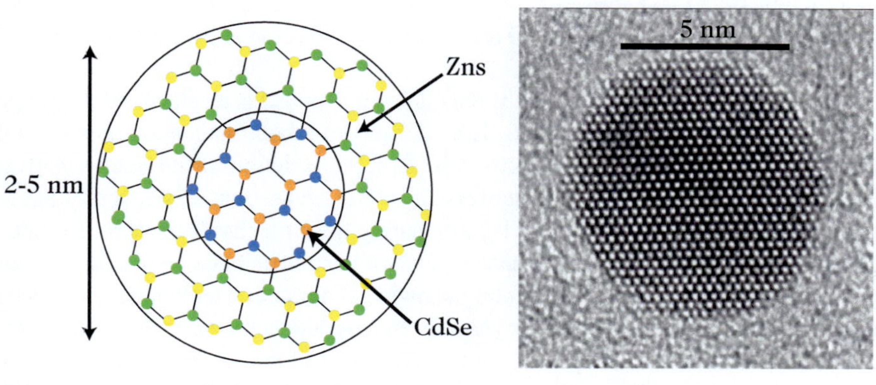

Fig. 2.10 Semiconductor nanocrystal (quantum dot) of CdSe surrounded by a ZnS shell. *Left* Schematic view. *Right* Transmission electron microscope image showing the atomic arrangement in the crystal. Courtesy of B. Dubertret, ESPCI, Paris

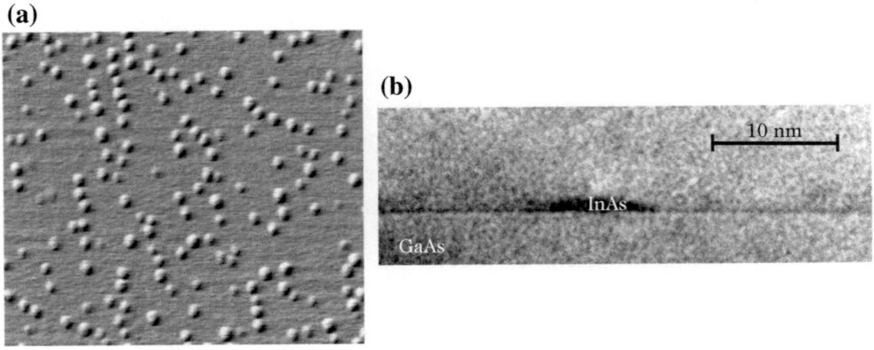

Fig. 2.11 **a** Atomic force microscope image of a plane of InAs quantum dots on a GaAs substrate. **b** Cross-section by transmission electron microscope of a single InAs quantum dot. Courtesy of J.-M. Gérard, Inac-CEA, Grenoble

Metal Nanoparticles and Clusters

Apart from processes reducing bulk matter to a powder state that were already used in ancient times, metal nanoparticles can be obtained equally well by colloidal synthesis (see, for example, [18]) or techniques involving deposition and etching of thin films (see Sect. 2.2.2). Small clusters are also produced by supersaturation of a metal vapour which is cooled suddenly, just as air saturated with water vapour will deposit frost on a very cold surface. Sputtering of metals under intense laser beams is another widely used technique [19]. Since the middle of the 2000s, there has been renewed interest in the development of metal nanostructures, in fact on a quite unprecedented scale, because they can be used in truly subwavelength optoelectronics (see Sect. 2.6).

Fullerenes and Carbon Nanotubes

A fullerene is a large spherical or ellipsoidal molecule. They were synthesised for the first time in 1985 by vaporisation of a piece of graphite at 2 700 °C. The best known fullerene is C_{60}, which contains 60 carbon atoms. It has the exact same structure as a football (see Fig. 2.9b).

It was in 1991, while analysing the dark deposit obtained in an electrical discharge between graphite electrodes, that S. Iijima observed tubular nanostructures several microns long using the electron microscope [20]. It was doubtless the sudden interest in nanoscience inspired by the discovery of the fullerenes that subsequently led to the discovery of carbon nanotubes by stimulating the curiosity of other researchers. Since this time and up to the last decade, carbon nanotubes have been without contest the most widely studied molecular nanowires. Fabrication techniques have been developed, novel physical properties have been reported, and many possible avenues of application have been explored. In the year 2008 alone, studies of carbon nanotubes gave rise to 7 000 scientific publications and 2 500 patents around the world. Electronics is one of the main areas of application. We shall return to this when we come to discuss nanoelectronics in Sect. 2.5.

VLS Synthesis and Functionalisation of Nanowires and Nanotubes

Among the various methods for making nanowires and nanotubes, vapour–liquid–solid (VLS) synthesis, or chemical vapour deposition in the presence of metal catalysts, is currently the most widely used, since it guarantees precise control over the growth of the resulting nanostructures. The basic idea is illustrated in Fig. 2.12 for the case of silicon nanowires. As for conventional methods using vapour phase deposition of thin films, chemical precursors (here silane) are introduced into the reaction chamber where they are decomposed in such a way as to deposit the desired element (here silicon). However, instead of depositing it directly onto the crystal substrate, this element is dissolved in metal nanoparticles (gold in this example) in the liquid state, previously deposited on the surface, which will play the role of catalysts. The concentration of the element in the particles grows until it exceeds the saturation threshold. This then leads to crystallisation upon contact between the particle and the substrate. The 'seed' for this growth, whose size is defined by the diameter of the particle, will thus determine the diameter of the growing nanowire. It is precisely because we can now reliably produce well calibrated nanometric metal particles, usually gold, iron, or nickel, that we can apply the VLS process to the growth of nanowires in all known semiconductors, viz., silicon, germanium, compounds like GaAs, InP, InAs, and so on, doped or otherwise, not to mention carbon in the case of nanotubes. In short, developments in nanoscience are beneficial to further such developments!

The VLS process just described provides a highly flexible way of making semiconductor nanowires. By varying the nature of the gas precursors during growth, the composition can be sharply modulated along the nanowire while maintaining perfect epitaxy, since the stresses are easily relaxed by the free surfaces of the nanowire. These modulations in the composition along the nanowire yield longitudinal heterostructures. Coaxial heterostructures, called core–shell structures, can also be obtained by altering temperature conditions in such a way as to favour direct crystal growth on the lateral faces of the nanowire, to the detriment of catalytic growth. It has even been shown that arborescent structures can be made by deposit-

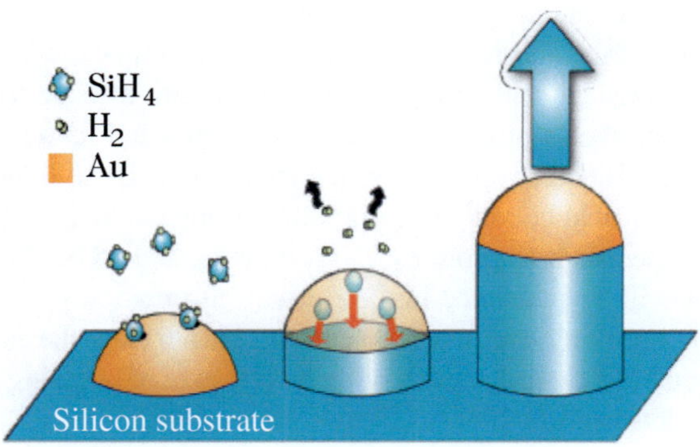

Fig. 2.12 Schematic description of the VLS process for synthesis of a silicon nanowire. *Left* to *right* Supply of chemical precursors, decomposition of precursors, incorporation of the silicon in a gold droplet, and growth of the nanowire by deposition of the silicon at the bottom of the droplet. Courtesy of L. Vincent, IEF, Orsay

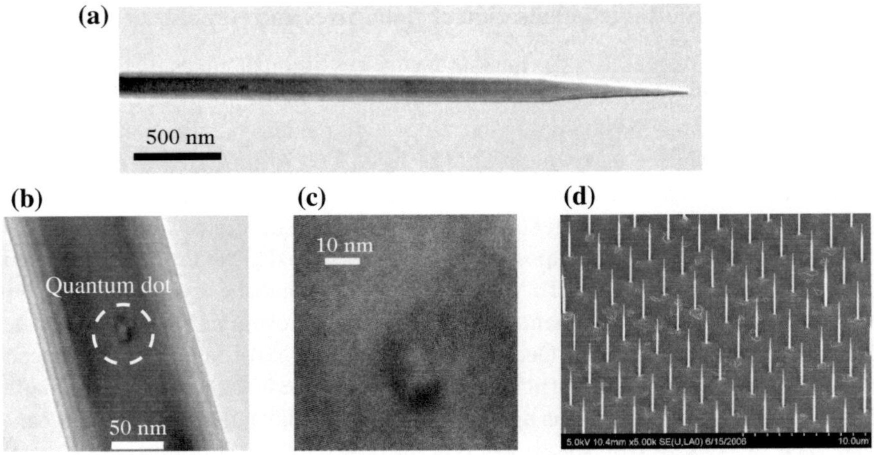

Fig. 2.13 Electron microscope images of InP nanowires ($L = 4\,\mu m$, $d = 150\,nm$) incorporating a single InAsP quantum dot [21]. **a** Image of a tapered nanowire. **b** Transmission electron microscope image of the region containing the quantum dot. **c** Enlarged image of this region. **d** Image of a regular array of nanowires. Courtesy of J.-C. Harmand, *Laboratoire de photonique et de nanostructures*, Marcoussis

ing catalytic nanoparticles on the nanowires. It would be difficult to imagine making such structures, either coaxial or arborescent, using top–down techniques.

Just the fact that one can overcome stress problems in nanowires is a small technological revolution in itself, dating from the 2000s. The crystal quality of the incorporated heterostructures is clearly a positive feature of these nanowires, for both electronics and optics. Furthermore, customised VLS growth opens the way to quite novel structures. One example among many is the nanowire incorporating a single quantum dot. An illustration is provided in Fig. 2.13 for an InP wire incorporating an indium and arsenic phosphide (InAsP) quantum dot emitting in the near infrared. Such devices could be used as sources in quantum cryptography, where photons can be emitted one by one (single-photon sources).

In addition to all this, and in contrast to carbon nanotubes, which are relatively unreactive, the surface of semiconductor nanowires can also be treated and functionalised to obtain precise physicochemical properties and favour surface interactions with various molecular species. When they occur, these interactions modify in their turn the electrical conduction properties of the nanowires. Hence, the presence of biomolecules such as proteins, DNA strands, and so on, bound chemically to the nanowires will be translated into electrical information [22], and this can be used for specific antibody–antigen recognition. This makes the nanowire into an ultra-miniaturised biochemical sensor, with the prospect of reaching the threshold of single-molecule detection.

Graphene

When one considers the relatively complex arrangements of carbon atoms found in fullerenes and nanotubes and their variants (see Fig. 2.9), one might think that a more elementary arrangement as in a graphene ribbon must have been found well before. In fact, although graphite, which is formed by superposing many sheets of graphene, weakly bound together, has been known since ancient times, particularly as a material for writing, it was not until 2004 that the physicists A. Geim and K. Novoselov were able to isolate a few graphene sheets and measure their electrical properties.

The somewhat crude technique they used to do this was simply to peel off pieces of graphite using sticky tape, until only a small number of monatomic layers remained, or even just one layer, of effective thickness of the order of a tenth of a nanometer [23]. The samples obtained in this way were then transferred to an appropriate substrate to facilitate characterisation (see Fig. 2.14).

Clearly, the crude exfoliation method used by A. Geim and K. Novoselov does not allow sufficient control over the material to be used for large scale production. Other methods are currently under investigation, including vapour phase deposition of hydrocarbons on metal, or surface graphitisation of a silicon carbide (SiC) substrate

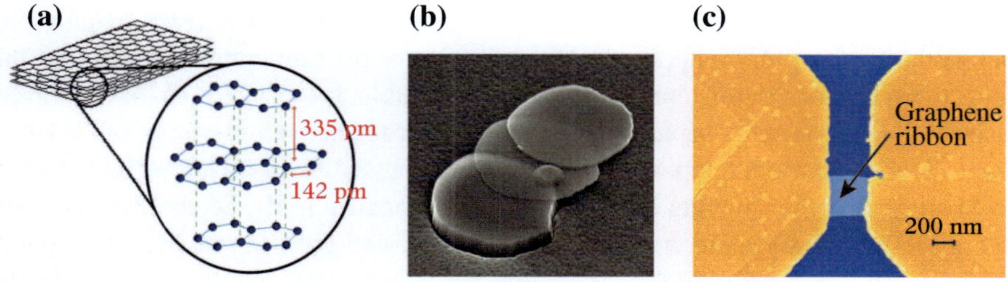

Fig. 2.14 **a** Schematic view showing a stack of graphene sheets in a piece of graphite. **b** Exfoliation method used by A. Geim and K. Novoselov to prepare graphene samples. **c** Graphene ribbon for electrical measurement of the Hall effect [23]. Courtesy of Konstantin Novoselov

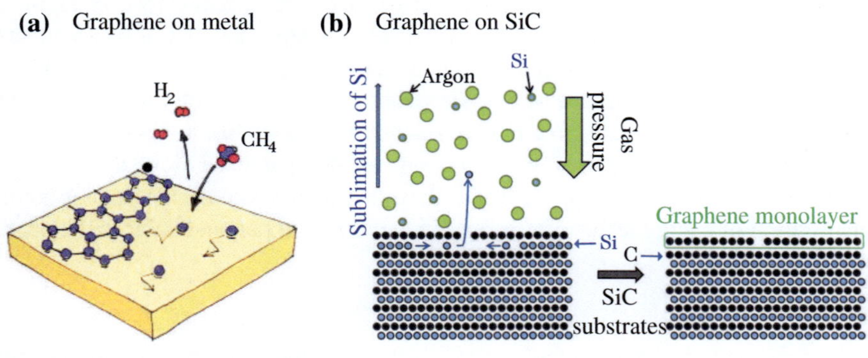

Fig. 2.15 **a** Deposition of a graphene monolayer on a metal substrate. **b** Producing a graphene layer on a SiC substrate by sublimation of silicon. Courtesy of D. Mailly, *Laboratoire de photonique et de nanostructures*, Marcoussis

(see Fig. 2.15). In the case of deposition on a metal, the hydrocarbon molecules, e.g., CH_4, are cracked at the surface in such a way as to leave only the carbon, which then arranges itself into the graphene structure. The other method, graphitisation on SiC, involves sublimation of the silicon. Since the sublimation energy of silicon is less than that of carbon, the silicon evaporates when the SiC surface is subjected to a high energy flux of argon atoms, while the carbon remaining on the surface can then rearrange itself to form graphene.

2.3.2 Physical Properties of Nano-Objects

Quantum Size Effects

In a metal, the wave functions associated with electrons which take part in conduction typically have wavelengths in the range 0.5–2 nm. It is thus easy to understand that specific quantum effects will occur whenever the size of the metal cluster is of the order of a few nanometers. In particular, it can be shown that the cluster will exhibit an electronic shell structure, just like a single atom. In the case of a semiconductor, characteristic wavelengths may exceed 10 nm. Quantum size effects thus arise whenever one of the dimensions of the semiconductor is of this order of magnitude. While there is no precise upper bound on the size of the semiconductor nanocrystal for observation of discrete energy levels to be possible, it is easy to see that too large a size will lead to energy separations below the thermal activation energy of the electrons (25 meV at room temperature). As a consequence, it will no longer be possible to 'distinguish' the separate levels. To see this, recall that the separation between energy levels for a system confined in a quantum well goes as one over the square of the width of the well. So for semiconductor nanocrystals measuring several tens of nanometers across, quantum size effects will begin to fade out and their physical properties can then be interpreted using a classical approach.

At this point, it is important to bear in mind that nano-objects fall into different categories, viz., nanoparticles (dots, clusters, islands, etc.), nanowires, nanotubes, or nanoribbons. It is only in 3D nanoparticles that energy levels can be quantised in the same way as in atoms and molecules. However, regarding the other nano-objects, the dimensions in one or two directions of space are such that the electron energies will vary classically, in a more or less continuous manner, in these directions. So for example, the energy levels of electrons are not quantised along the axis of a nanowire or a nanotube, but they will be in the transverse directions. For a nanoribbon, quantisation of energy levels occurs only in the direction perpendicular to the plane of the nanoribbon.

Nanotubes

A single-walled carbon nanotube is in fact a graphene ribbon that has been rolled into a cylinder (see Fig. 2.9c). The carbon nanotube is thus characterised by its diameter and its chirality, i.e., the angle at which the graphene sheet is rolled up relative to its structure. Depending on the values of these parameters, the single-walled nanotube will be either metallic or semiconducting. These properties have been confirmed, as has the surprising capacity of nanotube ropes to become superconducting through the proximity effect when they are connected to another superconductor [24]. Nanotubes also offer remarkable mechanical and optical properties, and in a more unexpected way, a potential for applications in nanofluidics. For example, they are a hundred times stronger than steel, while being six times lighter. They can emit particularly intense luminescence signals in the near infrared [25]. Water flows in nanotubes up to 10 000 times faster than one would predict given for such a narrow channel [26].

The shape of a nanotube lends itself well to use as a field emission cathode in devices able to operate at very high frequencies (see Fig. 2.16). This said, the main industrial application of carbon nanotubes remains the mechanical strengthening and/or hardening of materials by incorporation of nanotubes.

Transport in Nanotubes and Nanowires

Since the middle of the 2000s, carbon nanotubes and, even more so, semiconductor nanowires have become strategic elements for the integration of nanotechnologies. In a certain sense, we may say that, owing to their narrowness, the gate and channel of field effect transistors currently being developed by the semiconductor industry already belong to the category of nanowires. On a more fundamental level, nanowires are more or less ideal objects for studying the physics of heat or electrical transport and examining the interactions between particles such as electrons, phonons, and photons, in the highly specific case of a quasi-1D regime. This regime in fact obtains whenever the diameter of the wire is small enough to ensure confinement of the electrons and quantisation of the energy levels in the two transverse dimensions. This requires a diameter less than about 40 nm for a semiconductor nanowire. This is once again the quantum size effect described earlier. For a nanowire with no defects, the transfer of an electron along the wire does not obey Ohm's law, because quantum mechanics shows that it can occur without dissipation. The electrons are involved in no collisions, either elastic or inelastic, and are said to be ballistic. Only the ends of the nanowire will be a source of energy dissipation. It follows that the use of carbon

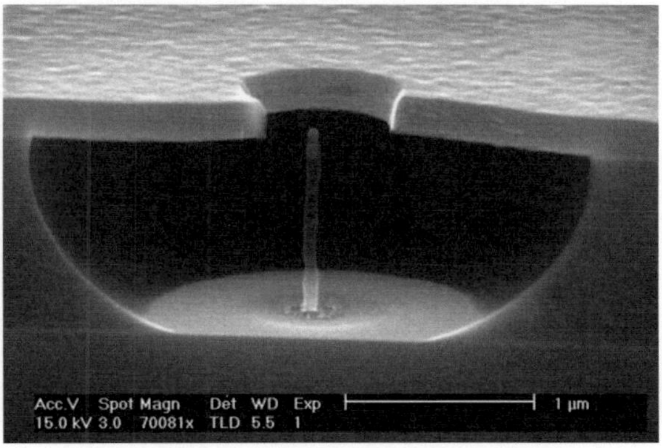

Fig. 2.16 Application of carbon nanotubes to electronics. Scanning electron microscope image of an emission cathode device. Courtesy of P. Legagneux, Thales TRT, Palaiseau, adapted from [27]

nanotubes as a conduction channel may allow very high frequency operation of MOS transistors (see Sect. 2.5).

Exceptional Conductivity of Graphene

The electronic properties of graphene were predicted theoretically over fifty years ago. However, the practical problems involved in obtaining a graphene layer of atomic thickness were considered more or less impossible to overcome. The exceptional properties of graphene follow from the dynamical behaviour of conduction electrons which, in the ideal case, behave like massless particles and can propagate without scattering in a perfect two-dimensional crystal. This is the field of application of quantum electrodynamics. In practice, this situation leads to record performances in terms of electron conductivity, with mobilities exceeding $200\,000\,\mathrm{cm^2\,V^{-1}\,s^{-1}}$ at room temperature [28], as compared with a mere $1\,200\,\mathrm{cm^2\,V^{-1}\,s^{-1}}$ in silicon. The electron spin diffusion length exceeds $2\,\mu\mathrm{m}$ at room temperature, with lifetimes of several hundred picoseconds, suggesting that it may be possible to make spin polarised currents circulate over long distances.

In addition to the electronic properties just described, graphene also displays several surprising optical, thermal, thermoelectric, and mechanical properties. Thanks to its transparency and excellent electrical conduction, it is a choice material for the electrodes in many electro-optical devices, and in particular touch screens. Record values of thermal conductivity of $5\,\mathrm{kW\,m^{-1}\,K^{-1}}$, ten times the value for copper, have been measured in a graphene monolayer [29]. Its mechanical strength is estimated to be close to two hundred times that of construction steel. Moreover, it is flexible and can be transposed to many different supports, such as flex circuits in flexible electronics. This long list of promising results thus led to a great deal of interest in graphene, some suggesting that it would soon compete with silicon. However, while it remains a good example of the kind of key discovery that nanoscience and nanotechnology can bring to the fore, the transition to innovative applications is a very different matter. Work is being carried out to characterise and model the role of structural defects and unwanted dopants, in order to get a better understanding of the physical properties of the material as a function of the number of layers or the nanoribbon geometry and describe spin injection and charge distribution when a potential is applied. The next few years will tell whether the discovery of graphene fulfills its industrial promise.

2.3.3 Lycurgus Cup

To end this overview of nano-objects, their fabrication processes, and their main physical properties, let us turn to an example of the use of nanotechnology that dates back to the fall of the Roman Empire. The Lycurgus cup (see Fig. 2.17 upper), a Roman glass cup dating from the fourth century AD and conserved at the British Museum, exemplifies the synthesis of metal particles of nanometric dimensions. Illuminated from within, it appears to be red, but illuminated from the outside, it looks green. The colour is due to the presence of a small number of gold and silver nanoparticles (40 parts per million) incorporated in the glass. These give rise to resonant absorption. Owing to the small size of the metal particles, the surface charges of the metal oscillate at specific resonance frequences when illuminated. These oscillations, known as plasmons by reference to the charge plasma contained in the metal [30], lead to absorption of light at the relevant frequencies. The change in the colour of the Lycurgus cup comes from the fact that we observe by transmission in one case, making it appear red, and by reflection in the other, when it appears green, the two colours being complementary.

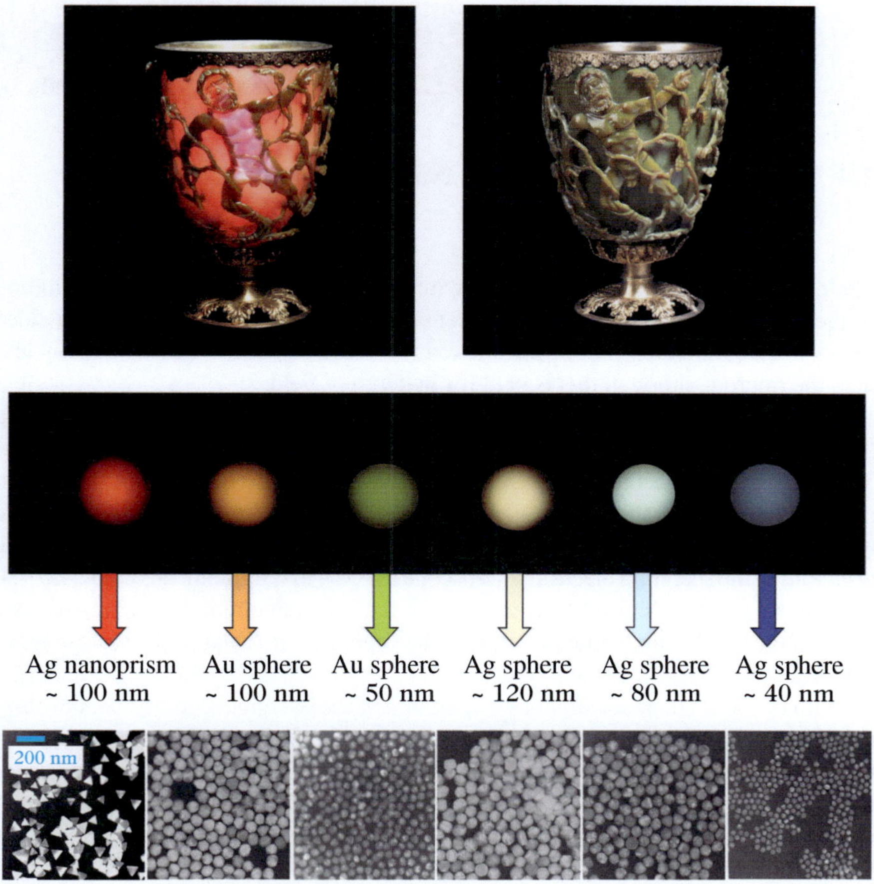

Fig. 2.17 *Upper* Lycurgus cup at the British Museum in London when lit from the inside (*left*) or by daylight from the outside (*right*). *Lower* Scattering of light by gold and silver nanoparticles of different shapes and sizes, obtained by colloidal synthesis. Courtesy of C.M. Mirkin, International Institute for Nanotechnology, Northwestern University

Fig. 2.18 SEM image showing the miniature architecture of metallization levels in an integrated circuit. © Artechnique, STMicroelectronics

2.4 Viewing and Manipulating Nanostructures: A Diversity of Microscopes

The observation of very small objects witnessed a first revolution at the beginning of the seventeenth century with the invention of the optical microscope. This idea has evolved continually ever since, and today we can obtain resolutions slightly less than the micron, although this is paid for in depth of field.

During the first part of the twentieth century, scientists had the idea of replacing the visible light beam of optical microscopes by a beam of high energy electrons (1–30 keV). When the surface of a sample is scanned with such a beam, it emits different particles in response, and in particular, secondary electrons which, after detection, are used to reconstruct an image showing the relief of the surface. This is the underlying principle of the scanning electron microscope (SEM), first implemented around 1930 by Max Knoll and Manfred von Ardenne. Today, we can obtain typical resolutions in the range 0.4–20 nm, while the depth of field suffices to visualise the surface relief (see Fig. 2.18).

With the boom in nanotechnology at the end of the twentieth century, the development of near-field microscopes made it possible, not only to view nanoscale objects, but also to move single atoms and gain access to other physical characteristics by measuring forces and mechanical properties, or to study biological samples in liquid media. In this section, we shall describe how these microscopes work and some of the possibilities they offer for physical characterisation of objects studied in the development of nanotechnology where they are very widely used. We shall also mention extensions of these systems that can be used to manipulate atoms or nanostructures.

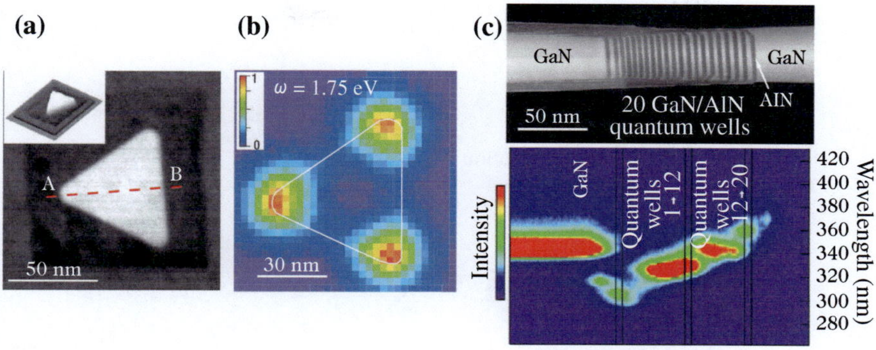

Fig. 2.19 **a** STEM image of a silver nanoprism, taken from [31], © Nature Publishing Group, 2007. **b** EELS measurement of a resonant absorption mode of the nanoprism at 1.75 eV. **c** *Upper* STEM image of a GaN nanowire in which several GaN/AlN quantum wells have been inserted, taken from [32], © American Chemical Society, 2011. *Lower* Spectral evolution of light emitted by the different quantum wells from one end of the nanowire to the other

2.4.1 Very High Resolution Transmission Electron Microscopy

The scanning transmission electron microscope (STEM), which has seen significant development with the advent of nanotechnology, differs from the scanning electron microscope in that it detects electrons transmitted through the sample, rather than secondary electrons. Originally investigated by Max Knoll and Ernst Ruska,[6] the STEM can reconstruct an image of the material traversed by the electrons. However, in order to be sure that the incident electron beam can actually get through the sample, it must first be thinned down to a thickness of the order of a few nanometers. The best performance so far from a STEM system is a resolution of around a tenth of a nanometer.

Figure 2.19 shows several examples of STEM images. The first on the left is a triangular silver nanoprism of side about 75 nm and thickness 10 nm. The second image, top right, shows a gallium nitride (GaN) nanowire about 200 nm long, in which twenty quantum wells have been incorporated. These are a few nanometers thick and are visible by their transparency. The GaN wells are separated from one another by layers of aluminium nitride (AlN), which plays the role of a potential barrier.

Apart from this use, recent developments with STEM systems have widened their horizons. Recall that electrons that go through matter lose energy by interacting with atoms in the target, and that this energy loss may be increased if there is resonant absorption by the target. By measuring the energy losses of the electrons in the method known as electron energy loss spectroscopy (EELS), one can carry out spatially resolved absorption spectroscopy on the sample. The image in Fig. 2.19b corresponds to a resonant absorption mode of the silver nanoprism described above, with an energy

[6]Ruska received the Nobel Prize for Physics in 1986.

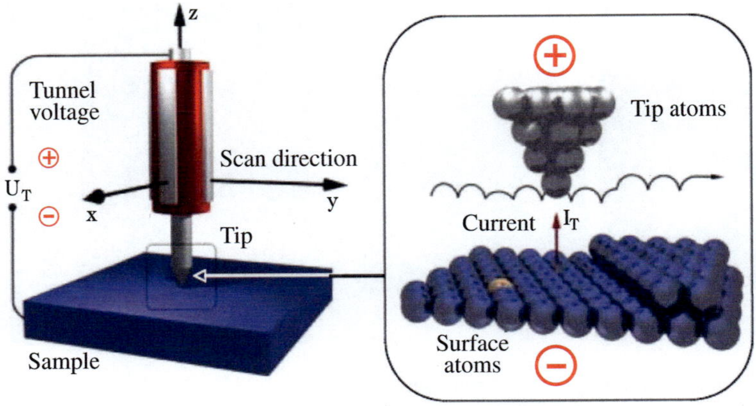

Fig. 2.20 Scanning tunnelling microscope. *Left* Macroscopic view. *Right* What is happening on the atomic scale. The STM is obtained by displacing the metal tip and measuring at each point the weak current between the tip and the region of the surface a fraction of a nanometer away. The current, which results from the quantum tunnel effect, reveals the electron density at the point of the surface just beneath the tip

loss of around 1.75 eV for each electron in the beam [31]. The yellow and red colours at the three corners of the triangle attest to the localisation of this absorption and allow identification of the resonance.

The energy lost by the electron beam can also give rise to photon emission when the sample is luminescent. This is known as cathodoluminescence, precisely the effect used to produce images on the screens of our old televisions sets, with their cathode ray tubes. The image to the bottom right of Fig. 2.19 shows the evolution of the luminescence wavelength produced by the different quantum wells from one end of the GaN nanowire to the other. According to theory, the broader the quantum well, the longer the emission wavelength [32].

2.4.2 Scanning Tunnelling Microscopy: Imaging and Manipulation

The scanning tunnelling microscope (STM) was an emblematic development of nanotechnology. Its discovery, which won the Nobel Prize for Heinrich Rohrer and Gerd Binnig in 1986, was the starting point for observation, analysis, and manipulation at length scales rarely attained before that time. As can be seen from Fig. 2.20, the idea is to measure the weak current circulating between a metal tip and a surface lying only a fraction of a nanometer from it when a voltage is applied between the two. This current results from the quantum tunnel effect and reveals the electron density in the region of the sample surface just beneath the tip (see the discussion in Sect. 2.1). The tip is displaced in the three space directions by piezoelectric transducers, with a

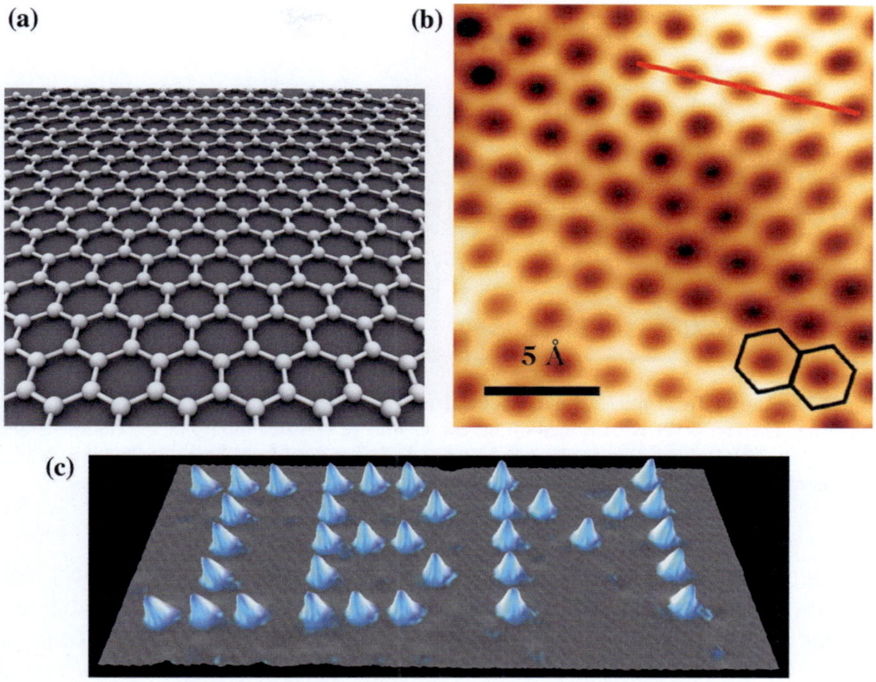

Fig. 2.21 The *top* two figures show graphene. *Left* Artist's view, as already shown schematically in Fig. 2.9, where each sphere represents a carbon atom and the rods joining them are bonds between neighbouring atoms. *Right* STM image [33], courtesy of G. Dujardin, *Institut des sciences moléculaires d'Orsay*, France. It shows the hexagonal lattice of carbon atoms and the higher electron densities on the bonds. *Lower* IBM logo written by manipulating 35 xenon atoms (*blue*) on a nickel substrate. The logo measures 17 nm across. © IBM, Eigler, IBM Almaden Visualization Lab

positioning accuracy that allows the system to image surface details with a resolution close to a tenth of a nanometer.

Figure 2.21b shows an image of graphene obtained by STM. One can make out the hexagonal array of carbon atoms, as already shown schematically in Fig. 2.9, the distance between neighbouring atoms being around 0.14 nm. One can also make out the high electron densities on the bonds.

Remarkably, individual atoms can be manipulated with a scanning tunnelling microscope. When the STM tip is brought near an atom adsorbed onto a surface and a potential difference is applied, there will be a force between the tip and the atom, which may be either attractive or repulsive (see the introductory chapter to this book). If the force is attractive, the adatom could then be stuck onto the tip or slid across the surface by displacing the tip. If the force is repulsive, the atom can be slid across the surface by pushing it. It is thus possible to manipulate single atoms at the surface of a sample. Figure 2.21 (lower) shows the first experiment of this kind, which has since become famous. In 1989, two research scientists at IBM, D.M. Eigler and E.K. Schweizer were able to spell out the logo of their company by

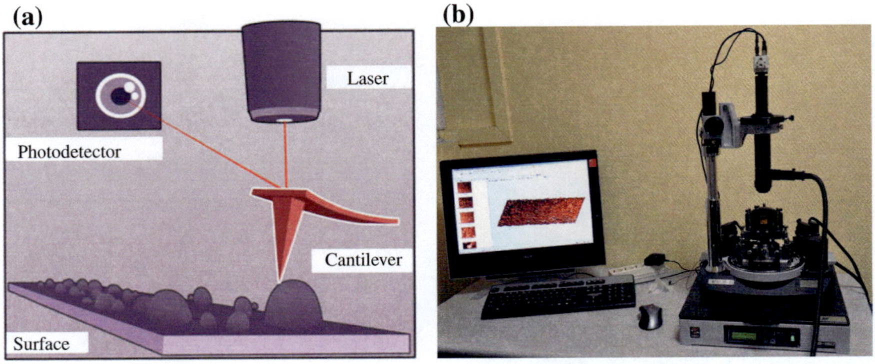

Fig. 2.22 Atomic force microscopy. **a** Schematic view, © CNRS/Sagascience, illustration Éric Vanneste. The cantilever is deformed by the forces arising between the tip of the cantilever and the surface. This deformation is measured by the deflection of a laser beam. **b** Photograph of a commercial microscope. Credit: *Université de Picardie-Jules Verne*

manipulating 35 xenon atoms on a nickel surface. The technique required very high vacuum conditions and also very low temperatures, obtained by cooling with liquid helium. This result, which was described in an article in the New York Times in 1990 [34], opened the way to mechanosynthesis with molecule-by-molecule control of chemical reactions. As has been shown recently, it is even possible to change the conformation of a single molecule and set it in motion by carefully selecting the point at which to excite it with the STM tip and the energy of the injected electrons [35].

2.4.3 Atomic Force Microscopy

The atomic force microscope (AFM) is another emblematic device in the rapid expansion of nanotechnology and has probably become one of the most widely used instruments for observing at the nanoscale. These systems exploit interatomic forces of the order of the nanonewton or less (see the orders of magnitude of various forces described in the introductory chapter to the book) exerted between a very fine tip and a surface only a fraction of a nanometer away (see Fig. 2.22). These forces are attractive van der Waals type forces at long range (>0.5 nm), but become repulsive at short range, whenever the electron clouds of the two approaching entities begin to interpenetrate. This means that there is a balance point which tends to keep the tip at a well defined distance from the surface. When the sample is moved relative to the tip, the latter will follow the relief of the surface just below it. The very slight deformation of the cantilever holding the tip is measured by deflection of a laser beam which a photodetector then converts into an electrical signal.

In fact there are several operating modes for the AFM, depending on the average separation imposed on the tip and surface, e.g., with the possibility of contact or

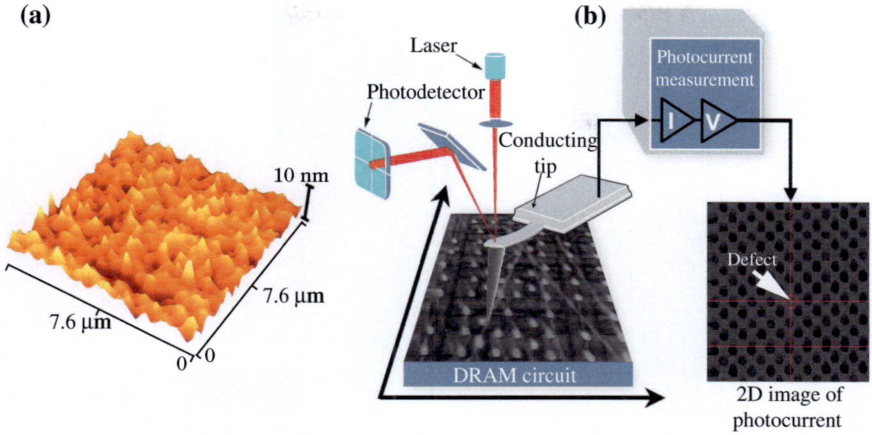

Fig. 2.23 Images obtained by AFM. **a** Three-dimensional measurement of the rugosity of a polycrystalline silicon surface. The average value is typically 1.2 nm, whereas it would be <0.2 nm for monocrystalline silicon. © CNRS/Institut d'électronique fondamentale, Orsay. **b** Analysing defects in an integrated circuit, in this case a dynamic random access memory or DRAM, by mapping picocurrents measured with a conducting AFM tip. Adapted from [36]. The missing black spot at the center of the *right-hand image*, indicating an absence of picocurrent, reveals a defective circuit element

friction, and depending on whether the cantilever is made to oscillate at its resonant frequency. Depending on the mode, one can obtain the surface topography with atomic resolution, measure the forces present to within a few tens of piconewtons, test the elasticity and molecular adhesion, or characterise the plasticity of materials by nanoindentation. If the tip is conducting, one can analyse defects in electronic components, probe the electric field associated with surface charges, and assess local doping values in semiconductors. If the tip is magnetic, one can analyse the domain structure of alloys or magnetic nanostructures, just as one can visualise vortex lattices in superconductors.

This brief inventory illustrates the many possibilities made available by AFM. Some examples are shown in Fig. 2.23. Note that, in contrast to scanning tunnelling microscopy, it is not necessary here to have a conducting surface. This means that one can analyse insulating samples like biological materials, which generally fall into this category.

2.4.4 Scanning Near-Field Optical Microscopy: Observation and Manipulation

The resolution of the best conventional optical microscopes, i.e., their ability to distinguish two distinct points of the same object, is essentially limited by the light

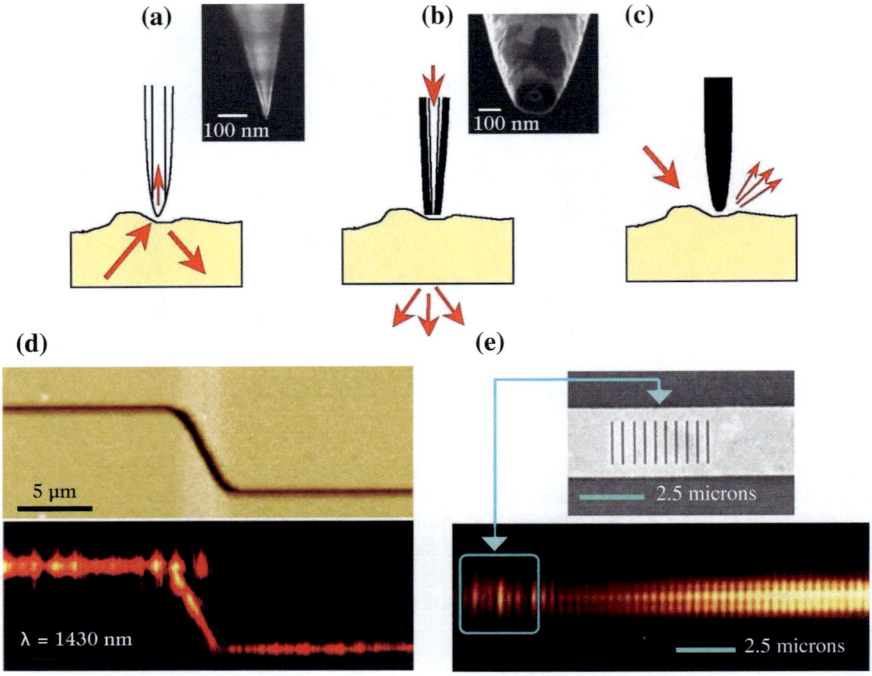

Fig. 2.24 **a** Schematic of scanning tunnelling optical microscopy and image of a tapered fibre used in detection. **b** Schematic of aperture near-field microscopy and image of a hollow fibre used in detection and emission. **c** Schematic of apertureless near-field microscopy. All three courtesy of Frédérique Fornel, *University of Burgundy*. **d** Light bent in a nanogroove hollowed from a metal surface [38]. **e** Interference pattern between a surface wave and the wave reflected by a metal nanoslit array. From [39], © AIP Publishing LLC, 2003

wavelength, which lies in the range 400–800 nm for visible light. To go beyond this limit, the only option is to move closer to the object and directly detect the so-called evanescent waves. Recall that these waves propagate in a sample when it is illuminated with light grazing the surface at such a low angle that it is totally reflected. These are surface waves sensitive to variations in the relief that are much smaller than the wavelength of the incident light. Although it was already envisaged as early as the 1930s, as attested by correspondence between E.H. Synge and A. Einstein [37], the advantages of scanning near-field optical microscopy (SNOM) only became a reality with the advent of micro- and nanotechnology and the development of other forms of microscopy. Indeed, most near-field optical microscopes share features with other microscopes. For example, a probe in the form of a tip is brought near the sample, while piezoceramic transducers ensure its exact positioning. The difference is that in this case photons are detected.

Various configurations are available depending on the type of tip and the way the sample is illuminated. In the optical tunnel effect configuration (see Fig. 2.24a), light is confined in the sample by total internal reflection or remains localised in the

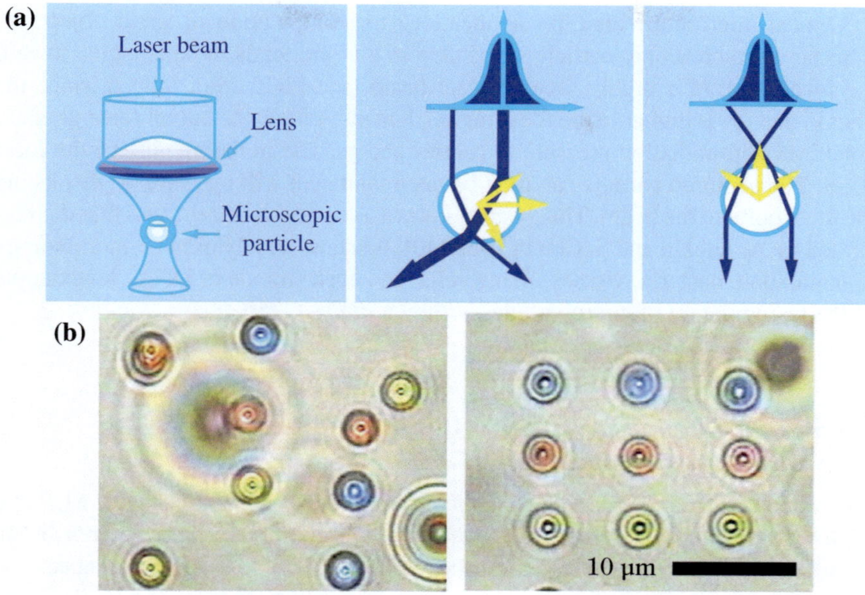

Fig. 2.25 **a** Optical tweezers. The object is captured at the focal point of a laser beam (*left*). If the object moves from this position (*center*), the forces due to momentum transfer from the photons will bring it back to the equilibrium position where they cancel one another (*right*). **b** Ordering of microparticles in terms of colour, using optical tweezers (*left* to *right*)

vicinity of the surface. This is what happens in optical waveguides or metal-coated samples across the surface of which light can surf, as it were (surface plasmons). The microscope tip, usually obtained by thinning one end of an optical fibre, picks up the evanescent waves as soon as it comes close enough to the sample. This process is the photon analogue of the tunnel effect described for electrons in Sect. 2.1. A tiny fraction of the light can then be detected at the other end of the fibre. In the configuration known as aperture near-field microscopy (see Fig. 2.24b), which is the most widespread commercial system, the sample is illuminated from above by a hollow tip used as a nanosource. Detection is carried out by the same tip. In the configuration known as apertureless near-field microscopy (see Fig. 2.24c), the sample is lit from above and a solid metal-coated tip converts a fraction of the evanescent waves by diffraction into propagating waves that can be detected in free space. Here the tip plays the role of a nanoantenna.

Apart from the very high resolutions, of the order of a few tens of nanometers, near-field optical microscopy offers different possibilities to other kinds of microscope. For example, one can visualise light propagation at length scales well below the wavelength, as illustrated by the bend in the light path shown in Fig. 2.24d and the interference pattern of Fig. 2.24e, obtained by reflection of a surface wave on an array of nanoslits.

Optical microscopy also has applications to manipulation of small objects. A transparent microscopic particle suspended in a liquid tends to be attracted toward the focal point of a highly focused laser beam (see Fig. 2.25a). Indeed, since the photon density is higher in the focal region than elsewhere, the lateral force gradient combines with radiation pressure to capture the particle in the vicinity of the focal point. The captured particle can then be manipulated at will by 'simply' displacing the focal point of the beam. This system, known as optical tweezers, was first demonstrated by A. Ashkin and S. Chu in 1986 [40]. It is now widely used in microbiology to manipulate bacteria, viruses, living cells, and even strands of DNA. An example of the manipulation of microparticles is shown in Fig. 2.25b.

2.4.5 A Few Remarks

Today, few research centres in the field of nanotechnology are without an AFM, STM, or SNOM system, or several of these, among their equipment. The high resolution scanning transmission electron microscopes (STEM) are less common because so much more expensive, and appeal must often be made to specialists who use them regularly in order to interpret results, especially when nanoscale processes play an important role. In fact, students doing a physics MSc or a doctorate in some field of nanoscience are now compelled to study these observational methods.

Going beyond the research laboratory, the various microscopy systems just described now constitute an integral part of production and characterisation in the microelectronics industry. For example, it has almost become a matter of routine to view, or rather probe surfaces and objects on the nanometric scale, and this relative trivialization of high resolution microscopes, for both the physicist and the biologist, as for the research scientist and the engineer, is in itself a significant achievement of nanotechnology.

The manipulation of adatoms using an STM tip is certainly a very sophisticated technique, but it is commonly used in the laboratory. As far as optical tweezers are concerned, diffraction phenomena currently limit manipulations to micrometric particles. The development of a new generation of optical tweezers able to trap and manipulate single objects of nanometric size therefore remains a considerable challenge.

2.5 Tomorrow's Nanoelectronics. The Challenge of Zero Consumption

While the development of new measurement tools (see Sect. 2.4) marked the beginning of nanotechnology in the early 1990s, the race toward miniaturisation in the electronics industry also played its role in accelerating their development. The main motivation of the electronics industry in promoting nanotechnology was to maintain the rate of development of integrated circuits and hence fulfill the predictions of Moore's law (see Sect. 2.1). However, innovation was also needed for new nanocomponents that would eventually replace the MOS transistors currently in use. There can be no doubt that nanoscience has succeeded in the first aim, with the prospect for the 2020s of an industrial production of MOS transistors with gate lengths around 10 nm. On the other hand, none of the new components presents itself as a successor to the MOS transistor.

In this section, we shall begin by discussing today's integrated circuits based on MOS transistors. We shall then describe spin electronics, which is on the point of leading to a genuine technological breakthrough, not only for microprocessors, but also for memory. We shall end with a brief overview of some alternatives, including quantum components, single-electron components, molecular electronics with carbon nanotubes and graphene, and neurally-inspired architectures.

As a preamble to this section, let us recall some general considerations concerning communications and information processing. There are two main tasks: data processing and data storage. The former is handled by the microprocessor. Its architecture today follows the traditional scheme first formulated by von Neumann [41], with four clearly distinguished units:

- A logic and arithmetic unit that processes the data.
- A memory unit that stores intermediate results.
- A set of circuits ensuring the interface with the outside world.
- A command unit that controls the whole system.

All these units are incorporated into the same integrated circuit, a so-called microchip on which the MOS transistors are etched to constitute switches, inverters, or memory cells. For its part, data storage is dealt with by optical or magnetic hard disks or by electronic circuits such as the storage registers of processors, USB keys, and so on.

Since the same components often handle both processing and storage, we shall discuss them in parallel.

2.5.1 MOS Transistors and Current Technology

Today's integrated circuits, microprocessors, and memories are all based on MOS transistors (see below). To improve the performance of integrated circuits in terms of speed, computational power, and storage capacity, they must be designed with more and more transistors, which are themselves faster and faster. Referring to Moore's law as discussed in Sect. 2.1, while today's integrated circuits comprise around a billion transistors with gate lengths of 30 nm, those of 2020 should reach 100 billion transistors with gate lengths of 10 nm. Operating frequencies of microprocessors are currently of the order of 2 GHz, but by 2020, they should reach 5 GHz.[7] Such ultimate levels of performance raise two kinds of problem: dispersion in the transistor characteristics and power consumption.

Dispersion in the transistor properties results mainly from size reduction. This is best seen through an example. The control voltage of switch transistors is adjusted by the number of fixed ions in the part of the substrate constituting the channel (see below). In today's transistors, the relevant volume is of the order of 18 000 nm^3 (30 nm × 30 nm × 20 nm).[8] With 10^{18} fixed ions per cubic centimeter, this implies 18 ions, and a fluctuation of one or two in this value would be acceptable. For the transistors of 2020, the surface area will be divided by a factor of ten, implying a factor of three in the dimensions, and the number of ions will be statistically somewhere between one and two. It is thus easy to understand that a fluctuation of one or two in this value would be disastrous. Put another way, not all the transistors will function as desired, and if we wish the overall operation to be satisfactory, we must design a circuit architecture that can tolerate such dispersions and if necessary allow the use of a learning phase.

The Ultimate MOS Transistor

The MOS transistor switch consists of a semiconductor channel, controlled by a gate voltage, in which charges pass from a source S to a drain D. The metal gate is separated from the semiconductor channel by an oxide. This explains the acronym MOS, which stands for metal–oxide–semiconductor. There are two complementary types of MOS transistor:

- the nMOS, where the charges circulating are negative electrons (n),
- the pMOS, where the charges circulating are positive holes (p).

In the first case, a control voltage expels the charges of the doped substrate (p) under the gate and creates a channel for the electrons. In the second case, it is the charges (n) of a reservoir previously added to the substrate that are expelled, allowing the holes to circulate freely. This complementary nMOS–pMOS pair constitutes the basic cell of all integrated circuits: the CMOS inverter (C for complementary), in which a logical input (0 or 1) leads to the complementary output (1 or 0).

Scale reduction of MOS transistors reduces the size of all their component elements, including in particular the gate which controls the conduction channel. Successive generations of transistors can be identified by their corresponding gate lengths.

---

[7] These are the frequencies of the clocks in the central processing unit which determine the number of on–off cycles made by the transistors every second.

[8] This is the volume for a channel with a surface area of 30 × 30 nm^2 and a thickness of 20 nm.

The electronics industry also refers to the technological node, which is half the minimal distance between two lithographed motifs (two metal lines, for example). The nodes are typically 1.5–2 times longer than the gate lengths. An important step was taken when the 100 nm node was passed in the 2000s, because the structure of the MOS transistor then had to be revised.

Indeed, the shorter the conduction channel, the less the control via the gate bias is efficient and independent of the voltage applied across the drain and source. To get round these difficulties, the gate can be brought closer to the channel by reducing the thickness of the insulating oxide. However, if this layer is too thin, of the order of a monatomic layer, the current may leak through the oxide. In order to limit the unwanted current which goes from source to drain through the substrate instead of going through the channel, one idea is to reduce the doped regions of the source and the drain. But if these regions are too small, that increases contact resistance. Compromise is essential here. Figure 2.26 illustrates the kind of structure that has been adopted over the past few years. Silica, the traditionally used gate oxide, has been replaced by

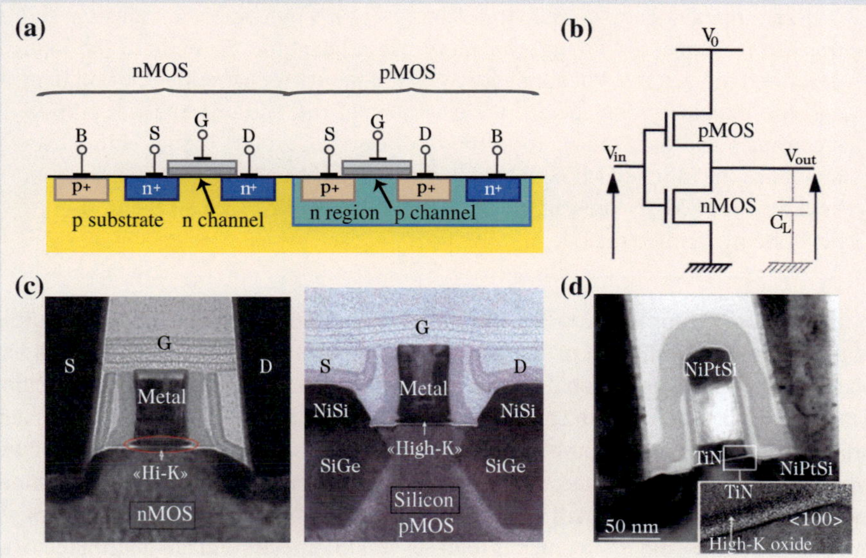

Fig. 2.26 MOS transistors. **a** Schematic cross-section of n-channel and p-channel complementary MOS transistors. **b** Circuit diagram of the CMOS inverter. **c** Transmission electron microscope images of nMOS and pMOS transistors made by Intel (45 nm node, production 2007), courtesy of Chris Auth, © Intel Corp. **d** Image of an nMOS transistor made by PanaSonic (32 nm node, production 2009), with zoom on the gate of the transistor. © Chipworks.com

insulators with higher electrical permittivity, known as high-K oxides, of sufficient thickness to maintain gate control efficiency. In the case of the pMOS, it is preferable to use a channel made from a composite SiGe material or tensile-strained silicon rather than a standard silicon channel, since the holes in these materials have higher mobility.

We may wonder in which direction the MOS will evolve and what will be its ultimate structure, if such a thing should really exist. According to the ITRS road map up to 2020, the date at which the 8 nm technological node should be reached in production, the materials used in transistors will continue to be optimised to ensure gate control efficiency and at the same time improve the mobility of the charges in the channel. Another objective will be to maintain a high contrast between the blocked and conducting states of the transistor. This means that, at the conduction threshold, all the positive (negative) charges must be expelled from the n (respectively, p) channel. We speak of a fully depleted channel, as opposed to a partially depleted channel. We are thus seeing structures where multiple gates almost completely cover an ever thinner channel, and where the latter may be insulated from the substrate (silicon-on-insulator or SOI).

Since the tendency is to confine charges to a single space dimension, the transistor channel is looking ever more like a quantum nanowire of the kind described in Sect. 2.3. Put another way, given the sizes becoming relevant here and the structures considered, there is no real discontinuity between MOS electronics and molecular electronics. In this situation, it is thus difficult to announce the ultimate MOS transistor for 2024. The challenges of fabrication, however, are clearly present, as are the challenges involved in the design of new circuit architectures.

Apart from the problems raised by extreme miniaturisation, another major hurdle is the energy consumption of these circuits. The size reduction of transistors has been accompanied by a global increase in the current consumption of the circuits. The smaller the transistor, the greater the risk of current leakage between source and drain, to the extent that a non-negligible current can pass through the transistor, even when it is in the blocked state. Moreover, the higher the transistor density, the more metal interconnects there will be between blocks of transistors. In one of today's integrated circuits, the interconnects total several kilometers and are distributed over a dozen or so levels (see Fig. 2.18). This huge increase in the amount of metal is obviously a source of Ohmic losses and consequent heat dissipation. Finally, to control a switch transistor, a given amount of energy is required. The faster the command, the higher the requisite power. This explains why integrated circuits need to consume ever more energy. It also explains why there will only be a very slight increase in the operating frequencies of microprocessors between now and 2020. Circuit speed will be improved by connecting processors in parallel, just a few for the moment (as advertised by Intel), but hundreds by 2020.

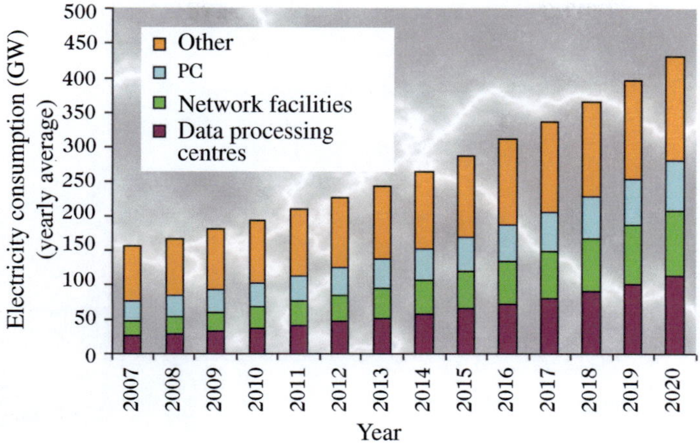

Fig. 2.27 World electricity consumption by information and communications technology. Adapted from [42], © IEEE 2008. This consumption currently represents about 8 % of world electricity consumption, but by 2020, this should rise to around 14 %. Note that an average power of 500 GW corresponds to an energy of 4 500 TW h

In 2012, data processing centres, computers, mobile phones, and other information and communications systems represented 5–10 % of the global electricity consumption (see Fig. 2.27).[9] Forecasts for the 2020s suggest values between 10 and 20 %. This further strengthens the motivation to reduce energy dissipation in tomorrow's circuits. Several solutions are already under development and others are being studied to overcome the inherent difficulties: the design of multicore processors, the use of non-volatile memory, the introduction of optics (see Sect. 2.6), the prospect of neural logic, the development of quantum computers, and others. All these solutions correspond to rather profound changes in circuit architecture, hence moving away from the traditional von Neumann architecture. Many of them involve innovations procured by nanotechnology. Failing a radical breakthrough, although that will happen at some point, nanotechnology is indeed in the process of engineering profound changes in the world of electronics.

2.5.2 *Spin Electronics and Nanomagnetism*

While for many of us electronics is synonymous with the use of charge currents to process and carry information signals, the developments brought about by nanotechnology are gradually introducing a new idea, namely spin electronics, according to which the electron spin could be used to supplement, or even replace, electri-

---

[9]Note that a hundred Google searches require as much energy as the operation of a low energy bulb during 1 h (30 W h).

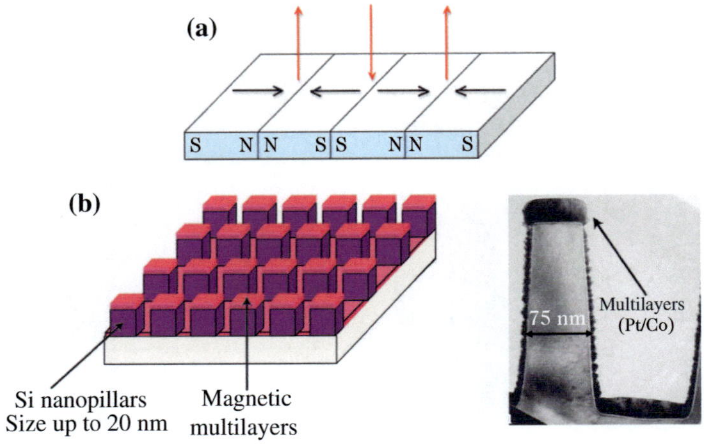

Fig. 2.28 **a** Track on a 2D magnetic recording medium. The area of one magnetisation domain is of the order of $200 \times 40\,\text{nm}^2$. Compare with the high density of magnetic pillars **b** prefiguring nanostructured media: pillar array (*left*) and single pillar (*right*). © CEA/CNRS, Bernard Dieny, LETI-Spintec/LTM

cal charge as information carrier. 'Spin electronics' and 'spintronics' have already become accepted terms in the English language. The connection with nanomagnetism is clear, because local magnetisation necessarily acts on electrons in their guise as elementary magnetic compasses (see "Spin and Magnetic Domains" hereinafter). The interaction between an electron current and local magnetisation imposed by magnetic films is already exploited in magnetic memory systems, where progress has been spectacular over the past few years. Indeed, these advances already promise the development of new circuit architectures. Apart from magnetic memory systems, spin electronics opens the way to new concepts such as the manipulation of magnetisation by an electric field, the manipulation of single spins, and the use of 'pure' spin currents.

Magnetic Memory

Changes in the orientation of the Earth's magnetic field have been recorded in the polar ice over many centuries. It is thus no surprise that magnetism should be exploited to produce 'non-volatile' memory systems, hard disks, and random access memory.

A hard disk is a magnetic system for recording data that can be both read and written. Information is stored in a thin magnetic film and coded by transitions between domains of opposite magnetisation (see Fig. 2.28). During writing, the orientation is forced one way or the other by the magnetic field produced by a small electromagnet (the write head), placed very close to the surface. During reading, the magnetisation thus recorded acts on the magnetoresistance contained in the read head. A zero or one will be read, depending on the value of the magnetoresistance.

Spin and Magnetic Domains

On the atomic scale, electrons and nuclei can be treated as spheres rotating about their own axis. They thus have an intrinsic angular momentum, known as spin, and with it an associated spin magnetic moment. These particles thus act as tiny compasses.

So what happens when different materials are subjected to a magnetic field? There are three main families of magnetic materials with different types of behaviours:

- diamagnetic,
- paramagnetic,
- ferromagnetic.

Most materials are diamagnetic or paramagnetic. The directions of the nuclear and electron spins are distributed randomly and uniformly, in such a way that the resulting magnetisation, i.e., the total magnetic moment per unit volume, is zero. If we now apply a magnetic field, diamagnetic materials acquire a magnetisation proportional to the field but in the opposite direction, whence the effect of the applied field tends to be cancelled out and, from the outside, nothing appears to happen. In contrast, in an external magnetic field, paramagnetic materials acquire a magnetisation in the same direction as the applied field, although this magnetisation disappears when the external field is removed.

Ferromagnetic materials behave quite differently. These are made up of tiny domains in which the spins tend to line up spontaneously, even when there is no external field. These domains are bounded by walls that can move under the effect of an external field, and such displacements allow the magnetisation of each domain to line up with the field, in the same direction, thereby strengthening its effect. This phenomenon is much more significant than in paramagnetic materials, so ferromagnetic materials form a family on their own. A good example is iron, hence the name for this kind of material.

Let us cite two direct applications of these phenomena in nanoelectronics (see also the remainder of Sect. 2.5):

- When an electron moves through a ferromagnet that is magnetised in a certain direction, this motion will be significantly perturbed if its spin is oriented in the direction opposite to that, while it will be unaffected if its spin is oriented in the same direction as the magnetisation. One can thus modulate conduction in spin valves and magnetic tunnel junction devices.
- The domains in a ferromagnetic film are used for the purposes of magnetic recording. Each domain constitutes a tiny independent magnet that can be oriented by an applied field.

Note that the displacement of the domain walls under the effect of an external field explains how permanent magnets work. If an external magnetic field is applied to a ferromagnet, the walls move in such a way as to line up the magnetisation in the direction of that applied field. When the external field is reduced, the walls tend to return to their initial state, but not completely. This leaves what is known as a remanent magnetisation. In the technical jargon, there is a hysteresis cycle when the magnetic induction B is graphed as a function of the magnetic field H.

Nanotechnology has brought decisive progress in the development of hard disks, since the discovery of giant magnetoresistance in 1988 led to the development of efficient electronic nanodevices exploiting this phenomenon. The discovery itself won the Nobel Prize for Albert Fert and Peter Grünberg in 2007. The structure, shown schematically in Fig. 2.29a, is a spin valve comprising a normal metal layer (M) sandwiched between a soft ferromagnetic layer (F_1), whose magnetisation can be reversed by an applied magnetic field, and a hard ferromagnetic layer (F_2) with fixed magnetisation. When the layers F_1 and F_2 have the same direction of magnetisation, electrons with spin parallel to this magnetisation can easily cross the spin valve,

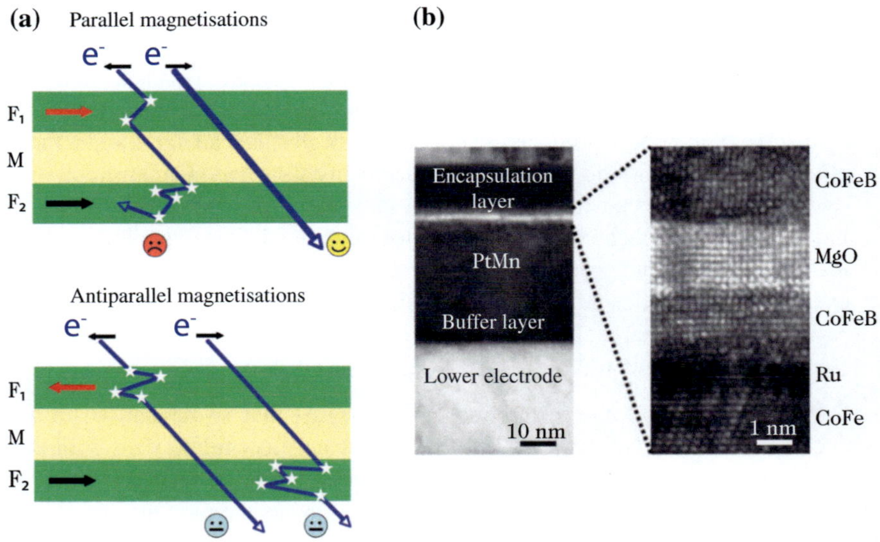

Fig. 2.29 **a** Schematic of magnetoresistance effect. Spin valve made from a metal layer M sandwiched between two ferromagnetic layers F_1 and F_2. The level of magnetoresistance is determined by the change in resistance when going from parallel magnetisations to antiparallel magnetisations in the two ferromagnetic layers. Courtesy of C. Chappert, *Institut d'électronique fondamentale*, Orsay. **b** Transmission electron microscope image of a magnetic tunnel junction (MTJ). Here the cobalt–iron–boron (CoFeB) alloy magnetic films are separated by a layer of magnesium oxide (MgO). Courtesy of H. Masnori, © Sony Corp

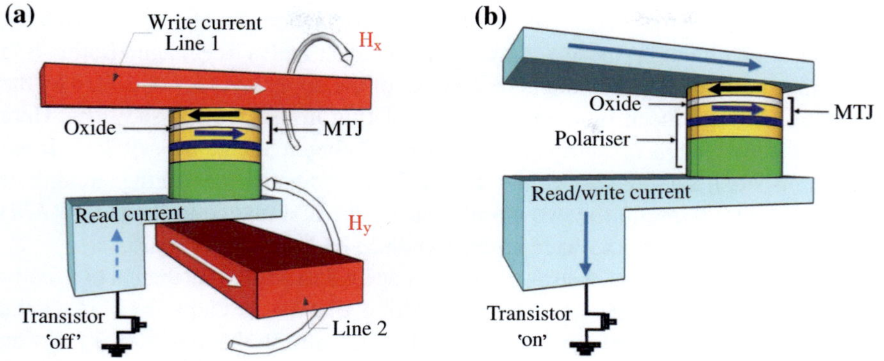

Fig. 2.30 **a** Schematic of a magnetic tunnel junction (MTJ) MRAM, where magnetic reversal of the upper layer is brought about by the magnetic field produced by the current circulating in the two crossed lines. The blocked transistor is in write mode. **b** Schematic of STT-RAM, where writing is achieved by injecting a polarised current directly through the MTJ (spin transfer). The polariser maintains the magnetisation direction of the lower layer. Reading and writing are achieved by the same current lines, although writing requires a stronger current than reading

whence the resulting resistance is relatively low. But when the magnetisations of F_1 and F_2 are antiparallel, electrons are strongly scattered in one or other of the two layers, whatever their spin orientation, and the resistance of the spin valve is thus very high. The relative change $\Delta R/R$ in resistance between the two states, which can be associated with the logic states 1 and 0, characterises the level of magnetoresistance. We speak of giant magnetoresistance whenever the ratio $\Delta R/R$ comes close to 100 %. Values exceeding 500 % are now obtained in magnetic tunnel junction devices (see Fig. 2.29b) [43], where a thin film of insulating oxide replaces the metal of the spin valve.

Hard disks have undergone even more spectacular development than microprocessors. Data storage capacity has been increasing by 150 % per year since 2007 to reach 0.5 terabyte/inch^2 in 2013. Current magnetic recording technology, based on continuous media with perpendicular anisotropy, is now very close to its physical limit, estimated at 1 terabyte/inch^2. Among the technological developments aiming to go beyond this limit by 2015, nanostructured media should provide storage capacities considerably greater than 10 terabyte/inch^2. Nanotechnology can fabricate nanoscale pillars which considerably reduce the area needed to record one bit of data (see Fig. 2.28b). Spin valves were introduced by IBM in 1997 as the basic element of the hard disks in our computers, and the magnetoresistive effect is still used today whenever there is a need to archive data at high densities.

Another challenge has been to use magnetic devices for random access memory (RAM). The problem was to demonstrate their compatibility with MOS technology while at the same time maintaining a capacity to switch rapidly between logic states. The basic principle here is not the same as for transistor-based RAM. A memory cell is shown schematically in Fig. 2.30a. A single MOS transistor plays the role of switch in series with a magnetic tunnel junction (MTJ) device. The logic state 0 or 1 results from the respective magnetisations of the two ferromagnetic layers, one with a fixed magnetisation direction (the lower layer) and the other with a free magnetisation (the upper layer). When writing, magnetic reversal of the upper layer is controlled by the magnetic field produced by the currents circulating in the two crossed lines. Here the transistor switch is open (blocked transistor). When reading, the switch is closed (pass transistor), and the state of the device (0 or 1) is read off from the strength of the current circulating in the two parallel lines. The first magnetic RAM (MRAM) systems using these devices were commercialised by Freescale in 2006.

With this configuration, a strong current is needed to write. It is thanks to a major discovery, the mechanism of spin-transfer writing, that the development of magnetic memory has been pursued down to sizes below 100 nm. In the new tunnel junction configurations (see Fig. 2.30b), it is the spin-polarised current of electrons through the junction which itself imposes the orientation of the free magnetic layer, if it is strong enough. The number of read and write cycles, i.e., the endurance of the memory, is as high as one would like, and switching times can be less than the nanosecond. The commercialisation of competitive spin-transfer memory systems (known as STT-RAM for spin transfer torque RAM) is expected in the forthcoming years, replacing semiconductor RAM.

As an extension of the spin transfer idea, there is further interest today in exploiting the displacement of domain walls by a spin-polarised current. A new kind of hard disk without mechanical parts would thus serve in sequence to read and write logical information on very large circuit areas. The potential for these devices was confirmed by major steps taken in 2011 [44]. Indeed, the propagation of domain walls induced by weak spin-polarised currents was demonstrated in various magnetic tunnel junction devices, and a memory system exploiting this phenomenon was integrated for the first time into a CMOS circuit of generation 90 nm by IBM (race track memory). Commercialisation is expected in 2020.

A New Paradigm: Magnetic Logic

The von Neumann architecture was the standard in computers for more than half a century, with a single memory unit holding all the instructions and data required or produced by the central processing unit (CPU). But today, this architecture has reached its limits, because access to the memory to read instructions or at the same time to read and write data limits the operating frequency and leads to very high power consumption. To get around these difficulties, today's computers run with a hierarchical memory structure containing several levels of cache memory.[10] But these in turn, so much faster and with such increased capacity as compared with a few years ago, have become preponderant sources of power consumption since the 90 nm technological node. This concerns in particular the static regime, where memorised data must be continually refreshed.

In contrast, magnetic logic offers the prospect of an integrated view of computation and memorisation with non-volatility of the memory. In such an architecture, non-

[10] In computing, a cache memory is one that temporarily records copies of data coming from other sources, such as DRAM random access memory, in order to reduce the time required by the processor to access this data (in read or write).

volatile memory elements would be dispersed among the computation operators, eventually making the processor itself non-volatile. Static consumption would then be totally suppressed, either in standby mode or during electricity cuts. The power consumption due to data transfers would be reduced too. The recent design of adders using magnetic tunnel junctions, compatible with CMOS technology and with very low consumption [45], already illustrates the progress that has been made in this direction.

Spin Electronics Beyond CMOS Technology

Apart from the new memory and logic concepts exploiting the spin-transfer process in CMOS technology, more exploratory ideas are also being developed, no doubt prefiguring the spintronics of the future. Two approaches are beginning to stand out:

- Manipulation of magnetisation by an electric field rather than a charge current, which tends to use too much energy.
- Replacement of charge currents by 'pure' spin currents, without displacement of electrons.[11]

These approaches are based mainly on the development of multifunctional materials that are simultaneously both ferromagnetic and ferroelectric and metal–insulator or semiconductor–insulator hybrid structures. Some studies have already observed magnetisation reversal and magnetic domain formation by application of an electric field in a ferroelectric material. Others report the propagation of a pure spin current along a magnetic track several millimeters long merely under the effect of a temperature gradient. A new discipline known as spin caloritronics has thus come into being, with possible applications for a new generation of electronic components.

2.5.3 Quantum Components

The extreme miniaturisation of components naturally favours the appearance of quantum effects by increasing the confinement of electrons, reducing their number within the component itself, and reducing their interactions with the surrounding medium. While the quantum tunnelling effect is commonly used in devices like the flash memories of our USB keys or spin transfer magnetic memory systems, other manifestations of quantum confinement, such as the discretisation of energy levels or conduction channels, or coherence effects associated with electron wave functions, are not yet directly exploited as such in electronic circuits. A wise strategy here is to assess the influence of these effects on ultra-miniaturised versions of existing components using models with differing degrees of complexity. The investigation of new components remains primarily a laboratory activity.

Some rather unexpected opportunities present themselves for quantum dots and wires. One of these exploits their thermoelectric properties to recover local heat losses

[11]The existence of such currents is a consequence of the spin Hall effect, but a serious discussion would go beyond the scope of this book.

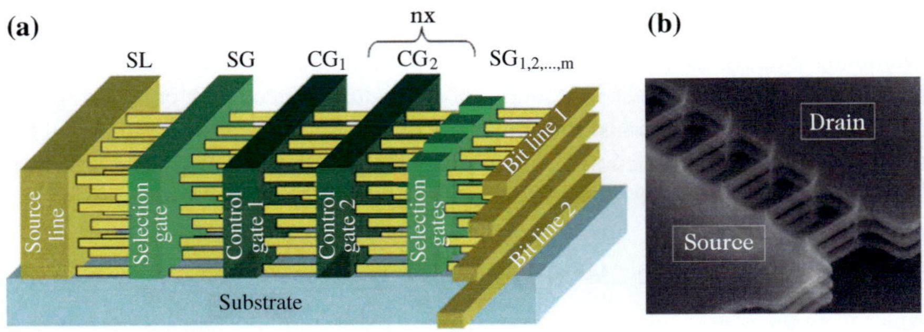

Fig. 2.31 **a** Schematic of a 3D nanowire memory device. **b** Three-level stack of nanowire arrays made using silicon technology. From [46], © IEEE 2008

in circuits in the form of electrical energy. Here silicon nanowires have been used in some remarkable 3D stacks of high-performance transistors (see Fig. 2.31). Although the technology implemented in this top–down approach remains very sophisticated, an industrial application may not be so far ahead.

2.5.4 Components Involving Few Electrons

Ten years ago, a great deal of enthusiasm was raised by the prospect of single-electron electronics (although in practice, involving small numbers of electrons), which makes use of the phenomenon known as Coulomb blockade[12] in quantum dots. Apart from the technological difficulties involved in making single-electron transistors (SET) that could operate at room temperature, and in particular the need for very low gate capacitances to ensure that the storage energy of an elementary charge remain greater than the thermal energy, a major obstacle was the fact that the command to open the SET depends not only on the charge stored in the quantum dot under the gate, but also heavily on the drain–source voltage. Attempts were then made to realise hybrid systems combining the advantages of the SET, such as low consumption and the possibility of multiple logic levels, with those of the MOS transistor, including voltage gain, control of the transistor output, and so on. Various systems now attest to a good level of fabrication control and a high reproducibility of operation of the SET at room temperature, giving impetus to further research in this field (see Fig. 2.32).

[12]Coulomb blockade is a mechanism that can control the passage of charge carriers one by one. In a quantum dot, the electrostatic repulsion between electrons leads to an increase in the confinement energy with the number of electrons. This means that, each time another electron is confined within the dot, more energy must be supplied.

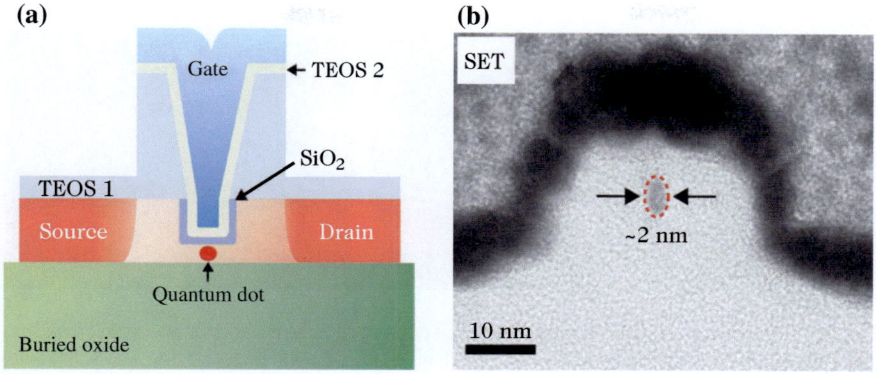

Fig. 2.32 **a** Schematic cross-section of a single-electron transistor (SET). From [47], © AIP Publishing LCC, 2010. **b** Transmission electron microscope image of the SET with a silicon quantum island of radius around 2 nm. Note that the tetraethyl orthosilicate (TEOS) used here is a chemical precursor. It forms silica (SiO_2) by hydrolysis

2.5.5 Molecular Electronics

Molecular electronics is usually understood to mean the fabrication by the bottom–up approach of electronic components involving just one or a few molecules, and by extension, just one or a few objects of comparable size to a molecule, i.e., measuring typically a few nanometers. These nano-objects include organic molecules, carbon-based nanostructures such as nanotubes and graphene, metal or semiconductor nanoparticles, and nanowires. Naturally, these include some of the nano-objects discussed in Sect. 2.3, except that there we were viewing them as quantum components, generally fabricated using a top–down approach. In the following, we shall consider only carbon-based nanostructures and their possible association with other molecules.

Significant progress has been made with transistors using one or more carbon nanotubes (see Fig. 2.33). These now have cutoff frequencies above 80 GHz [48]. However, nanotubes do raise a difficulty for large scale use in electronic circuits. The main problems are selective filtering of nanotubes in the case of transistors using nanotube 'carpets' and manipulation in the case of single-nanotube transistors. For these reasons, but also due to the unusual benefits promised by graphene, research effort has now swung heavily in favour of this other molecular arrangement of carbon atoms.

As mentioned in Sect. 2.3, graphene has remarkable electron conduction properties. Cutoff frequencies of 300 GHz have been demonstrated for graphene transistors on an SiO_2 substrate, where the graphene ribbon was self-aligned on a nanowire gate. The downside of its high electron mobility performance is that graphene is difficult to use in logic circuits, because it is hard to produce a blocked state that is clearly distinct from the conducting state. Having said that, in parallel with this

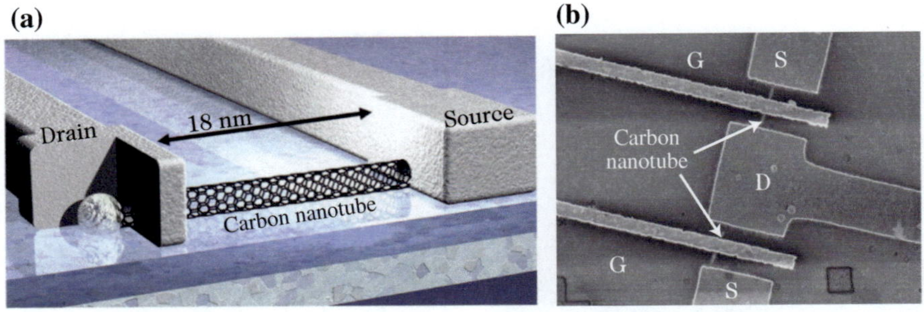

Fig. 2.33 Application of carbon nanotubes to electronics. **a** Schematic view of a field effect transistor using a carbon nanotube as conduction channel. © Infineon Technologies AG. **b** Scanning electron microscope image of a CMOS inverter made from two carbon nanotube transistors. The symbols D, G, and S stand for the common drain of the two transistors, the two gates, and the two sources, respectively. © CNRS Photothèque, Bernard Plaçais, *laboratoire Pierre-Aigrain*

research, graphene has stimulated unprecedented interest in flex circuits, or flexible electronics, an area where it can outdo everything so far achieved with polymer electronics. Particularly promising are the demonstration of transparent electrodes for large flexible screens and flexible transistors with very high cutoff frequencies [49].

In another line of research, it has been shown that a supramolecular spin valve can be made using a single-walled carbon nanotube upon which certain high spin molecules have been grafted. A giant magnetoresistance effect with $\Delta R/R = 300\,\%$ has been demonstrated at low temperature in a magnetic field of 1 T [50].

2.5.6 Memristors and Neurally-Inspired Architectures

There is nothing new about the idea of taking inspiration from neural mechanisms and the functioning of the human brain to design intelligent electronic circuits and systems capable of learning and self-correction. However, the development of such neuromorphic systems has come up against the problem of reproducing synapses in electronic circuits. Indeed, in our brain, more than 100 billion neurons are connected to one another by synapses with around 10 000 connections per neuron. Not only is this a remarkably high density of connections, but each synapse consumes only a tiny amount of energy and can memorise information over very long periods of time, even several decades. Here it is entirely thanks to progress in nanotechnology and the realisation of nanocomponents with a memory effect, and in particular the resistive memory elements known as memristors, that the study of neurally-inspired electronic architectures based on compact devices has received a new breath of life.

By definition, the electrical resistance of a memristor changes value each time an electrical information signal (current or voltage) is applied to it. A memristor is thus a variable resistor which, through the value of its resistance, reflects its own history.

First announced some forty years ago, memristors were only effectively demonstrated toward the end of the 2000s, by a research team at Hewlett-Packard [51].

A carbon nanotube transistor (see Fig. 2.33) coated with a light-sensitive polymer has been used in a recent version of the memristor [52]. Initial illumination of the polymer produces charges which are trapped along the interface between the gate and the nanotube. The conductance of the nanotube then changes steadily with each voltage pulse between the drain and the source, this modifying the number of trapped charges. This kind of memristor has been used as an elementary synapse in the construction of an artificial neural network with two inputs and one output. After specifying a threshold bias to distinguish the output logic levels 0 and 1, only seven learning cycles were needed to obtain an output signal agreeing with the expected signal. The development of much more sophisticated neurally-inspired architectures with many inputs and outputs is currently stimulating a great deal of interest for applications in image processing or shape recognition. Such developments will clearly require a concerted effort by nanotechnologists, electronicists, mathematicians, and neurobiologists.

2.5.7 Conclusion

Components involving small numbers of electrons, molecular electronics, and neurally-inspired architectures are alternative approaches to electronics that have been to a large extent encouraged by the development of nanotechnology. In contrast to CMOS logic, or even spin electronics, many investigations in these areas remain largely exploratory, without necessarily featuring on any particular road map. Considering the way things are going, should we conclude that the exploitation of quantum effects in dots and nanowires discussed above has now been shelved and that such effects are only relevant if we need to take them into account when modelling standard devices? The same question could be asked regarding ultimate molecular electronic systems. In reality, there will be no radical evolution toward a 'quantum' electronics until transport and storage processes can be made to involve a large number of electrons through their charges and/or spins, and until for example the wave functions of electrons confined in a nanowire can diffuse at its ends into the large reservoirs of energy levels constituted by drain–source Ohmic contacts. The price to pay for any radical development is that one must move away from the standard schemes of CMOS technology, and in particular, one must no longer restrict to dissipative silicon–metal Ohmic contacts, but make use of high barrier potentials as in tunnel junctions and seek out ways to 'get along with' quantum fluctuations.

2.6 Is Nanophotonics an Exact Parallel of Nanoelectronics?

Although the term 'nanophotonics' has become commonplace today, it remains more difficult to specify what it involves than the term 'nanoelectronics'. There is no doubt that optoelectronic components such as laser diodes, modulators, and detectors are getting smaller, especially to satisfy the needs of optical telecommunications, but there is no equivalent of Moore's law in optics, of the kind so commonly referred to in electronics. The most intuitively obvious limiting scale that can be imagined for optics is the wavelength, i.e., a scale close to the micron for visible or near-infrared waves.

Nanotechnology has been able to make nanometric light sources, just as it can carry out optical analysis of objects on the nanoscale thanks to the development of near-field microscopy. Furthermore, there is the prospect of nanomanipulation using optical tweezers, not to mention the development of metal and metal–dielectric devices which exploit localised or surface plasmons and bring photonics straight into the subwavelength domain. Even though it would appear that more elaborate components such as lasers or modulators must remain for the moment on the scale of the micron or ten microns and that they are unlikely yet to witness the same kind of evolution as the transistor, the structures used to make them are often specified to a comfortably submicronic accuracy. To see this, just note that the quantum well structures commonly used in semiconductor components are multilayer systems where the thickness of any individual layer is of the order of 10 nm. As a further example, consider an optical resonator of micrometric dimensions. In principle, a fabrication accuracy of less than a nanometer is required if the resonance wavelength is to be fixed with tolerance better than one in a thousand. So in a certain sense, while the progress in nanophotonics owes much to the various nanotechnologies, the latter have in their turn greatly benefited from the challenges raised by nanophotonics, not to mention the new concepts they have brought on the scene.

Among these new concepts, two of the most significant over the past decade have been the photonic crystal and metamaterials. Photonic crystals, obtained by periodic structuring of materials, have led to the idea of photonic circuits on a par with integrated electronic circuits [53, 54]. Even though the prospect of an optical computer has been abandoned for some time now, the joint integration of photonic and electronic components remains on the cards for optical links between electronic chips or between functional units within the same integrated circuit, or for light signal processing in telecommunications. Once again, one of the aims in combining nanoelectronics and nanophotonics is to reduce energy consumption. Extending in some sense the concept of the photonic crystal, metamaterials open the prospect of novel physical effects such as negative refractive indices and electromagnetic invisibility.

For these reasons, the field of research encompassed by nanophotonics is every bit as broad as the one embraced by nanoelectronics. Since it would be impossible to cover every aspect of it, we shall focus here on the most emblematic among its components, viz., optical micro- and nanosources such as lasers, single-photon sources, and so on, and new concepts such as photonic crystals and metamaterials, as well as the prospects for important applications in the field of photovoltaic conversion and optical interconnects.

2.6.1 *Optical Micro- and Nanosources, Lasers, and Controlled Photon Sources*

A light source can be characterised in many ways, e.g., by its efficiency, power performance, excitation and emission modes, spectral qualities, emission directivity, fields of application, and others. Regarding micro- and nanosources, the main aim is obviously size reduction, but with a view to preserving the efficiency of light emission. In this respect, semiconductor quantum dots represent choice nanoscale light sources because they can emit strongly with a spectral response that can be adjusted from the infrared to the ultraviolet depending on the material used and the size of the dot. However, if we consider isolated quantum dots, the resulting luminescence is generally dispersed in all space directions, whence only a small fraction of the emitted photons can be captured. To increase the directivity of this emission, the solution is to place the dot in an optical microcavity. The size of the whole system is then inevitably micrometric, but use of a cavity nevertheless opens up a range of possibilities.

From a source comprising a single quantum dot in a photonic crystal microcavity, it has been possible to produce, one by one and with high efficiency, a series of absolutely indistinguishable photons [55]. Such sources are key elements in quantum computing. Pushing the concept of the single quantum dot in a microcavity even further, extremely bright sources have been developed, emitting twin photons [56]. These are entangled photons, with the characteristic that, if we measure the properties of one, we instantaneously know the properties of the other, whatever the distance between them at the time of measurement. This feature has a huge potential for application in quantum teleportation, quantum cryptography, and more generally, quantum information processing. Figure 2.34a shows an example of a twin photon microsource, comprising a system of two twinned microcavities.

There is another way to focus the light produced by a quantum dot into a directive beam, namely to insert it in a nanowire which thus constitutes an optical waveguide, like an optical fibre. By thinning the nanowire at one end, a large fraction of the light propagating within it can be recovered (see Fig. 2.13a).

In all the examples of micro- and nanosources we have just described, photons are emitted spontaneously, although this emission is nevertheless controlled and considerably reinforced relative to the what would be produced by an isolated quantum dot. The idea of concentrating all the spontaneous emission in a single cavity mode gave rise in the 1990s to the prospect of a thresholdless laser, or at least to the possibility of a very low threshold (see below). This is the conclusion from experiments on quantum dot microlasers, where the spontaneous emission rate in the principal mode of the microcavity lies between 20 and 80 %. In optical pumping, the pumping power at

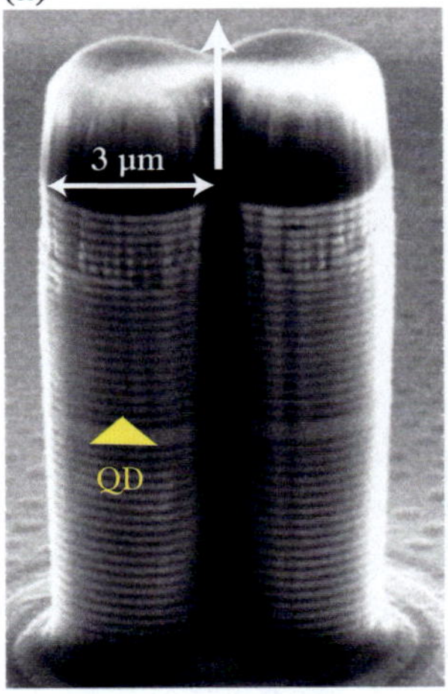

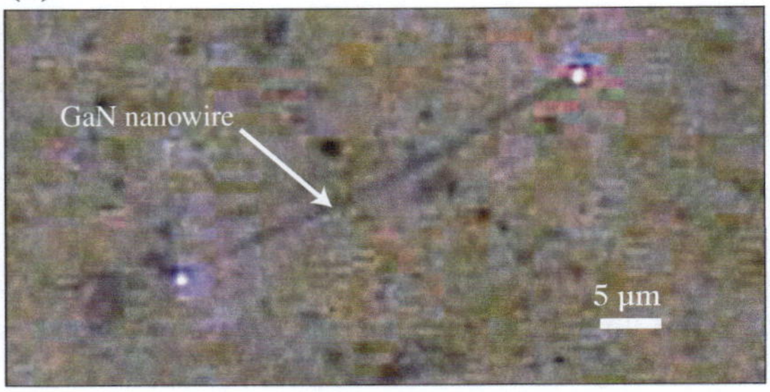

Fig. 2.34 Micro(nano)sources. **a** Electron microscope image of a microsource emitting twin infrared photons (wavelength $\lambda \sim 1.48\,\mu\text{m}$). The source comprises a system of two twinned vertical microcavities [56]. The mirrors of the two cavities, made from Bragg multilayers of thickness $\lambda/4$, are located on either side of the quantum dot (QD) active layers. © CNRS Photothèque, Pascale Senellart, *Laboratoire de photonique et nanostructures*, Marcoussis. **b** Optically pumped gallium nitride (GaN) nanowire microlaser, emitting in the ultraviolet ($\lambda \sim 0.375\,\mu\text{m}$). From [57], © Nature Publishing Group 2002. The wire is several tens of microns long and has nanometric diameter

the threshold of laser emission is just a few nanowatts and the modulation frequency of the laser can exceed 100 GHz. In electrical pumping, record threshold currents of 0.15 and 7.8 μA have been obtained at 150 and 300 K, respectively [58, 59]. On the other hand, with the exception of a first significant result in 2013 [59], the emitted power remains less than the microwatt. Further nanotechnological developments will thus be needed to improve the performance of these microlasers.

In parallel with studies of ultimate quantum dot sources, semiconductor nanowire lasers have seen remarkable development. Under optical or electrical pumping, a nanowire several tens of microns long 'naturally' constitutes a laser in which the amplifying medium is the semiconductor core and the optical cavity is bounded by the ends of the nanowire (see Fig. 2.34b) [57].

Semiconductor Lasers

When it is subjected to electrical, optical, or chemical excitation, any medium, whether it be a gas, liquid, or solid, may restore all or part of the excitation energy in the form of light emission. In the most common case, the photons are emitted spontaneously and we speak of a luminescent medium.

However, another process may occur if the photon density is high enough. This is stimulated emission, which underlies the laser effect, where each photon stimulates the emission of another photon. In order to actually obtain light amplification, however, this stimulated emission must predominate over the opposing absorption process. In other words, there must be more particles in the excited state than in the ground state, a situation known as population inversion.

These very general underlying principles of laser operation are also relevant in semiconductor lasers, where excitation is produced by injection of an electric current. The current transports electrons from an n-doped layer and holes (missing electrons) from a p-doped layer. Recombination of the electrons and holes in a central non-doped layer (the emitting layer) produces the light emission (see Fig. 2.35). The potential applied between the upper electrode and the substrate is used to control the current, i.e., the density of injected electrons and holes, and hence also the population inversion. It should be stressed, however, that only direct bandgap semiconductors like GaAs, InP, and GaN[13] can produce laser emission, and this is unfortunately not the case for silicon.

As for other lasers, an optical cavity made from two mirrors is used to store the photons and hence allow stimulated emission. The difference here lies in the geometry of the cavity, which is truly minuscule compared with the cavities in other lasers. The most common is the waveguide geometry, where the emitting layer itself serves to guide the light and the mirrors are simply the cleaved facets at the ends of the semiconductor crystal. Typical lengths range from a few tens to a few hundred microns. Another geometry is the vertical cavity, where optical emission occurs in a direction perpendicular to the emitting layer. The cavity mirrors are then made from Bragg multilayers of thickness $\lambda/4$ with very high reflectivity ($R > 99\,\%$). The length of the cavity is less than a micron.

Apart from their miniaturisation, semiconductor lasers differ from other lasers by their high efficiency (up to 50 %), their broad spectral coverage from the mid-infrared to the ultraviolet, and their compatibility with electronic circuits. Nanostructures, wells, wires, and quantum dots can be used today to confine charge carriers (electrons and holes) in ever smaller volumes, thereby limiting the number of carriers and the current required to trigger the laser effect. New geometries with photonic crystal cavities can also concentrate photons in very small volumes. For these new cavity geometries, the laser threshold current is close to a tenth of a microampere, the threshold itself thus becoming less significant among the laser characteristics.

[13]In a direct bandgap semiconductor, the process of electron–hole recombination can occur without supply of extra energy or momentum. In an indirect bandgap semiconductor, the process involves the energy and momentum associated with an elementary vibration of the crystal lattice called a phonon. In this case, the probability of a three-body process is clearly much lower.

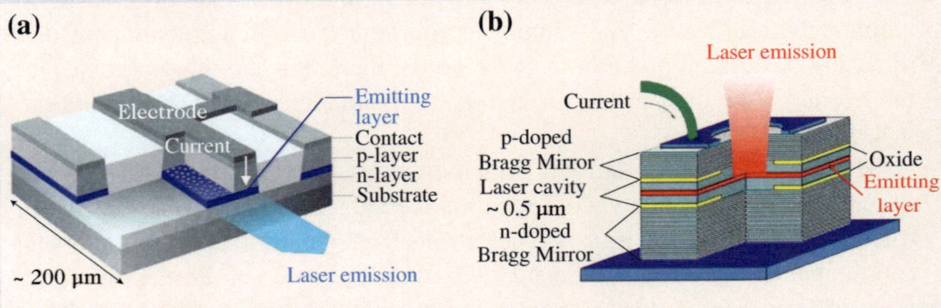

Fig. 2.35 Main types of laser. **a** Waveguide laser where the emitting semiconductor layer is also the layer that guides the light. **b** Vertical cavity laser where laser emission is produced in a direction perpendicular to the emitting layer

2.6.2 Photonic Crystals and Metamaterials

The motivation for research on photonic crystals since the middle of the 1990s has been to control the propagation of photons by endowing materials with a periodic structure, just as electron conduction is governed by the periodic structure of solid crystals. The 1D photonic crystal is nothing other than a Bragg mirror made from alternating layers of thin films with different optical indices and effective thickness 'lambda over four'. The extension to two and three dimensions raises a genuine technological challenge as far as fabrication is concerned. It is through progress made in etching that 2D planar structures have been fabricated for potential applications in guided optics. Figure 2.36 shows a waveguide made in a semiconductor membrane suspended in air. With similar geometries and by judicious positioning of the etched holes, microcavities can be achieved with truly impressive quality factors, exceeding 1 000 000 [60]!

Since then, many groups have gone back to and improved these miniaturised waveguides and optical cavities to develop innovative components, such as quantum dot microlasers, ultrafast all-optical photonic crystal modulators, photonic crystal components, light bends, superprisms, demultiplexers, interferometers, and beam splitters and combiners. Here we can only list them by name, referring the reader to the first volume in this series [61].

One of the most surprising properties a photonic crystal can have is without doubt a negative refractive index at certain wavelengths. The incident wave and the wave refracted by the crystal are then located in the same half-plane specified by the normal to the crystal, exactly as though the photonic crystal had a negative refractive index according to Snell's law. In actual fact, the value of the negative index of the photonic crystal depends on its interface with the surrounding medium, and the crystal cannot therefore be considered as a homogeneous medium with negative index. It is in this

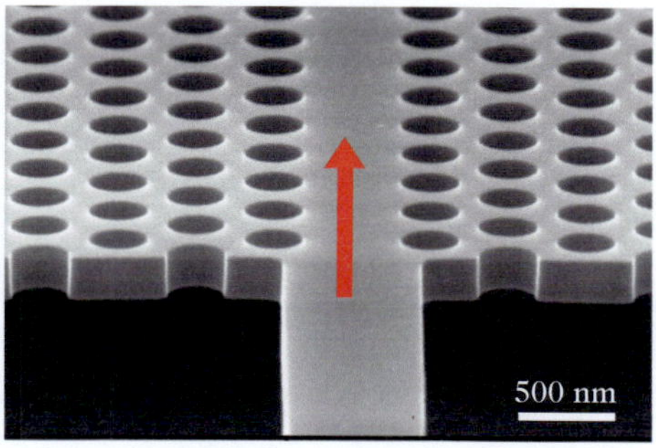

Fig. 2.36 Membrane-type photonic crystal waveguide [62]

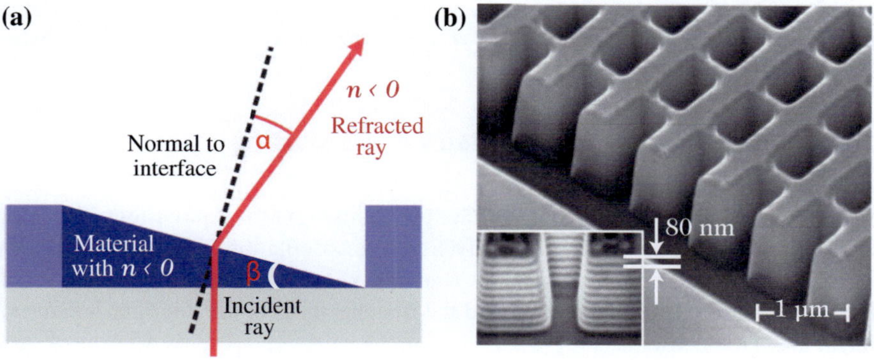

Fig. 2.37 **a** Negative refraction. **b** Metamaterial with negative refractive index made by superposing alternating silver (30 nm) and magnesium fluoride (50 nm) grid structures. From [63], © Nature Publishing Group 2007

context that metamaterials have been developed. These are composite structures in which the basic motifs, much smaller than the relevant wavelength, provide ways to approximate a homogeneous material in which one can simultaneously control the dielectric permittivity and the magnetic permeability, and hence also the refractive index, which depends on these two parameters. An example is shown in Fig. 2.37. The possibility of controlling these parameters opens up quite novel prospects. As a first example, if we could obtain a negative index metamaterial with index close to -1, we might be able to make flat lenses which would produce, after focusing, an ideal image of a point source. Another emblematic example is the invisibility cloak, which would force electromagnetic waves to avoid an object without otherwise modifying their propagation elsewhere in space (see Fig. 2.38).

The first demonstrations of negative index superlenses or invisibility cloaks have been reported in the microwave range. Extension to the infrared and visible remains a technological challenge given the very small size of the motifs that would be involved (nanoscale). Significant progress has nevertheless been made in the fabrication of guided optic invisibility cloaks (see Fig. 2.38).

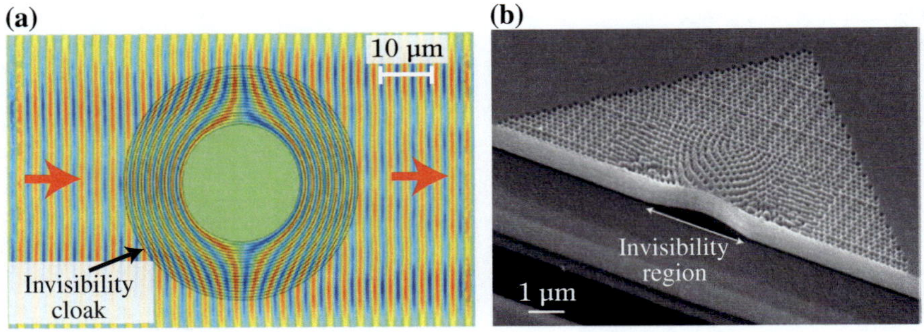

Fig. 2.38 **a** Model of the propagation of a wave going around an invisibility cloak in the infrared [64]. **b** Realisation in silicon for operation at a wavelength of 1.55 µm. From [65], © Nature Publishing Group 2009

2.6.3 Plasmonics for Photovoltaics or Biosensing

In the field of optics, metals exhibit surface plasmons or plasmon polaritons (SPP), the creation and propagation of which involves resonant oscillation of surface charges. By extension, the term 'plasmonics' refers to attempts to exploit this resonant interaction between electromagnetic radiation and the free electrons at the interface between a metal and a dielectric.

The efficiency of plasmonic nanoantennas (or nantennas) comes from the extreme confinement of the electromagnetic field in the vicinity of metal nanostructures. The strong fields there can be used to enhance the fluorescence of nano-emitters and also nonlinear optical processes such as Raman scattering. If we consider a nanosphere placed at nanometric distance from a metal plane, the electromagnetic field in the gap between sphere and plane may be more than a hundred times stronger than that of the exciting field (see Fig. 2.39a).

The enhancement phenomenon can be used not only to strengthen light emission, but also to increase the absorption of materials in the reverse situation where the light comes from an external source. This is what happens in photovoltaic conversion, where sunlight must be absorbed by the material used for conversion. One of the main problems so far encountered relates to the fact that significant thicknesses of material are needed to absorb all the light from the ultraviolet to the infrared. Since the idea is to deploy large areas of solar cells, the cost in terms of materials, including silicon, can soon become prohibitive. In this context, metal nanoparticles deposited on the surface of the photovoltaic element could scatter the incident light and thus lengthen the absorption path in lateral directions, thereby increasing the probability of absorption. In a different way, the insertion of metal nantennas within the conversion material may also provide an interesting solution for increasing the local absorption by the material and thereby reducing its useful thickness. Finally, light can also be trapped by nanostructuring the metal of the electrical contact underneath the photovoltaic element, thus exploiting the diffractive nature of the metal grating. All these proposals are illustrated in Fig. 2.39b.

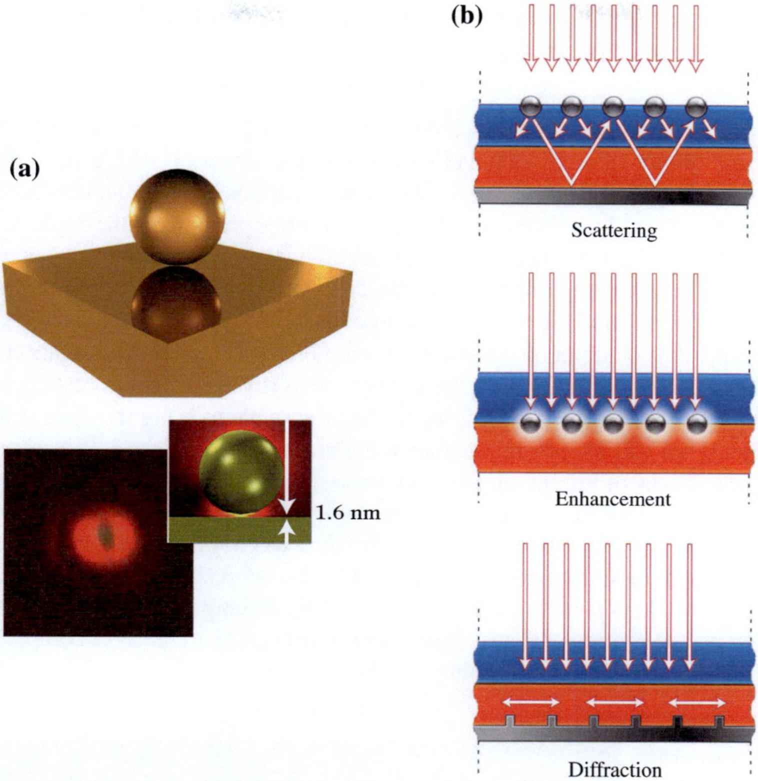

Fig. 2.39 **a** Enhancement of the electromagnetic field by a gold nanosphere coated with polymer (thickness 1.6 nm) on a gold plane and measured image of the scattered field. Adapted from [66], © American Chemical Society 2010. **b** Metal nanostructures and processes exploited to optimise absorption and hence photovoltaic conversion. Adapted from [67], © Nature Publishing Group 2010

In a different context, enhancement of the near field by metal nanostructures can also be used in the detection of very small amounts of biological molecules, in the field known as nanobioplasmonics. The idea is to detect the spectral shift of the plasmonic resonance when the biomolecules attach themselves to the nanostructures. To achieve this, specific molecular species must be grafted onto the nanostructures. These are species that the target biomolecules will recognise and fix onto. Not only is it possible to sense the chosen biomolecules, but one can also monitor their evolution within an artificial membrane, just as one would within a cell membrane.

2.6.4 Telecommunications and Optical Interconnects: Links Between Photonics and Electronics

Since the development of the optical fibre in the second half of the twentieth century, optics and photonics have joined electronics at the very heart of the telecommunications industry. Although wireless telecommunications has seen an unprecedented development recently, with the proliferation of mobile phones and the deployment of wifi communications protocols, fibre optics still has some major advantages. A fibre network can carry many communications channels in parallel, leading to throughputs in excess of 100 Tbits/s. A single fibre can itself carry several channels, each with a typical throughput of 10 Gbits/s. These fibres suffer lower propagation losses than their electrical counterpart, the copper cable (by three orders of magnitude). Even though for reasons of cost it would be impossible to bring the optical fibre to every household, fibre optic interconnects are becoming the norm in the professional world as a means of linking up the systems and subsystems of data acquisition and processing. An illustration is given in Fig. 2.40, which shows the steady progress of point-to-point optical links to the detriment of their electrical counterparts over recent years. This progression concerns not only connections between relatively distant systems, but also rack-to-rack, card-to-card, and chip-to-chip interconnects in servers, calculators, and routing systems. The trend toward optics is clearly accompanied by an increase in the throughput of numerical data.

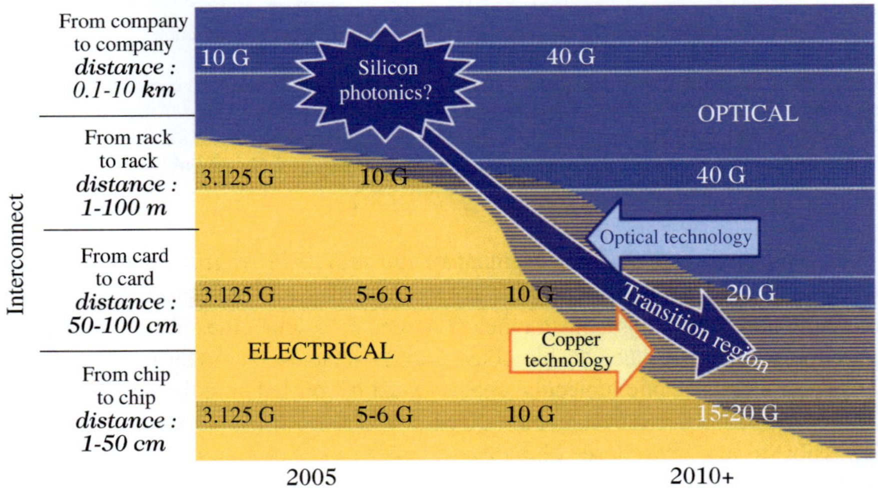

Fig. 2.40 Gradual transition from electrical to optical interconnects between entities communicating at short range (<10 km). Communication throughputs and their evolution during recent years are indicated in Gbits/s (G) for each kind of interconnects. Optical and copper interconnects coexist in the so-called transition region. Credit: CEA-Leti

While optical interconnects based on fibre optics have undeniable advantages over copper, the fact that the electronic and optical parts of systems remain strictly separated from one another is still a drawback for the shorter range links. Compactness is still a problem and production costs are too high. In this context, photonics in general and photonics on silicon in particular seem to be extremely promising options for finally superseding electrical interconnects made from copper. With silicon photonics, electrical signals can be converted directly to optical signals in the input connector, just as the reverse conversion from optical to electrical can be achieved directly in the output connector. In a certain sense, the electrons and photons follow the same path! This is the idea behind active optical cables, which can be connected to network devices without prior conversion of the signal, since everything is dealt with in the cable connector. The basic components of the connector are usually a light source, a modulator to convert the electrical signal to an optical signal, a detector for the opposite conversion, and some optical bits and pieces such as waveguides, beam splitters, light bends, optical gratings, and so on. All these components can be made using silicon technology, with the exception of the laser source, which is generally a III–V semiconductor microlaser positioned on the silicon circuit.

The possibility of exploiting photonic components in silicon circuits (between or within electronic chips) arises naturally enough when the circuits are used to process information that is already in the form of optical signals. This is the case in the material architectures of tomorrow's Internet, currently under preparation. Having said this, as emphasised in Sect. 2.5, reduction in energy consumption is essential for all future electronic circuits with their high-density integration, and any improvement to which photonics may contribute is of course welcome here. Figure 2.41 is a schematic view of a multicore processor, showing how it could benefit from optical interconnects. A first network of intra-chip interconnects provides the links between the processors and the cache memory units, while a second, inter-chip network links these temporary memory units with the random access memory, in this case dynamic random access memory (DRAM).

The development of new circuit architectures including optical interconnects has mobilised many research groups. Indeed, nanotechnology is essential to the task of combining photonics and electronics, which imposes stringent constraints on the dimensioning and positioning of photonic components. A laser is associated with each information channel from one of the electronic chips, and its beam is modulated in consequence for transmission to the other chip. A detector matrix on the latter converts the optical signals into electrical signals. We have indeed entered the world of nanophotonics.

2.6.5 Photonics at Extreme Wavelengths: Ultraviolet and Terahertz

In most of the topics discussed so far, from photonic crystals to plasmonics and optical interconnects, the relevant wavelengths lie mainly in the infrared and the

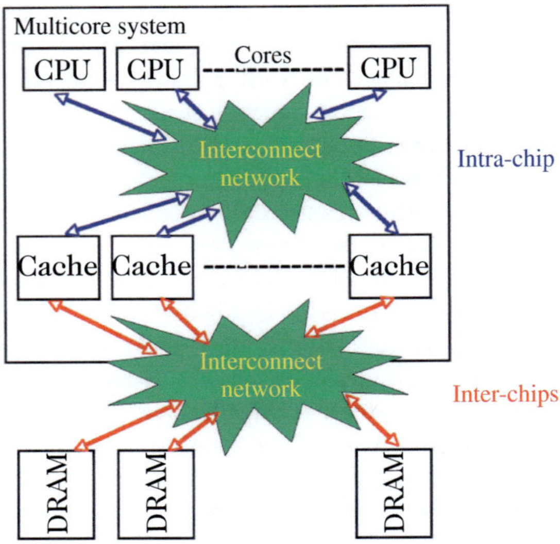

Fig. 2.41 Schematic view of a multicore processor using optical interconnects. A first network of intra-chip interconnects links the central processing units (CPU) to the cache memory units. A second network, this time inter-chip, links the cache memory units to DRAM-type random access memories

visible. However, much work is being done on the two extremities of this spectrum, with the help of nanotechnology.

In the visible to near-ultraviolet range, gallium nitride (GaN) and zinc oxide (ZnO) components are attracting a lot of interest due to both potential and actual applications. In particular, GaN lasers are used to read Blu-ray disks, while high efficiency GaN light-emitting diodes (LED) combined with phosphorescent processing now compete as lighting systems. We should also mention detection of UV radiation and applications to biology and biomedicine. One issue for all these applications is to obtain GaN and ZnO materials of high crystal quality, and nanotechnology already plays an important role in the growth and characterisation of such materials. It is still more important in the realisation of nanostructures such as quantum dots and nanowires. Furthermore, photonic crystals with submicron periods can provide a method of optimisation for extracting light produced by light-emitting diodes.

At the other end of the spectrum, the terahertz wave (or submillimeter) range presents other challenges and other prospects, particularly in terms of imaging. One of the main problems is to obtain a solid state laser source that is powerful enough and works close enough to room temperature, while emitting a beam with low angular divergence. Among the approaches to solid state sources, quantum cascade lasers are of some interest here. Their optimisation involves making quantum well multilayers with nanometric accuracy. Photonic crystal structures are useful for obtaining a low divergence output beam. With the help of such structures, the wave can be extracted from the largest surface of the device, at right-angles to the layers [68]. The crystal must then be etched with micrometric or even nanometric accuracy. Note also that, among the quantum well emitters, those made from GaN and AlGaN (gallium nitride and aluminium gallium nitride) are the most likely candidates for future terahertz lasers, thereby providing an unexpected link between the ultraviolet and long wavelengths.

2.7 Nanophysics: Genuinely Cross-Disciplinary Physics

By its very nature, nanoscience is cross-disciplinary, sharing ideas and methods of measurement, modelling, and fabrication on the nanometric scale and stimulating cross-fertilisation of every area from the physical sciences to chemistry and the life sciences. Indeed, many examples from the previous sections of this chapter illustrate the way physics has interacted with the other scientific disciplines. The growth of thin films, surface treatments, and etching of very small motifs (see Sect. 2.2) often require the knowhow of chemists. Innovative electronic components and promising circuit architectures are inspired by mechanisms observed in the living world, and in particular neural mechanisms from our growing understanding of the human brain (see Sect. 2.5). Going in the other direction, biologists and biochemists have appropriated the tools of high-resolution microscopy developed by physicists (see Sect. 2.4). Likewise, nanoelectronics and nanophotonics (see Sects. 2.5 and 2.6) play a key role in many analysis and diagnostic systems applied to medicine, the food industry, and climate and environmental science. There would be little point trying to list in this section all the interdisciplinary research in which nanophysics, nanoelectronics, and nanophotonics have become relevant. A few significant examples are discussed in the following, illustrating how nanoscience has influenced microsystems in ways other than the sole aim of miniaturisation, and also the boom in nanobiotechnology that has come about through methods and techniques born of the physical sciences.

2.7.1 Nanowires in Electromechanical Microsystems

Micro-electromechanical systems (MEMS) using silicon technology have seen continual development with an ever broader diversity of applications. Inertial sensors such as accelerometers, gyrometers, gyroscopes, inclinometers, etc., today represent a market of several billion euros. Used for several years now in many industrial applications, including the car industry, MEMS also made their appearance in mobile phones in 2007, breathing new life into the use of touch screens. Since then, motion sensors have been perfected and applications have proliferated. Apart from the accelerometer, which completely transformed the ergonomy of mobile devices, there is the gyroscope, which equips smartphones today and allows accurate control of motions in the three space directions. The gyroscope can deduce angular positions by integrating measured angular velocities, while the accelerometer can deduce linear velocities and relative positions by single and double integration, respectively.

The underlying idea of the accelerometer can be illustrated by thinking of a moving mass at the end of a spring, the other end of which is fixed to the object whose acceleration is to be determined (see Fig. 2.42). It is then by measuring the displacement of the moving mass relative to its support that the acceleration of the object can be ascertained. Usually, the displacement of the moving mass causes the displacement of a condenser electrode and measurement of this displacement amounts to

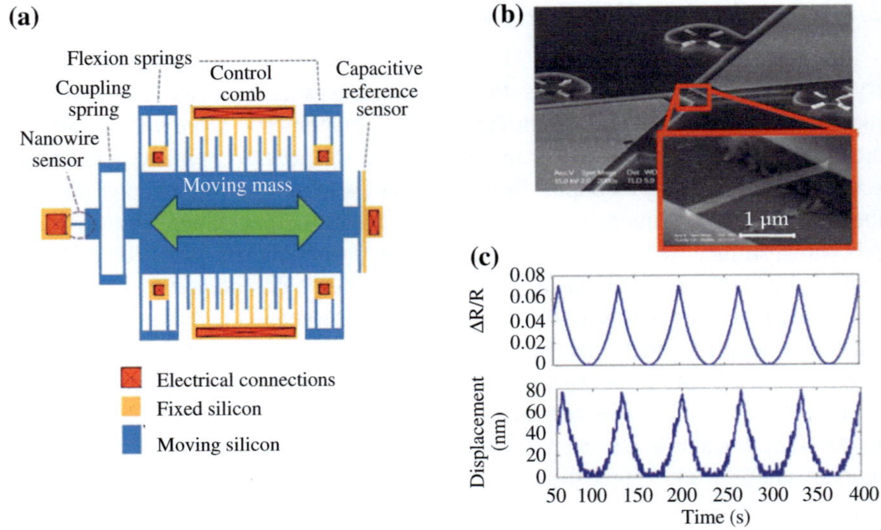

Fig. 2.42 **a** Test accelerometer using a piezoelectric silicon nanowire to measure displacements [69]. **b** Electron microscope image of the device, with a zoom on the nanowire. **c** Change in the piezoresistance of the nanowire (*upper*) as a function of the displacement of the moving mass measured by the capacitive effect (*lower*). Courtesy of A. Bossebœuf, *Institut d'électronique fondamentale*, Orsay

measuring a change in capacitance or voltage. Such devices are miniaturised using fabrication methods from silicon microelectronics, something that has been thoroughly mastered these days. Nanotechnology is important in processes used for the fabrication and integration of control and measurement electronics. Having said this, not all the component elements of these devices need be nanoscale, which is why we speak of microsystems. On the other hand, it may be that the insertion of nano-objects will prove useful for MEMS performance. This is the case, for example, with nanowires, whose piezoelectric properties can be used to measure the displacement of moving parts.

Due to their long thin shape, nanowires can be used as efficient converters of mechanical stress energy into electrical energy. In the diagram showing the test accelerometer in Fig. 2.42a, the nanowire is connected to the moving mass by a coupling spring and it is the stress transmitted by this spring that alters the electrical resistance of the nanowire. In the actual device made with silicon technology (Fig. 2.42b), the nanowire has the shape of a rectangular bar, several microns long and 250 nm wide. The measured changes in the resistance of the nanowire perfectly reproduce those recorded using more traditional methods based on the capacitive effect (see Fig. 2.42c). The key advantage of the piezoresistive measurement is the very low level of noise and the accuracy in the displacement, here 0.7 Å, the size of an atom [69].

2.7.2 Nanobiotechnology: Convergence Between the Physical Sciences and the Life Sciences

Nanobiotechnology is the area of nanoscience that is currently enjoying the fastest growth. Although there is no precise definition for this field of activity, the most widely accepted view includes everything at the interface between biology and nanotechnology, most of which results from the physical sciences. At this interface, nanomaterials and nanodevices are studied for the purposes of analysing biological systems (biology takes its cue from nanoscience), but the results can be used in return to bring innovation to nanotechnology (nanoscience takes its cue from biology). This duality is already illustrated by examples described in this chapter, whether it be quantum dots used for cell imaging (see Sect. 2.3) or neural architectures influencing electronics (see Sect. 2.5). In his discussion of nanomedicine in Chap. 9, Patrick Boisseau also considers many cases of nanobiotechnology in action, such as the lab-on-a-chip (LOC), sensors for medical diagnostics, nanodrugs, nanoimaging, and theranostics, among others. We shall thus restrict here to two examples where techniques from nanophotonics have led to innovation in biology.

Quantum Dot Sensing Platforms for Molecular Interactions

Fluorescent nano-objects, including proteins, dyes, rare earth complexes, quantum dots, nanowires, and so on, are now widely used to detect or monitor biomolecules in cell tissues, usually *in vitro*, but also *in vivo* [70]. Fluorescence is characterised by three parameters: its wavelength (its colour), its intensity, and the duration of emission. By using different colours, different levels of intensity, and different measurement times, it should be possible to control several biomolecular processes simultaneously, just as we can control several telecommunications signals on the same transmission line by the technique of multiplexing. If we compare with the familiar notion of the bar code, the detection of five colours, five intensity levels, and three measurement times would lead to the digitisation of almost 30 billion fluorescent codes!

In this context, among the fluorescent nano-objects, quantum dots offer a choice of wavelength from blue to red, 'simply' by adjusting their size and/or their composition. They have high luminescence efficiency and a broad absorption spectrum, so it is easy to separate the emission wavelength from the excitation wavelength. Furthermore, they have much shorter emission times than chemical complexes of the rare earths, typically between the microsecond and the nanosecond for the former, as compared with the millisecond for the latter. These features make it ideal for implementing the FRET detection of molecular interactions.[14]

In this setup (Fig. 2.43), the chemical complex of a rare earth, here terbium, and the quantum dot are first subjected to ultraviolet excitation. The quantum dot, which

[14]FRET is short for Förster resonance energy transfer. Theodor Förster (1910–1974) was trying to understand energy transport in plant photosynthesis. He was the first to provide a theoretical explanation for non-radiative energy transfer between molecules and its dependence on the intermolecular distance.

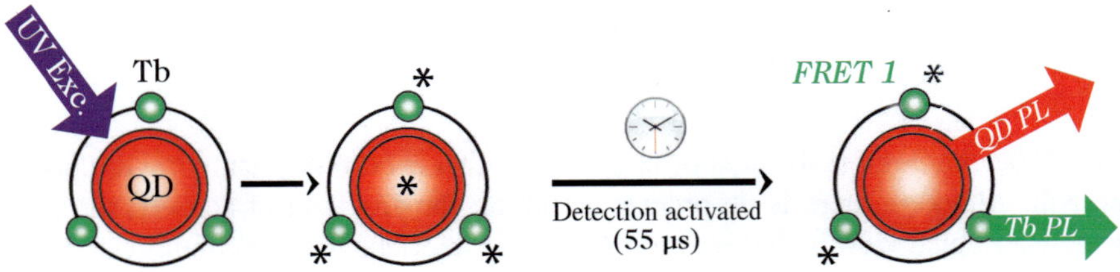

Fig. 2.43 Förster resonance energy transfer (FRET) between a quantum dot (QD) and terbium (Tb) complexes. Adapted from [70], © Wiley 2012. After excitation by a UV light pulse, the quantum dot de-excites by emitting a first (*red*) photon in less than 50 μs. The terbium complex transfers its excitation energy to the quantum dot (FRET 1) over a much longer time and any photon emission beyond ~55 μs will attest to the proximity of the quantum dot and the terbium complex, and hence also the proximity of the biomolecules associated with them, implying that they are in a position to interact

emits in the red, quickly sends out a first luminescence signal, in less than 50 μs, while the terbium complex, which emits in the green, transfers its excitation energy to the quantum dot before it even has time to emit. If the temporal detection window is fixed beyond 50 μs, any red luminescence signal measured will result from energy transfer from the terbium complex to the quantum dot. To be effective, however, this transfer requires the two elements to be within about twenty nanometers of one another. In other words, if the transfer does indeed occur and each of the elements is associated with a biomolecule such as a protein, peptide, enzyme, etc., detection of a luminescence signal after 50 μs will attest to the fact that the associated biomolecules are close to one another and thus able to interact together.

At this point, one can exploit the fact that the diameter of the quantum dot is generally one or two tens of nanometers, well above the size of organic or inorganic fluorophores (dyes or terbium complexes), to use the quantum dot as a platform for molecular interactions. Just as the terbium complex relinquishes its excitation energy to the quantum dot, so the latter can pass its excitation energy to a dye molecule, which will then emit luminescence shifted toward the dark red: the quantum dot thus serves as a donor rather than an acceptor. This double transfer mechanism has been used recently to control the hybridization of a nucleic acid and the activity of a protease. The relevant biomolecules (peptides and oligonucleotides) were grafted onto the terbium complex and the organic dye Alexa Fluor 647, respectively [71]. The double transfer is shown schematically in Fig. 2.44. This system, in which the quantum dot serves as a nanometric relay platform, is likely to become a practical tool for 'multiplexed' sensing of biomolecular interactions from time-resolved luminescence spectra.

Microdroplet Lasers for Opto-Microfluidics

The simultaneous, fast, and sensitive detection of one or more biological molecules in a biological sample of small volume (μl) is crucial in many applications, and especially for medical diagnostics. This explains the development of labs-on-a-chip or biochips which can be used to carry out immunoassays, enzyme reactions, DNA amplification, and many other tests besides. A biochip comprises a biocompatible

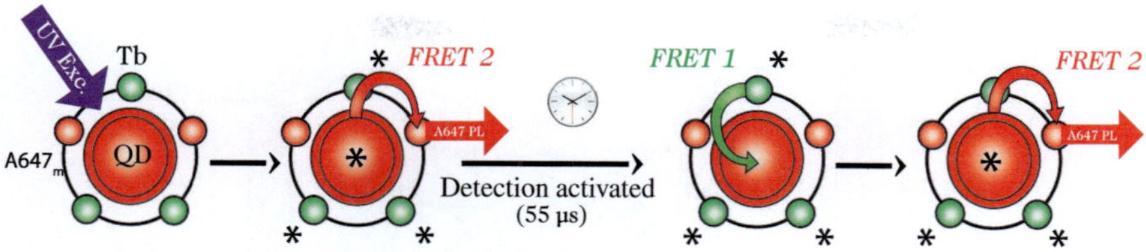

Fig. 2.44 FRET double transfer with the A647 dye, a quantum dot (QD), and a terbium (Tb) complex. Adapted from [71], © Wiley 2012. After UV excitation, the quantum dot quickly transfers its excitation energy to the A647 dye, which emits a first photon (*dark red*) (FRET 2). Beyond ~55 μs, the terbium complex can transfer its excitation energy to the quantum dot (FRET 1), which will subsequently transfer this energy to the A647 dye (FRET 2). Any emission from the dye after 55 μs (in fact, the only detected photon) will attest to the proximity of the terbium complex and the dye molecule, thus signifying that the biomolecules attached to them are also in close proximity. The quantum dot, which is both an acceptor and a donor of energy, serves as a nanoscale platform for the detection of biomolecular interactions

substrate such as glass or quartz with micrometric fluidic channels on the surface to convey the biofluid and carry out the desired detection and analysis. As we saw earlier, biomolecular detection often involves the use of optical techniques including some kind of light source. In Figs. 2.43 and 2.44, the light source was the source of UV excitation causing the luminescence of the quantum dot or dye. This source, often outside the biochip, increases the complexity of the analysis system.

To get round this difficulty, the solution is to incorporate the light source within the biochip by exploiting the fluidic architecture. Dye lasers lend themselves perfectly to this integration, emitting mainly in the visible range. Efforts to miniaturise and integrate them have moved quickly over the last few years to meet the demands of on-chip spectroscopy, and a new branch of microfluidics has come into being, known as digital optofluidics, which seeks to minimise the volume of liquid transported in the chip. The dye is thus introduced in the form of a microdroplet, carried along by another liquid whose sole role is as a fluidic carrier. A trail of droplets containing several kinds of dye rather than just one can also be used if we wish to switch the wavelength of the microlaser (see below). Figure 2.45 illustrates this case with the example of a particular microdroplet laser geometry [72].

The laser includes the system for producing microdroplets and the optical cavity. The droplets are produced from two T junctions placed in series and each containing a specific dye, here rhodamine 6G emitting at wavelength 565 nm and rhodamine B emitting at 586 nm. In this way, one can produce a series of droplets using just one dye, or using two dyes and alternating between them. This alternating droplet trail is then directed toward the laser cavity formed by two cleaved optical fibres coated with a gold film. The photograph at the bottom of Fig. 2.45 shows the cavity from above. The distance between the two ends of the fibres is small enough to obtain single-mode laser emission with a very narrow spectral width. Laser emission occurs only when a droplet crosses the cavity. When the dye droplets alternate, the laser switches from one wavelength to the other and the switching frequency can exceed the kilohertz. The range of emission wavelengths can in principle be extended to the whole visible spectrum simply be increasing the number of dyes.

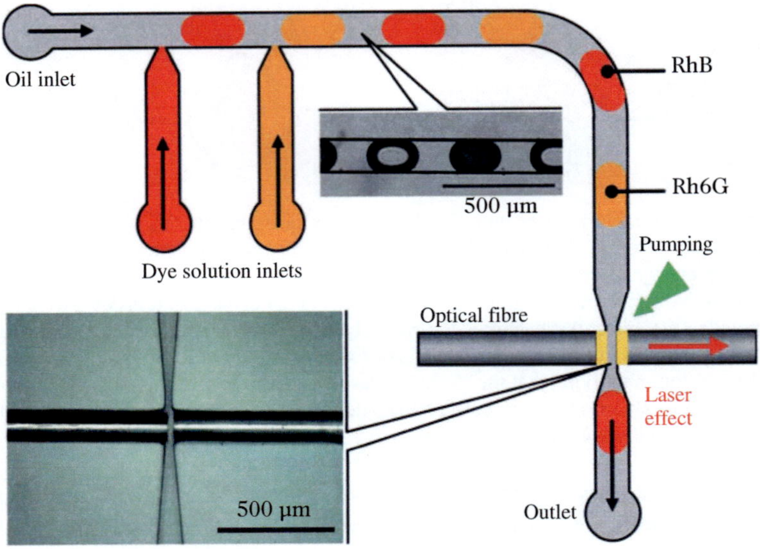

Fig. 2.45 Microdroplet laser. The dyes (rhodamine 6G and rhodamine B) are contained in glycol droplets, themselves carried in oil. *Upper* Droplets to scale. The mirrors of the laser cavity are made from two ends of metal-coated optical fibres placed opposite one another. *Lower* Photograph of the laser cavity to scale. Courtesy of A.-M. Haghiri, *Laboratoire de photonique et de nanostructures*, Marcoussis

This example, like the one before, shows how nano-optics and biology have become intertwined. Other examples would illustrate a similar interaction between nanoelectronics and biology, not only for acquisition, reading, and transmission of analysis data, but also for sorting and identifying biomolecules using electrofluidic or electrochemical processes. We should also mention the contribution of nanofabrication methods to micro- and nanofluidics and the resulting hydrodynamic studies, e.g., artificial microswimmers able to move through biological liquids like living micro-organisms [73]. There can be no doubt that the field of research at the interface between nanophysics and biology is one of the richest.

2.8 Nanophysics: What Future?

As we have seen in the different sections of this chapter, nanophysics is a particularly prolific field of research, with applications of the greatest importance in electronics and communications (what is known today as the digital world) and many openings in other disciplines. It is worth noting here that the microprocessor market on the one hand and the electronic communications market on the other each represent more than 40 billion dollars a year.

The historical record left by nanophysics, beginning with the advent of very high resolution microscopes in the mid-1980s, can now boast many discoveries regarding the properties and organisation of matter on very small length scales. Several of these have given rise to Nobel Prizes, the latest being the one awarded to Albert Fert and Peter Grünberg in 2007 for their work on giant magnetoresistance, and more recently still, the one awarded to Konstantin Novoselov and Andre Geim in 2010 for their work on graphene. The discovery of giant magnetoresistance has not only led today to data storage capacities never before attained, but also opens the way to a new form of electronics, spin electronics or spintronics, complementary to traditional charge electronics with the prospect of circuits with considerably reduced energy consumption. For its part, the more recent discovery of graphene opens up other prospects for electronics, including in particular the development of innovative circuits on flexible substrates.

While nanophysics represents an important share of the digital world of electronics and communications, it is also the perfect field for investigating quantum effects in matter. Several times throughout this chapter we have mentioned the widespread use of the quantum tunnel effect in today's electronic components. Several examples have been discussed where quantum physics plays a key role, including manifestations of electron interference (the quantum parallel of optical interference), systems involving small numbers of electrons or photons, and control of the spectral properties of emission, absorption, and conduction by nano-objects through their shape and size. And while quantum computers are not yet a reality, nanotechnology and nanophysics will play an essential part in the future of quantum information and computation, to the same extent as the work on atomic physics that recently saw the Nobel Prize for Physics awarded to Serge Haroche and David Wineland in 2012.

Looking beyond the major breakthroughs that have formed the springboard for research in nanophysics, it should be remembered that the association of a scientific field with a length scale, in this case the nanometer,[15] is quite unique. But while we may be justified in referring to nanoscience as a scientific discipline today, we may wonder what it will become tomorrow. After all, it might just turn into picophysics, that is, physics on the scale of the picometer, which is actually the scale of the atom (1 picometer = 0.01 Å).[16] The idea here would be to manipulate atoms in the same way as we currently build molecular edifices, using the bottom–up approach. On the other hand, it could be that nanophysics will disappear like a fashion which simply served to valorise the work of physicists at a propitious moment in time.

It should be said, however, that this scientific area is not just an unavoidable transition between two length scales, as one might think, for example, by considering the transition from microelectronics to nanoelectronics. The significant and radical change in the properties of materials at nanometric scales, sometimes but not always a consequence of the quantum effects discussed above, signals a boundary with other length scales which, although it is not particularly well defined, is real enough. The

---

[15]Historically, of course, the field of microsystems, itself characterised by a length scale (the micron), came before nanoscience and nanophysics. It too is situated at the meeting point of many scientific disciplines, but in contrast to nanophysics, it falls more or less completely into the field of applied science, while nanophysics lies at the frontier between fundamental science and applied science.

[16]In the Bohr model of the hydrogen atom, the Bohr radius which characterises the distance between the electron and the proton is about 53 picometers.

best known example is the nanoparticle itself, which inspires both curiosity and fear among the general public. The efficiency of its interactions with the surrounding medium is due to its very small size and the increased ratio of its surface area to volume. Another example is provided by electronic components, where the reduction to nanometric scales spells the end for any hope of 'zero defects' in electronic circuits. From now on, circuit designers will have to learn to live with the notions of defect tolerance, learning, and self-repair, just the kind of things we observe in biological systems and the living world.

The example of components and circuits is a good illustration of what one might refer to as the dual outlook of the nanophysicist: first to study matter at ever smaller length scales in order to extend our fundamental understanding of, and control over, substances at these scales, but also to design and create larger scale systems with a view to application, while exploiting the properties of matter on the smallest scales. Just as important is the cross-disciplinary attitude encouraged by nanoscience in general and nanophysics in particular. As we have seen from the examples described in this chapter, and as will be confirmed by others throughout this book, the interfaces between nanophysics, chemistry, biology, and medicine are the guarantee of future discoveries and innovations. Of course, the idea of cross-disciplinary research is nothing new, but it is on the nanometric scales of molecular constructs and biomolecules that the physicist most naturally encounters the preoccupations of chemists and biologists. For this reason, the development of nanotechnology and nanophysics has been, and will remain, a powerful driving force for cross-disciplinary research.

While it would be inappropriate to speak here of a scientific revolution, nanotechnology and nanophysics are nevertheless the source upstream of the industrial revolutions of our age, today in the digital industries, and tomorrow perhaps in the green industries. They also represent an important turning point in the approach of the nanophysicist within the physics community.

References

1. G. Fève, A. Mahé, J.-M. Berroir, T. Kontos, B. Plaçais, D.C. Glattli, A. Cavanna, B. Etienne, Y. Jin, An on-demand coherent single-electron source. Science **316**, 1169–1172 (2007)
2. F. Marquardt, S.M. Girvin, Optomechanics. Physics **2**, 40 (2009)
3. Le monde quantique, special issue of Pour la science, no. 68 (July-September 2010)
4. P.P. Naulleau, O.R. Wood, in *Extreme Ultraviolet (EUV). Lithography III. Proceedings of SPIE*, vol. 8322 (2012)
5. D. Winston, B.M. Cord, B. Ming, D.C. Bell, W.F. Di Natale, L. Stern, A.E. Vladar, M.T. Postek, M.K. Mondol, J.K.W. Yang, K.K. Berggren, Scanning-helium-ion-beam lithography with hydrogen silsesquioxane resist. J. Vac. Sci. Technol. B **27**, 2702–2706 (2009)
6. B. Schiedt, L. Auvray, L. Bacri, G. Oukhaled, A. Madouri, E. Bourhis, G. Patriarche, J. Pelta, R. Jede, J. Gierak, Direct FIB fabrication and integration of 'single nanopore devices' for the manipulation of macromolecules. Microelectron. Eng. **87**, 1300–1303 (2010)
7. A.M. Shevjakov, G.N. Kuznetsova, V.B. Aleskovskii, Chemistry of high temperature materials, in *Proceedings of 2nd USSR Conference on High Temperature Chemistry of Oxides*, 26–29 November 1965, Leningrad, USSR (Nauka, Leningrad, USSR, 1967), pp. 149–155 (in Russian)
8. T. Yatsui, K. Hirata, W. Nomura, Y. Tabata, M. Ohtsu, Realization of an ultra-flat silica surface with angstrom-scale average roughness using nonadiabatic optical near-field. Appl. Phys. B **93**, 55–57 (2008)

9. S.Y. Chou, P.R. Krauss, P.J. Renstrom, Imprint of sub-25 nm vias and trenches in polymers. Appl. Phys. Lett. **67**, 3114–3116 (1995)
10. T. Bailey, S. Johnson, D. Resnick, S. Sreenivasan, J. Ekerdt, C. Willson, Step and flash imprint lithography: an efficient nanoscale printing technology. J. Photopolym. Sci. Tech. **15**, 481–487 (2002)
11. A. Cattoni, E. Cambril, D. Decanini, G. Faini, A. Haghiri-Gosnet, Soft UV-NIL at 20 nm scale using flexible bi-layer stamp casted on HSQ master mold. Microelectron. Eng. **87**, 1015–1018 (2010)
12. S.H. Ahn, L.J. Guo, Large-area roll-to-roll and roll-to-plate nanoimprint lithography: a step toward high-throughput application of continuous nanoimprinting. ACS Nano **3**(8), 2304–2310 (2009)
13. T.J. Karle, Y. Halioua, F. Raineri, P. Monnier, R. Braive, L. Le Gratiet, G. Beaudoin, I. Sagnes, G. Roelkens, F. van Laere, D. Van Thourhout, R. Raj, Heterogeneous integration and precise alignment of InP-based photonic crystal lasers to complementary metal oxide semiconductor fabricated silicon-on-insulator wire waveguides. J. Appl. Phys. **107**, 063103 (2010)
14. M. Thiel, M.S. Rill, G. von Freymann, M. Wegener, Three-dimensional bi-chiral photonic crystals. Adv. Mater. **21**, 4680–4682 (2009)
15. A.P. Alivisatos, W. Gu, C. Larabell, Quantum dots as cellular probes. Annu. Rev. Biomed. Eng. **7**, 55–76 (2005)
16. I. Stranski, L. Krastanov, Sitzungsberichte d. Akad. d. Wissenschaften in Wien, Chap. IIb, vol. 146 (1937), p. 797
17. R. Rosales, K. Merghem, A. Martinez, A. Akrout, J.-P. Tourrenc, A. Accard, F. Lelarge, A. Ramdane, InAs/InP quantum-dot passively mode-locked lasers for 1.55 μm applications. IEEE J. Sel. Top. Quantum **1** (2011)
18. Y. Sun, Y. Xia, Shape-controlled synthesis of gold and silver nanoparticles. Science **298**, 2176–2179 (2002)
19. B. Baguenard, M. Pellarin, J. Lerme, J.-L. Vialle, M. Broyer, Competition between atomic shell and electronic shell structures in aluminum clusters. J. Chem. Phys. **100**, 754 (1994)
20. S. Iijima, Helical microtubules of graphitic carbon. Nature (London) **354**, 56–58 (1991)
21. J.-C. Harmand, L. Liu, G. Patriarche, L. Largeau, M. Tchernytcheva, N. Akopian, U. Perinetti, V. Zwiller, Potential of semiconductor nanowires for single photon sources. *Quantum Sensing and Nanophotonic Devices VI*. Book Series: Proceedings of SPIE, vol. 7222, article 722219 (2009)
22. J.I. Hahm, C.M. Lieber, Direct ultrasensitive electrical detection of DNA and DNA sequence variations using nanowire nanosensors. Nano Lett. **4**, 51–54 (2004)
23. K.S. Novoselov, A.K. Geim, S.V. Morozov, D. Jiang, Y. Zhang, S.V. Dubonos, I.V. Grigorieva, A.A. Firsov, Electric field effect in atomically thin carbon films. Science **306**, 666–669 (2004)
24. M. Kociak, A.Y. Kasumov, S. Guéron, B. Reulet, I.-I. Khodos, Y.-B. Gorbatov, V.-T. Volkov, L. Vaccarini, H. Bouchiat, Superconductivity in ropes of single-walled carbon nanotubes. Phys. Rev. Lett. **86**, 2416 (2001)
25. E. Gaufrès, N. Izard, L. Vivien, S. Kazaoui, D. Marris-Morini, E. Cassan, Enhancement of semiconducting single-wall carbon-nanotube photoluminescence. Opt. Lett. **34**, 3845–3847 (2009)
26. K. Falk, F. Sedlmeier, L. Joly, R.R. Netz, L. Bocquet, Molecular origin of fast water transport in carbon nanotube membranes: superlubricity versus curvature dependent friction. Nano Lett. **10**, 4067–4073 (2010)
27. P. Legagneux et al., Microwave amplifiers, in *Carbon Nanotube and Related Field Emitters: Fundamentals and Applications*, ed. by Y. Saito (Wiley-VCH Verlag GmbH & Co., KGaA, Weinheim, 2010)
28. M. Orlita, C. Faugeras, P. Plochocka, P. Neugebauer, G. Martinez, D.K. Maude, A.-L. Barra, M. Sprinkle, C. Berger, W.A. de Heer, M. Potemski, Approaching the Dirac point in high-mobility multilayer epitaxial graphene. Phys. Rev. Lett. **101**, 267601 (2008)
29. S. Ghosh, I. Calizo, D. Teweldebrhan, E.P. Pokatilov, D.L. Nika, A.A. Balandin, W. Bao, F. Miao, C.N. Lau, Extremely high thermal conductivity of graphene: prospects for thermal management applications in nanoelectronic circuits. Appl. Phys. Lett. **92**, 151911 (2008)

30. H. Raether, *Surface Plasmons*. Springer Tracts in Modern Physics, vol. 111 (Springer, Berlin, 1988)
31. J. Nelayah, M. Kociak, O. Stéphan, F.J. García de Abajo, M. Tencé, L. Henrard, D. Taverna, I. Pastoriza-Santos, L.M. Liz-Marzán, C. Colliex, Mapping surface plasmons on a single metallic nanoparticle. Nat. Phys. **3**, 348–353 (2007)
32. L.F. Zagonel, S. Mazzucco, M. Tencé, K. March, R. Bernard, B. Laslier, G. Jacopin, M. Tchernycheva, L. Rigutti, F.H. Julien, R. Songmuang, M. Kociak, Nanometer scale spectral imaging of quantum emitters in nanowires and its correlation to their atomically resolved structure. Nano Lett. **11**, 568–573 (2011)
33. H. Yang, A.J. Mayne, M. Boucherit, G. Comtet, G. Dujardin, Y. Kuk, Quantum interference channeling at graphene edges. Nano Lett. **10**, 943–947 (2010)
34. M.W. Browne, Two researchers spell IBM atom par atom. The New York Times (5 April 1990)
35. M. Lastapis, M. Martin, D. Riedel, L. Hellner, G. Comtet, G. Dujardin, Picometer-scale electronic control of molecular dynamics inside a single molecule. Science **308**, 1000–1003 (2005)
36. Y.-C. Yeh, C.-L. Lin, B.-J. Chen, Y.-W. Tseng, J.D. Russel, Application of atomic force probing on 90 nm DRAM cell failure analysis. *Proceedings of the 13th International Symposium on the Physical and Failure Analysis of Integrated Circuits* (2006) pp. 340–343
37. E.H. Synge, A suggested method for extending the microscopic resolution into the ultramicroscopic region. Philos. Mag. **6**, 356 (1928)
38. V.S. Volkov, S.I. Bozhevolnyi, E. Devaux, T.W. Ebbesen, Compact gradual bends for channel plasmon polaritons. Opt. Express **14**, 4494–4503 (2006)
39. E. Devaux, T.W. Ebbesen, J.-C. Weeber, A. Dereux, Launching and decoupling surface plasmons via micro-gratings. Appl. Phys. Lett. **83**, 4936–4938 (2003)
40. A. Ashkin, J.M. Dziedzic, J.E. Bjorkholm, S. Chu, Observation of a single-beam gradient force optical trap for dielectric particles. Opt. Lett. **11**, 288–290 (1986)
41. J. von Neumann, The principles of large-scale computing machines. Reprinted in Ann. Hist. Comput. **3**, 263–273 (1946)
42. M. Pickavet, W. Vereecken, S. Demeyer, P. Audenaert, B. Vermeulen, C. Develder, D. Colle, B. Dhoedt, P. Demeester, Worldwide energy needs for ICT: the rise of power-aware networking. *IEEE ANTS 2008 Conference*, Bombay (India), 15–17 December 2008
43. Y.M. Lee, J. Hayakawa, S. Ikeda, F. Matsukura, H. Ohno, Effect of electrode composition on the tunnel magnetoresistance of pseudo-spin-valve magnetic tunnel junction with a MgO tunnel barrier. Appl. Phys. Lett. **90**, 212507 (2007)
44. A. Chanthbouala, R. Matsumoto, J. Grollier, V. Cros, A. Anane, A. Fert, A.V. Khvalkovskiy, K.A. Zvezdin, K. Nishimura, Y. Nagamine, H. Maehara, K. Tsunekawa, A. Fukushima, S. Yuasa, Vertical-current-induced domain-wall motion in MgO-based magnetic tunnel junctions with low current densities. Nat. Phys. **7**, 626–630 (2011)
45. Y. Gang, W.S. Zhao, J.-O. Klein, C. Chappert, P. Mazoyer, A high-reliability, low-power magnetic full adder. IEEE Trans. Mag. **47**, 4611–4616 (2011)
46. T. Ernst, L. Duraffourg, C. Dupre, E. Bernard, P. Andreucci, S. Becu, E. Ollier, A. Hubert, C. Halte, J. Buckley, O. Thomas, G. Delapierre, S. Deleonibus, B. de Salvo, P. Robert, O. Faynot, Novel Si-based nanowire devices: will they serve ultimate MOSFET scaling or ultimate hybrid integration?, in *IEEE International Electron Devices Meeting (IEDM)* (2008), pp. 1–4
47. S.J. Shin, C.S. Jung, B.J. Park, T.K. Yoon, J.J. Lee, S.J. Kim, J.B. Choi, Y. Takahashi, D.G. Hasko, Si-based ultrasmall multiswitching single-electron transistor operating at room-temperature. Appl. Phys. Lett. **97**, 103101 (2010)
48. L. Nougaret, H. Happy, G. Dambrine, V. Derycke, J.-P. Bourgoin, A.A. Green, M.C. Hersam, 80 GHz field-effect transistors produced using high purity semiconducting single-walled carbon nanotubes. Appl. Phys. Lett. **94**, 243505 (2009)
49. C. Sire, F. Ardiaca, S. Lepilliet, J.W.T. Seo, M.C. Hersam, G. Dambrine, H. Happy, V. Derycke, Flexible gigahertz transistors derived from solution-based single-layer graphene. Nano Lett. **12**, 1184–1188 (2012)
50. M. Urdampilleta, S. Klyatskaya, J.-P. Cleuziou, M. Ruben, W. Wernsdorfer, Supramolecular spin valves. Nat. Mater. **10**, 502–506 (2011)

51. D.B. Strukov, G.S. Snider, D.R. Stewart, R.S. Williams, The missing memristor found. Nature **453**, 80–83 (2008)
52. W.S. Zhao, G. Agnus, V. Derycke, A. Filoramo, C. Gamrat, J.-P. Bourgoin, Nanotube devices based crossbar architecture: toward neuromorphic computing. Nanotechnology **21**, 175202 (2010)
53. J.D. Joannopoulos, R.D. Meade, J.N. Winn, *Photonic Crystals* (Princeton, New York, 1995)
54. J.-M. Lourtioz, H. Bénisty, V. Berger, J.-M. Gérard, D. Maystre, A. Chelnokov, D. Pagnoux, *Photonic Crystals. Towards Nanoscale Photonic Devices*, 2nd edn. (Springer, Berlin, 2008)
55. S. Laurent, S. Varoutsis, L. Le Gratiet, A. Lemaître, I. Sagnes, F. Raineri, A. Levenson, I. Robert-Philip, I. Abram, Indistinguishable single photons from a single-quantum dot in a two-dimensional photonic crystal cavity. Appl. Phys. Lett. **87**, 163107 (2005)
56. A. Dousse, J. Suffczynski, A. Beveratos, O. Krebs, A. Lemaître, I. Sagnes, J. Bloch, P. Voisin, P. Senellart, Ultrabright source of entangled photon pairs. Nature **466**, 217–220 (2010)
57. J.-C. Johnson, H.-J. Choi, K.P. Knutsen, R.D. Schaller, P. Yang, R.J. Saykally, Single gallium nitride nanowire lasers. Nat. Mater. **1**, 106–110 (2002)
58. B. Ellis, M.A. Mayer, G. Shambat, T. Sarmiento, J. Harris, E.E. Haller, J. Vuckovic, Ultralow threshold electrically pumped quantum dot photonic-crystal nanocavity laser. Nat. Photonics **5**, 297–300 (2011)
59. S. Matsuo, T. Sato, K. Takeda, A. Shinaya, K. Nosaki, H. Taniyama, M. Notomi, K. Hasebe, T. Kakitsuk, Ultralow operating energy electrically driven photonic crystal laser. IEEE J. Sel. Top. Quantum Electron. **19**, 4900311 (2013)
60. T. Asano, B.S. Song, S. Noda, Analysis of the experimental Q factors (> 1 million) of photonic crystal nanocavities. Opt. Express **14**, 1996–2002 (2006)
61. C. Dupas, P. Houdy, M. Lahmani (eds.), *Nanoscience* (Nanotechnologies and Nanophysics (Springer, Berlin, 2007)
62. S.J. McNab, N. Moll, Y.A. Vlasov, Ultra-low loss photonic integrated circuit with membrane-type photonic crystal waveguides. Opt. Express **11**, 2927–2939 (2003)
63. J. Valentine, S. Zhang, T. Zentgraf, E. Ulin-Avila, D.A. Genov, G. Bartal, X. Zhang, Three-dimensional optical metamaterial with a negative refractive index. Nature **455**, 376–380 (2008)
64. B. Kanté, A. de Lustrac, J.-M. Lourtioz, S. Burokur, Infrared cloaking based on the electric response of split ring resonators. Opt. Express **16**, 9191–9198 (2008)
65. J. Valentine, J. Li, T. Zentgraf, G. Bartal, X. Zhang, An optical cloak made of dielectrics. Nat. Mater. **8**, 568–571 (2009)
66. R.T. Hill et al., Leveraging nanoscale plasmonic modes to achieve reproducible enhancement of light. Nano Lett. **10**, 4150–4154 (2010)
67. H.A. Atwater, A. Polman, Plasmonics for improved photovoltaic devices. Nat. Mater. **9**, 205–213 (2010)
68. Y. Chassagneux, R. Colombelli, W. Maineult, S. Barbieri, H.E. Beere, D.A. Ritchie, S.P. Khanna, E.H. Linfield, A.G. Davies, Electrically pumped photonic-crystal terahertz lasers controlled by boundary conditions. Nature **457**, 174–178 (2009)
69. P.E. Allain, A. Bosseboeuf, F. Parrain, S. Maaroufi, P. Coste, A. Walther, Large range MEMS motion detection using integrated piezo-resistive silicon nanowire, in *Twenty-Fifth IEEE International Conference on Micro Electro Mechanical Systems (MEMS)*, Conference Digest (2012), pp. 1320–1323
70. B. Hötzer, I.L. Medintz, N. Hildebrandt, Fluorescence in nanobiotechnology: sophisticated fluorophores for novel applications. Small **8**, 2297–2326 (2012)
71. W.R. Algar, D. Wegner, A.L. Huston, J.B. Blanco-Canosa, M.H. Stewart, A. Armstrong, P.E. Dawson, N. Hildebrandt, I.L. Medintz, Quantum dots as simultaneous acceptors and donors in time-gated Förster resonance energy transfer relays: characterization and biosensing. J. Am. Chem. Soc. **134**, 1876–1891 (2012)
72. G. Aubry, Q. Kou, J. Soto-Velasco, C. Wang, S. Méance, J.-J. He, A.-M. Haghiri-Gosnet, A multicolor microfluidic droplet dye laser with single mode emission. Appl. Phys. **98**, 111111 (2011)
73. R. Dreyfus, J. Baudry, M.L. Roper, M. Fermigier, H.A. Stone, J. Bibette, Microscopic artificial swimmers. Nat. Lett. **437**, 862–865 (2005)

Chapter 3
Applications I. From Nanometers to Megawatts: Photovoltaic Applications

Bruno Ghyselen

Abstract Nanotechnology is omnipresent in photovoltaics. Here we illustrate with a particular technological example how an industrial sector can be created, starting from nanoscale objects and ending up with installations covering hectares of solar panels. These are solar farms, designed to generate large amounts of electricity in the multi-megawatt installations. The example chosen here is the sector known as concentrated photovoltaics (CPV).

Photovoltaic applications could not be omitted from such a book as this. On the one hand, renewable energies are very much in the news and a subject of much debate. On the other, they are full of examples of nanotechnology in action. Objects of nanometric dimensions have been omnipresent, whether it be to gain better control of light reaching us from the sun (i.e., to better channel, transmit, absorb, or trap it), to optimise the light conversion process and production of electricity, or to transport the resulting electrical energy. They can be found in the form of stacks of films of nanometric thickness, such as quantum wells and tunnel diodes (see the discussion of the tunnel effect in Sect. 2.1), or in the form of nanowires, quantum dots, and materials with nanostructured surface or bulk for better coupling with light.

Within the field of photovoltaic technology, many branches are competing, distinguished in particular by the material forming the active region of the solar cell: mono-, poly-, or microcrystalline silicon, amorphous silicon, cadmium telluride (CdTe), CIGS cells (made from an alloy of copper, indium, gallium, and selenium), III–V semiconductor materials, organic cells, pigment cells, and others. The aim is not to make an exhaustive list here. We have chosen to use a particular example to illustrate the way one can start with objects of nanometric size and end up with an industry generating installations that cover several hectares with solar panels. We shall speak here of solar farms, designed to produce electricity in a big way, of the order of the megawatt, as in the example shown in Fig. 3.1, and up to power outputs of several tens or hundreds of megawatts. The example we have selected is known as concentrated photovoltaics (CPV).

B. Ghyselen (✉)
SOITEC, Parc Technologique des Fontaines, 38190 Bernin, France
e-mail: bruno.ghyselen@soitec.com

J.-M. Lourtioz et al. (eds.), *Nanosciences and Nanotechnology*,
DOI 10.1007/978-3-319-19360-1_3

Fig. 3.1 CPV solar farm in Newberry Springs, California, producing 1.5 megawatt. © Soitec

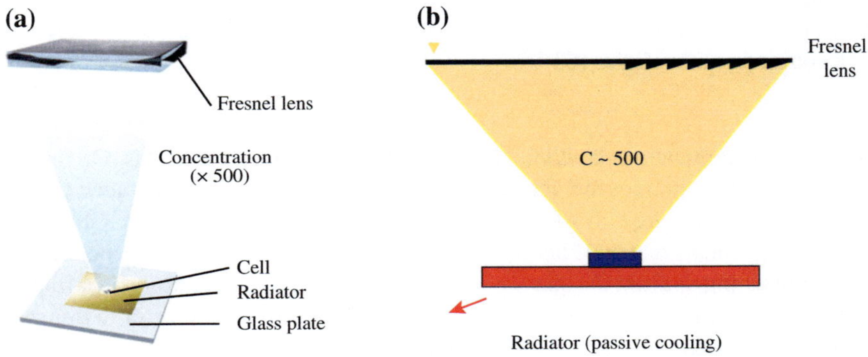

Fig. 3.2 **a** Concentrated photovoltaics (CPV). Concentration of solar radiation on a photovoltaic cell. **b** Fresnel lens. Reproduced with the kind permission of the Fraunhofer Institute for Solar Energy Systems ISE

In contrast to traditional technology, this approach involves an intermediate stage in which the incident light is concentrated optically before being sent to a smaller solar cell (in inverse proportion to the concentration factor C). Concentration factors range typically up to 300–500 in current systems, with a longer term objective of around 1000. Concentration can be achieved by reflection using mirrors or by transmission using lenses. There can be secondary optics to obtain a higher concentration or a more uniformly concentrated beam. Figure 3.2 shows how this works for the case of a Fresnel lens, while Fig. 3.3 shows schematically how these elements are then put together: cells are grouped in modules, then modules are grouped together on solar trackers which follow the sun as it crosses the sky. The solar farm thus consists of an array of these tracking systems. Depending on the model and the maker, each tracker generates a power from a few kilowatts to a few tens of kilowatts.

In such a system, the solar cell remains the central element, i.e., the place where light absorption takes place and the energy is converted to electricity. Traditionally,

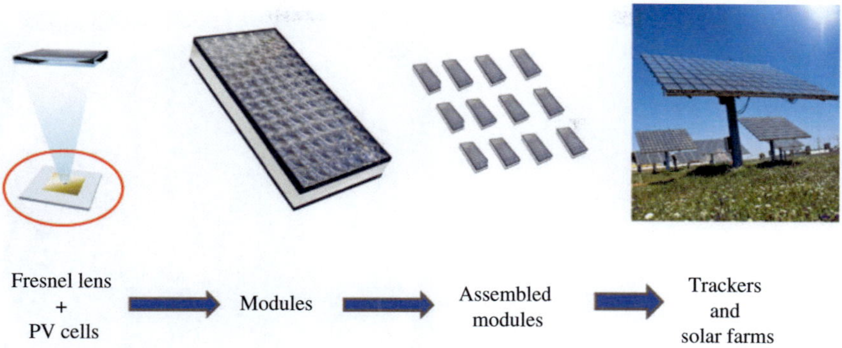

Fig. 3.3 Concentrated photovoltaics (CPV): from cell to solar farm. © Soitec

the cell is also considered to be the element contributing most to the cost of the modules, apart from installation costs. It is interesting to note that, in CPV systems, the area of the cell itself is considerably smaller than in traditional systems, by a factor close to the light concentration factor, i.e., a factor of several hundred. Figure 3.4 shows a top view of one of these cells, obtained by optical microscope. The dark regions are the absorbing active zones, while the lighter features are the interconnects, well spaced out to allow as much light to reach the cell as possible and uniformly distributed to recover the extracted current while limiting unwanted access resistance. Depending on the design and dimensions of the optical part, specific to each maker, the area of the cells can be as low as a few mm^2, or even less, while mono- or polycrystalline silicon cells typically have areas of the order of dm^2 (125×125 mm^2 or 156×156 mm^2). For reasons to be discussed below, CPV cells are commonly made on germanium (Ge) or gallium arsenide (GaAs) substrates of diameter 100 mm. A transition to 150 mm is expected in some cases, over the period 2012–2015. On each 100 mm substrate, almost a thousand concentration cells are then placed (see Fig. 3.5). After integrating these into concentrators, this leads to total power outputs of around a kilowatt, depending on the cell efficiency (see below).

In CPV systems, the cost of the optical parts used for concentration, combined with the need for a reliable solar tracking system, make it crucially important to improve performance at the cell level. One can envisage high performance cells, which can be relatively expensive, provided that the final economic equation in terms of the resulting kW h is favourable. At the present time, the cells with the highest efficiency are without question those combining III–V semiconductors with a so-called multijunction architecture.

As shown in Fig. 3.6a, a single cell made from a semiconductor with band gap[1] around 1.12 eV, as is typical in crystalline silicon, will only absorb in an optimal way that part of the solar spectrum above or close to this energy. Photons of lower energy are not picked up. Photons with much higher energy will be picked up, but with an energy efficiency that decreases as one moves away from the gap energy. A

[1]The band gap of the semiconductor and photon energy above which it will absorb light.

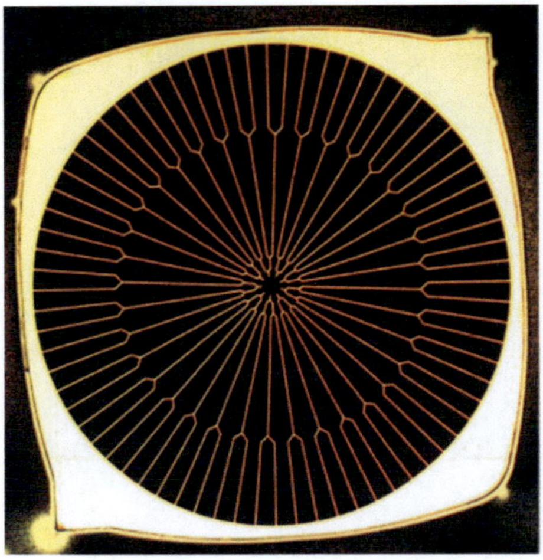

Fig. 3.4 CPV solar cell with a typical area of a few mm^2. Reproduced with the kind permission of the Fraunhofer Institute for Solar Energy Systems ISE

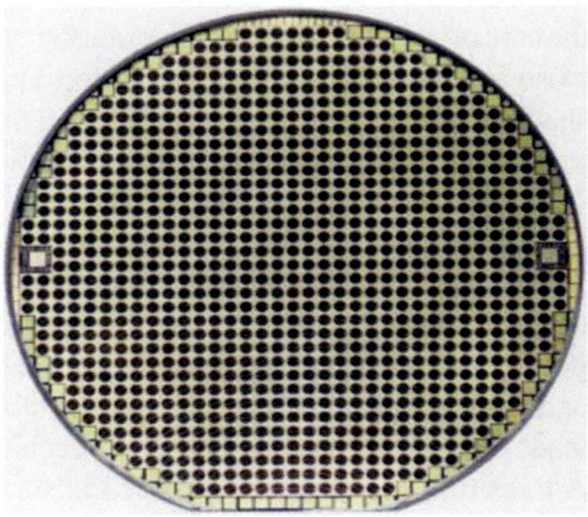

Fig. 3.5 Photograph of a 100 mm germanium substrate on which CPV cells have been placed. The latter are the hundreds of tiny black disks, on this scale. Reproduced with the kind permission of the Fraunhofer Institute for Solar Energy Systems ISE

non-negligible part of the energy collected thus tends to be dissipated in the form of heat, rather than electricity. By this principle, the energy efficiency of a CPV cell can be significantly increased by stacking materials up to form a multijunction cell. Depending on the gap of the semiconductor, each p-n junction[2] absorbs a certain range of the solar spectrum with high efficiency. The ideal thing would be to stack up as many junctions as possible to obtain the best efficiency. In practice, the number of junctions is in fact limited to three or four. This is to maintain a sufficient quality among the stacked materials, given the impact of the mismatches between their characteristic crystal lattices. There is a natural optical filtering in these architectures in which different junctions are piled on top of one another. Of course, one must respect the following order: the first absorption layer facing the sun has the largest gap and the gap diminishes as one moves to the lowest level.

[2]A p-n junction consists of two layers of a semiconductor material, one doped to give positve charges (p) and the other doped to give negative charges (n) (see Sect. 2.5).

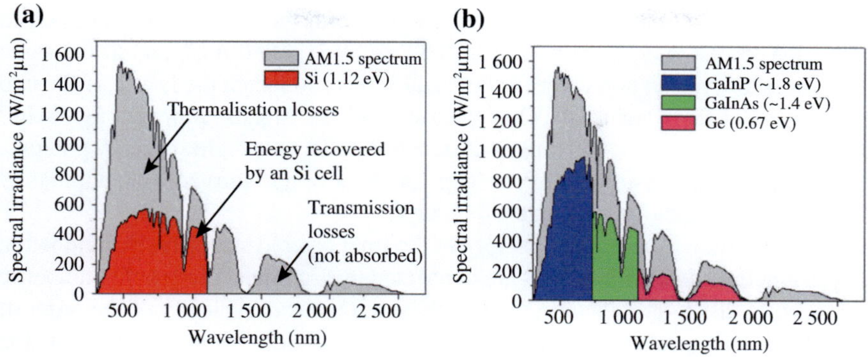

Fig. 3.6 Use of the solar spectrum arriving at the Earth's surface (the AM1.5 spectrum) for the case of a simple silicon cell (**a**) and for a multijunction cell with three levels (Ge and III–V materials) (**b**). Reproduced with the kind permission of the Fraunhofer Institute for Solar Energy Systems ISE

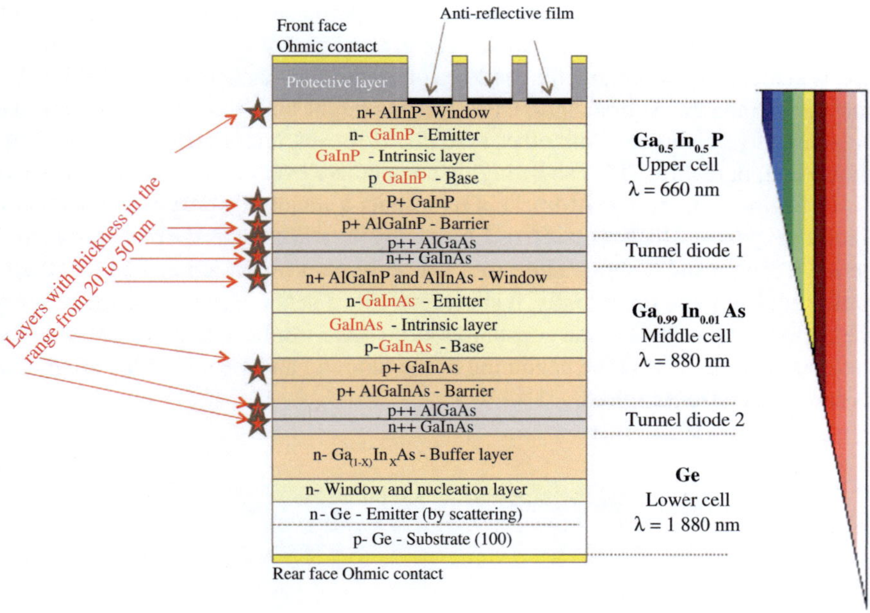

Fig. 3.7 Cross-section of a GaInP/GaAs/Ge triple junction, where GaInP is a ternary alloy of gallium, indium, and phosphorus. Reproduced with the kind permission of the Fraunhofer Institute for Solar Energy Systems ISE

Today, III–V semiconductors are used in both terrestrial and satellite CPV applications. The most widespread structure in production is a cell with three levels of junctions, built up on a bulk germanium substrate (see Fig. 3.7).

The germanium substrate forms the junction with the lowest gap, designed to collect the lowest energy photons (gap 0.67 eV, $\lambda = 1\,880$ nm), and at the same time serves as a nucleation layer for epitaxial growth to match the lattice of the two upper layers of III–V materials. One of these, with the highest gap, viz., the GaInP junction (gap 1.8 eV, $\lambda = 660$ nm), is designed to collect the highest energy photons. The middle junction using Ga(In)As (gap 1.4 eV, $\lambda = 880$ nm) collects photons of intermediate energies, in the range 1.4–1.8 eV.

Figure 3.7 shows the full complexity of the final layered structure. Each junction is a p-n junction, and the junctions are connected by n-p tunnel diodes of nanometric thickness. In 2012, the efficiencies of such junctions were typically of the order of 38–41 % in the optical concentration setup, with products commercially available in bulk, and of the order of 40–44 % for laboratory devices. Future generations of multijunctions are already under laboratory development. These should include (at least) one more junction level and, thanks to significant improvements in the materials, should achieve efficiencies of 45–50 %.

One consequence of optical concentration which might be considered a disadvantage is that the whole system must be kept permanently aligned with the sun. This alignment, which must be carried out to an accuracy within one degree of angle, is achieved by mounting the modules on multi-axis trackers that can follow the sun across the sky. At first sight, CPV systems might look somewhat complex and uncompetitive compared with other systems. However, a more detailed assessment shows that, in reality, CPV can be highly competitive over certain market segments.

To begin with, the use of trackers represents a major advantage. In contrast to traditional panels which remain fixed, following the sun extends the daily productive period of the installation, leading to a net daily gain in the amount of electricity produced. But more importantly, it also represents a better correspondence between the periods of production and consumption, so there can be a better response to the peaks in consumption at the beginning of the morning and the end of the afternoon, as can be seen from Fig. 3.8.

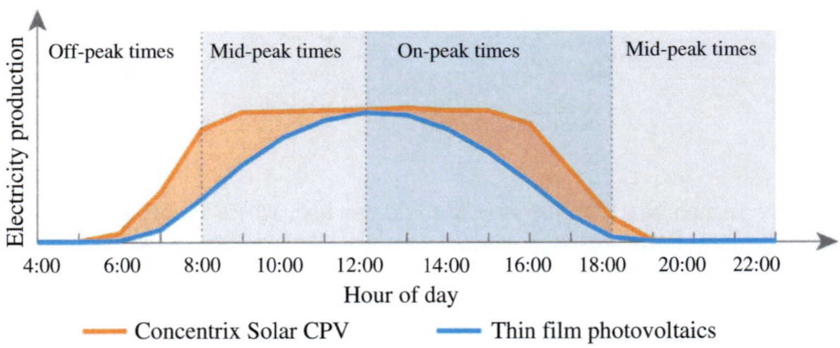

Fig. 3.8 Daily electricity production. Comparison between CPV technology (*orange*) and standard thin film technology (*blue*). © Soitec

Furthermore, as already discussed, the area covered by active semiconductor materials is significantly reduced, by at least two orders of magnitude compared with other systems. And in addition, the main materials used in this kind of system, such as glass or aluminium, are cheap and widely available, while the III–V semiconductors in ultrathin (nanometric) films represent only a small fraction of the total.

The main argument supporting the competitiveness of CPV systems is the record energy conversion efficiency achieved by these cells. Another point is that, in contrast to many of the traditional technologies, including systems based on silicon, the drop in efficiency when temperatures rise by several tens of degrees under conditions of use is considerably reduced for technologies based on III–V materials.

The fact that the solar panels must be aligned with the sun for concentration systems means that the dominant component of solar radiation used to generate the photocurrent is the one that is normally incident directly on the panels. Scattered light due to cloud cover or fog is barely used, since it is unlikely to reach the solar cell placed close to the focal point of the concentrator. Fortunately, the direct component of solar radiation predominates on clear days.

For all the above reasons, the market sector in which CPV systems are particularly competitive is solar farms in regions with high levels of direct insolation. As can be seen from Fig. 3.9, this includes the Mediterranean, Southern Europe, North Africa, the Middle East, Australia, the western part of the Americas, and so on.

Through the specific example of one type of photovoltaic system, viz., CPV, we have shown how stacks of nanolayers can be used to produce considerable amounts of energy in multi-megawatt solar farms.

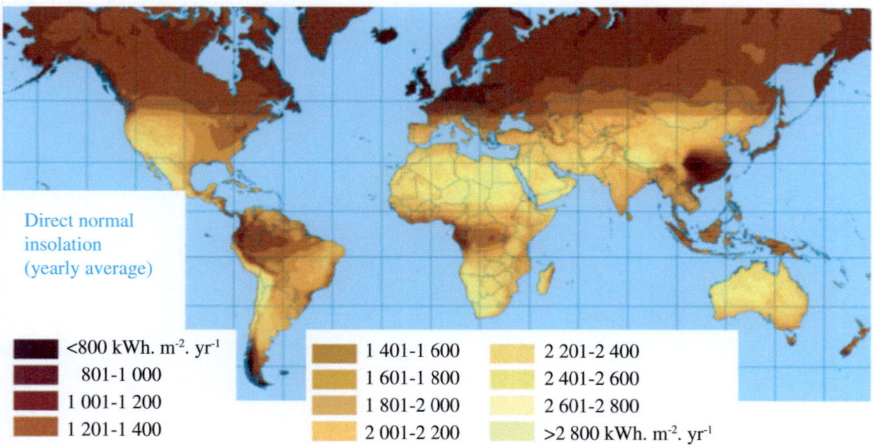

Fig. 3.9 World map showing direct sunlight (direct normal insolation DNI). *Colours* ranging from *yellow* to *dark brown* correspond to decreasing insolation. © Soitec

Chapter 4
Applications II: The Secret Revolution of Magnetic Sensors

Claude Fermon

Abstract Spin-electronics-based magnetic sensors, first developed for magnetic read heads, are now finding their way into many different technological devices from cars to medical systems. They combine very high sensitivity, great robustness, and very small size. Today they are integrated with CMOS electronics into a common process. In this chapter, we give an overview of that technology and describe some example applications, such as speed sensors, current sensors, and biomagnetic signal detection.

Every year almost four billion magnetic sensors are made for integration into various technological devices commonly used in everyday life, such as cars, washing machines, ovens, mobile phones, watches, and so on. Since the 1990s, nanotechnology has helped to developed a new type of sensor based on what is known as spin electronics, or spintronics. It is smaller and more sensitive than anything previously made. This magnetic sensor makes use of giant magnetoresistance (GMR), jointly discovered in 1988 by the research groups of Albert Fert and Peter Grünberg, and rewarded by the Nobel Prize in 2007.

The first direct application of these sensors was the development of read heads for hard disks. It is thanks to spintronics that we can have hard disks today capable of holding several terabytes. And this technology is beginning to find its place in many other areas, including the automobile industry, with a promise of greater safety, lower consumption, and greater automation. The basics of spintronics have already been described in Chap. 2. After a brief description of the key structure, the spin valve, we shall focus here on a few examples of its application in the field of magnetic sensors.

C. Fermon (✉)
CEA Saclay, Bât 772, 91191 Gif-sur-Yvette, France
e-mail: claude.fermon@cea.fr

J.-M. Lourtioz et al. (eds.), *Nanosciences and Nanotechnology*,
DOI 10.1007/978-3-319-19360-1_4

4.1 Spin Valves and Magnetic Sensors

In a ferromagnetic material such as cobalt, iron, or nickel, the internal magnetic field tends to orientate the electron spins parallel to itself. The electrons are then said to be polarised. This polarisation can vary between $+1$ when all the spins are lined up parallel with the internal field and -1 when they are all lined up antiparallel with it. To make a device that exploits spin, the polarisation of these electrons must remain intact outside the magnetic material, and this is where the nanometric scale becomes important. Indeed, when the electrons leave the magnetic material and move into a non-magnetic one like copper, for example, they lose their polarisation after just a few nanometers, owing to collisions with defects, internal atomic vibrations, or the presence of impurities. This means that the electron spin must necessarily be used at nanometric scales.

The spin valve is the basic building block in most magnetic sensors using spin electronics. It comprises two magnetic layers separated by a thin non-magnetic film which serves as a spacer (see Fig. 2.29a). The first layer, called the fixed or pinned layer, has completely blocked magnetisation and serves to polarise the electrons. The polarised electrons are then injected into the spacer, which is thin enough to ensure that they do not lose their polarisation before reaching a second magnetic layer. The latter orientates easily along an applied magnetic field, rather like a compass. This second layer is generally called the free layer. When the magnetisations of the two layers are parallel, electrons polarised by the first layer will easily enter into the second. However, when the magnetisations of the two layers are antiparallel, electrons polarised by the first layer have greater difficulty penetrating the second. The electrical resistance of the system is thus greater in the second case. The electrical resistance is thus modulated by means of a magnetic field, whence these systems are often referred to as magnetoresistive sensors. When the spacer is made from a metal, usually copper, the corresponding sensor is called a GMR sensor. When it is an electrical insulator, the electrons can cross this layer by the tunnel effect (see the discussion in Sect. 2.1), and the sensor is called a tunnel magnetoresistive or TMR sensor.

All sensors based on spintronics make use of giant magnetoresistance (GMR) or magnetic tunnel junctions (TMR) in the spin valve configuration. In fact, the spin valve is an indirect field sensor insofar as its resistance does not vary with the amplitude of the external field, but rather depends on the angle between the magnetisation of the pinned layer and that of the free layer orientated with this field. All the engineering aspects of sensors based on this technology consist in playing around with the shape and geometry of the sensor, the easy magnetisation direction of the free layer, and the specific architecture of the spin valve, in such a way as to transform the measurement of the angle between magnetisations into a highly sensitive measurement of the field, or linear over a broad range of fields, or effectively a measurement of the angular position of the field.

4.2 A First Application: Angle Sensors

A GMR (or TMR) spin valve is thus essentially an angle sensor, because its resistance depends on the cosine of the angle between the magnetisation of the free layer and that of the fixed layer, being maximal when they are opposite and minimal when they are parallel. If we build a spin valve in such a way that the free layer has a magnetisation that is effectively completely free to follow an external magnetic field, we can then precisely measure the rotation of an object carrying a magnet called the source magnet at the end. Accuracies of the order of 0.1° can be achieved for distances of several millimeters between magnet and sensor.

Up to the present time, angle sensors used in automobiles have mainly been based on another effect, known as the Hall effect. The point about these sensors, which serve in particular to measure the exact position of the steering wheel, is that they are intrinsically highly linear and rather cheap. Now GMR sensors are typically between a hundred and a thousand times more sensitive. In addition, their signal depends only on the angle and not the amplitude of the field, provided that the latter is greater than 1 mT, making them rather insensitive to distance fluctuations. GMR angle sensors are now produced by various manufacturers and will dominate the market for angle sensors within five years, because they will make it possible to replace the source magnets currently based on rare earths[1] by much cheaper ferrites.

4.3 An Essential Application for the Automobile

Magnetic sensors play an important role in automobiles because they measure the rotation of the engine and the wheels. Knowing the exact angular position of the drive shaft, one can trigger the ignition of the gas mixture at just the right moment and thereby reduce the fuel consumption. This is achieved by means of a cog wheel fixed on the main axis of the engine. The magnetic sensor, equipped on the outside with a magnet, measures when the teeth of the cog go by with accuracy better than 0.1° (see Fig. 4.1).

This increased accuracy was one of the key ingredients in reducing vehicle fuel consumption. The advantage with GMR sensors is not just that they increase accuracy still further, but also that they can be outside the oil bath, improving reliability and making it easier to replace them.

The angular speed of each wheel must be measured with some accuracy in order to control the anti-lock braking system (ABS). The use of the sensor is slightly different here. In this case, the wheel carries a magnetised ring with alternating north

[1]The strongest magnets we know are alloys of samarium and cobalt or of neodymium, iron, and boron. Samarium (Sm) and neodymium (Nd) are examples of rare earths, i.e., 4f elements of Mendeleev's periodic table, and have become very expensive these days. Efforts are being made to do away with them in magnetic systems.

Fig. 4.1 Measuring an angle of rotation without contact. A metal cog wheel rotates in front of the fixed sensor comprising a GMR and a strong permanent magnet. When a tooth of the cog wheel passes in front of the sensor, the field produced by the magnet is modified and a difference is detected in the signal by the sensor

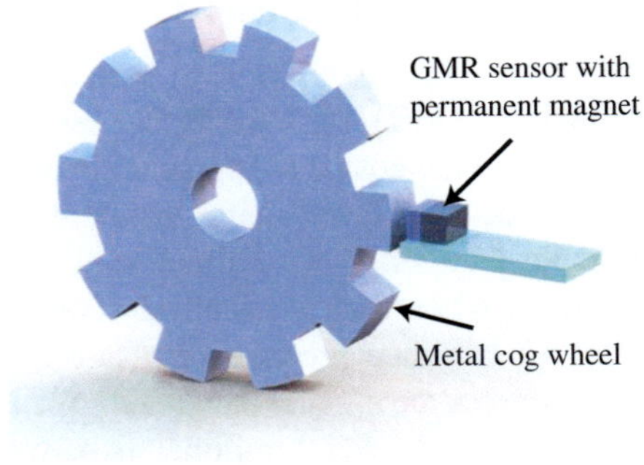

and south poles, while the magnetic sensor is placed just opposite and can detect them successively as they go by. In this measurement of the rotation angle, accuracies are of the order of 0.02°.

4.4 Detecting Very Weak Magnetic Fields

The sensitivity of GMR sensors can be exploited to measure very weak magnetic fields. Apart from electronic compasses, these sensors can be used for terrestrial or spatial magnetic mapping, non-destructive testing of metal parts, magnetic microscopy, and medical applications such as ultralow field magnetic resonance imaging (MRI) and magnetocardiography (MCG). In order to make an angle-sensitive GMR linear in the field, the trick is to arrange so that the magnetisation of the free layer is at 90° to that of the fixed layer, using an additional fixed field of high amplitude or a free layer with more sophisticated structure. In this case, the change in the resistance is linear for a small applied field in the direction of the fixed layer.

This linearity is very important in many applications, because one can then establish the true value of the measured field without sophisticated calculations. The typical response of a 'linearised' sensor is shown in Fig. 4.2, with the linear part in

Fig. 4.2 Response of a 'linearised' GMR sensor in an applied magnetic field

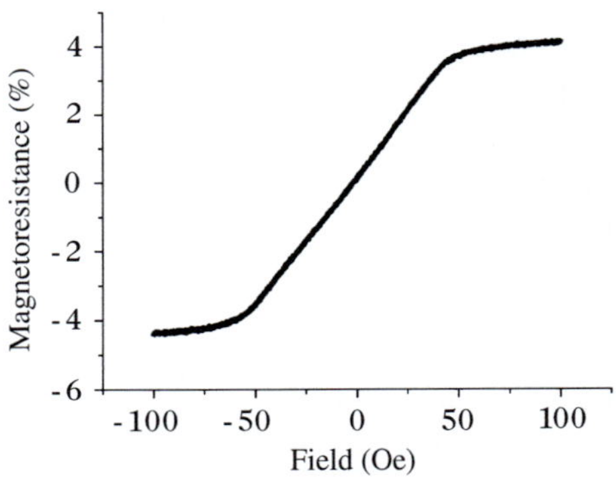

the middle and saturation for strong fields. For a GMR sensor, the resulting sensitivity, i.e., the percentage change in the resistance for one unit of field, is typically 2–4 %/mT.[2]

For a TMR sensor, the sensitivity is in the range 40–80 %/mT. These sensitivities are sufficient to detect magnetic fields of the order of 20 pT (20×10^{-12} T). This can be compared with the Earth's magnetic field, which is equal to 45 μT (45×10^{-6} T). TMR sensors have much higher sensitivity than GMR sensors, but they exhibit much greater electrical noise at low frequencies and are more difficult to make. For these reasons, they are mainly reserved for high frequency applications, such as hard disk heads, and applications requiring very low electricity consumption. However, an important application for these linearised sensors is the 3D electronic compass for mobile phones and GPS systems. In this field, in 2011, the company Freescale commercialised a 3D magnetometer for mobile systems that was based on TMR sensors.

4.5 Measuring Currents

Another major application of GMR magnetic sensors is the measurement of weak currents. When a current passes through a conductor, it generates a magnetic field in proportion to the strength of the current. This current can then be precisely measured without contact. Current sensors based on the Hall effect can resolve down to a few milliamperes. But using GMR sensors, one can have resolutions of the order of the micro-ampere. For example, this kind of sensor can very closely monitor the changes in the voltage of a battery or fuel cell (see the discussion of metal nanoparticles for fuel cells in Sect. 5.5.1). The voltage is converted into a local current using a metal line of well calibrated resistance. The point about using a magnetic sensor is to be able to carry out this measurement while guaranteeing a significant insulation (3 kV) between the battery which delivers up to 800 V and the electronics of the car under 12 V. In a project financed by the French national research agency, a monitoring circuit has been devised for fuel cells, integrating both CMOS electronic technology and GMR technology into a single integrated circuit (see Fig. 4.3).

4.6 Biomagnetism and Ultralow Field MRI

By combining a GMR sensor with a superconducting loop, sensitivity can be boosted still further, by a factor of several thousand, provided that the whole system is cooled using liquid nitrogen (−195.8 °C) or liquid helium (−269 °C). The superconducting loop serves to transform the magnetic field. It picks up a weak field on an area of

[2]The most common unit for the magnetic field strength is the tesla. The oersted (10^{-4} T) is often used for weak magnetic fields.

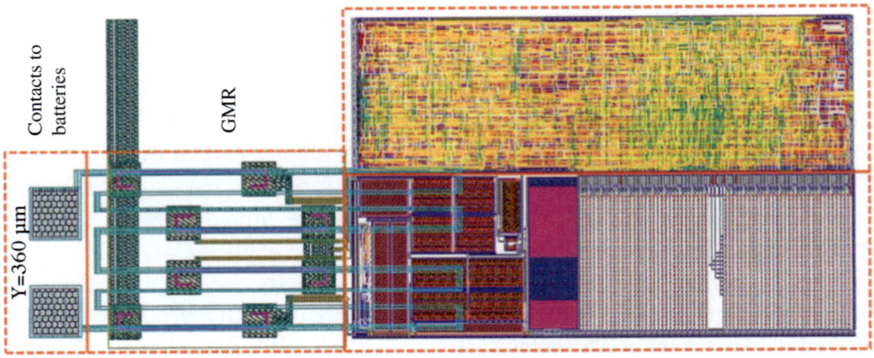

Fig. 4.3 Example of a complex circuit associating CMOS electronics and spintronics sensors to monitor a fuel cell. Source CEA/DSM (Saclay)–LIRMM (Montpellier)

order cm^2 and transforms it into a very strong local field, over a few μm^2, just as a microscope optics system would. This association of a magnetic sensor with a superconducting loop has been successfully applied to the detection of signals in the femtotesla range (10^{-15} T). Most applications of these sensors are in the medical domain.

One emblematic example among many is magnetic resonance imaging (MRI). This imaging technique is widely used today for medical diagnosis. The idea is to measure the magnetic resonance of the proton (a magnetic signal by the hydrogen atoms in biological tissues). The magnetic field detector is a current coil tuned to the resonance frequency. The coils used are very sensitive at high frequencies, but poorly sensitive at low frequencies. This is one reason why clinical MRI is carried out under very strong field conditions (1.5 or 3 T), in order to obtain a resonance at high enough frequency. Such systems are thus complex and expensive. With more sensitive sensors able to work at low frequency, one can envisage significantly cheaper and more accessible MRI systems, even if their performance is admittedly somewhat reduced. Many groups are currently investigating this possibility with different kinds of sensors able to detect fields in the femtotesla range.

Biomagnetism is another medical application of ultrasensitive magnetic sensors, and although it is less well known, it is just as promising. The idea is to detect the electrical activity of the body through the resulting magnetic field. The magnetic sensor can detect microcurrents in the neurons. The first application is in magnetoencephalography (MEG), which reconstructs the electrical activity of the brain in 3D using a large number of cooled sensors positioned around the head. The second application is magnetocardiography [1], which maps electrical activity in the heart, once again using the magnetic field it emits. Figure 4.4 shows an example of signals detected in this way.

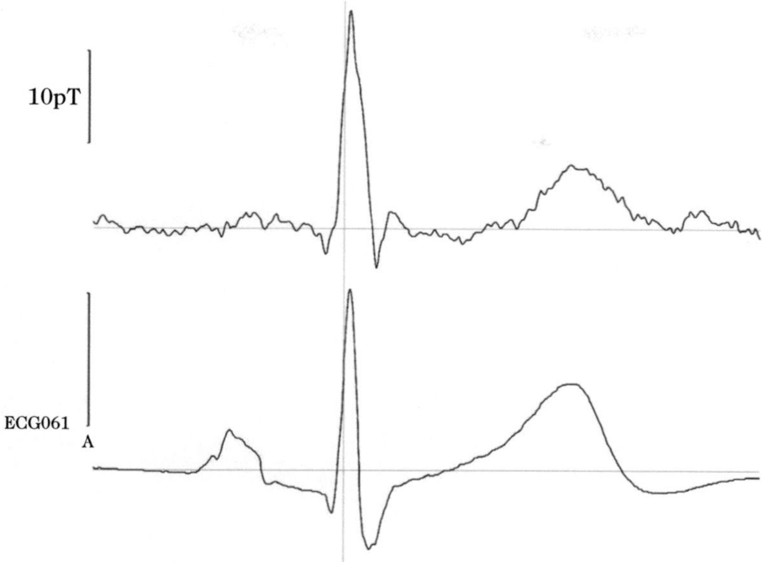

Fig. 4.4 Example of a magnetocardiography signal measured using a combined superconductor–GMR sensor (*upper*) and an electrocardiography signal (ECG) registered at the same time (*lower*)

4.7 Prospects

While hard disk read heads have long been based on GMR sensors, meeting the needs of high sensitivity and micrometric dimensions, the main manufacturers of magnetic sensors have just begun to mass produce GMR magnetic sensors for the automobile industry and other applications for the general public. What brought this on was the successful integration of GMR and TMR techniques in conventional CMOS electronics. Spin electronics thus becomes a natural complement to charge electronics, i.e., transistor electronics, for integration over large substrate sizes and hence at very low cost. We may estimate that, in seven years from now, one third of the four billion magnetic sensors manufactured each year will be based on spintronics. Further applications of these sensors are also coming to light, e.g., for stress or pressure measurements, or the fabrication of ultralow noise preamplifiers with high electrical insulation.

Reference

1. M. Pannetier-Lecœur, L. Parkkonen, N. Sergeeva-Chollet, H. Polovy, C. Fermon, C. Fowley, Magnetocardiography with sensors based on giant magnetoresistance. Appl. Phys. Lett. **98**, 153705 (2011)

Part II
Nanomaterials and Nanochemistry

Chapter 5
Overview of the Field

Christine Vautrin-Ul

Abstract Based on a comprehensive understanding of reactions at the nanoscale, nanochemistry is the art of building up nanomaterials via the bottom–up approach, as opposed to the top–down approach followed in electronics. In the first part, we use the examples of nanodiamond, carbon nanotubes, and graphene to illustrate the diversity of properties that can be procured by nanochemistry from the same element, viz., carbon, then describe their applications to energy production and storage. In the second part, we discuss soft nanochemistry, which appeals to the methods of molecular and supramolecular synthesis. The molecular route is exemplified by the so-called click chemistry, while the supramolecular route, molecular recognition, and self-assembly are illustrated by the cyclodextrins and self-healing rubbers. There follows a description of methods of functionalisation, key steps in controlling the properties of nanoparticles and nanomaterials. The last part deals with synthesis routes for preparing metal nanoparticles and nanostructured materials such as organic–inorganic hybrid nanocomposites. Finally, we present the bio-inspired approach in which the nanochemist produces hybrid nanomaterials with hierarchical structure in a single step.

5.1 Introduction: Nanochemistry

Chemistry is the science of matter, its nature, its composition, and its properties. It thus stands midway between physics, which studies the behaviour of matter from the laws of the universe, and biology, which seeks to understand the laws of living things. In fact, chemistry is by its very nature a nanoscience, since it manipulates and synthesises nanometric entities such as atoms, molecules, polymers, particles, and aggregates of varying levels of complexity (see Fig. 5.1). Nanochemistry is nevertheless a new subfield of chemistry, in which the skill of the nanochemist consists in guiding methods of synthesis either toward the production of highly functional

C. Vautrin-Ul (✉)
Centre de Recherche Sur la Matière divisée, Université d'Orléans,
1b rue de la Férollerie, 45071 Orléans Cedex 2, France
e-mail: christine.vautrin-ul@cnrs-orleans.fr

J.-M. Lourtioz et al. (eds.), *Nanosciences and Nanotechnology*,
DOI 10.1007/978-3-319-19360-1_5

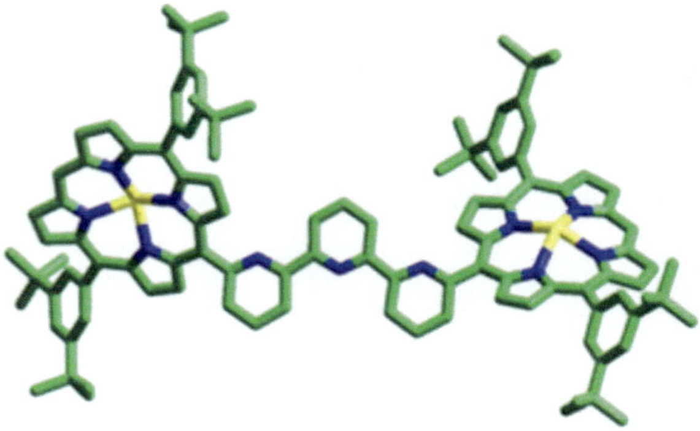

Fig. 5.1 Nanochemistry is a sub-discipline of chemistry that studies molecules or macromolecules with dimensions between 2 and roughly 500 nm

molecules, or toward the fabrication of nano-objects containing thousands of elementary molecules that have been suitably assembled to carry out specific functions. The nanochemist investigates ways of organising matter on the molecular scale, thereby determining specific physical, chemical, mechanical, or biological properties of nanomaterials.

Nanochemistry derives from an understanding of and control over the reactivity of matter on the nanometric, or even sub-nanometric scale. It thus involves the development of chemical syntheses leading to the construction of elementary nanometric building blocks of different size, shape, composition, surface structure and texture, and functionality.

These elementary building blocks can then be used as-is or to build more sophisticated architectures. Therefore, the nanochemist also designs them in such a way that they can be assembled, or they can self-assemble, to result in a given target nanomaterial. This is known as the bottom–up approach to the fabrication of nanomaterials, as opposed to the more traditional top–down approach widely used in microelectronics (see Fig. 2.3). The bottom–up approach, illustrated in Fig. 5.2, deals with

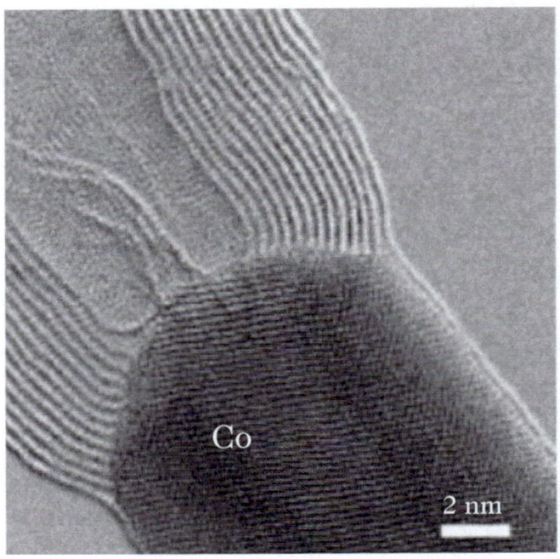

Fig. 5.2 Illustration of the bottom–up approach for producing nanomaterials. Growth of a multi-walled carbon nanotube on a cobalt crystal (image obtained by transmission electron microscope). © *Institut de physique et de chimie des matériaux*, Strasbourg

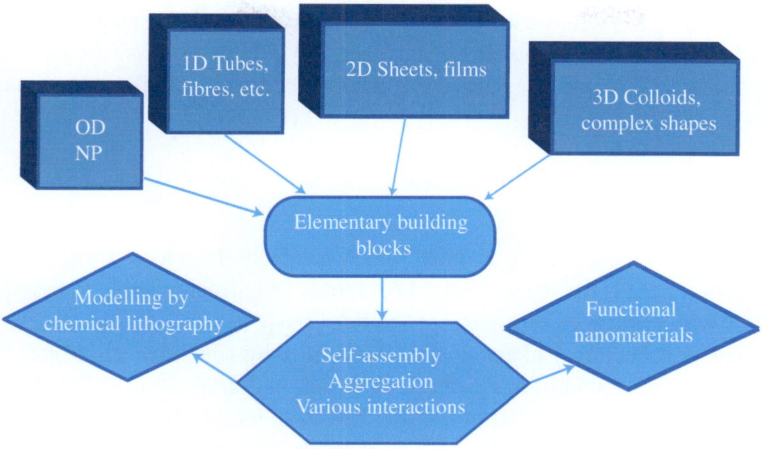

Fig. 5.3 Organigram showing the different stages in the elaboration of a nanomaterial. NP: nanoparticle

objects of increasing complexity, such as molecules, macromolecules, nanoparticles (NP), nanostructured and/or nanoporous materials, assemblies of biomolecules and nanoparticles, and so on.

The assembly of nanoscale building blocks, whether guided or spontaneous (self-assembly), is just as important as the synthesis of the building blocks themselves. It involves various interactions that work together to construct the desired functional nanomaterial, ready for its intended application (see Fig. 5.3). This kind of assembly can build upon units of very different types. Some associate organic and mineral building blocks, producing hybrid nanomaterials with applications in a broad range of fields, such as catalysis or energy conversion and storage. Others associate organic or inorganic building blocks with 'living' molecules in vitro or in vivo, notably in the development of nanomedicine, making it possible to release active ingredients in a targeted way within sick cells (see Part III of this book).

The aim in this chapter will be to demonstrate the importance of nanochemistry in the study and elaboration of nanomaterials, and also to demonstrate the almost infinite potential it makes available to the research chemist, although no attempt will be made to give an exhaustive account of the various methods of synthesis or indeed the whole range of their applications. We begin by illustrating the diversity of shapes and sizes of the elementary building blocks using the example of carbon, and taking the opportunity to discuss some of the many resulting applications. We shall thus be called upon to describe methods of synthesis which for the most part appeal to traditional techniques of chemistry, involving reactions that often require high temperatures and/or pressures. We shall then discuss how molecular and supramolecular chemistry can be used to develop a kind of 'soft' chemistry for the synthesis and functionalisation of nanomaterials. We end by presenting a few examples of syntheses exploiting this non-traditional chemistry to elaborate nanoparticles and nanomaterials. This will illustrate how nanochemistry can hold sway over the broad range of shapes, surfaces, functionalities, and physical properties of nanomaterials whose chemical compositions are nevertheless exceedingly similar to one another.

5.2 Carbon and Its Nanostates

Carbon is surely the chemical element which comes in the widest variety of forms, and also which leads to the most varied structures and an astonishing range of nanomaterials with ever more surprising properties [1–4].

Carbon is a chemical element that exists in nature in three allotropes on the macroscopic scale: amorphous carbon, diamond, and graphite (see Fig. 5.4). Each of these forms has exceptional properties that have long been put to use. In a first approach, we can see immediately that graphite can already be considered as a nanomaterial, built up as it is of nanometric crystal planes that confer upon it distinctly anisotropic properties, such as excellent electron conductivity and great strength under traction in the planes of its hexagonal structure. These properties are exploited in carbon fibre applications, for example.

Scientists have produced different nanometric forms of carbon, making use of its various allotropes, and this in one, two, or three nanoscale dimensions. Applications

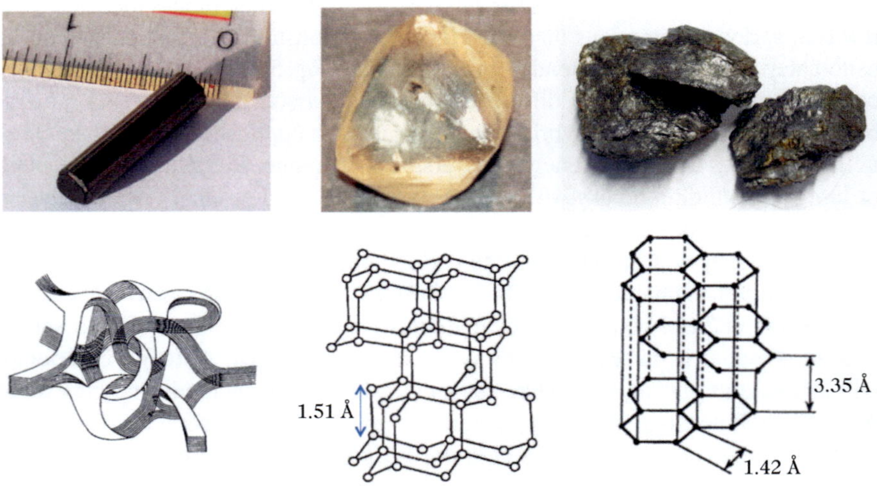

Glassy carbon:
entangled graphite ribbon
structure

Diamond:
face-centered cubic structure
sp^3 hybrid carbon atoms
Density 3.5 g. cm^{-3}

Graphite:
hexagonal sheet structure
sp^2 hybrid carbon atoms
Density 2.2 g. cm^{-3}

Fig. 5.4 Different allotropes of carbon and their structure. *Source* Wikimedia commons, Eurico Zimbres FGEL/UERJ, Daniel Schwen, and the structure of glassy carbon from [5] © Nature Publishing Group, 1971

concern fields as diverse as solar energy (organic photovoltaic cells), the removal of micropollutants, and supercapacitors. In the following, for each form of carbon, we shall describe one method of elaboration, the physicochemical properties attributable to its nanostate, and one example of application.

5.2.1 Carbon-Based Nanomaterials: Zero-Dimensional Structures or Nanoparticles

A Long Known Classic: Carbon Black

We begin with a classic nanomaterial that has been known for a long time, namely carbon black. These materials are aggregates of nanospheres obtained by a simple process of incomplete combustion. The carbon content is typically in the range 97–99 %. It is very cheap, costing 1–5 euros/kg, and occurs in the form of highly dispersed powders, usually produced by controlled pyrolysis of vapour-phase hydrocarbons. More precisely, there are in fact different industrial processes [1, 6] which produce different kinds of carbon black, including acetylene black, channel black, furnace black, lamp black, and thermal black. The mean diameter of the particles in most commercial carbon blacks lies in the range 10–300 nm, while the mean diameter of the aggregates (see Fig. 5.5) varies over 80–500 nm. Note that these aggregates can themselves form micrometric structures, or agglomerates, by interaction. Their specific surface area ranges from 10 to 300 m^2g^{-1}. Carbon blacks are mainly used as strengtheners in elastomers, especially in tyres. However, they also occur as black pigments in inks, paints, and mascara. Among the many other applications, we could also mention their use as percolating agent in batteries, ensuring electron conductivity in the electrodes.

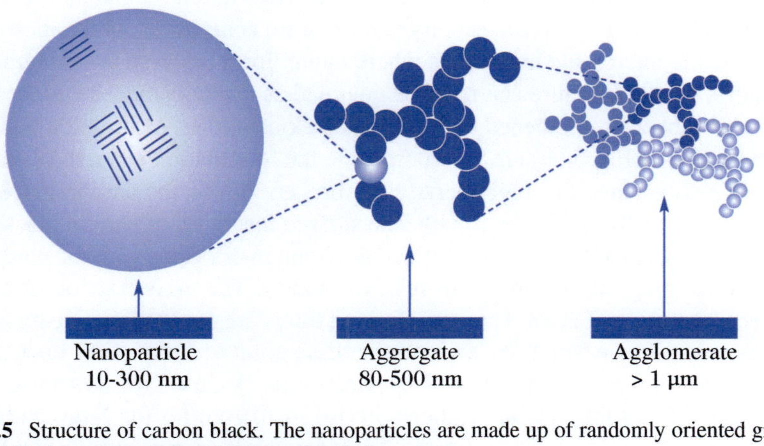

Fig. 5.5 Structure of carbon black. The nanoparticles are made up of randomly oriented graphite sheets [6]. *Source* International Carbon Black Association

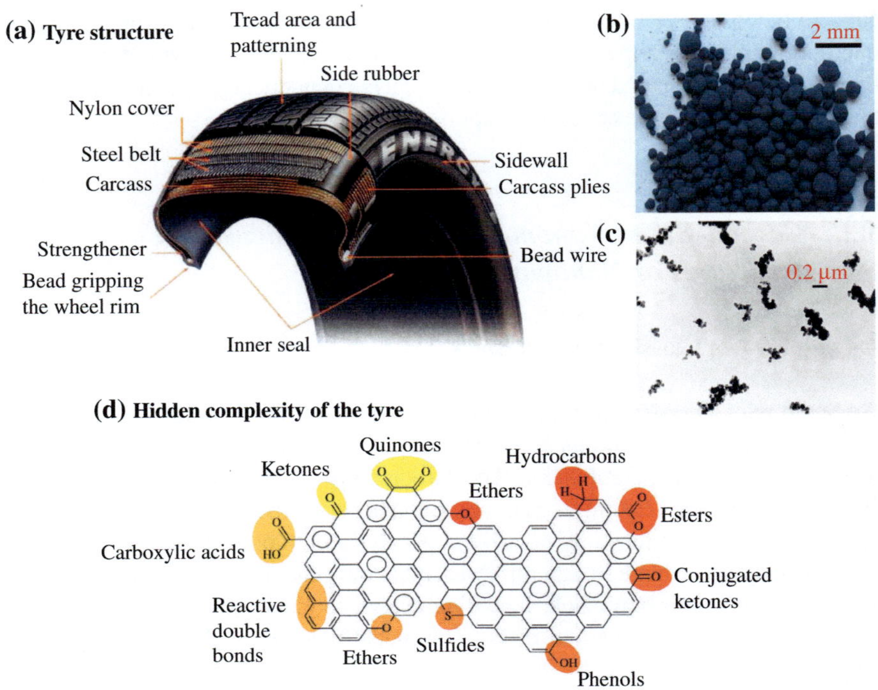

Fig. 5.6 The inner secrets of the tyre. **a** Schematic view of a tyre without air chamber. **b** Carbon *black* pellets before inclusion in the elastomer. **c** Nanoparticles and aggregates of carbon black. **d** Surface chemical functions of carbon blacks. Adapted from Francis Peters, Michelin [8]

A Well Known Nanocomposite Containing Carbon Black: Tyres

The tyre is an interesting example of a multicomposite structure made up of materials on different length scales [7]. The structure of a tyre without air chamber is shown schematically in Fig. 5.6a. The carcass and belts are composites on the mesoscale, or even on the microscale in the case where radial fibres are used as strengtheners. The matrix of this structure can itself be qualified as a nanocomposite, since it is a vulcanised rubber, strengthened by a carbon black nanofiller. Millimetric pellets of carbon black (see Fig. 5.6b) are incorporated in the formulation when the elastomer is vulcanised, whence the agglomerates are broken up and the carbon black ends up in nanometric form (see Fig. 5.6c). The surface reactivity of carbon blacks (see Fig. 5.6d) means that they can develop rather strong interactions with the elastomer, while maintaining the possibility of relative motion. The mass fraction of carbon black is of the order of 20 %. These reinforcing fillers are used to improve resistance to abrasion, reduce heating due to inelastic dissipation of mechanical energy, and contribute to the evacuation of such heat. Carbon blacks, used in tyres since 1910, are classified according to size: particles from 11 to 19 nm for the N100 and N200 standards and from 31 to 39 nm for the N400 standards (norms established by the

American Society for Testing and Materials or ASTM). The size range is selected with reference to the kind of rubber used in each part of the tyre. The smaller sizes improve heat dissipation effects, while the larger ones stand out for their effective role in adherence (road holding) and tear resistance.

It should be noted that, for the last twenty years, functionalised silica has also been used as a strengthening filler in rubber formulations for the tyre tread, the aim being to reduce roll resistance. At the present time, 'green' tyres are being developed, with a view to extending the lifetime of the tyre and reducing fuel consumption. These tyres are nanocomposites with hybrid fillers combining silica and carbon black nanoparticles [7].

New Carbon Nanomaterials: Fullerenes and Nanodiamonds

Fullerenes

The name 'fullerene' is a generic term for any carbon molecule of approximately spherical shape. This form of carbon was discovered in 1985 by Richard Smalley, Harold Kroto, and Robert Curl, and it won them the Nobel Prize for Chemistry in 1996 [9]. These molecules are named after the futuristic architect Richard Buckminster Fuller, who initiated many geodesic dome projects, such as *La Géode* in Paris. The fullerenes are built up from alternating hexagons and pentagons. The number of pentagons is strictly equal to 12, while the number of hexagons is arbitrary except for the fact that it must be even. The fullerenes satisfy Euler's relation

$$2F + V = E + 2 ,$$

where F is the number of faces, V the number of vertices, and E the number of edges.

The best known fullerene structure is C_{60} which has the shape of a football, with a surface made up of 20 hexagons and 12 pentagons (see Fig. 5.7 upper). In the molecule C_{60}, most of the carbon atoms have type sp^2 (generating double bonds). However, some of them must be sp^3 hybrids (generating single bonds), owing to the curvature of the molecule. In fact, there is a whole range of different fullerenes (see Fig. 5.7) [1, 2]. Many different structures such as C_{70}, C_{76}, C_{78}, C_{80}, etc., have been identified, in particular by mass spectroscopy. There are also fullerenes called nano-onions [1, 2] which are highly stable and sit inside one another (see Fig. 5.7c).

Fullerenes can be prepared either by combustion, or by methods based on the vaporisation and condensation of carbon in an inert atmosphere, such as helium or argon, e.g., electric arc or laser ablation techniques.

These nanostructures have inspired a great deal of interest in most industrialised countries, with applications in a whole range of fields, including chemistry, biology, electronics, materials science, optics, energy production, and others. Regarding their properties, the fullerenes are well known to be superconducting (below 40 K) when they contain, e.g., alkali metals such as potassium or rubidium. A resistivity as low as 5–10 mΩ cm has been measured at room temperature for a K_3C_{60} compound (to

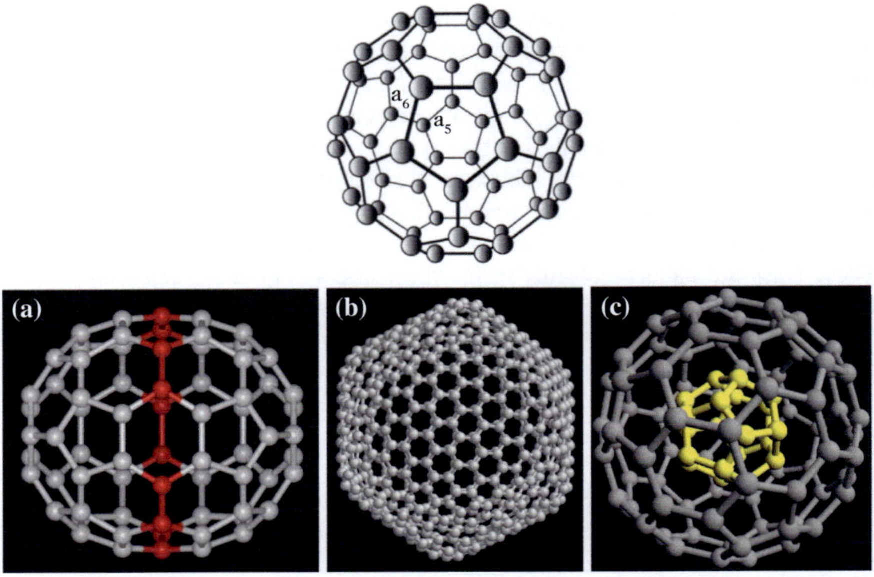

Fig. 5.7 *Upper* Fullerene C_{60} made from single bonds (a_5) in the pentagons and double bonds (a_6) in the hexagons. *Lower* Other fullerene structures: **a** C_{70}, **b** C_{540}, **c** carbon onion

be compared with 2×10^{-3} mΩ cm for copper and 1 mΩ cm for graphite or carbon fibres).

Nanodiamond

Discovered in the Soviet Union in 1960, nanodiamonds remained unknown to the rest of the world until the end of the 1980s [10, 11]. Since 1990, many techniques have been devised for making them, and three are currently used in industry: detonation, laser ablation, and high pressure–high temperature (HPHT) processes, with pressures up to 50 000 times atmospheric pressure and temperatures of around 1 500 °C, the diamonds being grown in industrial presses and then ground. Nanodiamonds are widely used today in industry as electronic or optical components, additives, and so on, and they are expected to become important in nanobiology as intracellular carriers of therapeutic molecules.

Synthesis of Nanodiamonds by Detonation. Nanodiamonds can be obtained by detonation of an explosive mixture of graphite (see Fig. 5.8). In general, two explosives are used, namely trinitrotoluene (TNT) $C_6H_2(NO_2)_3CH_3$ and hexogen $C_3H_6N_6O_6$ (see Fig. 5.8a). Monocrystals measuring 2–10 nm are formed and cluster together rapidly into strongly bound agglomerates measuring more than 20 nm.

The mechanism leading to the formation of nanodiamonds during detonation can be explained by referring to the phase diagram of carbon shown in Fig. 5.8b. It is clear that the stable form of carbon is graphite at low pressure and diamond at high pressure, the liquid phase of carbon existing for temperatures above 4 500 K. However, on the nanoscale, the size of aggregates must be taken into account in the

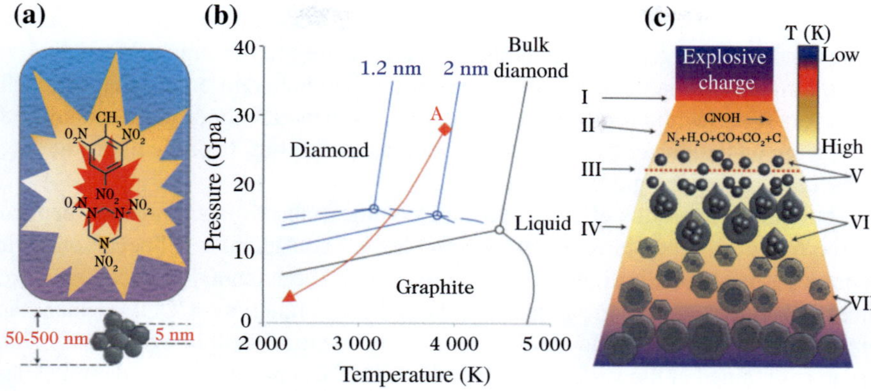

Fig. 5.8 Synthesis of nanodiamonds by detonation, taken from [10], © Nature Publishing Group 2012. **a** Example of an explosive charge used for the synthesis: 60 % TNT and 40 % hexogen. The explosion is carried out in a closed detonation chamber and in an atmosphere of N_2 and CO_2, in the presence of liquid water or water ice. After the detonation, the diamonds are recovered on the walls and at the bottom of the reactor. **b** Phase diagram of carbon. During the explosion, the pressure and temperature increase significantly, reaching the Jouguet point (point A), then fall isentropically (*red curve*), with the consequence that carbon nano-aggregates form, coalescing and crystallising in the form of diamonds, provided that the pressure is high enough for the system to remain in the stability region of diamond. For lower pressures, crystallisation leads to graphite. **c** Schematic view of the detonation: (*I*) shock wave caused by the explosion, (*II*) region of chemical reactions in which the molecules of the explosive charge decompose, (*III*) Chapman–Jouguet plane corresponding to the point A of (**b**), (*IV*) expansion region of detonation products, (*V*) formation of carbon nano-aggregates, (*VI*) coalescence of carbon nanodroplets, and (*VII*) crystallisation, growth, and agglomeration of nanodiamonds

phase diagram, because as the size decreases, so the liquid phase arises for slightly higher pressures and ever lower temperatures (of the order of 3 000 K for 1.2 nm carbon nanoparticles).

The temperature then becomes too low to form a continuous liquid carbon phase, but it is high enough to produce nanometric carbon droplets. This is effectively what happens in the detonation chamber. The nanodiamonds are then formed by homogeneous nucleation of supersaturated carbon vapours, followed by condensation and crystallisation of nanometric liquid carbon droplets (see Fig. 5.8c).

After synthesis, the nanodiamonds are filled with impurities, these being essentially other allotropic forms of carbon (25–85 % by mass), but also metals used to trigger the detonation (lead or copper azides, for example), and iron from the steel of the reactor (metal impurities represent 1–8 % by mass). Once extracted from the reactor, they are thus subjected to several chemical purification stages: an oxidising treatment to eliminate carbon impurities, e.g., using nitric acid or ozone, followed by an acid treatment to remove traces of metals.

There remains one more important step, namely dissociation of the agglomerates. This step, often necessary before the nanodiamonds can be used, can be carried out by means of different techniques such as grinding by ceramic (silica or zirconium oxide) microbeads, dry grinding by salts that are soluble in water and by sugars, or indeed ultrasound-assisted grinding methods. However, owing to the high surface area to volume ratio, the diamond nanoparticles tend to agglomerate once again to reach diameters of a few micrometers. To avoid this, the nanodiamonds can be stabilised by playing on their electrostatic and steric surface properties, in particular by adding surfactants during the grinding stages.

The chemical groups at the surface of commercialised nanodiamonds depend on the treatments they have undergone and have a considerable influence on their subsequent functionalisation (see Sect. 5.3) and on the intended applications. The most convenient surface group is the carboxylic acid function –COOH, which has high chemical reactivity and can facilitate many chemical coupling reactions. This group is obtained at the surface of nanodiamonds (ND) purified by oxidation using ozone or air (ND-COOH). Note that a hydrogen plasma treatment of these ND-COOH at 700 °C leads to a total reduction of the acid functions and produces surface-hydrogenated diamonds (ND-H).

Applications of Nanodiamonds. There are many potential applications of nanodiamonds in a wide variety of different fields. Figure 5.9 summarises their properties and related applications.

One of the most promising applications is the targeted delivery of therapeutic molecules in a way that can be monitored by luminescence. Thanks to their small size, nanodiamonds can cross biological barriers in the human body, such as cell

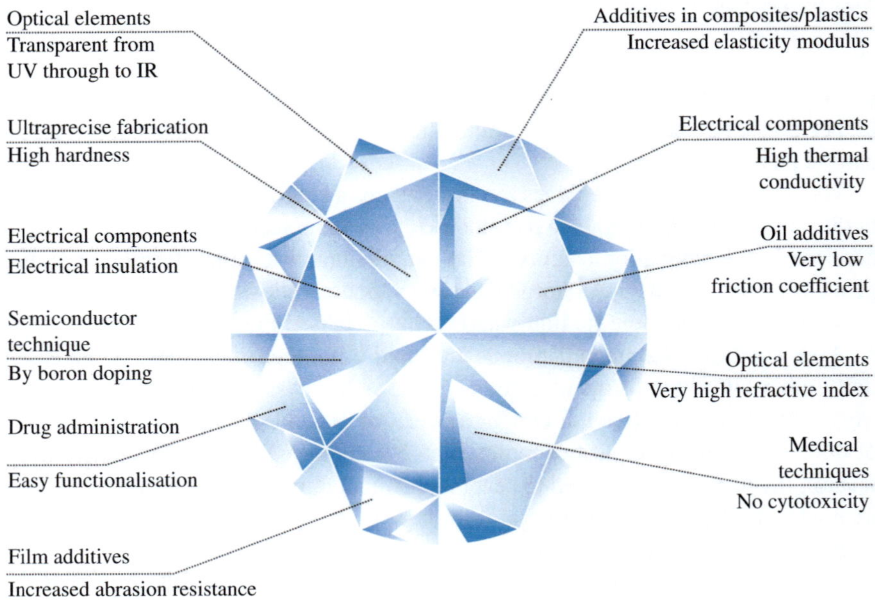

Fig. 5.9 Main properties and applications of nanodiamonds. *Source* Reishauer AG

membranes. Moreover, their chemical characteristics make them biocompatible and, after functionalisation, they can bind reversibly to therapeutic molecules. In this way, active principles can be delivered to specific targets in the very heart of the cell [12]. This promising strategy could be applied to the treatment of cancers that resist conventional chemotherapy. Recent studies [13] on liver and breast cancer models in mice have shown that, after injection of nanodiamonds bound reversibly to doxorubicin, an anticancer agent, the nanodiamond–doxorubicin complex remained ten times longer in the cell than the doxorubicin alone. In conclusion, the complex was significantly more efficient and less toxic than when the anticancer agent was injected independently.

Apart from drug delivery, biomarker functions can be associated with nanodiamonds. Indeed, if they are bombarded by an electron beam, extremely stable luminescent centers can be incorporated within them, providing possibilities for diagnostics or monitoring of tumours (see the third part of this book). Whether they are used for biomarking or drug delivery, the interactions between nanodiamonds and biological entities on the one hand or the drug on the other are essentially controlled by the surface chemistry of the nanoparticles, a key feature in their future exploitation.

5.2.2 Carbon-Based Nanomaterials. One-Dimensional Structures: Carbon Nanotubes

One-dimensional nanometric forms of carbon include nanotubes and nanofibres. The story of carbon nanotubes began in 1991 when they were discovered by the Japanese microscopist Sumio Iijima among the soots resulting from the electric arc synthesis of fullerenes (see Fig. 5.10a) [14]. Since then, the exceptional properties of carbon nanotubes have earned them a great deal of attention. There have been many books and reviews [15–18]. Some of their more important physical properties have already been described in Sects. 2.3 and 2.5 of Chap. 2. In the present chapter, we shall focus mainly on the structure and synthesis of these nano-objects.

Structure of Carbon Nanotubes

Carbon nanotubes are made by rolling up very long graphene sheets and closing them at the ends by half-spheres similar to fullerenes. There are two main families of carbon nanotube:

- Single-walled nanotubes (SWCNT) are made from a single graphene sheet which is rolled up as shown in Fig. 5.10a.
- Multi-walled nanotubes (MWCNT) consist in the perfect case of between two and a few dozen concentric graphene tubes placed inside one another like Russian dolls, or of a single graphene sheet rolled up several times about a central axis (see Fig. 5.10b).

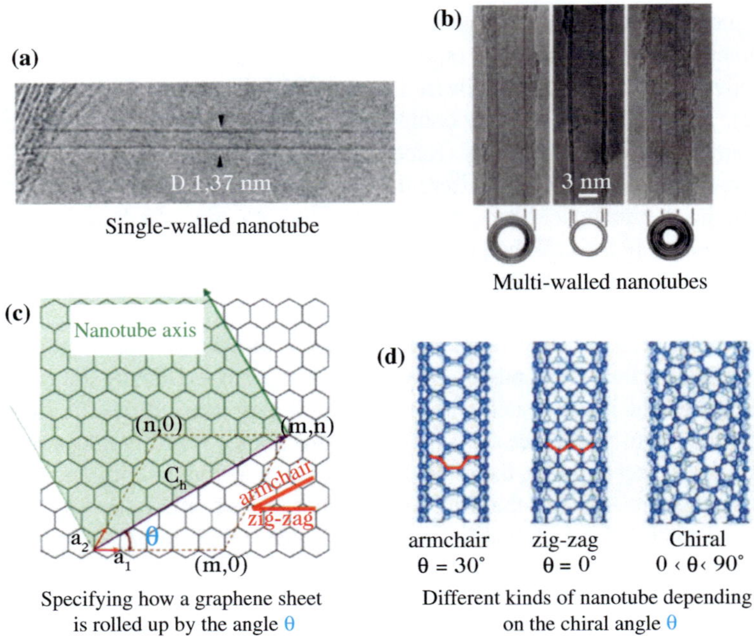

Fig. 5.10 Single-walled carbon nanotubes (SWCNT) and multi-walled carbon nanotubes (MWCNT). **a** Transmission electron microscope (TEM) image of an SWCNT [19]. **b** TEM image of MWCNTs [14]. **c** Different ways of rolling up a graphene plane as specified by the angle θ. **d** Various forms of carbon nanotube for different values of the chiral angle θ

Graphene sheets can be rolled up in several different ways, leading to different so-called helicities, i.e., different atomic structures, known as ziz-zag, armchair, or chiral (see Fig. 5.10c, d).

The presence of one or more sheets inside a nanotube creates nano-objects whose electronic, mechanical, and even chemical properties may differ. However, the helicity of carbon nanotubes can also affect their electronic properties. In particular, armchair nanotubes can have metallic behaviour, while some zig-zag or chiral nanotubes have properties making them more like wide bandgap semiconductors.

Note also that, in order to modify or amplify the properties of the generic nanotubes described above, or in order to facilitate their incorporation into nanomaterials by increasing their solubility, for example, they can be functionalised, doped, coated, or filled. These systems are referred to as meta-nanotubes [17].

Synthesis and Purification of Carbon Nanotubes

There are many ways of making carbon nanotubes [15]. These fall into two categories: methods based on sublimation of carbon in an inert atmosphere, by electric arc, laser ablation, or in solar furnaces, and chemical methods which are today expanding rapidly, including for example the catalytic decomposition of carbon-bearing precursors. The diversity of these methods produces nanotubes with a range of significantly different characteristics in terms of length, diameter, crystal structure of the walls, nature and amount of impurities, and so on. The industrial development of carbon nanotubes is also in rapid expansion, with prices dropping sharply over the last few years to less than a dollar per gram. In Europe, the market is shared between Bayer and Arkema, and the latter has just opened a production unit capable of producing 400 tonnes a year.

An important step before the nanotubes can be put to use is their purification, the aim being to remove soluble impurities, such as fullerenes or polycyclic aromatic hydrocarbons, and insoluble impurities which are essentially amorphous carbon, graphite particles, carbon onions, and catalyst particles. The soluble impurities are easily separated by washing with carbon sulfide (CS_2) or toluene.

Insoluble impurities pose a greater problem but many purification methods have been reported. Chemical methods based on the different reactivities of the nanotubes and the impurities, with the latter being generally more highly reactive, use oxidation reactions to degrade the impurities into soluble or gaseous products and hence remove any amorphous carbon and metal particles. Physical methods exploit either the difference in shape of the nanotubes, which are very long and highly anisotropic, and nano-impurities, which often have an approximately spherical shape, or the difference in density. These methods can remove catalyst, graphite, and carbon onion nanoparticles, but the amorphous carbon is too highly divided to be eliminated in this way. The purification techniques used in this case are mainly microfiltration, steric exclusion chromatography, or centrifugation. Since none of the methods just described, either physical or chemical, is 100 % selective, they must generally be used in combination.

An Example of a Synthesis Producing Different Types of 1D Nanomaterial: Thermal Plasma Synthesis

Thermal plasma synthesis was initially developed to produce carbon blacks and fullerenes, but it can be used to obtain a wide variety of different 1D forms of carbon. The idea is to create a large reaction volume at high temperature in order to vaporise or crack a carbon-bearing precursor in the presence of a catalyst. The processes operating in the gaseous phase are continuous and so can be extrapolated to the industrial scale for mass production. They are extremely efficient heat sources, whence carbon-bearing precursors can be used, including carbon black or graphite, for which the sublimation temperature of around 4 000 K is reached in the thermal plasma, or indeed various liquid or gaseous hydrocarbons, such as ethylene. The catalysts are transition elements, such as Ni, Co, Ni-Co, Y, Ni-Y, etc., which can be used in the form of a metal powder mixed with the carbon-bearing precursor. Nanomaterials produced in this way are of various kinds, as shown in Fig. 5.11, viz., SWCNT or MWCNT and, among them, ‘bamboo’ or ‘pearl necklace’ nanofibres.

The so-called bamboo nanofibres are in fact multi-walled carbon systems comprising some twenty graphene layers. Pearl necklace nanofibres have lengths exceeding a few tens of microns and an average diameter of around 50 nm. The ‘pearls’ are generally hollow and sometimes filled with catalyst. They are made up of highly ordered concentric layers of graphene. Bamboo nanofibres exhibit a high level of electro-

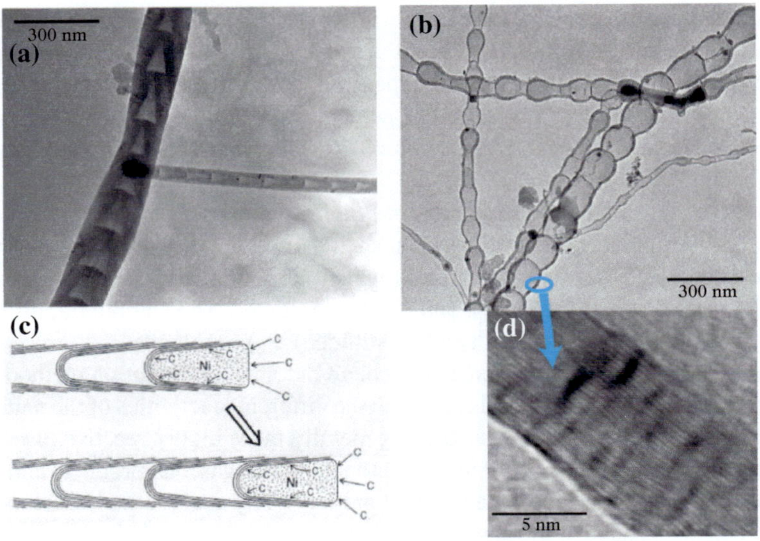

Fig. 5.11 Transmission electron micrographs of nanofibres produced by thermal plasma synthesis. **a** Bamboo type (helium plasma with Ni-Co catalyst). **b** Pearl necklace (nitrogen plasma, Ni-Co catalyst). *Black spots* are catalyst particles. **c** Growth mechanism of bamboo type nanofibres. **d** Zoom on the wall of a pearl necklace nanofibre. Courtesy of L. Fulcheri, *Université de Perpignan*, J.-C. Charlier, *Université catholique de Louvain*, and Yahachi Saito, University of Nagoya, Japan

chemical activity, greater than that of SWCNTs or MWCNTs, whence they are ideal for use in electrodes with applications as chemical or biochemical sensors [20].

Apart from the few examples described here, multi-walled nanotubes have a whole range of textures, depending on the general orientation of the graphene sheets relative to the axis of the nanotube, and also nanotextures, depending on the quality of the mutual orientation of the graphenes within the sheets taken as an ensemble. Their properties will depend very largely on this kind of texturing or nanotexturing [3]. Today, there are many prototypes incorporating these multi-walled nanotubes for use in high added value commercial applications, where post-processing overheads are easily absorbed. A case in point is surgical threads, tips for atomic force microscopes (see Sect. 2.4), and the like. There are also some large-scale applications of these nanotubes as catalytic supports or as a component in electrostatic paints.

5.2.3 Carbon-Based Nanomaterials. Two-Dimensional Structures: Graphene

Graphene, the last isolated allotropic form of carbon, is a monatomic carbon sheet which, when stacked up vertically under the action of van der Waals forces, forms graphite (see Fig. 5.4) [21, 22]. In reality, graphene is usually taken to refer to any materials with fewer than ten atomic layers [22]. Indeed, beyond this number, the electronic properties of the material begin to look more like those of graphite.

The key date for graphene was 2004, when Novoselov et al. published an article in the journal *Science* about the electronic properties of graphene sheets, in which they explained how they obtained them simply by repeated exfoliation of a piece of graphite until they ended up with a single monolayer of carbon [23]. The work carried out on this new nanomaterial won the Nobel Prize for Physics for its authors Andre Geim and Kostya Novoselov at Manchester university in 2010.

This 2D crystal with monatomic thickness has quite amazing properties: transparency, conductivity, mechanical strength and elasticity, and impermeability with respect to all gases, not to mention ease of functionalisation. With regard to these properties, it often does much better than other materials, thus suggesting many possible applications, particularly in the field of energy and portable systems, e.g., through the development of flexible electrodes. The physical and electronic properties of graphene for new components were outlined in Chap. 2. Easier to synthesise than nanotubes, graphene may eventually lead to a breakthrough technology in the electronics industry [24]. This is undoubtedly the opinion of the European Commission, which has just selected the graphene project as a Future and Emerging Technologies (FET) flagship, allocating a billion dollars over the next ten years (see Chap. 14).

Synthesising Graphene

Another reason for the rapid development of research on this nanomaterial is that it is relatively straightforward and rather cheap to obtain high quality graphene in the laboratory. Many of the most interesting properties have often been demonstrated using very high quality graphene films produced by mechanical exfoliation. However, there would never be a market for graphene applications without the development of industrial processes. The most promising ways to produce graphene industrially include vapour phase deposition on metal films, epitaxial growth by sublimation of silicon from a silicon carbide (SiC) wafer, but also mechanical or chemical exfoliation of graphite. Figure 5.12 illustrates the process of chemical exfoliation, while the deposition and epitaxial methods were discussed in Chap. 2. The size and quality of the resulting material depend on the process. In fact, the quality depends mainly on the type of defects and their concentration.

In graphite, the graphene sheets are bound together by van der Waals interaction forces that are proportional to $1/r^6$, where r is the distance between the sheets. The force required to separate them is estimated at $300\,\text{nN}/\mu\text{m}^2$ [22]. The idea behind chemical exfoliation methods is to reduce this force by modifying the chemical composition of the graphene planes by oxidation or by inserting molecules or ions, which tends to increase the intersheet distance and thereby facilitate exfoliation [22].

A relatively simple method is to set the graphite in contact with a polar solvent such as N-methyl-2-pyrrolidone (NMP) in the presence of ultrasound. The result

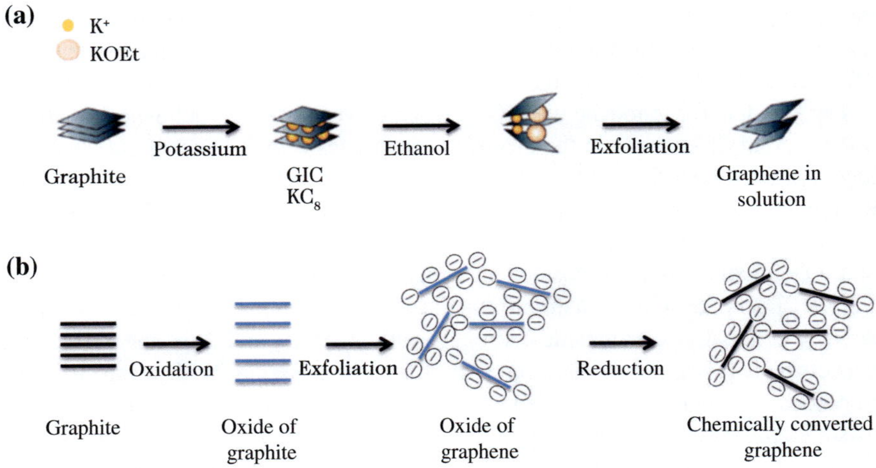

Fig. 5.12 Preparation of graphene by chemical exfoliation of graphite. **a** Graphene prepared from graphite intercalation compounds (GIC). KOEt potassium ethoxide. **b** Chemically converted graphene (GCC) obtained by reduction of oxidised graphite

is a colloidal solution of graphene sheets that can then by isolated by centrifuge. However, graphene concentrations remain rather low, of the order of 0.01 mg/mL.

A second way, the one shown in Fig. 5.12a, uses graphite intercalation compounds (GIC), which have been known since 1841. If graphite is heated in vacuum, one reaches a first intercalation stage with K^+ ions inserted between each layer of graphene. This creates the compound KC_8, a particularly powerful reducing agent. These alkali metal intercalation compounds are then exfoliated by placing them in an aqueous or alcoholic solution where they react vigorously to yield a bulky ethoxide or hydroxide, together with molecular hydrogen gas, which assists in exfoliation.

A third promising way of exfoliating graphene is shown in Fig. 5.12b. This is known as chemically converted graphene (CCG). It consists in oxidising graphite and stabilising the oxidised graphene planes in an aqueous solution using oxygenated functional groups which tend to form hydrogen bonds. The oxidised graphene sheets are then reduced by a reducing agent such as hydrazine, in a process that can be carried out thermally. This kind of preparation is simple and lends itself well to industrialisation. The graphene obtained in this way has too high a level of impurities to be used in microelectronics or photovoltaics, but the quality is nevertheless sufficient for the production of nanocomposites.

Fullerenes, Carbon Nanotubes, and Graphene: Development of Organic Photovoltaic Cells

What Is at Stake?

The advantages of solar energy as an alternative source of energy have by now been clearly demonstrated (see Chap. 2). Standard photovoltaic technology uses crystalline silicon and represents today about 85 % of world production. This option is still favoured, but the high cost of silicon cells, the high investments required for their production, and the presence of heavy metals such as cadmium and rare metals like indium, gallium, etc., in significant amounts in the thin inorganic films are all factors encouraging the development of innovative photovoltaic materials and products. Since the 1990s, organic photovoltaic (OPV) technologies have been developed in this context [25, 26]. Indeed, organic materials bring several advantages:

- Low cost of fabrication with regard to process, energy, and raw materials, which mean that organic cells could cost just 25 % as much as silicon technologies.
- They are extremely light.
- They are highly flexible, since they can be deposited on plastics.
- The basic compounds are easily available and highly innocuous, in contrast to certain metals used in second generation cells.

On the other hand, the performance of OPV technologies is limited by the lower electrical mobility of charges in organic materials as compared with inorganic semiconductor materials, together with a lower stability. Despite the fact that their efficiencies remain low, OPV technologies are beginning to reach the commercial phase, and many in research and industry consider them to represent a real solution for the future. They are already envisaged for use in portable electronic devices for the general public and flexible products for the building and leisure industries. Here one may think of applications in laptop computers, small electronic devices in cars such as GPS or air-conditioning, self-heating clothes, and others.

How it Works

As in conventional cells using inorganic semiconductors, current is generated in organic systems through dissociation of excitons, i.e., electron–hole pairs bound together by electrostatic interactions, created by absorption of photons. The first condition is thus that the molecules or polymers are able to absorb significant amounts of photons emitted by the sun over a broad range of wavelengths. However, in contrast to what happens in inorganic materials, the excitons do not dissociate easily, because the electron is always strongly attracted by the hole to which it is bound.

It is thus important to be able to dissociate the two charges in order to ensure the formation of free charges that can be collected at the electrodes (anode and cathode). Here we appeal to the heterojunction, an interface between an electron donor material and an electron acceptor material. The force attracting the electron to the acceptor material must overcome the one binding it to the hole. The excitons will therefore migrate toward the interface where they will dissociate. The electron donor material could, for example, be a π-conjugated polymer,[1] while the electron attractor could be a fullerene (see Fig. 5.13a).

The energy levels of the polymer must be precisely positioned relative to those of the acceptor compound, to ensure efficient dissociation of the excitons. The best way to do this is to combine chemically different aromatic heterocycles within the polymer chains (see Fig. 5.13b). This is indeed a positive point in using organic chemistry, since the electronic properties of such materials can be modulated!

Another important parameter is the diffusion length of the excitons, limited here to a few tens of nanometers. If the heterojunction is to be perfectly suited, the optimal phase segregation between donor material and acceptor material must therefore lie between 5 and 10 nm. In practice, one encounters planar (2D) heterojunction devices using evaporable molecules and 3D heterojunction devices using polymers. Once the charges have been created, the next task is to transport them to the electrodes. The ideal situation is thus to have high mobility holes in the donor material and high mobility electrons in the acceptor material.

Over the past few years, scientists have been able to synthesise organic materials with broad absorption ranges, from the visible to the near-infrared, and optimised charge transport properties. The best efficiencies, between 6 and 8 %, are obtained using low bandgap polymer materials combined with fullerene derivatives. These low bandgap materials are in fact copolymers possessing both electron donor and electron acceptor groups. To make them soluble and easier to implement, the monomers used to synthesise them generally carry alkyl groups, resulting in very long alkyl chains which improve solubility.

Further optimisation concerns the electrode material, which must be transparent so that the active layer receives maximal radiation (see Fig. 5.13a). The material most commonly used is a transparent conducting oxide, viz., indium tin oxide (ITO), a mixture of indium oxide and tin oxide. Apart from the fact that indium resources are not compatible with mass production of organic photovoltaic cells, the use of ITO in flexible photovoltaic devices requires a reduction in thickness, and this in turn leads to a drop in its conduction performance. Graphene and carbon nanotubes are excellent candidates for making

[1] A π-conjugated polymer is one that is rich in double bonds and aromatic rings, which delocalise π electrons over the whole structure, thereby bestowing semiconductor properties on the material (see Fig. 5.13b).

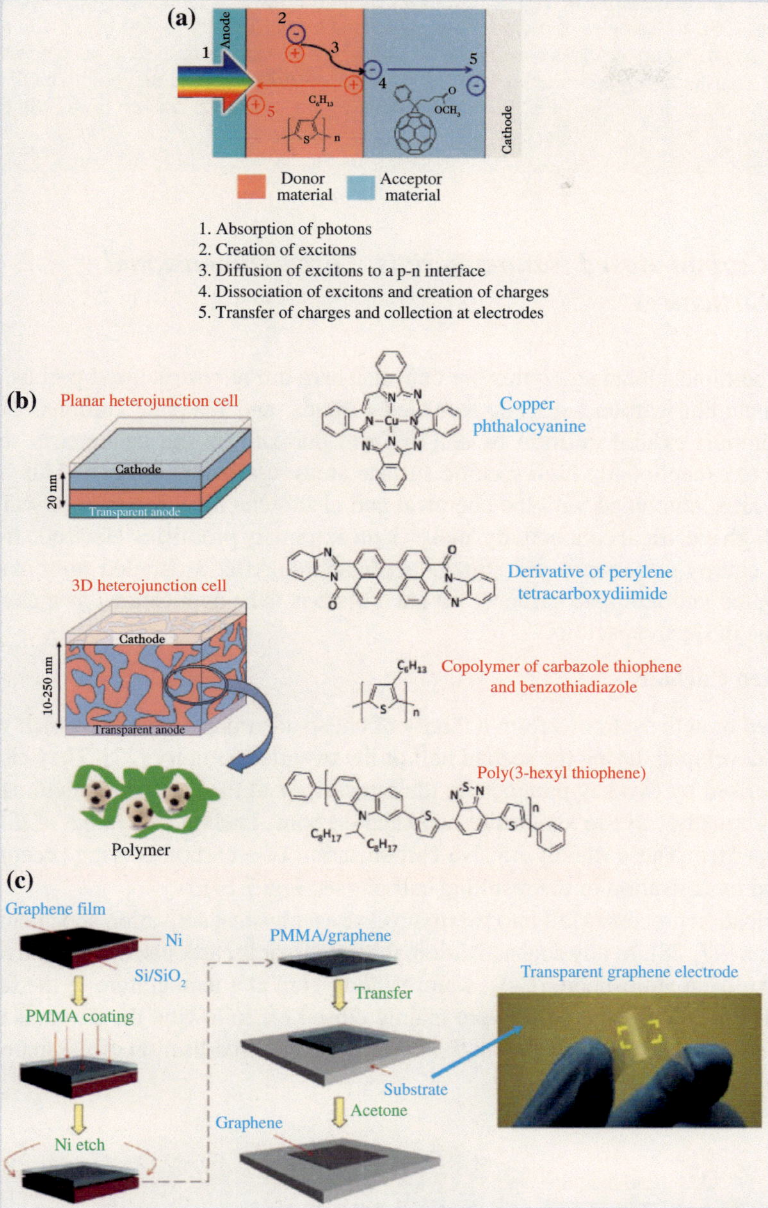

flexible and transparent electrodes that could replace ITO, in particular, in organic photovoltaic cells (see Fig. 5.13c). And this is not the only promising prospect for these carbon-bearing nanomaterials. Many studies are under way to apply them more widely in solar cells.

Fig. 5.13 **a** Principle of a heterojunction solar cell and various structures for heterojunction devices [25]. **b** Examples of p-type donor organic molecules in OPV systems [25]. *Top to bottom*: two evaporable molecules and two soluble polymers that can be used in wet processing. **c** Transfer of a graphene film onto a transparent substrate for construction of an OPV system [26]. **a**, **b** courtesy of R. Demadrille, CEA Inac, Grenoble. **c** is taken from [26], © Elsevier 2012

5.2.4 *Carbon-Based Nanomaterials. Three-Dimensional Structures*

The three-dimensional structures we consider here are nanostructured porous carbons, including carbon aerogels, activated carbons, and template carbons. These nanomaterials exhibit uniform or non-uniform porosity on the nanometric scale, leading to exceptionally high specific surface areas, up to $4\,000\,\mathrm{m^2 g^{-1}}$. This huge surface area, combined with the chemical and electrochemical inertness of carbon and its high electrical conductivity, makes it an extremely promising electrode material for energy conversion and storage applications. After a detailed presentation of activated and template carbons, we shall discuss their applications as electrode materials in supercapacitors.

Activated Carbons

Activated or active carbons form a family of cheap adsorbent materials which were mainly developed during the second half of the twentieth century [27]. They can be characterised by two key parameters: their porosity and their surface functionality. There are many ways to synthesise activated carbons, leading to a range of different properties. These mainly involve carbonisation of a carbon-bearing precursor, followed by activation of the resulting carbon (see Fig. 5.14).

Fabrication processes fall into two main classes: physical activation and chemical activation [27, 28]. In physical activation, the precursor is carbonised, then activated in an oxidising atmosphere (CO_2, water vapour, etc.) at a temperature in the range 900–1 100 °C. While the pores were mainly closed off following the carbonisation stage, they now become accessible thanks to the partial oxidisation of the material.

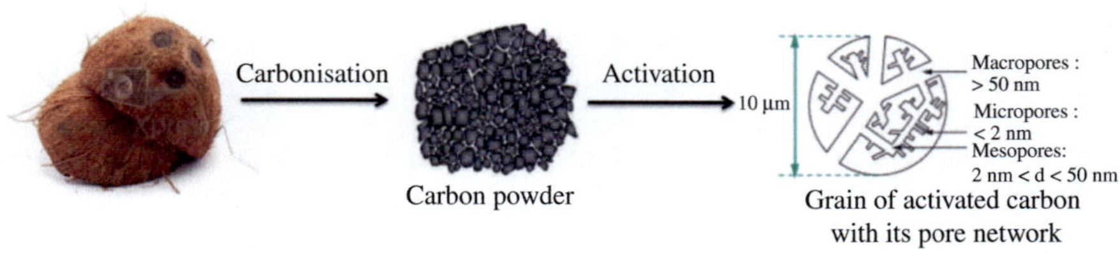

Fig. 5.14 Elaboration of an activated carbon and classification of the different kinds of pore. Courtesy of P. Simon, lecture given at the *Collège de France*, 2011

This process is generally implemented with coconut shells, phenolic resins, various plant residues, or carbon fibres.

Chemical activation is carried out at lower temperatures, of the order of 700 °C, in the presence of a chemical reagent such as phosphoric acid, zinc chloride, potash, or others. This causes dehydration, carbonisation, and structural reorganisation of the precursor (e.g., wood, fruit kernels, petroleum residues), thereby producing micropores and mesopores while retaining in part the initial pore structure of the precursor. The resulting activated carbon must then be flushed with water to eliminate the reagents or their decomposition products.

The size of the pores and specific surface area of the activated carbons is determined by the choice of carbon-bearing precursor, means of synthesis, and activation parameters. The specific surface area can exceed $2\,500\,m^2g^{-1}$. The IUPAC classification specifies three kinds of pores (see Fig. 5.14):

- macropores with diameters above 50 nm,
- mesopores with diameters in the range 2–50 nm,
- micropores with diameters less than 2 nm.

Activated carbons can exist in different forms, including powders, fibres, tissues, and granules. Fields of application are many and varied thanks to their specific features in terms of porosity and functionality, combined with the intrinsic properties of carbon. They can be used to trap gaseous or liquid chemical species, decontaminate smoke, and filter water or gases, as well as serving as a support for heterogeneous catalysis in organic chemistry or a material for converting and storing energy.

The different ways of producing activated carbons nevertheless lead to materials with a high dispersion in pore size. Such materials often contain both micropores and mesopores. Other methods of synthesis have thus been developed to provide better control over the nanostructure of the resulting carbon: the template method produces mesoporous or microporous materials, while the carbide-derived carbons (CDC) are essentially microporous.

Template Carbons

The template method [29, 30], illustrated in Fig. 5.15, involves the following steps:

- Impregnation of a microporous or mesoporous silica or zeolite substrate by a carbon precursor in gaseous, liquid, or solid form.
- Pyrolysis to transform the carbon precursor.
- Dissolution of the silica or zeolite template in a hydrofluoric acid solution to obtain a carbon material in which the pore network is an exact replica of the silica template.

Carbide-Derived Carbons

Carbide-derived carbons or CDCs are microporous carbons [31], with subnanometric porosity, i.e., less than 1 nm. They are obtained by chlorination of metal carbides (MC) at temperatures in the range 400–1 000 °C according to the reaction

$$MC + nCl_2 \longrightarrow MCl_{2n} + C_{solid} .$$

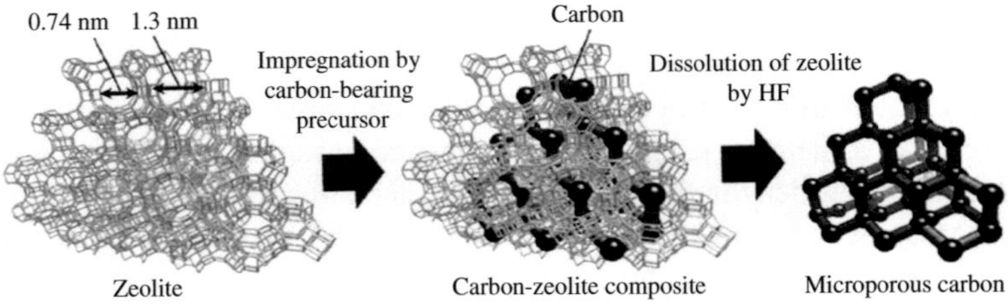

Fig. 5.15 Formation of a carbon from a zeolite using the template method. HF hydrofluoric acid. From [29], © Elsevier 2009

The point about this method is that the resulting characteristics can be perfectly controlled, including both the specific surface area and the average pore diameter. Indeed, these characteristics are determined by the choice of carbide (with perfectly well defined crystal structure) and the chlorination temperature. The most important point is that this method can be used to produce a much narrower pore size distribution than could be obtained by any of the conventional methods for making activated carbons.

Supercapacitors: An Application of Nanostructured Carbons

Uses

Electrochemical systems for converting and storing energy are classified into two categories:

- High energy density systems such as lead accumulators, nickel–metal hydride accumulators, lithium ion accumulators, and fuel cells. The specific energy is the amount of energy stored per unit mass ($Whr\,kg^{-1}$) or per unit volume ($Whr\,l^{-1}$).
- High power density systems such as electrical capacitors. The specific power is the power that the electrochemical storage system can deliver per unit mass ($W\,kg^{-1}$) or per unit volume ($W\,l^{-1}$).

The Ragone plot in Fig. 5.16 provides a way to compare electrochemical energy conversion and storage systems. This shows that supercapacitors are electrochemical systems with power and energy intermediate between dielectric capacitors and accumulators.

Supercapacitors are particularly useful for onboard systems which are characterised by high efficiency energy transfers over short time scales, typically of the order of the second. Their high performance in dynamic regime can be attributed to the very low internal resistance and the largely electrostatic nature of the energy storage. However, their relatively low specific energy may make it necessary to adjoin a main energy source, such as an electrochemical accumulator, to meet the electrical energy requirements of the relevant application.

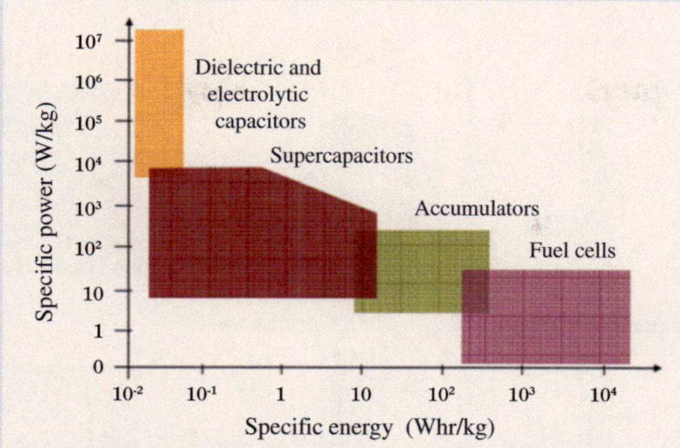

Fig. 5.16 Ragone plot showing the main systems for electrical energy storage. From the energy point of view, supercapacitors fall somewhere between dielectric capacitors and accumulators

Applications using supercapacitors as unique primary source include backup memories, transmission antennas, and so on, which involve a peak in power output, and devices that must be recharged quickly, such as toys, tools, etc. In the military domain, supercapacitors are also used in detonators, for controlling water levels in submarine ballasts, starting up vehicles such as tanks and lorries in tough climatic conditions, and others. They are also used in aeronautics, and in particular for the emergency opening system on the doors of the A380 aircraft. In the civilian domain and in land transport, supercapacitors provide the power needed for ignition and recover the energy supplied when the brakes are applied. For a vehicle stopping frequently such as a bus, tram, underground train, dustbin lorry, and so on, the energy recovered during stoppage can then be used to move on to the next stop. This combines the advantages of reduced energy consumption, reduced pollution, and lower operating costs.

In electrical transport systems, depending on the architecture, functionalities, and missions of the vehicle, electrical power requirements are often discontinuous in nature and effective at different levels and for different durations. To satisfy this kind of requirement, one solution is to use a hybrid electricity source which combines features such as specific power, specific energy, energy efficiency, and lifetime. A good solution for this kind of application would be the association of a supercapacitor with an accumulator or a fuel cell, for example.

Description

Like accumulators, supercapacitors have two electrodes separated by a liquid electrolyte wetting a porous separator. Here we shall only consider supercapacitors using activated carbon. Also known as double-layer electrochemical capacitors, they comprise two symmetrical carbon electrodes with very high

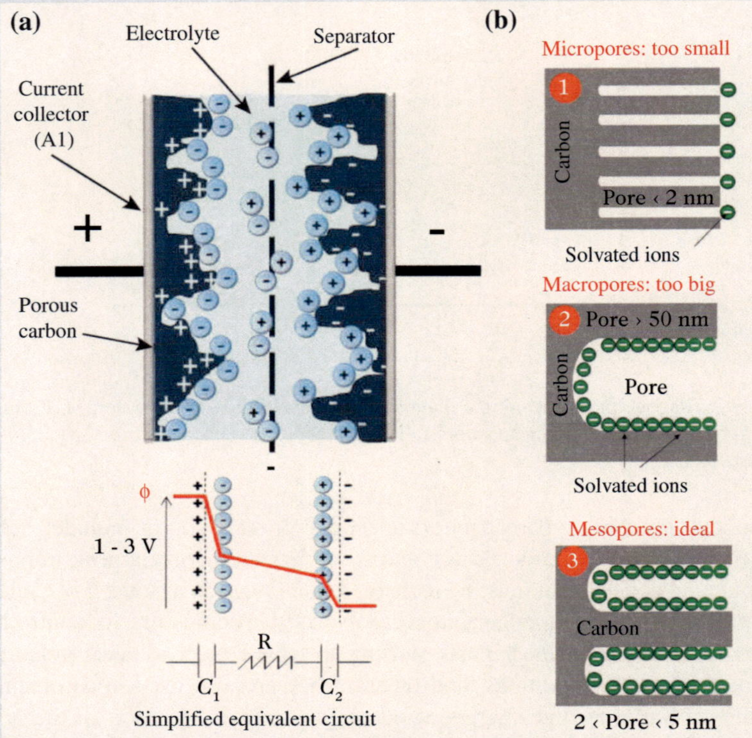

Fig. 5.17 Supercapacitor with nanostructured carbon electrodes. **a** Configuration of the supercapacitor showing the electrochemical double layer. **b** Role of the pores [32]. Courtesy of P. Simon, lecture given at the *Collège de France*, 2011

specific areas, separated from one another by a separator soaked with an organic liquid electrolyte. Storage is achieved by adsorption of ions at the interface between electrode and electrolyte, thanks to the formation of an electrochemical double layer (Fig. 5.17a).

During cycling, the phenomena occurring within the supercapacitor are fully reversible and in theory involve no alterations in the structure or composition of the materials. Their lifetime is thus theoretically unlimited. Even though degradation of the electrolyte and corrosion of the current collectors will in practice limit the lifetime, it will always last longer than an accumulator. Furthermore, the charging time of a supercapacitor is only a few seconds, while it will take at least a few minutes to charge up an accumulator.

Effect of Nanostructured Electrodes

The main characteristics of an electrochemical double layer supercapacitor are the specific energy $E_{\max}$, specific power $P_{\max}$, and maximum voltage $U_{\max}$ [32]. These characteristics are related as follows:

$$E_{\max} = \frac{C_{\mathrm{SC}} U_{\max}^2}{2m} , \qquad P_{\max} = \frac{U_{\max}^2}{4 R_{\mathrm{ESR}} m} ,$$

where C_{SC} is the capacitance of the supercapacitor (F), m is its mass (kg), and R_{ESR} is the internal resistance (Ω). The capacitance C_{SC} is given by

$$C_{\mathrm{SC}} = \frac{\varepsilon_0 \varepsilon_{\mathrm{r}} S}{\delta} ,$$

where ε_0 and ε_{r} are the permittivity of the vacuum and the relative permittivity of the electrolyte, respectively, S is the active surface area, and δ is the dielectric thickness. The key to understanding the supercapacitor is the high value of S/δ, whence the charges are localised on very large surface areas and at very small distances (see Fig. 5.17a).

The point in using activated carbon as the active material of the electrode is mainly that this greatly increases the specific surface area and hence the capacitance of the system, this leading to a corresponding increase in the specific energy. The pore size distribution is also a very important parameter. It would seem that the ideal size corresponds to small mesopores (2–5 nm) (see Fig. 5.17b), bearing in mind the size of the solvated ions, while micropores (<2 nm) would be too small for the solvated ions to penetrate. However, more recent work on CDC materials has shown that micropores contribute significantly to the capacitance and that ions adsorbed in the double layer are at least partly desolvated there [32, 33].

5.3 Nanochemistry in the Service of Nanomaterials: From Molecular to Supramolecular Chemistry

In the last section, we reviewed a broad range of nanomaterials in terms of shape, size, properties, and applications, using the example of carbon. However, for the main part, the methods for synthesising these carbon-based nanomaterials involved high temperature processes and hence what one might call conventional chemistry. But in parallel, nanochemists have been developing a ‘soft’ chemistry based on molecular chemistry, and often on supramolecular chemistry [34]. The aim in the present section is to define, then illustrate these two routes to the synthesis of nanomaterials.

5.3.1 Click Chemistry: Molecular Methods

For the molecular chemist, matter is made up of atoms and molecules in which atoms are held together by covalent bonds [35]. Quite generally, chemistry, and more precisely molecular chemistry, can be defined as the science describing the composition of matter, its properties, and its transformation reactions. As a route to the synthesis of nanomaterials, molecular chemistry is thus based on the covalent assembly of elementary building blocks.

Click chemistry provides a particularly good illustration of the way molecular chemistry can be used to synthesise nanomaterials. The basic idea, introduced in 2001 by Sharpless and coworkers [36], is to implement a string of highly efficient reactions, each meeting precise requirements. A click reaction must be open to modulation, produce high yields, generate only harmless byproducts, and be stereospecific, using a simple protocol and freely available basic materials and reagents. The example in Fig. 5.18 illustrates the simplicity of click chemistry, which uses a sequence of fast, reproducible, selective, and modifiable reactions to build molecular assemblies or to produce a wide range of different functionalisations in a great many materials.

The chemical transformations satisfying these criteria generally involve a high enthalpy of reaction (>20 kcal/mol) and fast kinetics in order to lead quickly to a single reaction product. Several classes of chemical transformation can thus be considered as click reactions, such as cycloadditions of unsaturated species, certain nucleophilic substitutions (opening of electrophilic heterocycles), or additions on C–C bonds (epoxidation, dihydroxylation, and others). Most of these click reactions involve the formation of energetically highly favorable carbon–heteroatom bonds. However, among the different click reactions, 1,3-dipolar cycloaddition between an azide and an alkyne, catalysed by monovalent copper (I), is by far the most commonly used, owing to its simplicity and efficiency, which make it the click reaction *par excellence*.

This cycloaddition reaction between an azide and an alkyne produces a 1,2,3-triazole unit (see Fig. 5.19). The thermal process has been known since the nineteenth century, but the reaction became more attractive when the catalytic process was

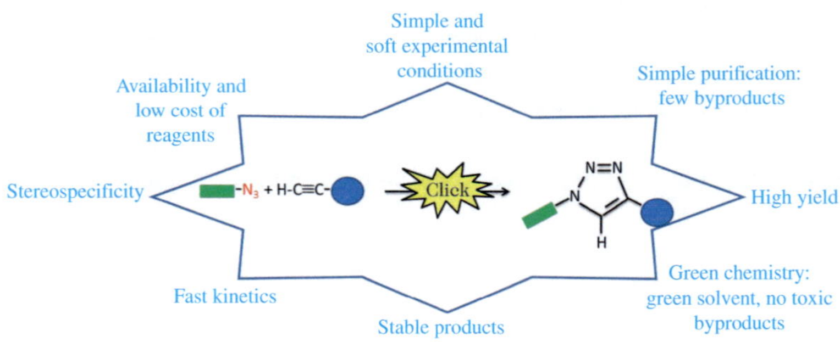

Fig. 5.18 Schematic view of click chemistry. Cycloaddition between an azide and an alkyne

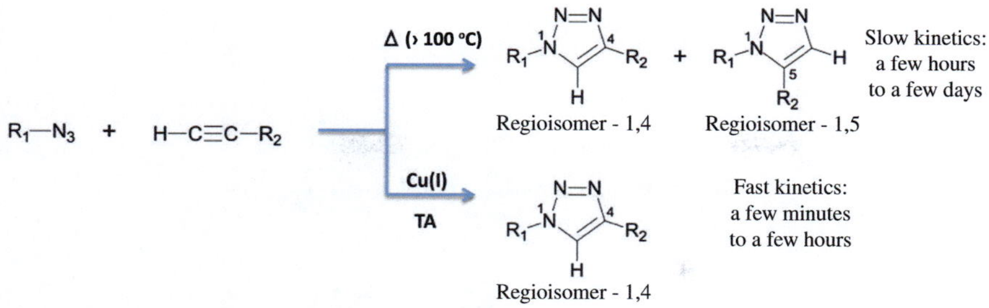

Fig. 5.19 1,3-dipolar cycloaddition between an azide and an alkyne using a conventional thermal process and using catalysis (click reaction)

discovered. The latter is regioselective, kinetically much faster, and can be carried out under soft conditions at room temperature. The triazoles produced by the reaction are five-atom heterocyclic compounds containing three nitrogen atoms. They have an aromatic structure conducive to π–π interactions (π-stacking). These are very stable compounds, generally chemically inert with respect to reducing or oxidising conditions, even at high temperatures. The triazoles also have a high dipole moment and a notable capacity to form hydrogen bonds (both protophilic and protogenic).

Compounds incorporating triazole groups in their structure have caught the attention of biochemists [37] because they have several important biological properties, viz., antifungal, antibacterial, and anti-inflammatory properties. They are used as analogues of amino acids, particularly when constructing nucleoside derivatives, with the triazole ring replacing the pyrimidine bases. This leads to new compounds with antiviral and antitumoral properties. The triazoles arise as building blocks in many reactions of organic chemistry, particularly in drug formulations.

Today, most research in click chemistry concerns 1,3-dipolar cycloaddition. This reaction has become an indispensable tool for organic chemists, materials chemists, and biochemists. From a range of azides and alkynes, the organic chemist can very quickly prepare a wide variety of organic molecules, avoiding all the usual tedious steps to protect and deprotect functional groups.

In materials chemistry, 1,3-dipolar cycloaddition proves to be an attractive way of functionalising surfaces or rigid core nano-objects like fullerenes and gold nanoparticles. Recently, Nierengarten and coworkers [38] have shown that click chemistry can be applied to produce highly functionalised C_{60} fullerenes: twelve click reactions have been carried out at the surface of the same C_{60} sphere. Gold nanoparticles functionalised by small organic or organometallic molecules, as well as polymers and DNA strands, have also been synthesised in this way.

Applications of Click Chemistry

Two examples in Fig. 5.20 illustrate functionalisation by click chemistry. The first concerns the grafting of liquid crystal dendrimers carrying alkyne functions onto gold particles that are themselves functionalised by azide groups. We shall see in Sect. 5.5 that many methods of synthesis are available today for the preparation of

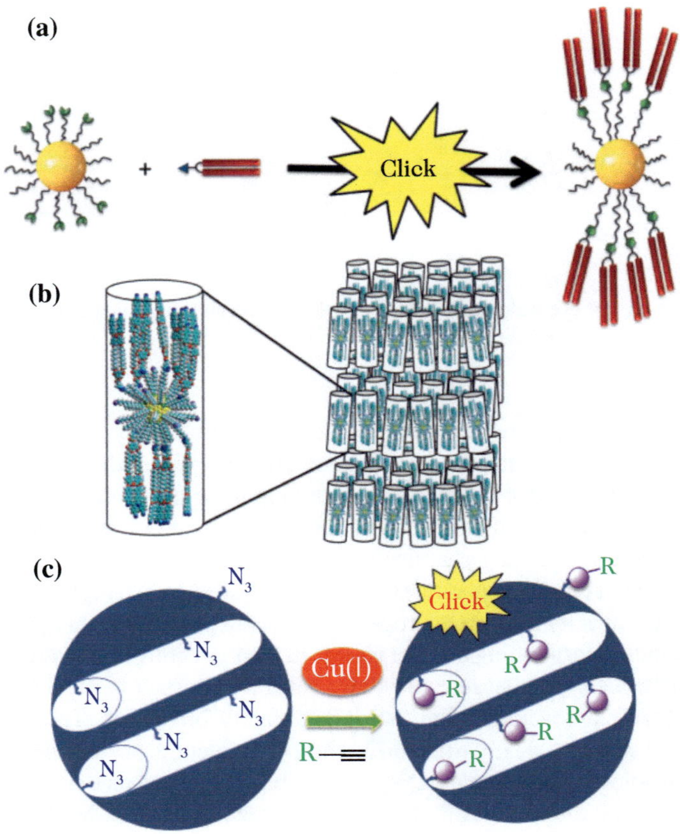

Fig. 5.20 **a** Synthesising gold–liquid crystal hybrid nanoparticles using click chemistry to graft liquid crystal dendrimers on the surface of gold nanoparticles. Courtesy of S. Mischler, Neuchâtel University, 2012. **b** Supramolecular organisation suggested for hybrid nanoparticles in smectic phase A [35]. Courtesy of S. Mischler, Neuchâtel University, 2012. **c** Functionalisation of mesoporous silica by click chemistry [39]. Courtesy of J.O. Durand, *Institut Charles-Gerhardt* and X. Cattoën, *Institut Louis-Néel*, from [39]

gold nanoparticles. However, organising them into a well determined 3D arrangement remains a major challenge. And as it happens, there is growing interest in 3D lattices of gold nanoparticles given the properties arising from their collective interaction. These lattices could be exploited in a variety of novel devices, such as high-density data storage systems, light-emitting diodes, plasmon waveguides, or charge transport systems. Theoretically, one can also conceive of materials with negative refractive index which could be used to produce very high resolution superlenses or invisibility cloaks making it impossible to observe certain objects from afar (see Chap. 2).

Three-dimensional arrangements of gold nanoparticles thus have a great deal of potential, but this is only the beginning of attempts to develop fabrication methods and analyse the remarkable phenomena associated with such systems. Click chemistry offers an elegant approach here, exploiting the intrinsic self-assembly properties of liquid crystals. By incorporating gold nanoparticles in a liquid crystal system, hybrid materials are produced which combine the organisational capacity and anisotropy of liquid crystals with the optical and electronic properties of metal nanoparticles. This is where click chemistry plays its role, making use of the liquid crystal system as an ideal vehicle for controlling the ordered arrangement of nanometric particles.

The second example of click chemistry, illustrated in Fig. 5.20c, concerns the functionalisation of mesoporous silica nanoparticles, grafted by pendant alkyne or azide groups. Lattices of mesoporous organosilica nanoparticles are deposited using an inkjet technique. They can then be functionalised by attaching fluorescent compounds to them using click methods, so that they can be used as biosensors, for example. Hybrid materials obtained in this way should also lead to progress in the synthesis of nanomachines for controlled delivery of active principles [39].

5.3.2 *Self-Assembly and Supramolecular Chemistry*

Supramolecular chemistry was first defined in 1978 by Jean-Marie Lehn as 'chemistry beyond molecular chemistry' [40–42]. The idea is to design supramolecular constructions by exploiting non-covalent interactions between chemical species. To achieve this aim, the chemist can appeal to a wide range of tools for controlling the connections between molecules involving the various kinds of non-covalent interactions (see Fig. 5.21).

Complex and ordered entities can be assembled from several molecules using these so-called weak interactions. Naturally, their architecture depends on the molecules chosen as basic building blocks. Supramolecular chemistry thus requires molecular chemistry in order to create or modify the molecules to be assembled. There are three main areas of study in supramolecular chemistry:

- The first focuses on molecular recognition, which depends on the nature of the relevant chemical entities, viz., polymers, cage molecules, metal cations, biomolecules, etc., but also their prior organisation and the information they contain and use. The information in question is stored structurally in the form of the covalent bonds at the molecular level. It is then transferred and used at the supramolecular level via specific non-covalent interactions. Molecular recognition thus involves the storage and restitution of structural molecular information.
- The second area of supramolecular chemistry concerns self-assembly processes. These depend mainly on the structure of the relevant building blocks.
- The third area concerns, on a larger scale, the materials that can be designed and created by exploiting molecular recognition and self-assembly of chemical entities.

Molecular Recognition: The Lock-and-Key Mechanism

Molecules are recognised by their shape, size, and functional groups. These are the primary characteristics involved in the formation of a given chemical bond. If two

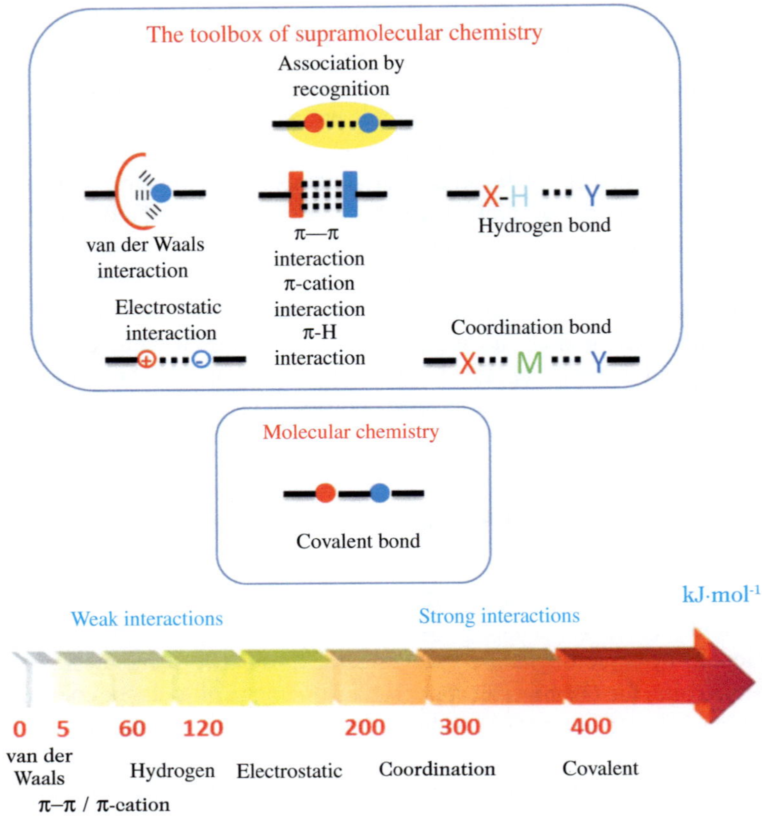

Fig. 5.21 Different kinds of intermolecular interactions exploited in supramolecular chemistry and their respective energies [43]. Adapted from W. Hosseini, *Introduction à la chimie supramoléculaire*, MSc course, 2011

molecules are to form a supramolecular structure, they must therefore be outwardly compatible, like a lock and key (see Fig. 5.22) [44]. This idea was defined for the first time by Fischer in 1894 to describe enzyme activity.

A further principle of supramolecular chemistry, directly related to the lock-and-key effect, is selectivity. Just as one can try several keys before finding the one that fits a given lock, a molecule may also try to form bonds with different partners before bonding to the one that fits it best. It is easy to understand this situation by thinking of the human genetic heritage, which is itself recorded, transmitted, read, and applied by pairs of molecules held together by hydrogen bonds in DNA (see Fig. 5.23). It is the combined effect of these low energy hydrogen bonds that provides for the stability of such structures, while ensuring flexibility and reactivity, because these bonds can evolve quickly with changing conditions of temperature, pH, ionic strength, and so on.

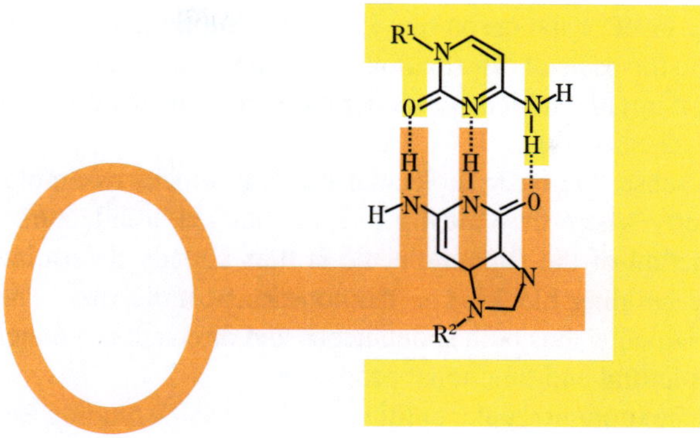

Fig. 5.22 Molecular recognition and the lock-and-key mechanism. Two molecules are compatible in the same way as a lock and key, an idea first formulated more than a hundred years ago by Emil Fischer, Nobel Prize for Chemistry, to describe enzyme activity. Courtesy of Marcel Mayor, from [43]

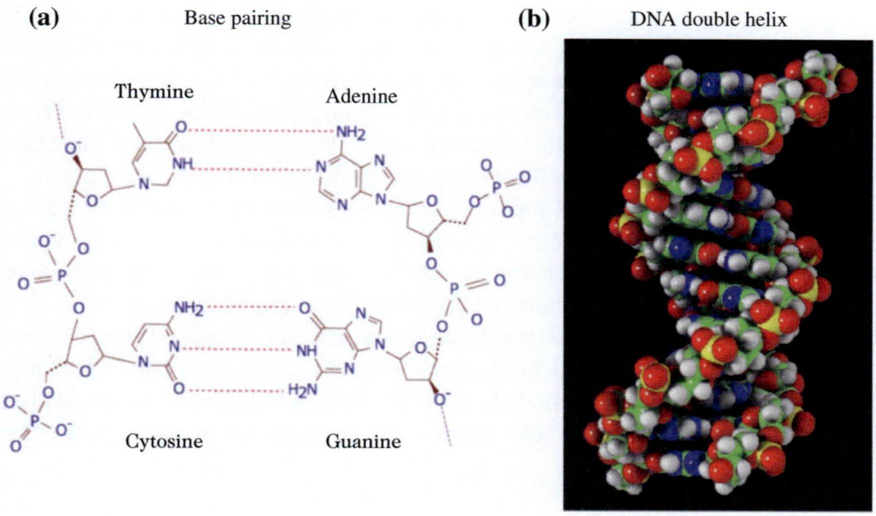

Fig. 5.23 **a** Pairing of two polynucleotide chains of DNA by means of hydrogen bonds between the four bases C, T, G, and A. **b** Double-helix structure of two paired DNA molecules. Courtesy of J.-M. Lehn, *Année de la chimie*, 2011

Molecular Self-Assembly or Self-Ordering

From the simple to the complex, matter assembles thanks to interactions between molecular entities. Self-assembly occurs when ordered structures are obtained from elementary building blocks which bind spontaneously under the effect of physical or chemical interactions, generally forming weak chemical bonds like hydrogen bonds or hydrophilic–hydrophobic interactions. Complementary shapes and charges may also play a role. Ordered and spontaneous assembly of molecules corresponds to a minimal energy level of the ensemble. In short, a supramolecular architecture will self-assemble on the basis of information stored at the molecular level (the chemical groups making it up) and molecular recognition mechanisms (the interactions it can engage in with its partners).

In a certain sense, the basic molecular building blocks assemble on their own, 'knowing' exactly where they must go in the planned architecture. The blueprint for the construction of the supramolecule is thus already determined by the very choice of these building blocks, i.e., the constituent molecules. The result of this self-assembly, which is thus both spontaneous and directed, is a material entity with well defined structural and functional properties.

Concerning the supramolecular entities, the fact that the binding forces are weak is an advantage. If damage occurs, the damaged building block is automatically replaced by a new one in good working order. This natural healing process is just one example of the fascinating benefits one may hope to gain from supramolecular materials. An oft-cited illustration is the tobacco mosaic virus, described in Fig. 5.24. To build this, 2 130 simple proteins self-assemble to form a helical capsid structure. The starting protein, the basic building block, thus possesses all the information needed for the resulting ordered construction. The molecular recognition is so precise that the programmed architecture is spontaneously realised in space. Furthermore, the assembly reaction ceases at just the right moment. This is brought about through the intervention of the ribonucleic acid (RNA) located inside the construction. This biological self-ordering process occurs in three steps: recognition, growth, and termination.

The chemist can also build supramolecules by exploiting the self-ordering tendencies of basic building blocks. Consider the example of double or triple helix structures elaborated by association of suitable metal ions with linear bipyridine-containing molecules, which contain two to five bipyridine groups (see Fig. 5.25a). This can be done by simply mixing these strands and adding copper ions to yield a spontaneous double strand helical assembly. Replacing the copper ions, which bind tetrahedrally, by nickel ions, which bind octahedrally, a triple helix is obtained (see Fig. 5.25b). In a similar way, self-ordering can be exploited to assemble a cylin-

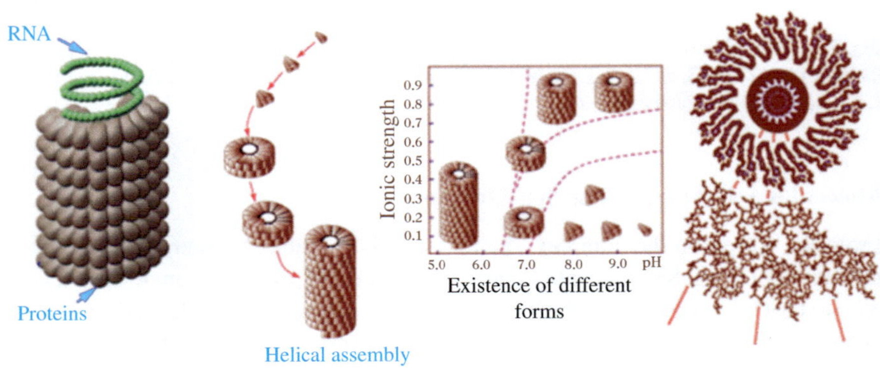

Fig. 5.24 Tobacco mosaic virus [40]. Adapted from J.M. Lehn, *Année de la chimie*, 2011

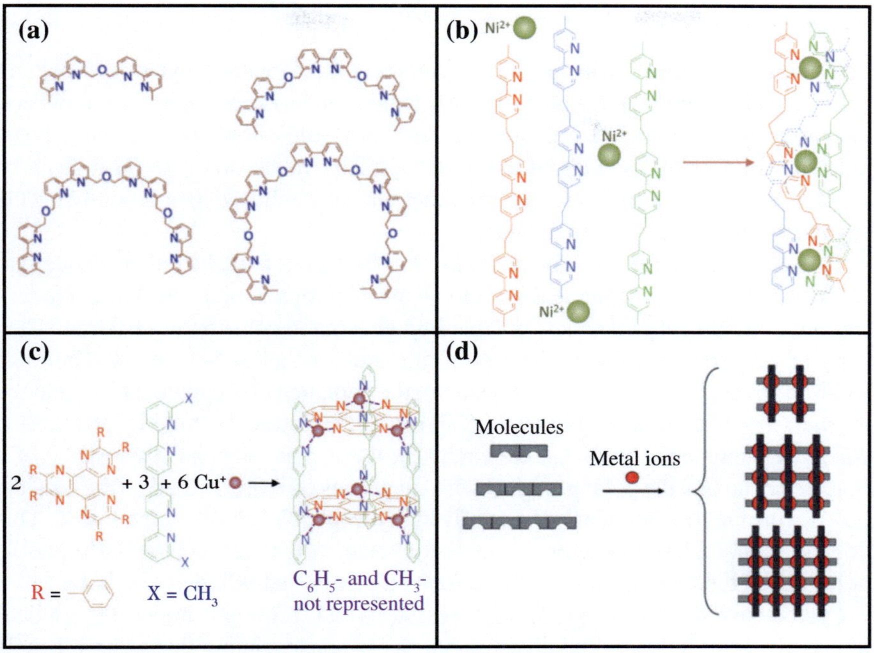

Fig. 5.25 Supramolecular structures obtained by associating suitable metal ions with organic molecules. **a** Linear bipyridine-containing organic molecules with 2, 3, 4, and 5 bipyridine groups. **b** Triple helix obtained by associating nickel ions with these molecules. **c** Cage obtained by associating copper ions with linear molecules and planar polycyclic molecules. **d** Construction of ladder structures. Adapted from J.M. Lehn, *Année de la chimie*, 2011

drical cage from two planar polycyclic molecules and three linear molecules (see Fig. 5.25c). Ladder structures can be obtained by similar methods (see Fig. 5.25d).

In practice, it is not hard to obtain such supramolecular structures. No sophisticated equipment is required and it is much easier to synthesise molecular building blocks capable of self-ordering than it would be to synthesise the final structure. The hard part here lies in the initial design of the components that will be brought into play and elaboration of the strategy or program that will lead to the desired result. It should be stressed that, in the rapidly expanding field of nanoscience and nanotechnology, self-assembly processes offer a particularly powerful alternative approach to fabrication and manipulation: rather than having to do something, we just let it happen!

Properties of Supramolecular Structures

Supramolecular chemistry provides a wealth of solutions for elaborating materials from the nanoscale to the macroscale. The variety of structures, shapes, scales, and properties brought within reach of the chemist, combining supramolecular and molecular chemistry, is truly immense. In the first part of this section, we chose to present two examples that illustrate the prospects opened up by supramolecular nanochemistry, one based on molecular recognition and the other resulting from self-assembly.

Cyclodextrins

Many organic molecules, including crown ethers, cyclophanes, cyclodextrins, and others, contain within them a kind of cavity which can be exploited to form inclusion complexes [45]. In such a process, the host molecule allows one or more guest molecules into its cavity without the need for any covalent bonds to form. So here we have a genuine supramolecular construct whose stability depends solely on the way the partners adapt to one another.

These 'molecule traps' can be put to use in many areas to make ion exchange systems or catalysts, or to separate the components of complex mixtures. The cyclodextrins are natural host molecules, obtained by enzymatic conversion of starch. They occur in the form of cyclic oligomers of glucose containing 6–12 units. The most common are the hexamers (α-cyclodextrin), the heptamers (β-cyclodextrin), and the octamers (γ-cyclodextrin). Figure 5.26 shows the structure of these cyclodextrins, which are nanometric entities. Schematically, these molecules have the shape of a truncated cone. All the polar groups (hydroxyl groups OH) are located on the outside, thereby forming the boundary of a relatively hydrophobic cavity on the inside. This globally amphiphilic character means that cyclodextrins can include hydrophobic molecules in their cavity to form inclusion complexes that will dissolve in water.

Cyclodextrins are biodegradable, a feature which promises important applications in the food industry and pharmaceutics. By encapsulating them, these fragile molecules can be protected until needed, then released in a controlled way. Moreover,

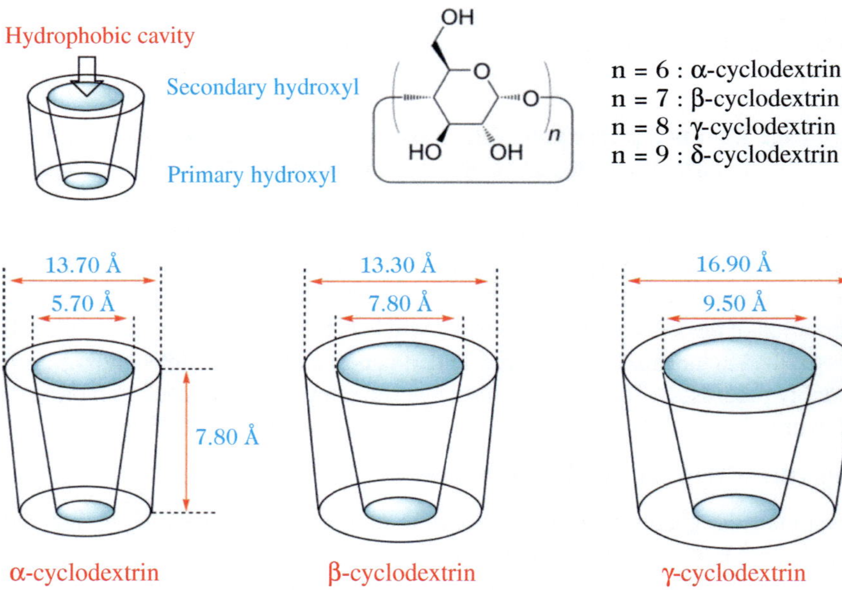

Fig. 5.26 Cyclodextrins. Chemical structure and spatial structure in the shape of a truncated cone, capable of trapping molecules [43]. Courtesy of M.W. Hosseini, Strasbourg University, *Introduction à la chimie supramoléculaire*, MSc course, 2011

drugs that are insoluble in water can be solubilised in the form of inclusion complexes in cyclodextrins, whence injectable preparations become possible. Finally, anti-odour textiles have been commercialised, using cyclodextrins to trap evil smelling molecules produced by perspiration.

Supramolecular Elastomers: Self-Healing Rubbers

In 2008, Leibler and coworkers [46] found a way to synthesise a new kind of elastomer made up of small molecules which join up by means of hydrogen bonds to form a network. This synthesis was carried out using molecules produced from fatty acids in vegetable oils and functionalised by chemical groups able to join together by multiple hydrogen bonds. There are two steps in the process (see Fig. 5.27), leading to a supramolecular assembly that is basically a mixture of molecules associated by hydrogen bonds between carbonyl and amine groups.

The first novelty in this approach is the simple means of synthesis and the use of fatty acids that are widely available in many different kinds, since they are found in renewable vegetable products, such as pine, sunflower, maize, and colza. The material has many interesting properties, starting with its elastomeric behaviour. However, the most remarkable property of this material is its ability to repair itself when damaged. Indeed, in contrast to conventional elastomers, if we cut a sample, we can simply hold the two parts together for a few minutes, without external effects such as heating or exerting pressure, and the material will return to its original state, with the same properties, even though it has absolutely no adhesive property. It can then be put under stress once again, deforming and stretching, and if the healing time was long enough, it will not break in the place where it was cut (see Fig. 5.28). The first tests on this material show that, the longer the healing time, the better the repair.

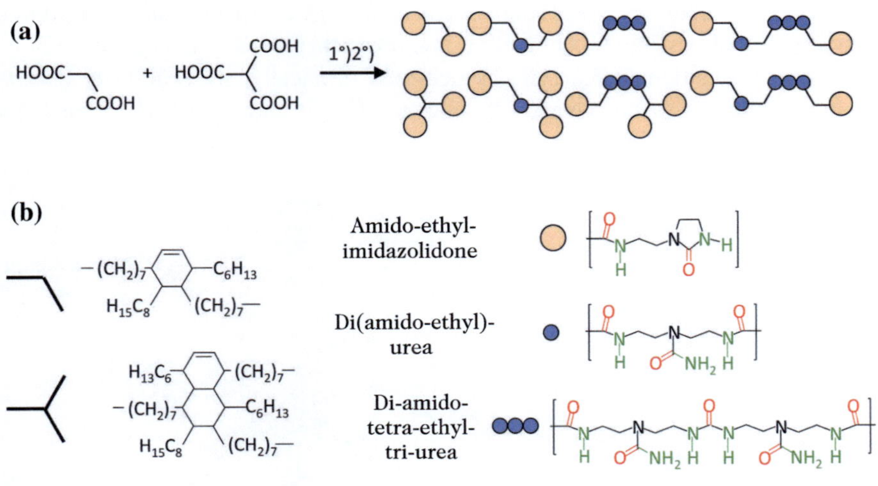

Fig. 5.27 Two-stage synthesis of a self-healing elastomer. **a** A mixture of fatty acids is condensed with diethylenetriamine. **b** This then reacts with urea. Acceptors (Hydrogen bonds) are shown in *red* and donors in *green* in the chemical formula. From [46], © Nature Publishing Group 2007

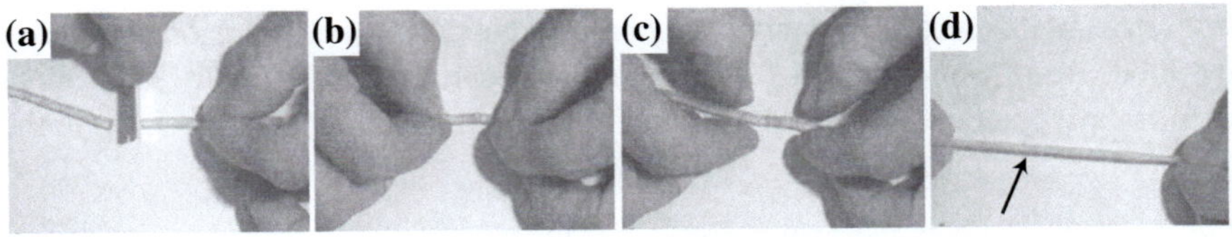

Fig. 5.28 Self-healing properties of a supramolecular elastomer. **a** Cut. **b** Repair by contact. **c** Repair completed. **d** Traction test [47]. Courtesy of L. Leibler and J.-L. Halary, ESPCI Paris

The rupture–repair–stretch cycle can be reiterated several times and repair can be carried out several hours after rupture!

Note that, less than one year after this discovery, a partner company in the project called Arkema began industrial production at a rate of 100 tonnes per year. The field of application of these self-healing rubbers is potentially unlimited. Indeed, they have all the virtues of conventional elastomers, and in particular the same elongation properties, but with another advantage on top of that: their ability to repair themselves. The potential applications are thus many and varied, from the building industry to high technology.

5.4 Functionalisation of Materials

For most potential applications of nanomaterials, whether in molecular electronics, biosensors, or organic photovoltaic cells, the nano-objects must be chemically or physically modified to endow their surface with functionalities such as solubility or wettability, and/or to facilitate their assembly into more complex architectures with other materials. To illustrate the last point, one might want to fix nanoparticles onto the surface of electrodes, for example. In these cases, we speak of functionalisation. The diversity of possible functionalities is shown schematically in Fig. 5.29.

Although functionalisation can be achieved physically, we shall only discuss chemical methods here. Chemical functionalisation is carried out in one or more

Fig. 5.29 Possible functionalisations of a nanoparticle (here, gold), leading to a considerable diversity of applications. Courtesy of S. Mischler, Neuchâtel University, 2012

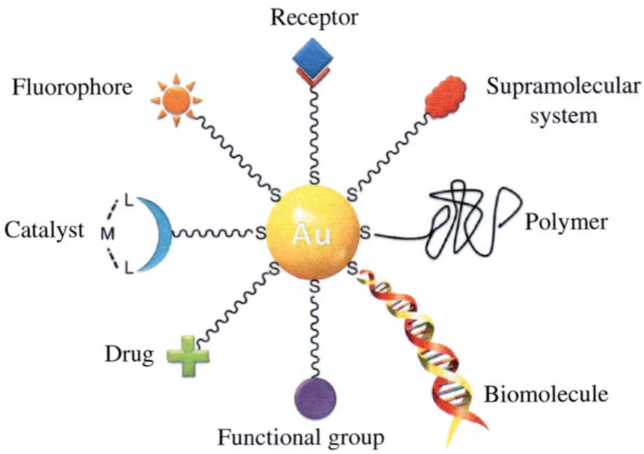

steps. After giving an overview of the many possible methods, we shall consider two approaches to functionalisation of nanomaterials in more detail. These concern chemical or electrochemical grafting. Note that, when a macroscopic material is functionalised by grafting a molecular nanolayer onto its surface, it becomes *ipso facto* a nanomaterial.

5.4.1 The Many Roads to Functionalisation by Chemical Methods in Gas Phase or in Solution

Figure 5.30 exemplifies the many ways of functionalising a material by considering the case of nanodiamonds, whose synthesis was already discussed in Sect. 5.2.1. The nanodiamonds in this figure are taken to have been synthesised with well controlled initial surface properties. They have been 'decorated' with carboxylic acid groups (–COOH).

Gas phase functionalisation methods require high temperatures and reactive gases like ammonia, chlorine, or hydrogen, with each gas leading to different functionalities [48]. For the same treatment and the same gas, several chemical groups can be obtained on the surface. For example, treatments using ammonia can introduce nitrile groups (–C≡N), amine groups ($-NH_2$), or double bonds. Softer methods at room temperature and in solution are also possible. Although numerous, the examples presented are not exhaustive. Chemical reduction by hydrides like $LiAlH_4$ can produce nanodiamonds with a surface made hydrophilic by the presence of hydroxyl groups

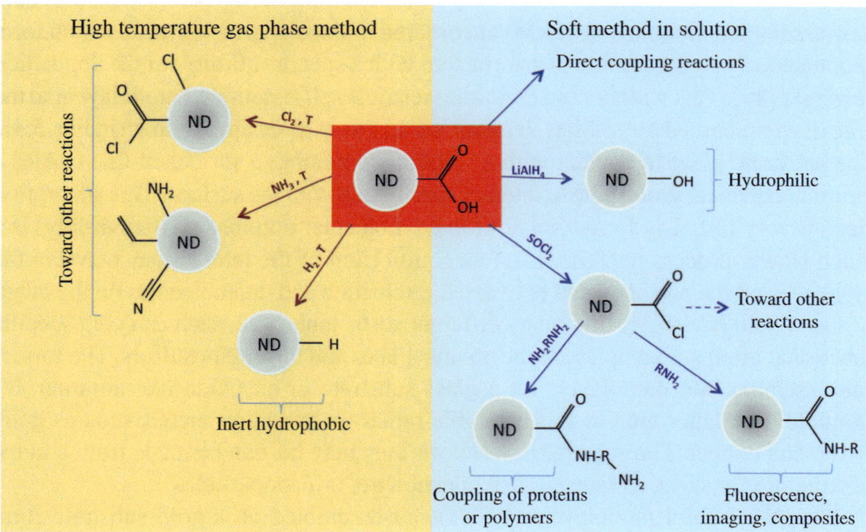

Fig. 5.30 Different ways to functionalise nanodiamonds (ND) by chemical methods

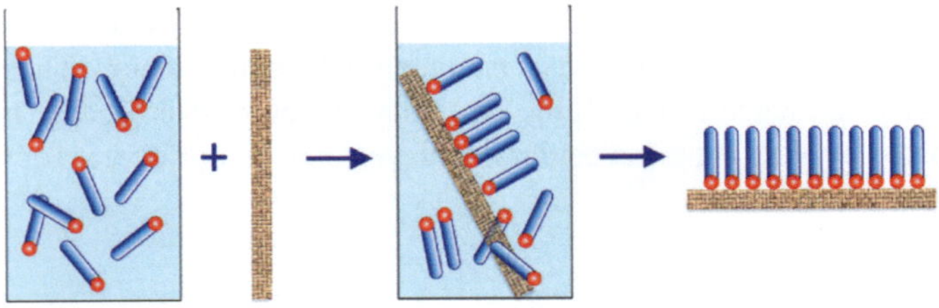

Fig. 5.31 Technique for producing self-assembled monolayers

(–OH). Functionalisation using thionyl chloride $SOCl_2$ will leave an acyl chloride on the surface, and this can be used to implement a further reaction from organic chemistry. Nanodiamonds can thus be functionalised with many and varied chemical groups, bringing an equal variety of surface properties, including fluorescence, molecular recognition, and so on. The chemist's task is then to achieve the required functionality after this first stage.

5.4.2 Example of Chemical Functionalisation: Self-Assembled Monolayers

Self-assembled monolayers (SAM) are ordered molecular arrangements which form spontaneously by adsorption of a surfactant with a specific affinity for the deposition substrate [49]. This widely used definition restricts self-assembled monolayers to the case of chemisorbed assemblies. It is particularly easy to set up, as shown in Fig. 5.31. The substrate is set in contact with a solution containing a surfactant that carries a terminal chemical group able to interact specifically with the surface. This adsorption stage is very fast. The formation of ordered molecular domains (self-assembly) is a much slower process resulting from an equilibrium of the interactions between the molecules on the one hand and between the substrate and the molecules on the other.

One could envisage using many different surfactants here, each carrying specific functional groups. Examples are the organosilanes and the organosulfurs. The former can assemble into monolayers on a glass substrate or an oxide like alumina, for example. The latter tend to self-assemble rather on transition metals such as gold, silver, and copper. The substrates, whatever they may be, can be made from a range of structures, such as a plane surface, a nanowire, or nanoparticles.

The alkane thiol monolayers (RS–H) self-assembled on a gold substrate form the most widely studied class of such nanomaterials. Gold and sulfur easily form covalent bonds ($40\,\mathrm{kcal\,mol^{-1}}$) by the reaction

$$\mathrm{RS{-}H} + \mathrm{Au}_n^0 \longrightarrow \mathrm{RS^-Au^+} \cdot \mathrm{Au}_{n-1}^0 + \frac{1}{2}\mathrm{H_2}\,.$$

By adjusting the length of the chain (radical R) or incorporating specific groups into it, e.g., conjugated double bonds, electron transfer across the film can be controlled. The presence of an appropriate functional group at the end of the chain can be used to modulate surface properties such as wettability or implement a subsequent chemical reaction to integrate the functionalised nano-objects into a more complex architecture.

There are many applications of SAM. We may cite their role as active layers in sensors or their use for studying interface reactions and as analogues of cell membranes.

5.4.3 Electrochemical Functionalisation: Electrografting by Reduction of Diazonium Salts

Over the past ten years or so, electrografting by reduction of diazonium salts has become the method of choice for electrochemical functionalisation of a great many materials. Discovered in 1992 by Jean Pinson [50], this method has since met with considerable success because it facilitates covalent bonding between aryl groups, substituted by one or more organic functions, and many different substrates, including carbon, noble or oxidisable metals, semiconductors, polymers, and inorganic dielectrics. One of the main advantages of this method is the simplicity of synthesis of diazonium salts, whose aryl group can carry many different substitutes, including –COOH, –Cl, –Br, –OH, $–NH_2$, –CN, alkyl or perfluorinated chains of varying lengths, etc. The electrografted layers are particularly robust and can resist many solvents, ultrasound, temperatures as high as 150 °C, and a wide range of electrochemical potentials.

Functionalisation is also very easy to implement. It is achieved by polarising the substrate with a sufficiently negative potential at the cathode for a few seconds or minutes to reduce the diazonium salt. The latter is in an organic or aqueous solvent at a concentration of the order of $1–10\,\mathrm{mmol\,l^{-1}}$.

The mechanisms in the grafting process are summarised in Fig. 5.32a. This involves the formation of an electrogenerated aryl radical which reacts with the surface. Dense monolayers can thus be obtained on certain substrates, but multilayers made up of disordered polyphenylenes or involving azo bonds (–N=N–) can also be formed (see Fig. 5.32b). Control of the grafting is a delicate matter, achieved by optimising the experimental conditions. Key factors are the quality of the substrate, its surface state, its polarisation, and the nature of the substituent carried by the aryl ring of the diazonium salt.

Finally, it should be noted that grafting can occur spontaneously by simply immersing the substrate in a solution containing the diazonium salt [51]. This approach, particularly useful for functionalising nano-objects like carbon nanotubes,

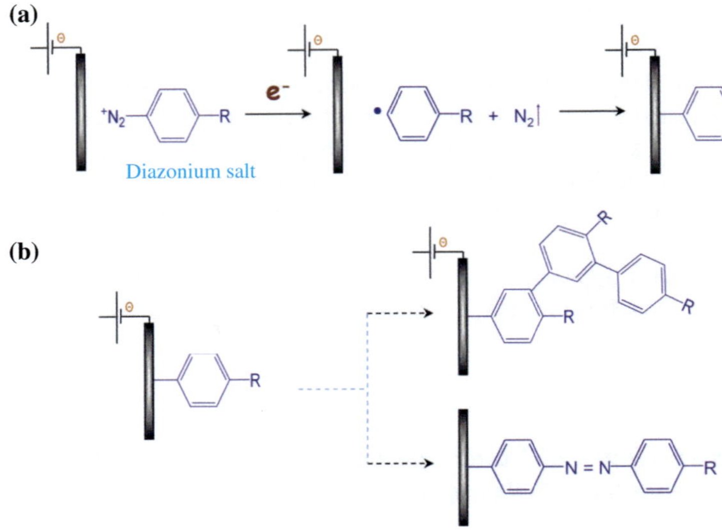

Fig. 5.32 **a** Electrografting mechanism producing molecular monolayers. **b** Spontaneous or electrogenerated evolution toward molecular multilayers

mesoporous carbons, and various nanoparticles, has been implemented on carbon, oxidisable metal, gold, and platinum substrates, among others.

There is keen interest in this approach to functionalisation due to the broad potential of methods using diazonium salts. The resulting applications concern many different fields. A first example is shown in Fig. 5.33a. This is a molecular electronic device with variable conductance, comprising a thin film of 4-nitro-azobenzene grafted onto carbon. Voltage pulses induce a modification in the structure of the film, accompanied by a spectacular drop in the film resistance [52].

The two other examples in Fig. 5.33 concern specific sensors for biological or chemical analysis. Figure 5.33b shows a biosensor that can detect several entities by electroaddressing. This is made by grafting antibodies modified by diazonium groups onto screen printed microelectrodes [53]. Figure 5.33c describes a sensor for detecting metal pollutants (U^{VI}, Pb^{II}, Cu^{II}, Cd^{II}, etc.) occurring as traces (ppb) in natural waters by direct measurement on site [54]. This is made by grafting a diazonium salt onto screen printed carbon electrodes. The salt introduces a function complexing metal cations at the electrode surface. Other industrial applications concern the functionalisation of carbon blacks used in printing inks or coatings for automobile bodywork.

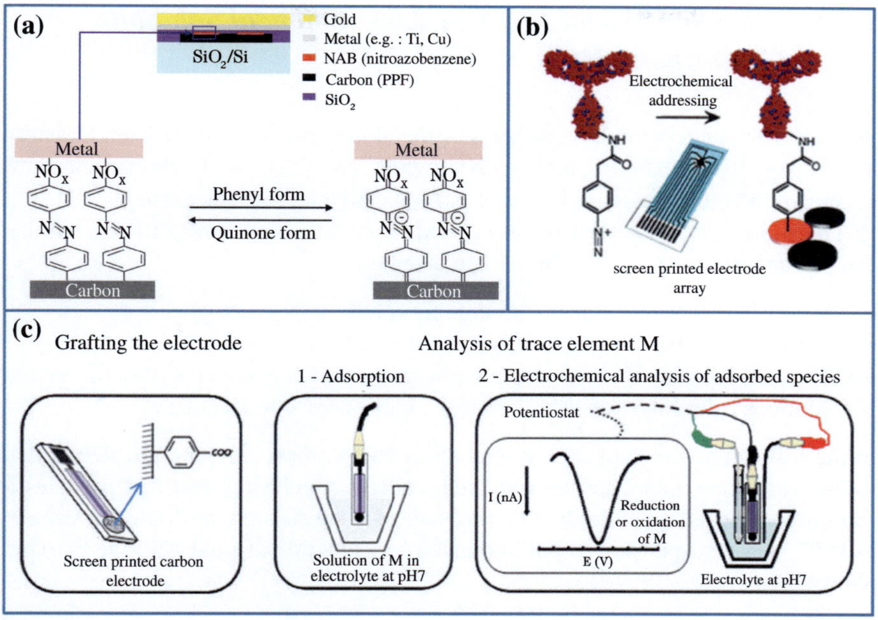

Fig. 5.33 Applications of functionalisation of diazonium salts by electrografting. **a** Conductance-changing electronic device. The initial 'phenyl' structure displays high resistance. The low resistance 'quinone' structure results from application of a 4 volt potential. Courtesy of P. Allongue, *Ecole Polytechnique*, and J. Pinson, *Alchimer – L'Actualité Chimique*, 2009, no. 327–328. **b** Immobilisation and direct addressing of proteins. Antibodies modified by diazonium groups are selectively grafted onto screen printed carbon microelectrodes, which can be individually addressed (*red electrodes*). Adapted from [53], © American Chemical Society 2005. **c** Chemical sensor for detecting trace elements in natural waters. The screen printed carbon electrode is electrografted by 4-carboxybenzene diazonium. At a pH greater than 5, the grafted carboxylate function complexes the metal cations which are then detected by electrochemical analysis (reduction here). The detection limit is of order ppb

5.5 Different Approaches to Chemical Synthesis of Nanomaterials

In the last two sections, we have described soft chemical methods for functionalisation and assembly of basic building blocks such as nanoparticles, monolayers, and nanoporous materials. In the present section, we shall discuss several methods associating molecular and supramolecular chemistry and which lead to nanomaterials that differ considerably in form, nature, and composition, and hence also in their properties. We shall not attempt to give an exhaustive review. The aim is to show that, apart from the synthesis and association of basic building blocks, the nanochemist also plays a major role in the elaboration of the final nanomaterial. Even at the design stage, she or he must conceive of the kind of building blocks that could be assembled, or self-assemble, to form the target nanomaterial, something which clearly depends on the desired characteristics of that nanomaterial.

5.5.1 Synthesis of Metal Nanoparticles: Colloidal Solutions of Transition Metals

We begin by considering the synthesis of metal nanoparticles made from transition metals like silver, platinum, nickel, cobalt, gold, etc. [55]. Gold nanoparticles have generated the most abundant literature, so we shall take them as example.

There are a great many chemical methods for elaborating nanoparticles, but they can be considered to fall into three main categories:

- Chemical synthesis in aqueous or organic solution.
- Chemical synthesis on specifically designed surfaces.
- Chemical deposition in the vapour phase, carried out under ultrahigh vacuum conditions (ultrahigh vacuum chemical vapour deposition or UHVCVD).

In the following, we shall describe methods for synthesis in solution. Recall that, in the solid state, nanoparticles are often called nanoclusters and that particles in solution can be called colloids. The first colloidal suspension, attributed to Faraday in 1857, was a suspension of gold obtained by reduction of gold chloride by white phosphorus.

Choice of Process

Before choosing a method of synthesis, various parameters must be assessed, including the shape of the nanoparticles, their size, their size dispersion (monodisperse or polydisperse), and the solvent in which they are to be dispersed, viz., aqueous or organic. Table 5.1 presents the broad range of gold nanoparticles produced by different means of synthesis in a liquid medium.

Whichever method is selected, nanoparticles in solution tend to cluster together spontaneously in the form of precipitates at the bottom of the container. For this reason, a crucial parameter for obtaining single nanoparticles is the choice of dispersing agent or surfactant.

Stabilising the Nanoparticles

The stability of particles in solution results from a balance between the attractive van der Waals forces and repulsive electrostatic forces (see the introductory chapter to this book). In the absence of repulsive forces to oppose the van der Waals forces, colloids aggregate to form the bulk metal. There are three kinds of stabilisation for nanoparticles, illustrated in Fig. 5.34: electrostatic stabilisation by anions and cations adsorbed on the surface, steric stabilisation by large and bulky molecules like polymers or ligands, and finally, stabilisation combining both steric and electrostatic effects [35, 55]:

- **Electrostatic Stabilisation.** Ions adsorbed on the metal surface, like halides, carboxylates, or polyoxoanions in aqueous solution and their respective counterions form an ionic double layer around the particle, and this is what gives rise to electrostatic stabilisation (see Fig. 5.34a). Indeed, it leads to a Coulomb repulsion between the metal nanoparticles, preventing them from aggregating. Methods for

Table 5.1 Main characteristics of gold nanoparticles synthesised in a liquid medium [56]

Method	Shape	Size	Stabiliser	Medium
Turkevich–Frens reduction	Spheres	$\Phi = 15\,\text{nm}$ (10–80 nm)	Sodium tricitrate	Aqueous
Anisotropic nucleation	Rods	$\Phi = 3.5\,\text{nm}$ $L = 15\,\text{nm}$	CTAB	Aqueous
Self-assembly	Ribbons	$\Phi = 10\,\text{nm}$ $L = 40\,\text{nm}$	Mercaptopropionic acid	Aqueous
Growth in membranes	Filaments	$\Phi = 150\,\text{nm}$ $L = 1\text{–}3\,\mu\text{m}$	None	Al_2O_3 membrane
Brust synthesis (MPC)	Spheres	1–3 nm	Thiol	Toluene
Au_{55}	Spheres	1.4 nm	PPh_3 ligand (exchangeable)	Aqueous/organic
Gamma-ray or X-ray radiolysis	Spheres, rods	3–100 nm (depending on dose)	PVA, CTAB, etc.	Aqueous

CTAB cetyltrimethylammonium bromide, *MPC* monolayer protected clusters, *PVA* poly (vinylalcohol)

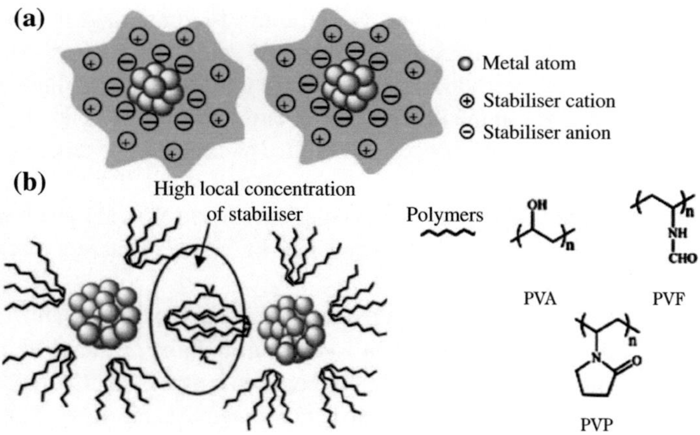

Fig. 5.34 **a** Schematic view of electrostatic stabilisation of nanoparticles. **b** Schematic view of steric stabilisation of nanoparticles and examples of polymeric stabilising agents. Courtesy of S. Monserrat-Gomez, PhD thesis of S. Sanhes, *Université de Toulouse*

synthesising electrostatically stabilised particles mainly use salts as stabilising agents.

- **Steric Stabilisation.** The adsorption of macromolecules such as polymers, oligomers, or dendrimers on the particle surface provides a protective layer (see Fig. 5.34b). This prevents particle aggregation in two different ways. Firstly, motion is restricted in the interparticle space, and secondly, the high concentration of stabilising agents in this space prevents the two layers from interpenetrating.

Polymers often used to create a steric barrier between particles are poly(*N*-vinyl-2-pyrrolidone) (PVP), poly(vinylalcohol) (PVA), poly(vinylether) (PVE), and the cyclodextrins.

Methods of Synthesis

Among the chemical methods for preparing colloids in suspension reported in the literature, we shall discuss only those involving reduction of metal salts, these being the most widely used for obtaining transition metal colloidal solutions. The main reducing agents are hydrides, citrates, alcohols, molecular hydrogen, and carbon monoxide. Standard stabilising agents are polymers, and more recently, dendrimers. Reduction can be carried out either in an aqueous or an organic medium.

In the Turkevich–Frens method [56], which derives directly from Faraday's way of synthesising gold nanoparticles, the starting product is a gold salt such as chloroauric acid ($HAuCl_4$) or potassium tetrachloroaurate ($KAuCl_4$), in which the gold is in oxidation state +III. A reducing agent is then added to reduce the gold ions to gold metal (oxidation state 0). The solution becomes supersaturated in gold atoms and these begin to aggregate and precipitate out. The relative amount of stabilising agent is then used to control the particle size: the more stabilising agent, the smaller the nanoparticles. In this method, the roles of reducing agent and electrostatic stabilising agent are played by the same reagent, viz., sodium tricitrate, and in principle, the particle size can be adjusted over a wide range of values from 10 to 80 nm. However, it is commonplace to produce spherical gold nanoparticles of diameter 15 nm, with very little variability in the diameter (standard deviation of order 1 nm). These nanoparticles, clusters of smaller gold crystallites measuring 3–4 nm, are characteristically ruby red and negatively charged, being coated with a layer of citrate. While this is undoubtedly the simplest method to implement, we are only just beginning to understand the particle growth mechanisms and influence of factors like the pH or the citrate concentration.

Another method for synthesising gold nanoparticles is the Brust–Schiffrin method shown in Fig. 5.35. This uses steric stabilisation through the presence of a thiol (R–S–H end group) with a chain (R) of different possible lengths which serves to protect the nanoparticle core. The synthesis is carried out in an organic solvent and consists once again in reducing a gold salt, but this time very quickly preventing aggregation of the nanocrystals thanks to the ligands which form a protective molecular layer firmly fixed to the gold by covalent bonds (Au–S–). The nanoparticle size can be approximately controlled by suitable choice of the reduction conditions and the proportions of gold salt and thiol. However, it is generally necessary to carry out a post-synthesis selection.

Properties of Gold Nanoparticles

Three main properties of gold nanoparticles explain the interest in these objects over the past few decades: their catalytic properties, their optical properties, and their biocompatibility.

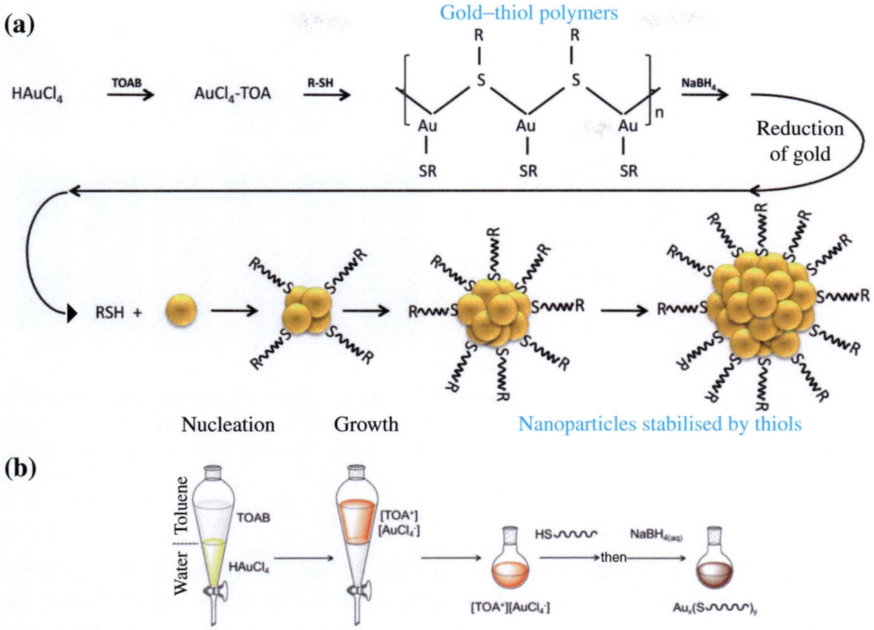

Fig. 5.35 Brust–Schiffrin synthesis of gold nanoparticles. **a** The relevant chemical reactions. **b** Simplicity of implementation [35]. TOAB tetraoctylammonium bromide (phase transfer agent). Adapted from S. Mischler, Neuchâtel University, 2012

Catalytic Properties

While gold in its bulk state is well known for being chemically inert, it turns out that gold nanoparticles have excellent catalytic properties, on a par with or even superior to those of other materials commonly accepted as good catalysts. They have demonstrated these catalytic properties in many chemical reactions, such as the oxidation of carbon monoxide (CO), the oxidation of alcohols into acids or aldehydes, the formation of hydrogen peroxide (H_2O_2), and selective hydrogenation of C–C and N–O bonds [57].

Optical Properties

The optical response of gold nanoparticles is one of their best known properties [35, 56]. According to Mie's theory, electromagnetic waves induce a polarisation of the nanoparticle surface, creating collective oscillations (resonances) of their conduction electrons (see Fig. 5.36a). At a certain frequency, the oscillations of the electron clouds will themselves modulate absorption of the incident light wave and generate a characteristic absorption band, the plasmon band, in the UV–visible range.

The shape, intensity, and spectral position of this absorption band depends on the dielectric permittivity of the medium, the electronic interactions between the stabilising ligands and the metal core, and above all the morphological characteristics

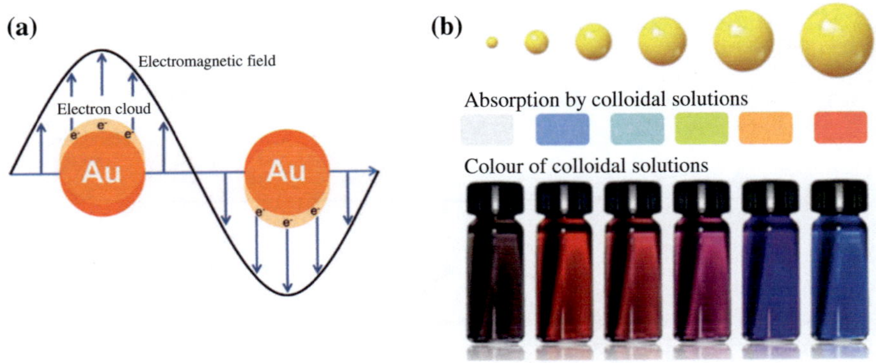

Fig. 5.36 **a** Schematic view of the displacement of the electron cloud of a nanoparticle. **b** Examples of the colours produced by different gold nanoparticle solutions, viewed in absorption and in transmission, for different values of the nanoparticle diameter [35]. Courtesy of S. Mischler, Neuchâtel university, 2012)

of the particles, viz., shape, size, and size dispersion. A change in one of these parameters will lead to a change in the electric field distribution at the surface of the particles. The electrons will thus oscillate at another frequency, thereby displacing the plasmon band.

Particles of different shapes have different oscillation modes, resulting in the formation of distinct plasmon bands. The more complex the particle geometry, the greater the number of oscillation modes. Hence, spherical particles are characterised by a single plasmon band, while rod-shaped particles have two, one corresponding to transverse oscillations and the other to longitudinal resonances.

Nanoparticles of the same shape but different sizes will also have distinct absorption spectra. This leads to differences in the apparent colour of the particles (see Fig. 5.36b). Hence spherical gold nanoparticles with diameters of 20 nm or more exhibit strong resonant absorption for wavelengths in the range 520–570 nm. An increase in diameter shifts the absorption band toward the red with an increase in intensity. A reduction in diameter shifts toward the blue and the absorption diminishes in intensity to become practically imperceptible for very small particles (<2 nm). Note that the apparent colour of samples observed by transmission rather than reflection changes in the opposite way, from blue to red as the diameter of the spheres decreases (see Fig. 5.36b and recall also the Lycurgus cup discussed in Sect. 2.3).

Biocompatibility

The chemical inertness of gold nanoparticles ensures their neutrality in the presence of living tissue, just as it ensures that they will not degrade in agressive environments. A further advantage for biological applications is that gold nanoparticles are easy to functionalise. Indeed, a whole panoply of molecules can be grafted onto them, some of which have therapeutic properties.

Metal Nanoparticles for the Fuel Cell

The fuel cell illustrates the key role played by chemistry in the production of nanoparticles for large scale applications, in this case the elaboration of electrode materials for energy production. The idea of the fuel cell has been around for almost two centuries now. It converts the chemical energy contained in hydrogen and oxygen into electrical energy (see Fig. 5.37a), and this in a continuous manner. The electrical energy supplied in this way is clean and renewable, since the only things that come out of the fuel cell are water and heat, while the fuel (mainly H_2) can be produced from many sources, but in particular the electrolysis of water, even though efficiencies must still be further improved. In a fuel cell, oxidation and reduction reactions take place at the anode and the cathode, respectively, these being separated by an electrolyte which ensures the transfer of charge by ion conduction. On the other hand, the electrolyte blocks the transfer of electrons, these being forced to circulate via an external circuit, thereby setting up an electric current that can be put to work.

There are several kinds of fuel cell, distinguished by their operating temperature, the electrolyte, or the fuel. The proton exchange membrane fuel cell (PEMFC) is the energy solution chosen by many industrial actors, and in particular by the automobile industry. The PEMFC offers undeniable advantages, among which the low operating temperature ($<100\,°C$), insensitivity to carbon dioxide, and fully robust technology. In its final configuration, the PEMFC has the form of a stack of elementary cells, each comprising an electrolyte and two electrodes: an anode and a cathode supplied with reactive gases by either porous distributers (foams) or channelled distributers (embossed metal or graphite/polymer composite plates) (see Fig. 5.37b).

Figure 5.37 shows the complexity hidden behind the simple idea of the PEMFC. These systems involve elements on different scales and several different physical processes, whence the technology exploits a wide range of different materials. One of the bottlenecks today relates to the oxygen reduction kinetics at the cathode, which requires the use of a platinum catalyst. The cost and availability of this noble metal are a hindrance to the development of the PEMFC for mass production. The viability of the device depends on the discovery of electrocatalytic systems that do not use platinum, or at least contain very low levels of platinum, and yet remain effective and stable. Currently preferred solutions consider the second option.

Electrode Materials Used in PEMFCs

Due to its slow kinetics, the reduction of oxygen at the cathode tends to limit the efficiency of the cell, reducing the potential difference across the cell terminals. With platinum as catalyst, a potential difference of 0.8 V can be achieved, the theoretical value being fixed at 1.23 V by thermodynamic considerations.

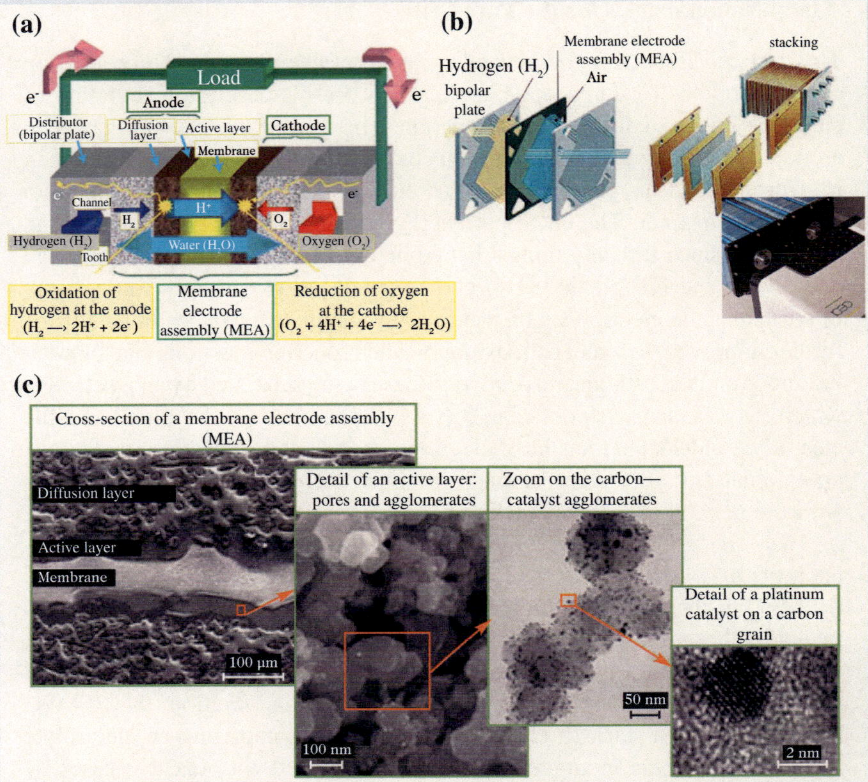

Fig. 5.37 **a** Working principle of the elementary cell in a PEMFC. **b** *Left* Stack of elementary cells. Each one contains two elements: (i) bipolar plates (anode, cathode) to distribute gas and to collect the electric current, and (ii) the membrane electrode assembly (MEA) element. *Right* 80 kW cell developed for the automobile industry by the French atomic energy authority (CEA) in association with PSA Peugeot-Citroën as part of the Genepac project (*générateur électrique à pile à combustible*). **c** Electron microscope image of an MEA and detail of an active layer showing the different physical scales coming into play [58]. Courtesy of J. Pauchet, CEA Liten Grenoble, Clés CEA, 2010, 59, 51–59

The electrodes are thus made from an electron-conducting material, usually carbon black, and the catalyst. Two possibilities have been investigated for reducing the amount of platinum: nanostructuring platinum or platinum-containing particles and deposition of platinum on nanostructured substrates (see Fig. 5.38). In both cases, the resulting increase in the surface to volume ratio provides a way to reduce the mass of platinum in the electrode while maintaining a good level of electrocatalytic activity:

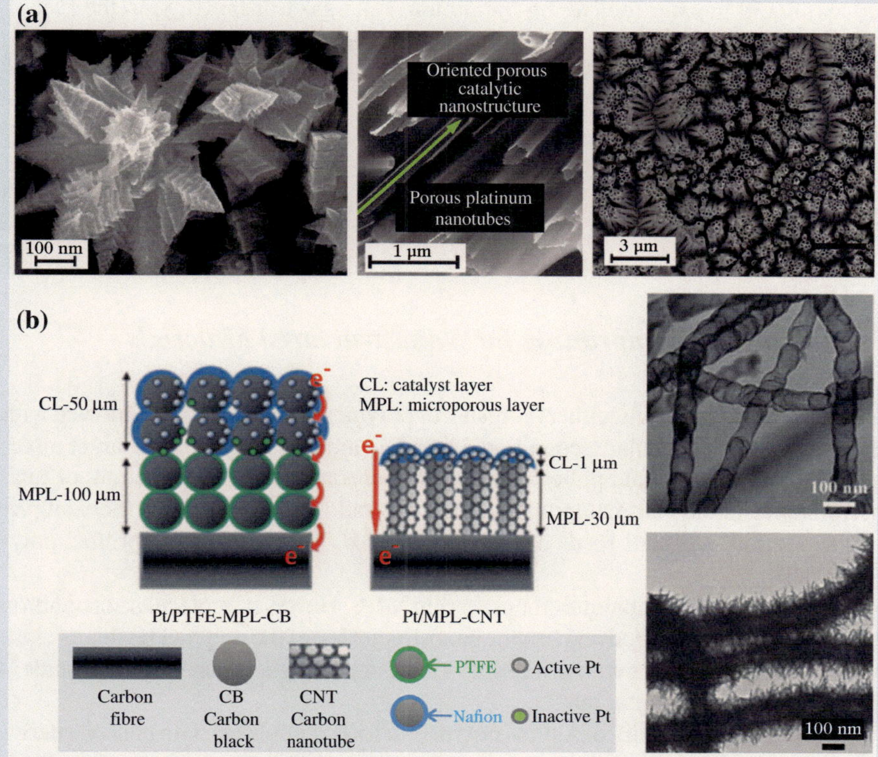

Fig. 5.38 a Electron microscope images illustrating the diversity of platinum nanostructures [58, 59]. Courtesy of J. Pauchet, CEA Liten Grenoble, Clés CEA, 2010, 59, 51–59. **b** *Left* Cathode for PEM, deposition or growth of platinum nanostructures on carbon nanotube electrodes. From [60], © Royal Society of Chemistry 2011. *Right* Electron microscope images of nitrogen-doped carbon nanotubes before and after deposition of platinum nanostructures. From [61], © Royal Society of Chemistry 2009

- **Nanostructuring Platinum Particles.** This is illustrated in Fig. 5.38a [58, 59]. The methods used to synthesise metal nanoparticles (here platinum) are discussed in the first part of this section. The electrocatalytic activity of platinum nanoparticles is sensitive to their shape and also the orientation of the crystal planes in the platinum. Another line of investigation is the synthesis of bimetal nanoparticles such as Pt–Co [59]. The amount of platinum is then very small thanks to the presence of the other metal. Some bimetal nanoparticles have a core–shell configuration, where only the shell is platinum, while the core is some other metal. All these nanoparticles are then incorporated in some way into an electron-conducting support.
- **Deposition of Platinum on Nanostructured Supports.** This is illustrated in Fig. 5.38b [60, 61]. This can be a good solution to relieve problems relating

to liquid transfer (delivering the protons) and gas transfer (O_2) in the fuel cell. In Fig. 5.38b, the electrode is made from carbon nanotubes, thereby maintaining electron conduction while facilitating the transfer of protons and molecular oxygen. A platinum nanofilm is formed by chemical vapour deposition. In the same way, platinum nanowires can be grown on carbon nanotubes or nanospheres.

5.5.2 Molecular Imprinting for Nanostructured Materials

Nanostructured materials, either inorganic or polymer, contain cavities endowed with the capacity for molecular recognition when they are fabricated around target molecules [62, 63]. Potential applications include sensors, selective extraction, or high performance liquid chromatography for the separation of enantiomers, for example. In this section, we shall focus on the special case of molecularly imprinted polymers (MIP).

There are several techniques for preparing MIPs. However, molecular recognition turns out to be more effective when the nanostructuring of the polymer begins as close as possible to the target whose imprint is to be made, rather than when the latter is buried within the polymer matrix.

Historically, the ability of a silica polymer to form complexes with benzene derivatives used during its synthesis was first observed in 1931. The method of synthesis was taken up again by Dickey in 1949 to produce a polymer selecting for methyl orange [62]. He introduced a technique called specific adsorption, which subsequently became known as molecular imprinting, used to create synthetic receptors. The materials produced are resins with an ability for molecular recognition analogous to a similar ability in biological receptors. However, because they are synthetic, MIPs are extremely robust, more stable, and can be used under more severe conditions (in acidic or basic environments, etc.). Quite generally, these MIPs are obtained by polymerisation of a mixture containing a crosslinking agent, a porogenic agent, functional monomers, and the template molecule whose imprint is required.

The simple and cheap process used for this synthesis has three main steps (see Fig. 5.39):

- Polymerisable compounds with a presumed chemical affinity for the target molecule are pre-ordered in a porogenic solvent in the presence of this molecule.
- The mixture is then co-polymerised by a crosslinking agent.
- The target molecules and the solvent are extracted from the resulting polymer.

After extracting the target molecules, cavities form in the polymer. These cavities will be able to selectively recognise the chosen molecule. The polymer matrix, which is both solid and porous, thus contains cavities whose conformation is predetermined for

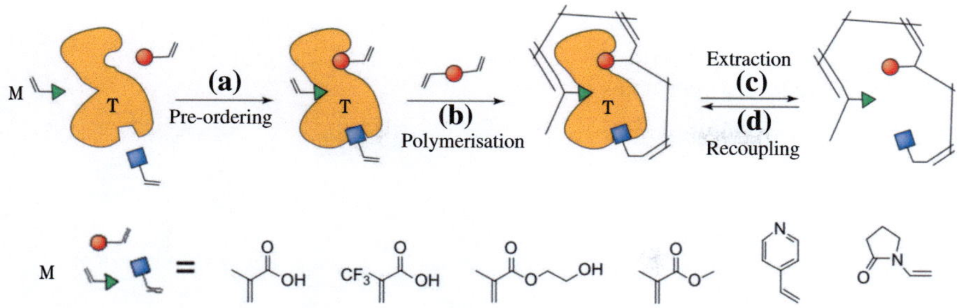

Fig. 5.39 Reactions leading to the synthesis of molecularly imprinted polymers (MIP). **a** Pre-polymerisation complex between the template molecule T to be imprinted and the functional monomers M. **b** Co-polymerisation with an excess of crosslinking agent. **c** Extraction of the template molecule T from the complexing sites. **d** Recoupling with the template molecule T by the imprint sites when the MIP is used. From [64], © Elsevier 2003

selective recognition of molecules identical to the target. This selectivity comes not only from the spatial geometry of the cavity, but also from the specific supramolecular interactions of its surface functions with the target.

5.5.3 Multifunctional Hybrid Nanomaterials Produced by Soft Chemistry

The chemistry of hybrids is by nature soft, ecological, and respectful of the environment [65, 66]. It is based on inorganic polymerisation reactions, better known by the name of sol–gel processes, carried out at rather low temperatures with molecular precursors such as metal salts or metal alkoxides in aqueous or hydro-alcoholic media (see Fig. 5.40). It can be used to combine organic or biological components, which are generally rather fragile from a thermal point of view, with the world of inorganic compounds. This chemistry of hybrid materials makes full use of the flexibility and diversity inherent in both molecular and supramolecular chemistry.

The methods of soft chemistry were first developed to prepare entirely inorganic systems such as glasses, ceramics, and inorganic–inorganic composites. They can now be used to produce inorganic and organic components simultaneously in the same material to end up with hybrid nanomaterials. The latter can be defined as genuine nanocomposites on the molecular scale in which at least one of the components is organic (or biological) and the other inorganic, with a size of a few tenths to a few tens of nanometers.

The inorganic component of the material can play several roles: strengthening the mechanical or thermal behaviour of the material, modulating its refractive index, or contributing to its magnetic, electronic, electrochemical, or chemical properties, and in the case when this component is a porous matrix, making the material accessible to chemical reagents. The organic component for its part serves to extend the

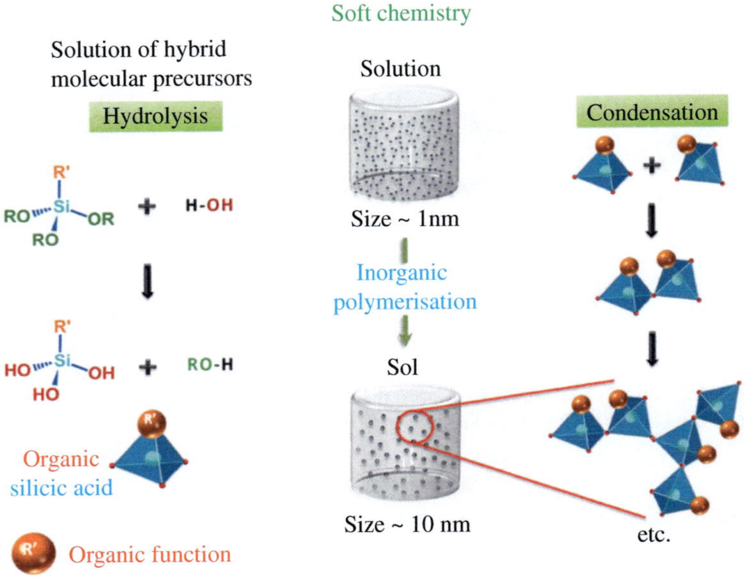

Fig. 5.40 Schematic representation of inorganic polymerisation reactions (hydrolysis–condensation) or sol–gel processes [66]. Courtesy of C. Sanchez, *Leçon inaugurale au Collège de France*, 2012

range of possible matrices, to obtain various geometric structures by simple moulding, to control the porosity and connectivity of the resulting networks, and to adjust the physical, chemical, and electrochemical properties. However, the many different functionalities of hybrid materials do not result from the simple addition of individual contributions from each of their components, but also from the considerable synergy produced by an extensive hybrid interface. Indeed, by the very nature of the interactions between inorganic and organic entities, and by the energy and lability of the bonds, this interface plays a determining role in the modulation of optical, mechanical, and catalytic properties, but also in the chemical and thermal stability of the material.

Apart from the physicochemical properties, the association of soft chemistry with the structuring processes of soft matter, such as dip coating, sputtering, spin coating, extrusion, microemulsions, and others, provides a straightforward way to elaborate hybrid nanomaterials in a broad range of forms, either as a thick or thin film ($5 <$ thickness $< 50\,\mu$m), a fibre, a powder, a foam, or a monolith. This wealth of methods is illustrated in Fig. 5.41.

Classification of Hybrid Nanomaterials

Hybrid materials fall into two main classes depending on the kind of interface which associates the organic component with the inorganic component. Class I contains hybrid systems in which the organic and inorganic components are associated by weak bonds of van der Waals type, hydrogen bonds, or electrostatic interactions (see

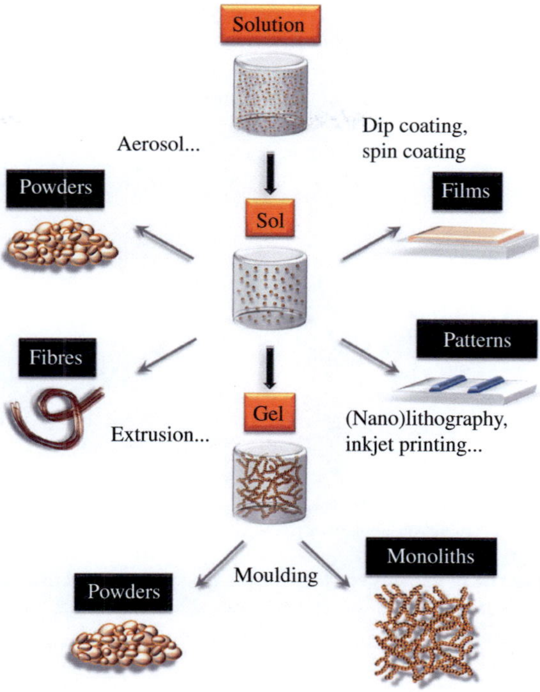

Fig. 5.41 Examples of different structuring processes used in the chemistry of hybrid materials [65]. Courtesy of C. Sanchez, *Leçon inaugurale au Collège de France*, 2012

Fig. 5.21). Here we are dealing mainly with supramolecular assemblies. Class II contains materials in which these components are bound by covalent or ionic–covalent chemical bonds. The development of class II nanomaterials is growing in importance today. In contrast to class I materials, the covalent grafting of organic functions onto the inorganic substructure helps to avoid the loss of organic components when these materials are in use. The strength of the organic–inorganic binding within these composites opens the way to novel possibilities for synthesis by minimising phase separations. Better definition of the organic–inorganic interface also allows a better understanding of the material and of the relation between its properties and its microscopic structure. This in turn allows a better optimisation of the relevant properties, whether they be optical properties or ones relating to the hydrophilic or hydrophobic nature of the material, for example.

The Wealth of Strategies for Synthesis

It is important to emphasise the wealth of different strategies for synthesis and processes for elaborating and structuring hybrid nanomaterials. In fact there are three main strategies.

Sol–Gel Strategies

The standard sol–gel synthesis labelled (A1) in Fig. 5.42 is based on hydrolysis and condensation reactions (see also Fig. 5.40), initiated by alkoxide molecular precur-

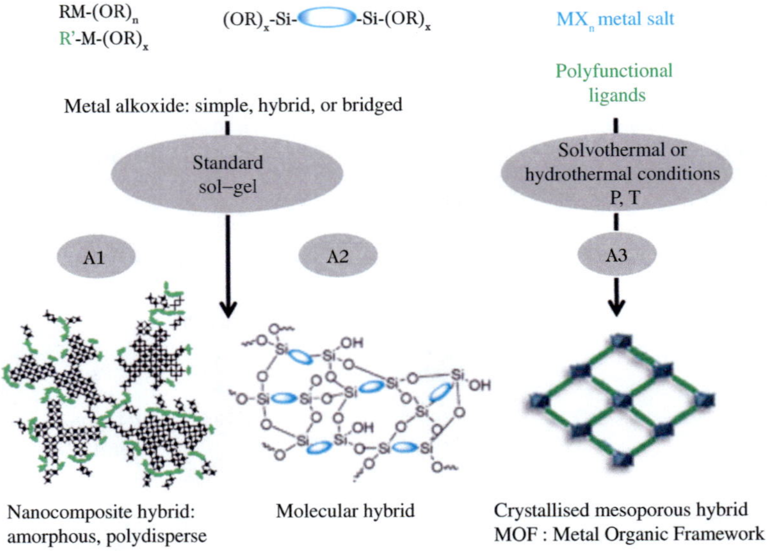

Fig. 5.42 Variants on standard sol–gel synthesis. From C. Sanchez, *Le nano-monde de la chimie. La lettre de l'Académie des Sciences*, 2008, 23, 10–19

sors (metal salts in the presence of polyfunctional complexing agents, etc.). It is easy to implement, cheap, and easy to adapt, producing hybrid nanomaterials that are then easily structured. However, the resulting nanomaterials are amorphous and polydisperse with regard to size, so it is more difficult to relate their properties to their microscopic structure.

In the second scheme, labelled (A2) in Fig. 5.42, one starts with bridged silsesquioxane precursors [X_3-M-Rp-M-X_3, where M is a metal (Si or Sn), Rp represents the organic bridge, and X the hydrolysable group, generally Cl or OR]. The reaction path then produces hybrid materials that are homogeneous on the molecular scale.

The third scheme, labelled (A3) in Fig. 5.42, is based on hydrothermal or solvothermal processing of solutions in a polar solvent that contains both metal cations and organic structures carrying metal-complexing functions, e.g., carboxylic di- or tri-acid. This produces crystallised microporous hybrid solids called metallic–organic frameworks (MOF), with perfectly defined structure and exhibiting very large specific surface areas (1 000–5 000 m^2/g). These MOFs have very promising properties for catalysis or gas absorption, but also magnetic or optical properties depending on the chosen metal cations.

Synthesis by Legochemistry

The legochemistry approach consists in assembling monodisperse, preformed nano-objects with an inorganic core and organic functions over the surface (see Fig. 5.43). These nano-objects may be clusters, metal oxide, metal, or alloy nanoparticles, core–

shell composite nanoparticles, or nanosheets of lamellar compounds such as nanoclays. They must be stable under the chemical conditions imposed by the assembly stage. For this reason, they are often previously functionalised by organic groups, such as polymerisable ligands, organic spacers, telechelics, functional dendrimers, biomolecules, etc. In this way, legochemistry can construct tailor-made hybrid nanomaterials, built up step by step, as it were. Since the inorganic compounds are monodisperse, this leads to better defined structures than the standard sol–gel synthesis. This in turn makes it easier to characterise the final material.

Two examples of this approach are illustrated in Fig. 5.43. The first is the synthesis of 3D covalent networks from titanium oxide aggregates $Ti_{16}O_{16}(OEt)_{32}$, in which we may selectively modify the ethoxy groups at the surface through exchange reactions with alcohols. A first stage of legochemistry consists in coupling these aggregates in a common solvent with polyfunctional dendrimers whose dendrons terminate in alcohol functions. After evaporation of the solvent, we then obtain a nanostructured glass in which the dendrimers are crosslinked by the aggregates. Another reaction substitutes an alcohol carrying a polymerisable function (hydroxyethyl methacrylate HEMA) for each titanium aggregate and co-polymerises the whole system to form a hybrid network that is easily prepared in the form of monolithic blocks. The network will display greater or lesser permeability to gases or liquids depending on which reaction is used.

The second example in Fig. 5.43 involves the chemical intercalation of clays, with clay nanosheets reacting with organic monomers or polymers. A first approach to synthesis uses polar solvents in which the clay is mixed with the polymer, so that the latter can intercalate between the nanosheets. The solvent is then eliminated by vacuum evaporation. Although weakly polar polymers can be used here, this approach has the disadvantage of employing large amounts of solvents. Another kind of synthesis carries out polymerisation in situ by introducing a liquid monomer or a monomer in solution. The latter migrates into the intersheet gaps and polymerisation is triggered by heating, radiation, or an initiator that is also intercalated. Nanocomposites made in this way have a relatively ordered lamellar inorganic structure and exhibit excellent impermeability and fireproofing properties.

Self-Assembly of Amphiphilic Molecules

Amphiphilic molecules have a hydrophilic polar head and a hydrophobic nonpolar tail (see Fig. 5.44). This configuration is conducive to molecular self-ordering which exploits hydrophilic–hydrophobic interactions. The self-assembly properties of amphiphilic molecules have been used to produce supramolecular motifs controlling the texture and morphology of systems undergoing growth. This is a particularly useful form of self-assembly because it can be used to build a continuous range of nanocomposites from ordered dispersions of inorganic nano-objects in organic matrices to the controlled segregation of organic polymers in inorganic matrices. Figure 5.44 provides a perfect example of this kind of self-assembly.

In the first step, alkoxide and surfactant (amphiphilic) precursor molecules are brought into contact and a cooperative sol–gel self-assembly process leads to the formation of hybrid liquid crystals. Inorganic polymerisation then occurs at the inter-

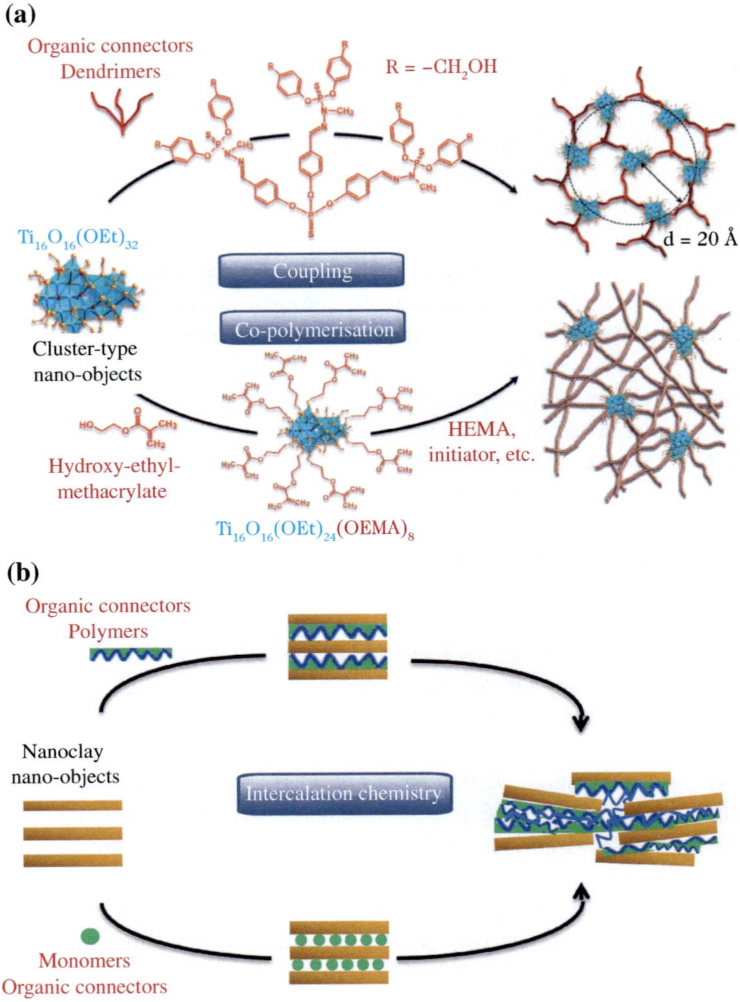

Fig. 5.43 Synthesis of hybrid nanomaterials by assembly of preformed nano-objects. **a** The nano-objects are titanium oxide aggregates which are assembled in a 3D network by coupling or co-polymerisation reactions. Courtesy of C. Sanchez, *Leçon inaugurale au Collège de France*, 2012. **b** The nano-objects are clay nanosheets, between which are intercalated polymers or monomers that will subsequently be polymerised. From C. Sanchez, *Le nano-monde de la chimie. La lettre de l'Académie des Sciences*, 2008, 23, 10–19

face between the water and the polar heads of the amphiphilic molecules which form cylindrical micelles. An intermediate hybrid phase with a well-defined structure (hexagonal in Fig. 5.44) is thereby obtained, with a 'soft' core made up of micelles and a continuous envelope that hardens during inorganic polymerisation. In a second step, the micelles can be eliminated simply by rinsing, thus leaving perfectly calibrated pores in an inorganic framework. The pore size, in the range 2–20 nm,

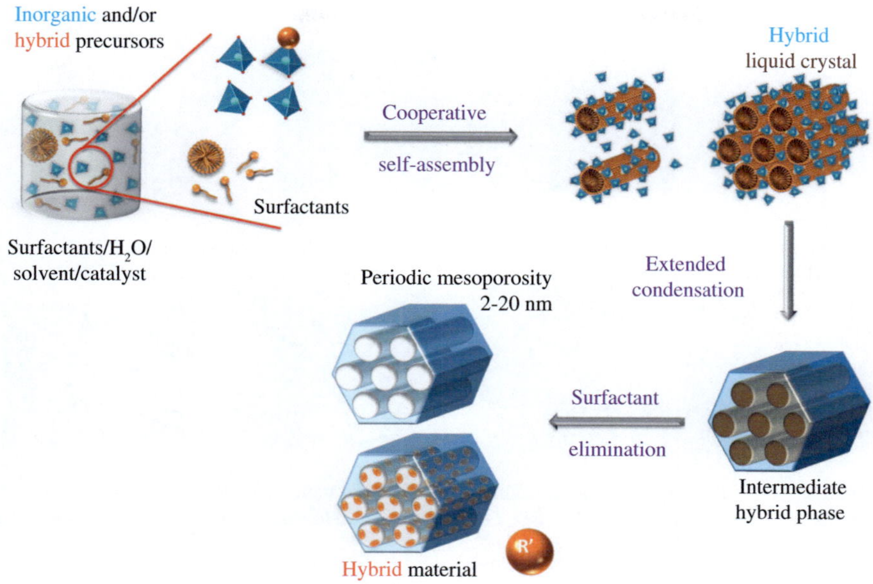

Fig. 5.44 Synthesis of hybrid nanomaterials by self-assembly of amphiphilic molecules. Courtesy of C. Sanchez, *Leçon inaugurale au Collège de France*, 2012

corresponds approximately to the diameter of the micelles, which itself depends on the size of the amphiphilic molecule selected to build the nanomaterial.

The final system can be entirely inorganic, e.g., silica or a metal oxide. Having said that, one only needs to change the molecular structure of the original precursor to obtain a hybrid mesoporous material. Quite generally, the properties of the resulting nanomaterials are determined by the nature of the inorganic or hybrid compound forming the framework of the mesoporous material. Thanks to the particularly large specific surface area, of the order of 1 000 m^2/g, and thanks to a high surface reactivity with greater possibility for exchange of matter, these nanoporous materials have considerable potential for applications in the environment and energy sectors, where they can be used as membranes, sensors, catalysts, or adsorbers.

The Coupling of Chemistry and Process: Toward a Bio-Inspired Approach

A material is not simply a compound or a chemical composition. If we take the example of silica, with chemical composition SiO_2, made up of chains of SiO_4 tetrahedra, we have the main constituent of sand, quartz crystals, and window glass, but also of the frustule, the silicified cell wall of a diatom, which is a single-celled alga, abundantly common in plankton. This diversity of materials for just one given chemical composition results from the coupling between the chemistry and the elaboration process. The properties of the material in use and its robustness depend on the quality of this coupling between chemistry and process. But it is also through a well chosen and well controlled coupling that one can produce a whole range of hybrid materi-

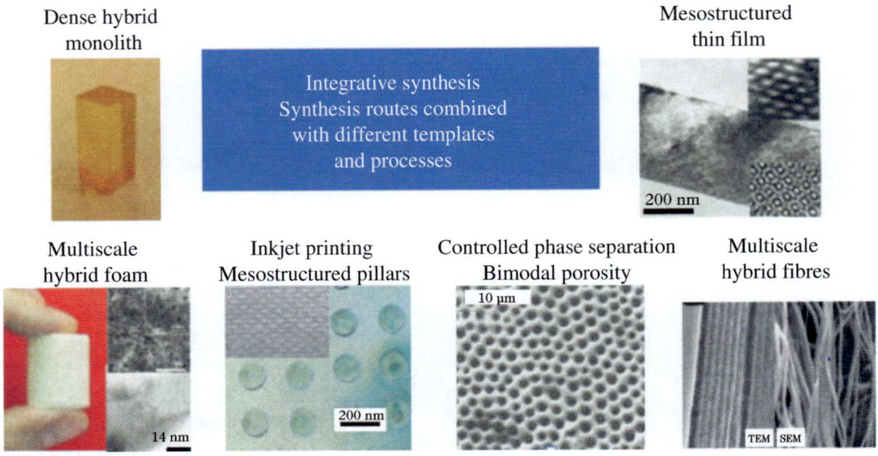

Fig. 5.45 Integrative approach to the elaboration of hybrid nanomaterials. The coupling between chemistry and process. From C. Sanchez, *Le nano-monde de la chimie. La lettre de l'Académie des Sciences*, 2008, 23, 10–19

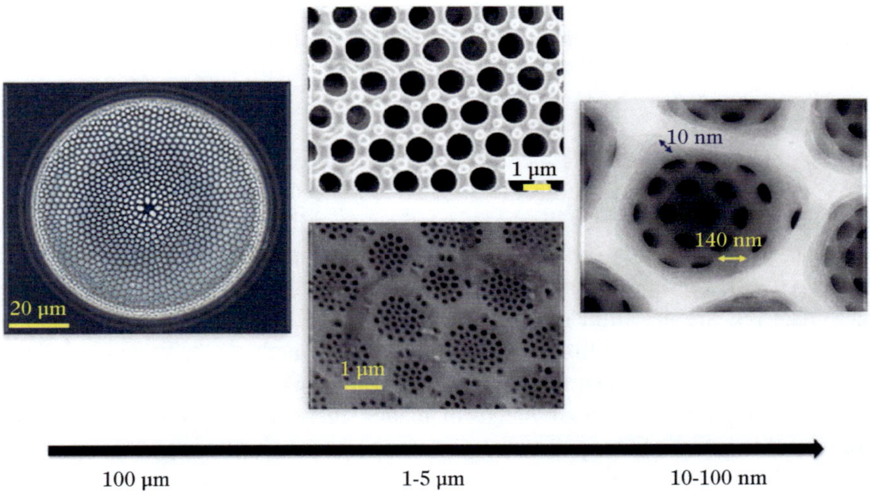

Fig. 5.46 Multiscale structure of diatom frustules. Courtesy of C. Sanchez, *Leçon inaugurale au Collège de France*, 2012

als (see Fig. 5.45). The integrative approach thus combines the arsenal of synthetic routes discussed above with the physicochemical processes of soft matter (repeated replications and nanomoulding, reactive extrusion, dip coating, etc.) to make structures and motifs of variable sizes and shapes in materials differing enormously from one another (latexes, inorganic colloids, organogels, etc.).

Having said this, while these strategies can produce hybrid materials with hierarchical structure in a single step, they do not yet lead to the kind of hierarchies observed in vivo, which are not only extremely complex, but also highly selective. Taking the example of diatoms, one observes a porous structure on length scales from a few microns to a few nanometers (see Fig. 5.46). Clearly, living tissues are not only organic–inorganic hybrid materials *par excellence*, but they also exhibit hierarchical and multifunctional structures. These natural materials can thus serve as models to inspire the development of new ideas and new strategies. This is the field of bio-inspired materials.

5.6 Conclusion

As we have shown in this chapter, nanochemistry is important at every level of the bottom–up approach to making nanomaterials. Even at the design stage, the nanochemist decides like an architect which strategies to use for synthesis in order to build nanomaterials with specific properties. She or he then prepares the basic building blocks required for the construction, and programs the assembly of these 'nanobricks'. At each stage in the elaboration of the nanomaterial, the nanochemist controls the evolving matter, and this is indeed one of the most important tasks. Finally, she or he must also plan the way the nanomaterial will be integrated into whatever systems are to be used for specific applications.

On the nanoscale, the nanochemist can imagine novel materials starting out from very different kinds of nanobricks, often to be bound together by covalent bonds, the strongest chemical bonds available, but also by the non-covalent bonds that form the basis for supramolecular chemistry. In this nanoworld, organic and inorganic get along fine together, binding closely to give rise to hybrid nanomaterials.

Today nanochemistry is a growing industry, but it is far from having reached its full potential. In fact, the possibilities for creating nano-objects are almost infinite. It remains only for the nanochemist to suggest the most suitable methods for assembly and integration. In the interdisciplinary world of nanoscience, nanochemistry forms a bridge between physicists and biologists. It plays a central role because it can build directly from basic matter. Of course, this is not the first time that chemists and physicists have worked together to develop processes or materials, but the nanoscale now offers possibilities for designing novel systems for electronics and optics by making use of molecular chemistry.

This is the case for magnetic nanosystems in particular, where the magnetic properties of nanoparticles or giant molecules, e.g., high spin molecules, can be custom built. During the last few years, there has been increasing interest in molecular nanomagnets in the form of organometallic clusters for which the ground state is characterised by a high spin moment ($S > 10$), with a predominantly uniaxial anisotropy. The main interest here is in potential applications to quantum information systems and spintronics devices (see Chap. 2). Nanochemists play a key role here in the synthesis of these clusters, bringing to bear the impressive resources of supramolecular chemistry.

Likewise, chemists and biologists have long cooperated to develop biological molecules and analyse their properties, but here too, the nanoscale opens up novel prospects for the imaging of living systems, drug delivery, and therapeutic treatments in general (see Part III of this book). It seems highly likely that, over the coming years, some of the most innovative features will arise from the strengthened cooperation of chemists and biologists, going beyond the simple association of chemical and biological molecules. In fact, this revolution is already under way, even if it is only just beginning.

References

1. P. Delhaes, *Science et technique des carbones: de l'énergie aux matériaux* (Hermes et Lavoisier, Cachan, 2013)
2. A. Krueger, *Carbon Materials and Nanotechnology* (Wiley-VCH Verlag GmbH & Co., Weinheim, 2010)
3. Special Issue: Les matériaux carbonés. L'Actualité Chimique **295–296** (2006)
4. Website of the Groupe français d'étude des carbones. www.gfec.net/
5. G.M. Jenkins, K. Kawamura, Structure of glassy carbon. Nature **231**, 175–176 (1971)
6. Carbon Black User's Guide, Safety, Health, & Environmental Information. International Carbon Black Association. www.carbon-black.org/
7. C. Bathias et al., Matériaux composites, Chap. 19, *L'Usine Nouvelle*, Mécanique et matériaux, 2nd edn. (Dunod, Paris, 2009)
8. www.dgcis.gouv.fr/files/files/archive and www.industrie.gouv.fr/colloque/nanomateriaux/francis-peters-michelin.pdf
9. H.W. Kroto, J.R. Heath, S.C. O'Brien, R.F. Curl, R.E. Smalley, C60: Buckminsterfullerene. Nature **318**, 162–163 (1985)
10. V.N. Mochalin, O. Shenderova, D. Ho, Y. Gogotsi, The properties and applications of nanodiamonds. Nat. Nanotechnol. **7**, 11–23 (2012)
11. O.A. Shenderova, D.M. Gruen, *Ultracristalline Diamond: Synthesis, Properties and Applications*, Micro- and NanoTechnologies Series, 2nd edn. (Elsevier, Amsterdam, 2012)
12. E. Osawa, D. Ho, Nanodiamond and its application to drug delivery. J. Med. Allied Sci. **2**, 31–40 (2012)
13. E.K. Chow, X.Q. Zhang, M. Chen, R. Lam, E. Robinson, H. Huang, D. Schaffer, E. Osawa, A. Goga, D. Ho, Nanodiamond therapeutic delivery agents mediate enhanced chemoresistant tumor treatment. Sci. Transl. Med. **3**, 73ra21 (2011)
14. S. Iijima, Helical microtubules of graphitic carbon. Nature **354**, 56–58 (1991)
15. S. Yellampalli, *Carbon Nanotubes. Synthesis, Characterization, Applications* (InTech, Open Book, 2011). doi:10.5772/978
16. D.M. Guldi, N. Martín, *Carbon Nanotubes and Related Structures* (Wiley, New York, 2010)
17. M. Monthioux, *Carbon Meta-Nanotubes: Synthesis, Properties and Applications* (Wiley, New York, 2012)
18. M.F.L. De Volder, S.H. Tawfick, R.H. Baughman, A.J. Hart, Carbon nanotubes: present and future commercial applications. Science **339**, 535–539 (2013)
19. S. Iijima, T. Ichihashi, Single-shell carbon nanotubes of 1-nm diameter. Nature (London) **363**, 603–605 (1993)
20. S. Shanmugam, A. Gedanken, Electrochemical properties of bamboo-shaped multiwalled carbon nanotubes generated by solid state pyrolysis. Electrochem. Commun. **8**, 1099–1105 (2006)
21. K.S. Novoselov, V.I. Fal'ko, L. Colombo, P.R. Gellert, M.G. Schwab, K. Kim, A roadmap for graphene. Nature **490**, 192–200 (2012)
22. J.H. Warner, F. Schaffel, M. Rummeli, A. Bachmatiuk, *Graphene: Fundamentals and Emergent Applications* (Elsevier, Amsterdam, 2013)

23. K.S. Novoselov, A.K. Geim, S.V. Morozov, D. Jiang, Y. Zhang, S.V. Dubonos, I.V. Grigorieva, A.A. Firsov, Electric field effect in atomically thin carbon films. Science **306**, 666–669 (2004)
24. C. Berger, Z. Song, X. Li, X. Wu, N. Brown, C. Naud, D. Mayou, T. Li, J. Hass, A.N. Marchenkov, E.H. Conrad, P.N. First, W.A. de Heer, Electronic confinement and coherence in patterned epitaxial graphene. Science **312**, 1191–1196 (2006)
25. R. Demadrille, Élaborer des architectures moléculaires conjuguées pour le photovoltaïque organique. Clés CEA **60**, 26–28 (2011)
26. G.V. Dubacheva, C.K. Liang, D.M. Bassani, Functional monolayers from carbon nanostructures—fullerenes, carbon nanotubes, and graphene as novel materials for solar energy conversion. Coord. Chem. Rev. **256**, 2628–2639 (2012)
27. H. Marsh, F.R. Reinoso, *Activated Carbon* (Elsevier, Oxford, 2006)
28. P. Bernier, S. Lefrant, *Le carbone dans tous ses états* (Gordon and Breach Science Publishers, Amsterdam, 1997)
29. M. Inagaki, Pores in carbon materials importance of their control. New Carbon Mater. **24**, 193–232 (2009)
30. A. Walcarius, Electrocatalysis, sensors and biosensors in analytical chemistry based on ordered mesoporous and macroporous carbon-modified electrodes. Trends Anal. Chem. **38**, 79–97 (2012)
31. P.L. Taberna, P. Simon, Supercondensateurs carbone/carbone à haute densité d'énergie. Apport des carbones microporeux. Technique de l'Ingénieur (2011)
32. P. Simon, Supercondensateurs: principes et évolutions. Conférence du Collège de France, Chaire Développement durable, Environnement, Énergie et Société, (2010–2011). www.college-de-france.fr/media/jean-marie-tarascon/UPL19317_P_Simon_2F_vrier.pdf
33. J. Chmiola, G. Yushin, Y. Gogotsi, C. Portet, P. Simon, P.L. Taberna, Anomalous increase in carbon capacitance at pore sizes less than 1 nanometer. Science **313**, 1760–1763 (2006)
34. G.A. Ozin, A.C. Arsenault, L. Cademartiri, *Nanochemistry: A Chemical Approach to Nanomaterials*, 2nd edn. (RSC Publishing, Cambridge, 2009)
35. S. Mischler, La chimie click pour la conception de nanoparticules d'or liquides-crystallines. Ph.D. thesis, Neuchâtel University (2012)
36. H.C. Kolb, M.G. Finn, K.B. Sharpless, Click chemistry: diverse chemical function from a few good reactions. Angew. Chem. Int. **40**, 2004–2021 (2001)
37. A.M. Isloor, B. Kalluraya, P. Shetty, Regioselective reaction: synthesis, characterization and pharmacological studies of some new Mannich bases derived from 1,2,4-triazoles. Eur. J. Med. Chem. **44**, 3784–3787 (2009)
38. J. Iehl, R. Pereira de Freitas, J.F. Nierengarten, Click chemistry with fullerene derivatives. Tetrahedron Lett. **49**, 4063–4066 (2008)
39. N. Moitra, P. Trens, L. Raehm, J.O. Durand, X. Cattoën, M. Wong, Chi Man: Facile route to functionalized mesoporous silica nanoparticles by click chemistry. J. Mater. Chem. **21**, 13476–13482 (2011)
40. J.M. Lehn, *La chimie supramoléculaire: concepts et perspectives* (De Boeck Supérieur, 1997)
41. J.M. Lehn, *Chimie des interactions moléculaires: leçon inaugurale au Collège de France, 7 March 1980* (Collège de France, 1980)
42. J.W. Steed, J.L. Atwood, *Supramolecular Chemistry*, 2nd edn. (Wiley, Chichester, 2009)
43. W. Hosseini, Introduction à la chimie supramoléculaire. MSc course 2011. http://lcco.u-strasbg.fr/wp-content/uploads/2011/09/Cours-Supramol-copie.pdf
44. M. Mayor, PNR Matériaux fonctionnels supramoléculaires. Vision (Swiss magazine of science and innovation), special issue (2001)
45. E. Bilensoy, *Cyclodextrins in Pharmaceutics, Cosmetics, and Biomedicine* (Wiley, Hoboken, 2011)
46. P. Cordier, F. Tournilhac, C. Soulie-Ziakovic, L. Leibler, Self-healing and thermo-reversible rubber from supramolecular assembly. Nature **451**, 977–980 (2008)
47. J.L. Halary, L. Avérous, M.É. Borredon, S. Bourbigot, B. Boutevin, C. Bunel, S. Caillol, S. Commereuc, S. Duquesne, L. Lecamp, L. Leibler, É. Pollet, C. Soulié-Ziakovic, F. Tournilhac, C. Vaca-Garcia, V. Verney, Matériaux polymères et développement durable. L'Actualité Chimique **338–339** (2010)

48. A. Krueger, The structure and reactivity of nanoscale diamond. J. Mater. Chem. **18**, 1485–1492 (2008)
49. J.J. Gooding, F. Mearns, W. Yang, J. Liu, Self-assembled monolayers into the 21st century: recent advances and applications. Electroanalysis **15**(2), 81–96 (2003)
50. M. Delamar, R. Hitmi, J. Pinson, J.M. Saveant, Covalent modification of carbon surfaces by grafting of functionalized aryl radicals produced from electrochemical reduction of diazonium salts. J. Am. Chem. Soc. **114**, 5883–5884 (1992)
51. A. Adenier, E. Cabet-Deliry, A. Chaussé, S. Griveau, F. Mercier, J. Pinson, C. Vautrin-Ul, Grafting of nitrophenyl groups on carbon and metallic surfaces without electrochemical induction. Chem. Mater. **17**, 491–501 (2005)
52. J. Ru, B. Szeto, A. Bonifas, R.L. McCreery, Microfabrication and integration of diazonium-based aromatic molecular junctions. Appl. Mater. Interfaces **2**, 3693–3701 (2010)
53. B.P. Corgier, C.A. Marquette, L. Blum, Diazonium–protein adducts for graphite electrode microarrays modification: direct and addressed electrochemical immobilization. J. Am. Chem. Soc. **127**, 18328–18332 (2005)
54. S. Bouden, A. Chaussel, S. Dorbes, O. El Tall, N. Bellakhal, M. Dachraoui, C. Vautrin-Ul, Trace lead analysis based on carbon-screen-printed-electrodes modified via 4-carboxy-phenyl diazonium salt electroreduction. Talanta **106**, 414–421 (2013)
55. C. Altavilla, *Inorganic Nanoparticles: Synthesis, Applications, and Perspectives* (CRC Press, Boca Raton, 2011)
56. O. Pluchery, M. Carriere, Nanoparticules d'or. Dossier Nanomatériaux: élaboration, propriétés et applications. Techniques de l'ingénieur (2011)
57. A. Corma, H. Garcia, Supported gold nanoparticles as catalysts for organic reactions. Chem. Soc. Rev. **37**, 2096–2126 (2008)
58. J. Pauchet, A. Morin, S. Escribano, N. Guillet, L. Antoni, G. Gebel, La pile à combustible PEMFC: une solution très crédible. Clefs CEA **59**, 51–59 (2010)
59. H. You, S. Yang, B. Ding, H. Yang, Synthesis of colloidal metal and metal alloy nanoparticles for electrochemical energy applications. Chem. Soc. Rev. **42**, 2880–2904 (2013)
60. H.Y. Du, C.H. Wang, H.C. Hsu, S.T. Chang, S.C. Yen, L.C. Chen, B. Viswanathan, K.H. Chen, High performance of catalysts supported by directly grown PTFE-free micro-porous CNT layer in a proton exchange membrane fuel cell. J. Mater. Chem. **21**, 2512–2516 (2011)
61. S. Sun, G. Zhang, Y. Zhong, H. Liu, R. Li, X. Zhou, X. Sun, Ultrathin single crystal Pt nanowires grown on N-doped carbon nanotubes. Chem. Commun. **45**, 7048–7050 (2009)
62. K. Haupt, Molecular imprinting. Top. Curr. Chem. **325**, 1–110 (2012)
63. V. Dufaud, L. Bonneviot, Molecular imprinting, in *Nanomaterials and Nanochemistry*, ed. by M. Lahmani, C. Bréchignac, P. Houdy (Springer, Berlin, 2007), pp. 597–616
64. D. Batra, K.J. Shea, Combinational methods in molecular imprinting. Curr. Opin. Chem. Biol. **7**, 434–442 (2003)
65. C. Sanchez, *Chimie des matériaux hybrides. Leçon inaugurale au Collège de France* (Collège de France, Fayard, 2012)
66. C. Sanchez, Matériaux hybrides multifonctionnels: du champ d'investigation pluridisciplinaire aux applications. Le nano-monde de la chimie. La lettre de l'Académie des sciences **23**, 10–19 (2008)

Chapter 6
Applications I: Nanolubricants

Fabrice Dassenoy

Abstract The exceptional properties procured by nanoparticles in terms of reduction of friction and resistance to wear open new prospects in the field of tribology and lubrication. When used as lubricant additives in automobile applications, unrivalled performance levels are obtained, promising in the long term a significant reduction in fuel consumption and hence also in environmental pollution.

There has been growing interest in nanoparticles for tribological[1] applications over the past few years. The many studies on this subject have revealed the remarkable lubricating properties, viz., friction-reducing and anti-wear, of certain nanoparticles. This makes them potential candidates for replacing the lubrication additives currently used in automobile lubricants, especially since the latter are known to be pollutants. This has not gone unnoticed to professionals in the sector, including those commercialising these additives, the petroleum industry, and the car industry, all of whom are following this burgeoning area of research with a keen interest. Indeed, all of them are faced with the problem of providing lubricants that meet the needs of the technological evolution of engines while respecting ever stricter environmental norms. And this is no small challenge, especially when we realise that, in Europe alone, the annual consumption of liquid lubricants for cars is around 5 million tonnes, while reaching 40 million tonnes worldwide. This is by far the greatest use of liquid lubricants, all applications considered.

Among the nanoparticles with proven tribological performance are carbon nanotubes and onions, boron nitride nanoparticles, and inorganic fullerene-like (IF) metal disulfides (IF-MoS_2, IF-WS_2) (see Fig. 6.1). Without doubt, the latter exhibit the best friction-reducing and anti-wear properties ever observed. At the *Laboratoire de tribologie et de dynamique des systèmes* (LTDS) in the *Ecole Centrale de Lyon*, these nanoparticles have been the subject of detailed investigation for more

[1] Tribology is the study of friction, wear, and lubrication.

F. Dassenoy (✉)
Laboratoire de tribologie et dynamique des systèmes, École Centrale de Lyon, Bât H 110, 36 Avenue Guy de Collongue, 69134 Écully, France
e-mail: fabrice.dassenoy@ec-lyon.fr

J.-M. Lourtioz et al. (eds.), *Nanosciences and Nanotechnology*,
DOI 10.1007/978-3-319-19360-1_6

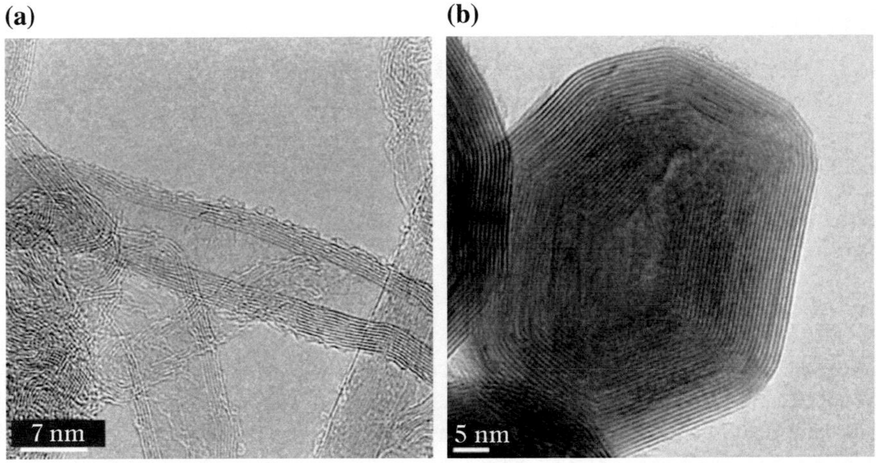

Fig. 6.1 High-resolution transmission electron microscope images of **a** a multi-walled carbon nanotube and **b** a fullerene-like molybdenum disulfide (IF-MoS_2) [1]

than ten years now, and during this time, many key issues have been tackled, such as the conditions leading to these properties, the lubrication mechanisms coming into play, and the influence of parameters such as size, structure, and morphology of the nanoparticles on their tribological properties and the lubrication mechanisms. In order to answer such questions, state-of-the-art characterisation techniques are required, often *in situ*, and sometimes extremely complex to set up. Some of these can even visualise the behaviour of a nanoparticle in real time while it is undergoing tribological testing. These years of research have given us a good understanding of the way these nanoparticles behave, and we can now identify the key parameters to be adjusted when optimising their lubrication properties.

6.1 Why Nano?

The key feature of nanoparticles is that they have 'closed' structures which considerably minimise the number of dangling bonds.[2] For this reason, they are much less chemically reactive than conventional molecular additives such as organic or organometallic sulfur-bearing or phosphorus-bearing compounds. Moreover, thanks to their small size, of the order of a few tens of nanometers, and their morphology, viz., spherical or cylindrical, these nanoparticles can easily penetrate the contact zone between two objects undergoing friction, and at the same time they display superior tribological properties to conventional additives. Considering the morphology

[2]Chemically active dangling bonds are much less common in closed systems like nanoparticles than in open systems where the molecules are distributed over the whole contact zone.

of these nanoparticles, it was conjectured that such structures could either roll or slide in the contact zone and thereby contribute to reducing the friction between the opposing surfaces. On paper, such nanoparticles look extremely promising because they meet all the requirements of today's lubricants (high performance, ecological, long-lasting). Furthermore, the first toxicological studies have shown the absence of any risk for their use in tribological applications, since the nanoparticles remain confined within the lubricant.

6.2 Tribological Properties of Nanoparticles

The inorganic fullerene-like metal disulfides IF-MeS_2, with Me = Mo or W, exhibit by far the best lubrication properties. Figure 6.2 shows the dependence of the coefficient of friction on the number of friction cycles for a lubricant (or dispersion) made from nanoparticles (single-walled carbon nanotubes, carbon onions, IF-MoS_2, and IF-WS_2) with a mass fraction of 1 % in a typical oil (poly alpha olefin PAO6). These dispersions were tested under particularly severe conditions, at the limit of the lubrication regime, where the lubrication additives take over from the oil, which is no longer able to fulfill its role of lubricant. The tests were carried out using an oscillating tribometer at room temperature and a contact pressure of 1 GPa. The figure also shows friction curves obtained for a dispersion containing a 1 % mass fraction of micrometric graphite and a dispersion containing a 1 % mass fraction of molybdenum dithiocarbamate (MoDTC), the friction-reducing additive currently

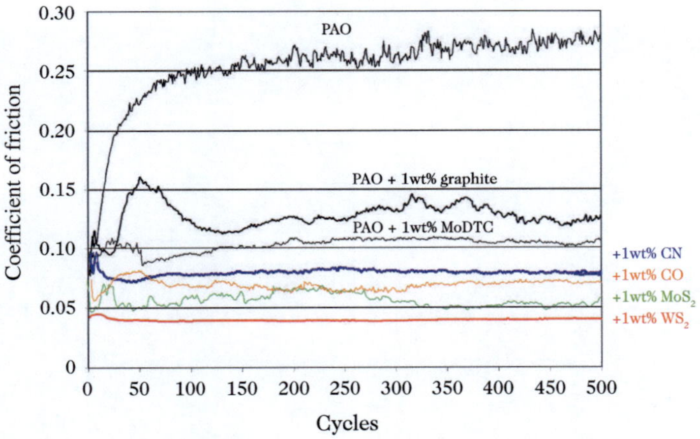

Fig. 6.2 Friction curves obtained for different dispersions of nanoparticles in a poly alpha olefin (PAO) oil. The tests were carried out at the limit of the lubrication regime with steel friction surfaces, a contact pressure of 0.83 GPa, a slip speed of 2.5 mm/s, and at a temperature of 25 °C [1]. CN: carbon nanotube, CO: carbon onion, IF-MoS_2 and IF-WS_2: inorganic fullerene-like metal disulfides, wt %: mass fraction

used in engine oils. Whichever nanoparticle system was studied, they were all found to considerably reduce the coefficient of friction in comparison to the oil alone or the oil containing current additives. However, it is with IF-MeS_2 nanoparticles that the reduction in friction is the most spectacular. The decrease can be as much as 70 % compared with the oil alone, confirming the superiority of these systems over all the others. Furthermore, it should be noted that the coefficients of friction are low right from the first test cycles, and this even at room temperature. This gives the IF-MeS_2 systems a further advantage over conventional molecular additives, which require an induction period and high temperatures to get them working properly. Nanoparticles, and in particular the IF-MeS_2 systems, might thus provide an interesting response to problems associated with cold starts, and notably the considerable wear generated during this operating phase.

6.3 Mechanism of Lubrication by IF-MeS_2 Nanoparticles

Following the friction tests, the main mechanism involved in the action of these nanoparticles was determined by characterisations carried out on opposing surfaces and on wear particles. It was shown that fullerenes entering the contact zone gradually

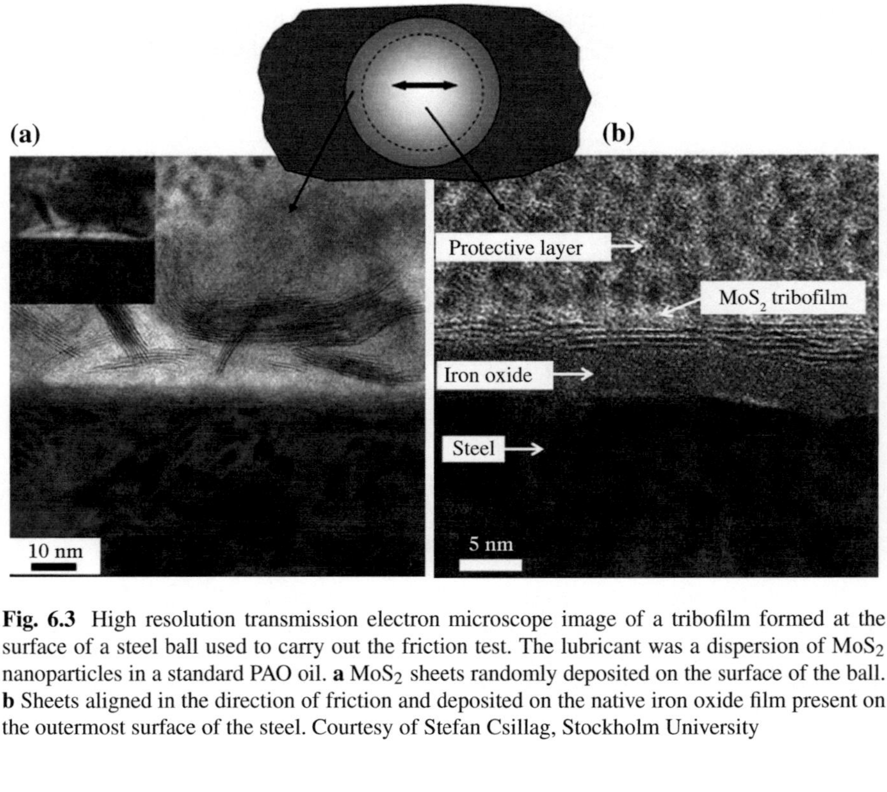

Fig. 6.3 High resolution transmission electron microscope image of a tribofilm formed at the surface of a steel ball used to carry out the friction test. The lubricant was a dispersion of MoS_2 nanoparticles in a standard PAO oil. **a** MoS_2 sheets randomly deposited on the surface of the ball. **b** Sheets aligned in the direction of friction and deposited on the native iron oxide film present on the outermost surface of the steel. Courtesy of Stefan Csillag, Stockholm University

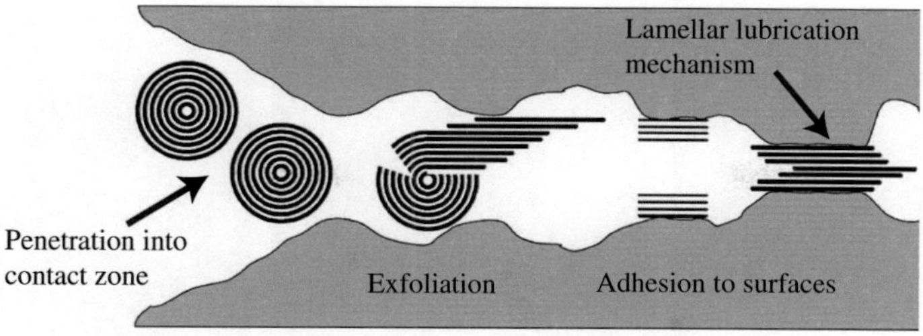

Fig. 6.4 Lubrication mechanism for IF-MeS_2 nanoparticles, Me = Mo or W

exfoliate under the action of mechanical stresses, giving rise to lubricating sheets of MeS_2 with a lamellar structure (see Fig. 6.3a). These sheets end up in the wear debris, mixed with intact particles, or are deposited on the friction surfaces (see Fig. 6.3b), leading to the formation of a very thin tribofilm, of thickness 5 nm, which adheres to the metal surfaces and protects them. It is therefore a lamellar lubrication mechanism that lubricates the system, as shown schematically in Fig. 6.4. This mechanism is extremely interesting in that the lubricating additive only acts in the contact zone and does not require any chemical interaction. This is an enormous advantage compared with conventional molecular additives, since these require an induction period lasting for a time that depends on the temperature and pressure. We thus understand why the fullerene-like metal disulfides are effective right from the first moments of frictional contact, and at room temperature.

Even though the main mode of action of fullerene-like structures operates through the exfoliation of the particles and the formation of a tribofilm made up of lamellar sheets on the friction surfaces, additional lubrication mechanisms such as rolling or slipping have often been put forward in the literature without ever having been observed. In fact, it is only recently that the behaviour of the nanoparticles in the contact zone has been visualised directly at the LTDS of the *Ecole Centrale de Lyon*, using nanocompression and shear tests of the nanoparticles carried out *in situ* in a high resolution transmission electron microscope equipped with a nanoindenter. The deformation under mechanical load was thus observed in real time. Different kinds of behaviour were observed depending on the nature of the particles, i.e., their structure, morphology, size, etc., and the test conditions. Critical values of the pressure were measured at which the particles, roll, slide, or exfoliate. Figure 6.5 shows a compression test carried out on a single particle of WS_2. The particle structure is disrupted under the effect of the pressure, thereby releasing its lubricating sheets (see Fig. 6.5f). Figure 6.6 shows that the same particles can, under certain load conditions associating pressure and slipping, roll in the contact zone and thus behave exactly like ball-bearings.

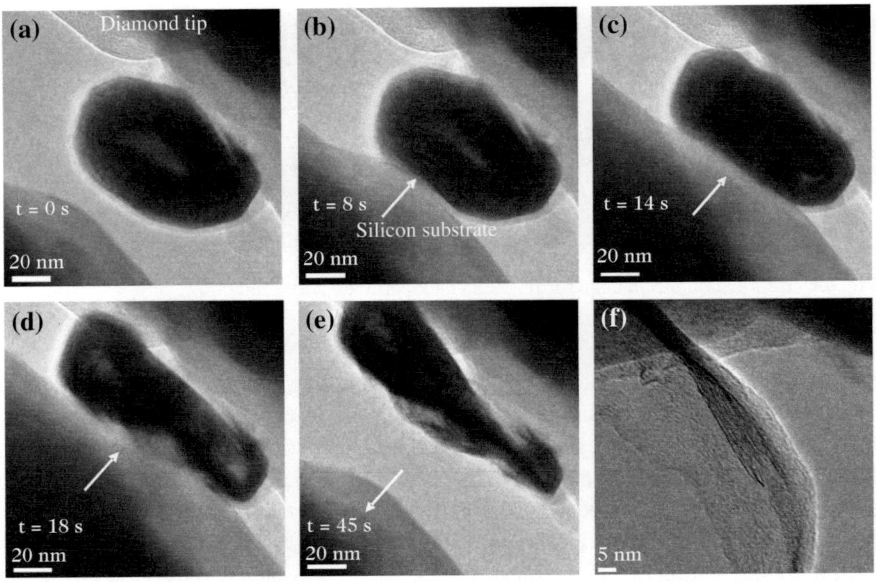

Fig. 6.5 Images obtained from a video recorded during a compression test carried out on a single WS_2 particle *in situ* in a high-resolution transmission electron microscope equipped with a nanoindenter [2]. **f** Lubricant sheets observed after the compression test. The final publication is [3]

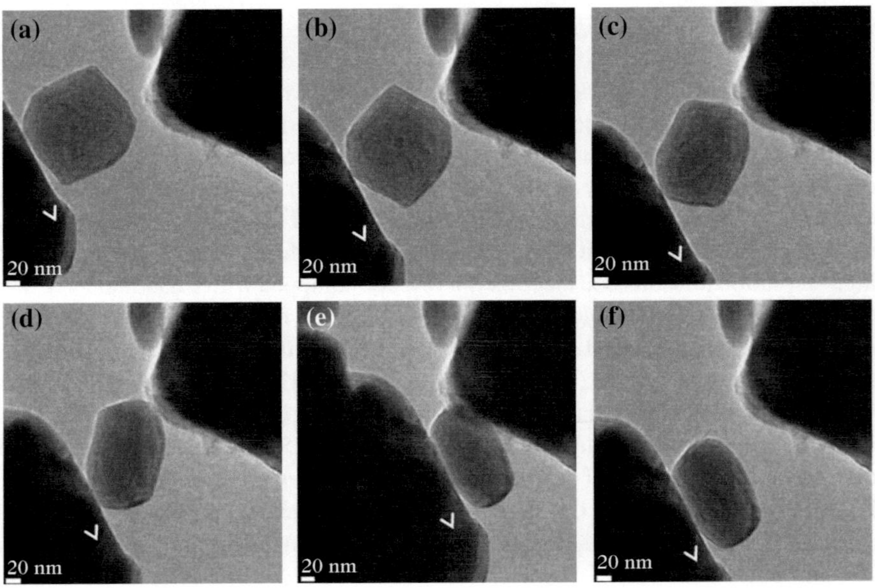

Fig. 6.6 Images obtained from a video recorded during a compression test carried out on a single MoS_2 particle *in situ* in a high-resolution transmission electron microscope equipped with a nanoindenter [2]. The final publication is [4]

6.4 Conclusion

Nanoparticles open up interesting prospects in the field of tribology and lubrication. They have considerable potential as lubrication additives thanks to anti-wear and friction-reducing performance never observed in other materials. The possibility of introducing these nanoparticles in fully formulated lubricants is currently under investigation. There remain many technological bottlenecks before we can proceed to large scale production while maintaining compatibility with other additives, but things are moving fast and there can be little doubt that these nanoparticles will soon be playing their role in our car engines. So there is much hope for these nanoparticles that some are already considering as the additives of the future!

References

1. L. Joly-Pottuz, Nanoparticules lubrifiantes à structure fermée. Ph.D. thesis, École Centrale de Lyon (September 2005)
2. I. Lahouij, Mécanismes de lubrification des nanoparticules à structure fullerène. Approche multi-échelle. Ph.D. thesis, École Centrale de Lyon (December 2012)
3. I. Lahouij, E.W. Bucholz, B. Vacher, S.B. Sinnott, J.M. Martin, F. Dassenoy, Lubrification mechanisms of hollow-core inorganic fullerene-like nanoparticles: coupling experimental and computational work. Nanotechnology **23**, 375701 (2012). http://iopscience.iop.org
4. I. Lahouij, F. Dassenoy, B. Vacher, J.-M. Martin, Real time TEM imaging of compression and shear of single fullerene-like MoS_2 nanoparticle. Tribol. Lett. **45**, 131–141 (2012). www.springerlink.com

Chapter 7
Applications II: Nanocrystallites in Cement: What Future After Two Thousand Years?

Abdi Feylessoufi

Abstract Cement nanocrystallites constitute the binding phase in concrete, a material that has been around for more than two thousand years and remains the most widely used of manufactured materials. Thanks to their size, their disordered nature, their reprecipitation after combination with water, and their rapid evolution when the concrete sets, these nanostructures have only been subjected to careful investigation quite recently. Nanotechnology is now able to control their formation and exploit their evolution (shrinking and swelling) during the setting phase to produce a concrete with a strength and ductility close to those of steels. Could this be the beginning of a new industrial revolution? Here we discuss the future of these nanocrystallites in new applications, considering not only the economic and industrial aspects, but also the issues of research and training.

7.1 A Material Full of Surprises

Cementitious materials, known mainly in the form of concrete, are without doubt the oldest and the most widely used manufactured materials in the world, with more than a tonne produced per year and per person. They comprise a granular assembly bound together by a nanostructured matrix, also called the binding phase, which is the common feature of all such materials.

In fact, this nanostructured matrix is made up of hydrated chunks, each measuring a few tens of nanometers. These are hydrated calcium silicates, but often also hydrated calcium aluminosilicates or ferroaluminosilicates, in proportions that are non-stoichiometric with respect to the oxide content (see Fig. 7.1). These oxides (CaO, SiO_2, Al_2O_3, Fe_2O_3, etc.) are already in the cement grain in crystallised form. Upon contact with water, part of these oxides dissolves, then diffuses under the strong concentration gradients and precipitates out again in the form of nanometric

A. Feylessoufi (✉)
Centre de Recherche sur la Matière Cohésive, Université d'Evry, Rue du Père Jarlan, 91025 Evry Cedex, France
e-mail: afeyless@univ-evry.fr

J.-M. Lourtioz et al. (eds.), *Nanosciences and Nanotechnology*,
DOI 10.1007/978-3-319-19360-1_7

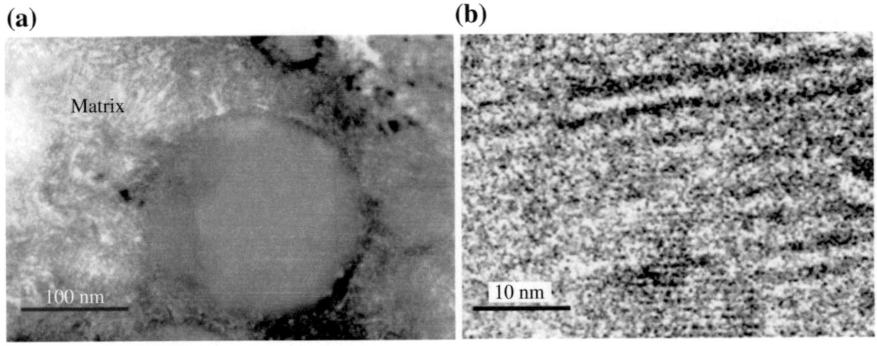

Fig. 7.1 Nanocrystallites in the cement matrix. Transmission electron microscope images (**a**) on the scale of the cement matrix and (**b**) on the scale of the nanocrystallites, showing their poorly ordered structure. From [1], © 2001 Springer New York LLC, reproduced with kind permission

hydrates, which are also called nanocrystallites owing to their size and the fact that they are poorly ordered, midway between the amorphous and crystalline phases [1].

As for any composite structure, it is this matrix that plays the determining role in the mechanical behaviour of the cementitious material. While the setting is under way, which may take a few hours or a few tens of hours, the cementitious material goes from a quasi-Newtonian liquid state to a viscoplastic state, then gradually acquires a visco-elasto-plastic behaviour, until finally it becomes a solid that is both very rigid, with Young's modulus as high as a few tens of GPa, and very strong, with a compressive yield stress between about ten and a few hundred MPa, depending on the formulation.

It is the exceptional reactivity of this nanostructured matrix, and at room temperature, hence very economical, which is really the strong point of these cementitious materials, because one can control this or that property either at the prefabrication stage or at the building site, depending on the use the material will be put to. Moreover, its very basic pH favours the use of passive or active reinforcement steel.

However, if the setting stage is not properly controlled, the cement matrix can quickly become of very poor quality, exhibiting various pathologies such as carbonatation, sulfation, or the alkali reaction, which may affect both aesthetics and durability. It is also through the matrix that radioactive atoms can penetrate concrete when it plays the role of barrier against radioactivity produced in nuclear power stations. And it is once again the matrix that is responsible in cases of shrinkage and consequent crack formation, with a possible loss of structural integrity.

7.2 The Roman's Were Here

It was the Romans who discovered cement serendipitously around the village of Pozzuoli near Naples in Italy. The place has given its name to the hydration of cement, which is still called the pozzolanic reaction. What they found was a powder

Fig. 7.2 Two thousand years went by between the construction of these concrete edifices. *Left* The Pont du Gard in France (photo credit Emanuele), assembled with a mortar using pozzolanic ash. *Right* The Footbridge of Peace in Seoul, built with ultra-high performance concrete (photo credit Philippe Ruault, Bouygues)

which, when it came into contact with water, turned into an artificial stone. In fact, it consisted of ash thrown out of the volcano. Naturally, the latter was able to boil up the essential ingredients of cement, which are limestone and clay.

One of the most remarkable edifices in which the Romans used concrete was the Pont du Gard, built near Nîmes in the south of France between 50 and 60 AD. A mortar made from sand and pozzolanic powder (an ancestor of concrete) was used as binder in this stone construction, some time before the famous dome of the Pantheon in Rome, proving that the Romans were the first to produce and use this material on an industrial scale (see Fig. 7.2). However, the recipe for making it was lost in the Middle Ages.

7.3 Late Rediscovery and a Burst of Development

Cement was reinvented at the beginning of the nineteenth century by the English bricklayer and businessman Joseph Aspdin who, in 1824, began to manufacture 'Portland' cement, referring to it as artificial stone and naming it by analogy with the Portland stone, an oolitic limestone quarried on the Isle of Portland in Dorset. Its composition differed only slightly from the product of the same name today. In 1855, the French engineer Louis Vicat built the bridge in the *Jardin des Plantes* in Grenoble, the world's first poured concrete construction. But it was only in 1887 that the French chemist Henri Le Châtelier devised the theory of cement hydraulicity, i.e., the body of knowledge that elucidates the nature of hydrates, until then unexplained. Cement was thereafter understood to be a salt with the exceptional property of being water stable!

To illustrate the rapid development of cementitious materials during the second half of the twentieth century, it is useful to consider the evolution of one of its properties, namely its compressive strength, i.e., the maximum stress a material can sustain without rupture under a compression test. Up until the 1930s, the standard compressive strength of concretes was 20 MPa. It reached 40 MPa in the 1960s, then 90 MPa in the 1970s with high-performance concretes (HPC), and finally 120 MPa with very high-performance concretes (VHPC) at the end of the 1980s.

Indeed, up until the 1930s, it was known that increasing the amount of mix water for a given mass of cement reduced the compressive strength of the resulting concrete, because the water associating with the cement in the hydrates gave way to a certain porosity in the material, and the strength of the concrete was thereby diminished in the same proportion. For the engineer, the amount of mix water should be just enough to obtain the rheology required for the application to hand. To reduce the amount of water needed, blood can also be used. This is a natural surfactant, i.e., a natural wetting agent, allowing better wetting of the cement grains for a given amount of water, and hence producing better hydration of the cement. So, off to the slaughterhouse!

In the 1960s and 1970s, sulfonated industrial surfactants were commonly used, including sulfonated naphthalene condensates and naphthalene-formaldehyde condensates or sulfonated melamine condensates and melamine-formaldehyde condensates. This made the concrete much stronger (HPC) and/or more fluid for a given amount of water (self-compacting or self-consolidating concretes). Fresh concrete could be pumped to a height of over 100 m, as for the construction of the *Arche de la Défense* in Paris, with a considerable gain in productivity. Note that the word 'performance' is used in preference to 'strength', because the issue here is not just optimisation of the strength. Other aspects of the performance are important, such as the rheological properties or durability, and these can be optimised through different but similar reformulations.

7.4 Ultra-High Performance Concretes

A considerable technological leap occurred during the 1990s and 2000s. Industrial levels of compressive strength reached values above 800 MPa with the ultra-high performance concretes (UHPC). At the beginning of the 1990s, university research centers began to advocate the use of large amounts of polymers (up to 20 %) to produce macro defect free (MDF) concretes, i.e., without macroscopic defects, using the simple idea of filling any residual porosity with hydroxypropyl methylcelluloses (HPMC) and endowing the composite with a certain ductility. However, despite their great strength, these concretes lost out with regard to durability because the polymer had to be water soluble in order to be incorporated in the mix, but would subsequently lose its properties under the effects of humidity. In addition, preparation of these materials required a process associating kneading and calendering which proved too complex and too costly, and hence was incompatible with industrial applications. This approach was quickly dropped from the agenda.

In the mid-1990s, thanks to the efforts of a visionary French industrialist who was scientific director at Bouygues, considerable resources were devoted to the development of new concretes. Several French research projects under the name *Saut technologique* or *Réactif*, and Brite-Euram European projects bringing together a network of industrial companies, such as Lafarge, Rhodia, etc., and about twenty research centers set out to develop the *bétons Pierre Richard* or *bétons de poudres réactives* (BPR), named after their main advocate and also known as reactive powder concretes (RPC), later commercialised under the name of Ductal. These concretes, which achieved strengths from 300 MPa to more than 800 MPa, were set to revolutionise the rules of architectural design. They were clearly beginning to compete with steels, having similar strengths, a specific mass three times lower, and a cost ten times lower! One might think that, with concretes ten times stronger, the mass of concrete used would be ten times less, but that would be forgetting that 80 % of the strength of a bridge is used only to support its own weight! Much lighter and more aesthetic architectures thus become possible, like the Sherbrooke Footbridge in Quebec and the Footbridge of Peace in Seoul (see Fig. 7.2), where the apron of the bridge is only 3 cm thick!

7.5 Toward Nanometric Sizes

Without reducing the size and fineness of the pore network, it would have been illusory to try to improve performance and strength by focusing on nanocrystallites. Indeed, fracture mechanics teaches us that strengths are inversely proportional to the square root of the typical size of a characteristic defect, and the defects here are the pores. There was thus no way of making further improvements while the pores remained at micrometric sizes. It was therefore essential to achieve nanometric pore sizes, without forgetting that, if the grains are too closely packed, the hydration reaction will lead precisely to a localisation of shrinkage, causing cracks to form.

The challenge was taken up on the basis of two in principle contradictory ideas. The first was optimal filling of intergrain gaps using a decreasing granulometry, down to nanometric size, and an Apollonian packing. This goal was achieved rather cheaply using colloidal silica with optimal sieving, from about ten nanometers to several hundred nanometers. The silica was available as a waste product from the silicon and steel industries, but today this precious waste product is no longer available in sufficient quantities, so it has to be made specially for the cement manufacturers!

The other idea was grain loosening, i.e., separating the grain sizes and ensuring that the shrinkage which occurs when the concrete sets is uniformly distributed throughout the material, without localisation. Indeed, any localisation would lead to harmful microcracks, quite the opposite of the desired effect. In addition, grain loosening facilitated the flow and spread of the fresh concrete into the required volumes by preventing the grains from jamming up! However, the use of nanoscale colloidal silica grains considerably increased the specific surface area and hence also the amount of water required. This problem was tackled using highly efficient new generation surfactants, viz., methacrylic polymers with long neutral chains of ethylene polyoxide. The problem there was that, at high concentrations, these polymers acted at the water–cement interface to perturb the hydrate formation mechanisms. A perfect balance had therefore to be found between the grain packing, grain loosening, and the proportion of polymers acting at the interfaces.

7.6 More Rigid Nanocrystallites

The introduction of colloidal silica into the formulation had some quite unexpected results. Forming by solid condensation and strictly spherical, they provided excellent lubrication during mixing and very good rheology in the fresh concrete. Furthermore, they diffused within the nanocrystallites, thereby reducing their Ca/Si ratio to values less than unity. In such proportions, their elastic moduli in the matrix were higher and close to those of the non-hydrated cement and silica grains remaining in the matrix, so that, during mechanical loading, there was a better distribution of stresses with less gradation from one phase to the other.

7.7 Progress in Nanocrystallite Thermodynamics

Another property of nanocrystallites with a low ratio of Ca to Si was their improvement in heat treatments at the prefabrication stage. Heat treatments of concretes had previously been poorly regarded because conventional hydrates evolve toward hydrates with inferior mechanical properties. However, the situation turned out to be quite the opposite for hydrates made from nanocrystallites with high silica content, because they transform into nanocrystallites with excellent mechanical quality, such as xonotlite or gyrolite. Further studies carried out with university research centers led to the specification of a heat treatment that perfectly controlled the thermodynamics involved in the formation of these phases [2], and this had industrial spinoffs for the heat treatment of prefabricated concretes.

7.8 Better Resistance to Wear and Tear

This is how concretes are obtained with nanoscopic pore networks. The typical pore size as found by high resolution porosimetry [3] for ultra-high performance concretes was only two or three nanometers (see Fig. 7.3). But apart from strength, there are other advantages in having such a fine nanometric pore structure. The first is durability, because pathogenic agents, or even radioactive atoms, are no longer able to enter such very small pores. This type of concrete has thus been used to store category B radioactive waste. The second advantage is resistance to freeze–thaw cycles. Resistance to freezing is generally the weak point of conventional concretes, because residual water in the capillary pores tends to freeze when temperatures are very cold and break the concrete from within. But inside a pore that is only two or three nanometers across, the water is bound to hydrates and does not begin to freeze until the temperature falls below $-20\,^{\circ}C$, a considerable advantage for concretes to be used in cold climates.

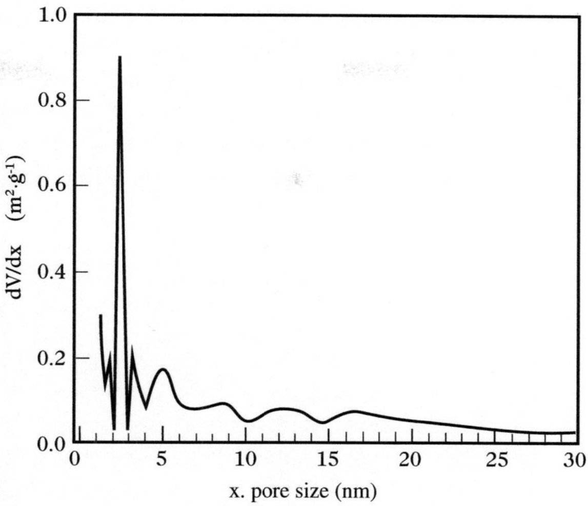

Fig. 7.3 Modern concretes are characterised by a predominantly nanoscopic porosity, which increases strength and makes it difficult for pathogenic agents to penetrate. Here the specific volume occupied per pore radius is analysed by high resolution nitrogen porosimetry. The predominant pore size lies between two and three nanometers. From [3], © 2001 Elsevier

7.9 The Beginning of the Twenty-First Century: A Clear Future

The development of experimental techniques and modelling at the end of the twentieth century and the beginning of the twenty-first has led to a better understanding of the genesis, aggregation, and structuring of the nanocrystallites during setting, as

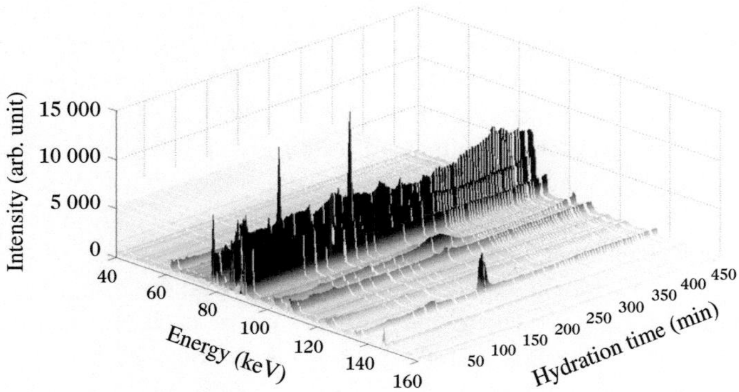

Fig. 7.4 Synchrotron radiation can be used to obtain almost instantaneous X-ray diffraction patterns as hydration proceeds [4]. The nanocrystallites, often poorly ordered, appearing and disappearing over periods of ten seconds or so, can be distinguished and analysed, something that was quite impossible using laboratory X rays about a million times less intense!

well as the gradual development of their textural and physical–chemical–mechanical properties (see Fig. 7.4). Whereas all the relevant mechanisms used to be encompassed as one under the general term ‘setting’, we can now identify the chemical and topological mechanisms occurring during this process [4–6]. We have thus entered into a phase where the aim is no longer to improve the product, but to develop a genuine ability to engineer these nanocrystallites, targeting specific properties.

Setting Concrete

Setting is an exothermic reaction whose progress can be measured in various ways. For a given sample of ultra-high performance concrete, two methods are considered here [6]: isothermal calorimetry and shear wave ultrasonic testing, both highly sensitive to the structuring of the cementitious matter.

Isothermal calorimetry tells us the dissipated power per unit mass of concrete (see Fig. 7.5a). As soon as the water has been introduced, heat power is emitted from the hydration of tricalcium aluminates, these being one of the components of the cement. This is followed by a dormant period which may last for several hours, whereupon the dissipated power climbs to a maximum. The concrete is then considered to have set. On the building site, it is at this moment that formwork is removed. However, hydration continues for several months. The integral of the power curve (see Fig. 7.5a) indicates the progress of the reaction. The degree of hydration corresponds to the heat emitted since the beginning of the setting process divided by the heat that will have been emitted when the whole sample has been hydrated. Note that, for modern ultra-high performance concretes, the cement is never fully hydrated. The degree of hydration never exceeds 20–25 %. The rest of the cement remains available and, in the case of nano- or microcracking it can combine with penetrated water or water provided by humidity to mend such cracks.

Shear (or transverse) wave ultrasonic testing also tells us how far setting has proceeded by revealing the ordering of the cementitious nanocrystallites (see Fig. 7.5b). In the early stages, there is no connection between the nanocrystallites, which means that the transmitted shear waves cannot cross the sample. The shear waves are thus reflected in their entirety and the modulus of the reflection coefficient is equal to one. As hydration proceeds, the matter percolates, leading to a decrease in this modulus, then orders itself into clusters of different sizes. This explains the dispersion of the reflection curves at different frequencies until they come back together again, indicating that the clusters have joined up into a single structure.

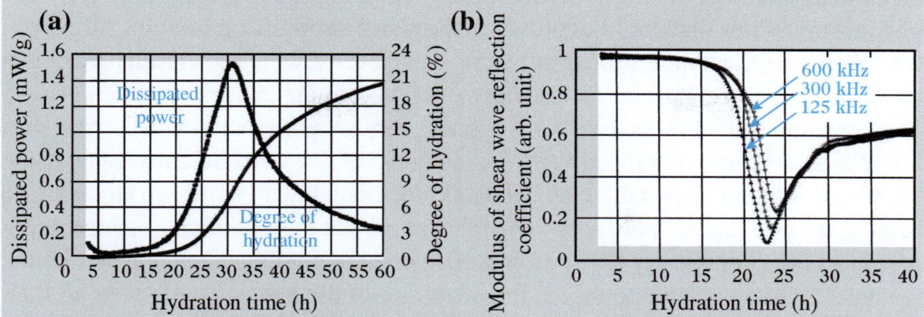

Fig. 7.5 a Dissipated power per unit mass of concrete as determined by isothermal calorimetry. **b** Shear wave reflection coefficient as determined by ultrasonic testing. Adapted from [6], © Elsevier

The most effective methods then used synchrotron X-ray radiation [4], e.g., at the European Synchrotron Radiation Facility in Grenoble, or very intense neutron beams, e.g., at the Laue-Langevin Institute in Grenoble, to identify the hydrates. Indeed, using conventional X-ray diffraction techniques in the laboratory, it was never possible to identify the hydrated calcium silicate nanocrystallites: they were too small (ten nanometers or so) and they were very poorly ordered. In addition, exposures had to be much too long, around ten hours, so it was impossible to monitor the development of these nanocrystallites during the setting process. With fluxes up to a million times more intense than laboratory X rays, the synchrotron source made it possible to detect the nanocrystallites as they formed and disappeared on time scales of around ten minutes (see Fig. 7.4).

Other spectroscopic techniques using nuclear magnetic resonance (NMR), silicon-29 NMR or aluminium-27 NMR, provided a better analysis of the inorganic polymerisation of the nanocrystallite networks, while proton NMR [5] was used to characterise the bonds between the nanocrystallites and water and the incorporation of water molecules into the nanocrystallites during hydration (see Fig. 7.6).

Imaging techniques such as ultra-high resolution transmission electron microscopy [1] have made it possible to measure lattice spacings within highly disordered crystallites. Atomic force microscopy was unable to reconstruct a clear image owing to the poor ordering of the nanocrystallites, but it proved useful for measuring surface and van der Waals forces on the level of the nanocrystallites.

The methods of high-resolution porosimetry, in near-equilibrium, using molecules with different shapes and properties, are crucial for measuring the size and distribution of nanopores, and even for producing an image of their shape. This in turn made it possible to model mechanisms of shrinkage and permeability with respect to pathogens.

However, it was the advent of high resolution ultrasound spectroscopy [6] that did most to improve our understanding of the mechanisms of percolation and aggregation in the evolving phases by providing precise information about the elastic moduli of the developing nanocrystallites (see Fig. 7.7). During the 2000s, this resulted in the first attempts to engineer the nanocrystallites. Formulations able to develop networks of nanocrystallites that could improve performance by blocking harmful shrinkage, the cause of microcrack formation, were patented. Other direct results from this technique were formulations facilitating extrusion forming of cement nanomaterials.

In parallel with the analysis of the basic nanoscopic reactions occurring during setting, and their modification by tiny amounts of polymer additives acting at the interfaces, numerical modelling techniques known as cellular automaton models have been developed and are now *potentially* able to predict the evolution of nanocrystallites in all their complexity [7–9]. However, in order to produce a genuine prediction, we must actually be able to model the evolution of the nanocrystallites in all their complexity, and we are a long way from being able to do that in practice, because not all the reactions occurring during setting are in fact known for the moment and because the computations are still too heavy.

7.10 The Scientific Challenge

The challenge here is to improve the nanocrystallites for specific needs, develop ways to engineer them, and 'program' their evolution during hydration in such a

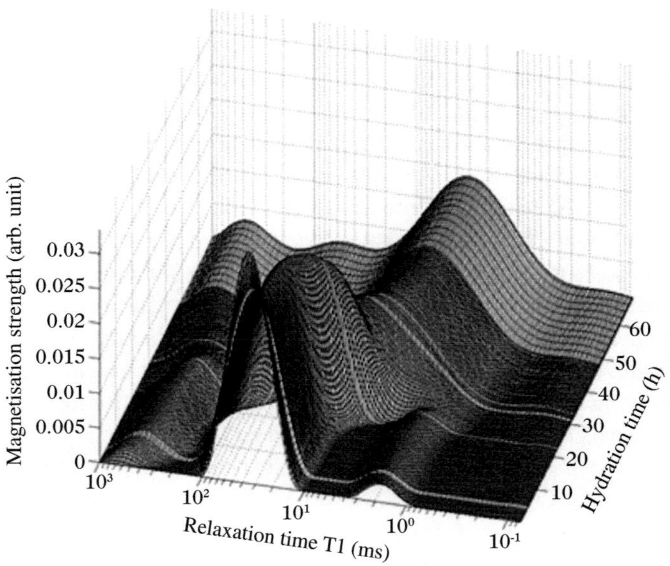

Fig. 7.6 After excitation in a magnetic field, proton spin moments undergo exponential relaxation until they recover the equilibrium magnetisation. This relaxation, known as spin-lattice relaxation, has a characteristic time denoted by T1, which depends on the level of bonding of the protons in the water molecule. The dependence of the relaxation time T1 on the extent of hydration informs us, on the scale of a few angstrom units, about the bonds between the water molecules and the nanocrystallites, until the former are entirely incorporated, thus forming cement hydrates [5]

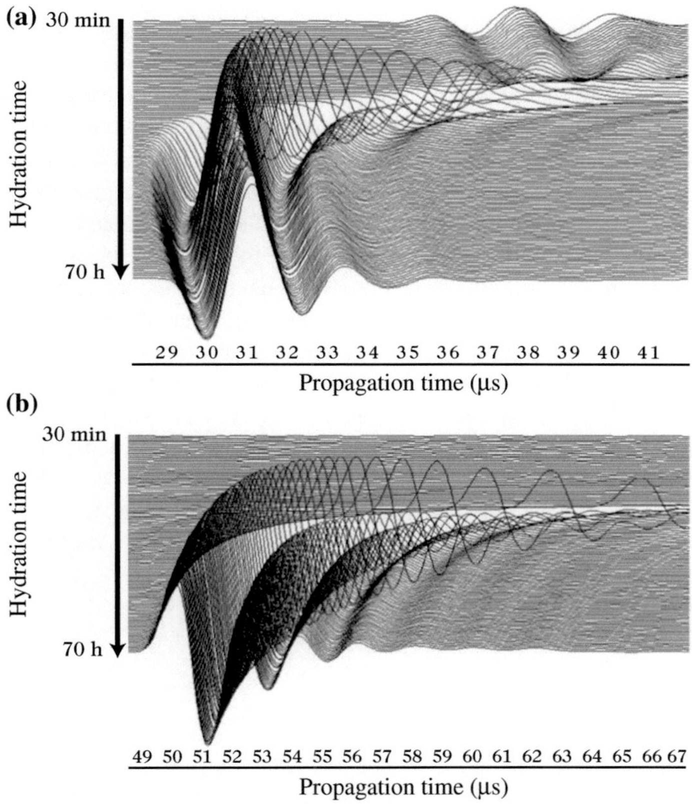

Fig. 7.7 Signals corresponding to compression waves (longitudinal component) (**a**) and shear waves (transverse component) (**b**) after different hydration times. The speed of propagation obtained by measuring the propagation time of acoustic compression and shear waves in a sample at different frequencies can be used to deduce the viscoelastic moduli of the developing crystallites. Adapted from [6], © Elsevier

way as to control their properties at various stages. However, our knowledge in this respect is extremely limited, and incomparably inferior to what is known in solid state physics, metallurgy, or biology for the development of nanomaterials. The local thermodynamic conditions of deposition and accumulation of nanocrystallites remains to be determined, and so too are the causes of local fluctuations in the composition and adhesive strength observed by electron microscopy and near-field microscopy.

It also remains to understand the morphological, structural, and crystallochemical development of the nanocrystallites if we are to be able to modify the surface properties and cohesive properties of the assemblies. This will require state-of-the-art knowledge of surface chemistry and crystal growth, because the growth of calcium silicate nanocrystallites must be optimised to obtain the most cohesive interface possible with the limestone or the slag.

Nanocrystallite assemblies are also controlled on the mesoscale. Intermediate levels of organisation over lengths greater than the micrometer would appear to have been observed between the nanocrystallite sheet and the continuous binding matrix [10–12]. They look like stacks of sheets measuring a few tens of nanometers laterally and a few nanometers in thickness. The question is whether an interstitial solution could modify their size, shape, and assembly. In plaster, the growth of hydrates (gypsum) is controlled by adding organic molecules which selectively modify the surface energies of the different crystal faces of the gypsum when adsorbed. One can therefore consider applying this approach to cement nanocrystallites, especially now that it can be supported by molecular modelling calculations.

Another possibility for modifying the mechanical properties of the assemblies would be to change the topology of the cement nanocrystallites, which are generally two-dimensional, by *in situ* synthesis of mixed hydrates of silicon and aluminium, which are three-dimensional. Industrial development would be achieved by the synthesis and control of protozeolitic structures.

We must also relate the hydromechanical properties to the morphology of the cementitious material on all the scales just mentioned. There is a long way to go, but it is nevertheless worth consideration.

7.11 The Industrial Challenge

The main industrial challenge is to convince the decision-makers that these new nanomaterials are economically advantageous. Recall that ultra-high performance concretes have been advocated for their mechanical strength and their ability to meet the technical challenges in an economically viable way. Their exceptional durability, their resistance in agressive environments, the fact that no maintenance is necessary, and the ease with which they can be applied are all arguments justifying their use. Their early age rheology has seen them pumped to a world record height of 606 m, thereby avoiding the need to transport prefabricated structures to this altogether exceptional height.

The construction of the Burdj Khalifa in Dubai, the tallest building in the world, culminating at 830 m, is a recent demonstration of the economic viability of these 'nanoconcretes'. A tower more than a kilometer high is under study. Energy could be generated by exploiting the air flow produced by the temperature difference between the top and bottom! For such constructions, absence of maintenance is a decisive economic argument when the comparison is made with metal structures. Nanoconcretes are the only materials able to resist both freeze–thaw cycles and pathogenic carbonatation and sulfation agents without the need for maintenance. When titanium dioxide (TiO_2) is incorporated in very small amounts, they even become self-cleaning under the Sun's UV radiation, displaying a smooth and aesthetic surface, and above all they are highly economical for the customer.

A further advantage of nanoconcretes is that they are very light for a given strength, in particular when compared with steel! It is a common mistake to assume that concrete is heavy. Its density of 2.0–2.6 t/m^3 is lower than the density of aluminium. There has even been a suggestion to use nanoconcretes to armour military helicopters! This low density can procure a considerable saving in material, and the cost of the material represents around a quarter of the cost of a construction, while profit margins are only a few percent in major building projects. The contractor or the owner can thus double their profit by using nanoconcrete.

Hence the greatest challenge lies outside the field of construction, industrialists wishing to set up in markets where they are currently absent. The aim is to exploit all the properties offered by the engineering of 'customised' nanocrystallites and to substitute nanoconcretes for metal materials or high-resistance polymers by commercialising premixes, i.e., optimised dry mixtures comprising cement, polymer additives modifying the setting mechanisms, sand, and ground quartz. This aim seeks to exploit the visco-elasto-plastic properties during setting (at room temperature) and the design of forming machines adapted to suit the nanoconcretes. The profits expected in these markets are truly enormous because the forming processes for the metal alloys currently used must be carried out at high temperatures, whence they are energy costly, and they also require considerable investment in the production unit.

Finally, nanoconcretes have some environmental advantages, an important factor in the eyes of the authorities. In addition to their low energy consumption during manufacture (1.4 GJ/t for concrete, compared with 30 GJ/t for steels) and their low greenhouse gas contributions (CO_2, SO_2, and NOx: on average 147, 0.2, and 0.6 kg/t for concrete, compared with more than 3 000, 3, and 5 kg/t for steels), they are easy to recycle because hydrated cements can simply be loaded into the cement oven to dehydrate them and recover the original cement.

7.12 The Many Remaining Obstacles

In the light of what has been said, an enormous potential has built up for the future development of cementitious nanomaterials. However, the latter have generated their own difficulties. These result from a lack of academic research, the fact that R&D has focused too much on specific applications, and the fact that industrial organisation involves large building companies that often play the role of project manager.

Regarding academic research on cementitious nanomaterials around the world, few research scientists work in this area compared with the number who work on nanomaterials in general. Referring to the publications in the two main journals for this area, viz., *Cement and Concrete Research* and *Cement and Concrete Composites*, this community represents fewer than 300 researchers. Moreover, this theme has traditionally derived from mechanics and civil engineering, whereas current progress in analysis and modelling can rather be attributed to nanochemists and nanophysicists. This leads to certain inconsistencies. For example, mechanicists investigating nano- or micro-heterogeneous materials focus mainly on long term constitutive laws, ignoring the nano- and microstructural complexity. Research chemists who wish to create nanocrystallites with novel mechanical properties generally have little knowledge of the problems of localisation or homogenisation of behaviour. The profession of physico-chemico-mechanicist combining the studies of nanomechanics and physicochemistry and working on cementitious nanomaterials remains to be created!

Finally, one hears researchers speak about 'complex media' or 'artificial geomaterials' as if research on cements was somehow degrading compared with more fashionable fields. Although France leads the world in cement manufacture and is one of the leaders in the construction industry, there are no international level research centers for cementitious materials in this country that could bring together skills as varied as nanocrystallite chemistry, nanomechanics, and molecular dynamics or cellular automaton modelling in the same geographical site, as one finds in the United States (MIT, Princeton, the University of Illinois), the United Kingdom (the University of Aberdeen, Imperial College), or indeed in Switzerland (*Ecole polytechnique fédérale de Lausanne*). Naturally, there are few training courses at MSc level and we find ourselves in a situation where it is the large companies that train young engineers or scientists by sending them abroad to follow courses, something the smaller companies are of course unable to do!

With regard to industrial R&D, current effort concerns mainly expert missions and technical consulting for clients using ultra-high performance premixes in specific applications. Researchers suffer from the dual nature of their work, on the one hand developing new cementitious materials and on the other advising clients about the use of new cement premixes. Unfortunately, as often happens, the most urgent is also the priority. This is a shame because genuine progress can be made when industry succeeds in isolating researchers for longer term work (see Sect. 7.4).

Most large projects are mounted with design and construction regulations, and often entirely entrusted to contractors who finance them themselves, then rent or lease them to the public or private sectors. In France, major concession projects like the Millau viaduct, under a 78 year concession, or the *Stade de France*, under a concession of 30 years, are witness to this. The result is a rather particular kind of organisation where the major construction companies act as project managers for large projects, while the actual work is entrusted to small and medium-sized enterprises (SME) under the technical control of the contractor. This was what made the use of the highly fluid self-compacting concretes so successful, because they do not require any special skills, while they lead to a considerable gain in productivity. But this is no longer the case when more specialised materials are to be used. The solution adopted by the major contractors is then to ask the design office to set up very strict quality control regulations. The idea is to sort out the slightest details of formulation and calculation in the design office, then set up a verification laboratory on site. This solution is problematic when there are deadlines to respect, and many SMEs have gone out of business due to difficulties supplying the product laid down in the specifications, whereupon they were simply not paid!

In parallel, during the 1990s, many competent SMEs set themself up to supply ready-to-use concretes. This activity, which generates a lot of jobs, consists in purchasing the cement and suitably dosed additives controlling the hydration reactions for specific applications, and selling these so-called ready-to-use concretes directly to nearby construction sites, thereby ensuring highly competitive development on a local or regional scale. However, such companies do not have the means or the infrastructure to file patents and maintain them. The development of complex formulations and the stringent requirements of European standards (Eurocodes), largely dominated by the large industrial groups, have put considerable pressure on these smaller entreprises. They must choose between revealing their methods or being swallowed up by these larger groups and disappearing. This effective monopoly, which is still tightening its grip in the name of quality requirements, has led to a loss of knowhow and jobs which is unfortunately backed by European regulations and the powers that be. This is a collateral consequence of the evolving technicality of cementitious materials, in which innovative micro-enterprises were able to establish a niche in the market by developing state-of-the-art products.

Finally, it should be noted that, in the field of cementitious materials, the small number of academic research institutes and the particular goals of industrial R&D centers have so far been an obstacle to the development of new start-up companies, something which is true for nanomaterials in general and is certainly a handicap for development in this sector. The problems discussed above must be overcome if there is to be a genuine development of cementitious nanomaterials in France. And this could be a source of considerable economic stimulus given the related market volumes.

References

1. L. Gatty, S. Bonnamy, A. Feylessoufi, C. Clinard, P. Richard, H. Van Damme, A transmission electron microscopy study of interfaces and matrix homogeneity in ultra-high performance cement based materials. J. Mater. Sci. **36**, 4013–4026 (2001)
2. A. Feylessoufi, M. Crespin, P. Dion, F. Bergaya, H. Van Damme, P. Richard, Controlled rate thermal treatment of reactive powder concretes. Adv. Cem. Based Mater. **6**, 21–27 (1997)
3. A. Feylessoufi, F. Villieras, J.-L. Michot, P. De Donato, J.-M. Cases, P. Richard, Water environment and nanostructural network in a reactive powder concrete. Cem. Concr. Compos. **18**, 23–29 (1996)
4. R. Soth, A. Feylessoufi, L. Cassar, Early age hydration of cement pastes by energy-dispersive synchrotron X-ray diffraction and neutron diffraction, in *Second Advanced Cementitious Materials Symposium (ACM'06)*, Evry, 4–5 September 2006
5. R. Soth, P. Faure, S. Rodts, A. Feylessoufi, Nuclear magnetic resonance and ultrasonic spectroscopy investigation for understanding setting mechanisms of a modified cement paste, in *Second Advanced Cementitious Materials Symposium (ACM'06)*, Evry, 4–5 September 2006
6. A. Feylessoufi, F. Cohen-Tenoudji, V. Morin, P. Richard, Early ages shrinkage mechanisms of ultra-high performance cement based materials. Cem. Concr. Res. **31**, 1573–1579 (2001)
7. J.-W. Bullard, A determination of hydration mechanisms for tricalcium silicate using a kinetic cellular automaton model. J. Am. Ceram. Soc. **91**, 2088–2097 (2008)
8. S. Bishnoi, K.L. Scrivener, A new platform for modelling the hydration of cements. Cem. Concr. Res. **39**, 266–274 (2009)
9. J.J. Thomas, J.J. Biernacki, J.W. Bullard, S. Bishnoi, J.S. Dolado, G.W. Scherer, A. Luttge, Modeling and simulation of cement hydration kinetics and microstructure development. Cem. Concr. Res. **41**, 1257–1278 (2011)

10. G.W. Scherer, J. Zhang, J.A. Quintanilla, S. Torquato, Hydration and percolation at the setting point. Cem. Concr. Res. **42**, 665–672 (2012)
11. G.W. Scherer, J. Zhang, J.J. Thomas, Nucleation and growth models for hydration of cement. Cem. Concr. Res. **42**, 982–993 (2012)
12. A.C.J. de Korte, H.J.H. Brouwers, Random packing of digitized particles. Powder Technol. **233**, 319–324 (2013)

Chapter 8
Applications III. Nano-TiO_2

François Tardif

Abstract Titanium dioxide in the form of perfectly crystallised nanometric particles (<100 nm) is used for its semiconductor properties in anti-UV and self-cleaning films, but also for air purification and in certain photovoltaic cells. More than three million tonnes of white titanium dioxide pigments are produced every year, comprising particles with sizes typically in the range 250–350 nm but some also of nanometric size, for applications in paints, paper, polymers, and inks. At the present time, there is no formal proof of toxic effects being induced by titanium dioxide nanoparticles, either for humans or for ecosystems, and the many studies carried out on sun creams show that cutaneous exposure does not allow transfer to the systemic route when these products are used on healthy skin. As a precautionary measure, the US NIOSH nevertheless recommend limiting concentrations in the air in case there should be carcinogenic effects.

Since the beginning of the thirteenth century, titanium dioxide (TiO_2) has been widely used in industry in two crystal forms, namely rutile and anatase, in particular for their property as white pigments.[1] This is due to their high refractive indices of 2.75 and 2.54, respectively. Indeed, these high optical indices mean that there is significant scattering of incident light, whence about 96 % of light in the visible part of the spectrum, with wavelengths from 400 to 700 nm, gets reflected, with no absorption zone.

The current market for titanium dioxide concerns above all particles with sizes typically in the range 250–350 nm. French production is around 250,000 tonnes per annum, while world production is more like three to four mega tonnes per annum depending on the source.

Out of 1 015 nanoproducts identified across the board, the Woodrow Wilson Institute has listed almost 10 % that contain nanoscopic TiO_2! In the following, we shall review a broad cross-section of these applications, focusing more particularly on

[1]According to the *Société chimique de France*, many forged paintings were exposed as such by the presence of TiO_2 pigments, since this indicates that the painting was done after 1920.

F. Tardif (✉)
CEA Liten, 17 Rue des Martyrs,
38054 Grenoble, France
e-mail: francois.tardif@cea.fr

J.-M. Lourtioz et al. (eds.), *Nanosciences and Nanotechnology*,
DOI 10.1007/978-3-319-19360-1_8

those containing the smallest particles (measuring a few tens of nanometers), which are still only produced in relatively small amounts (in France, only a few thousand tonnes per annum), but which offer promising prospects while nevertheless raising some doubts over their potential toxicity.

8.1 Applications of 250–350 nm TiO_2 Particles: From Paints to Foods

Titanium dioxide is mainly obtained from the mineral ilmenite, which has the overall chemical formula $FeTiO_3$ and comprises a mixture of titanium oxide (30–70 %) and iron oxide. These are separated chemically by two industrially used processes: a sulfuric acid process for ilmenites with a low TiO_2 content and a chlorine process when the TiO_2 content is higher. In the sulfuric acid process, the $FeTiO_3$ is first dissolved, then calcined at 800 °C. In the hydrochloric acid process, the titanium is first purified by conversion to titanium tetrachloride, then reoxidised by exposure to oxygen at more than 1000 °C. There are also mines of rutile titanium dioxide (bulk TiO_2 with only traces of iron), but this ore is much scarcer than ilmenite.

Used as pigments, TiO_2 particles with sizes in the range 250–350 nm serve above all as the white basis in paints (60 %), polymers (25 %), and paper (13 %). There are also applications in inks, floor coverings, glasses, and others.

Thanks to their chemical inertness, white TiO_2 pigments can also be used in certain cosmetic products, soaps, drugs, and toothpastes. They also turn up under the code E171 in foodstuffs, and in particular in industrial cheeses, cakes, and other sugary foods such as brightly coloured sweets where a thin layer of white TiO_2 on dark coloured chocolate can reduce the amount of coloured coating needed to conceal the chocolate.

In a quite different register, the refractory nature of TiO_2, which has a melting temperature of 1850 °C, makes it particularly interesting in ceramics for electronic applications and as support for catalysts. Its biocompatibility also makes it interesting for the manufacture of prostheses.

It should be noted that, even for an average size distribution of TiO_2 pigments centered around 250–350 nm, there is sometimes a certain fraction of particles smaller than 100 nm, which therefore come under the heading of nanoparticle.

8.2 Titanium Dioxide Nanoparticles (<100 nm): From Anti-UV Films to Elimination of Pollutants

Industrially, titanium dioxide nanoparticles are mainly produced by high temperature processes (>1000 °C), particularly by pyrolysis of aerosols. The idea is to make a fine spray of a liquid (or gaseous) titanium precursor and then oxidise it to titanium dioxide. But there are other ways to make TiO_2 nanoparticles, e.g., by sol–gel syn-

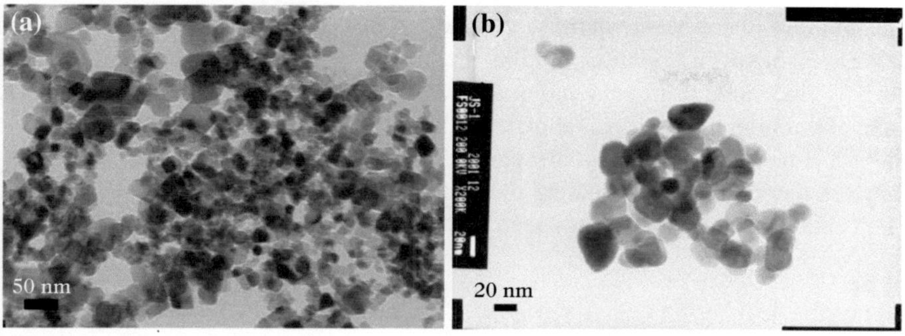

Fig. 8.1 Commercial TiO_2 nanoparticles of diameter 20 nm produced by aerosol pyrolysis and observed by transmission electron microscopy. Scales are 50 nm (**a**) and 20 nm (**b**). Image CEA

thesis (see Chap. 5) or by hydrothermal reactions, possibly under supercritical water conditions[2] or using microwaves to increase the crystallinity of the nanoparticles.

In the form of perfectly crystallised nanometric particles (<35 nm), agglomerated to various degrees (see Fig. 8.1), nano-TiO_2 is used for its semiconducting properties. Anatase, which is optically the most active form in terms of cross-section, has band gap energy 3.23 eV, implying absorption at wavelengths shorter than 384 nm. The band gap energy of rutile is 3.02 eV, with absorption at wavelengths shorter than 411 nm.

The three main applications are in anti-UV films, photocatalysis and the elimination of organic pollutants (air quality improvement, self-cleaning surfaces), and certain kinds of photovoltaic cell. Photons with shorter wavelengths than those corresponding to the band gap energy are absorbed and converted into electron–hole pairs. Depending on the application, one can then favour recombination of these pairs to produce heat, or the production of oxidising reactive species, or again the creation of a current in a photovoltaic cell.

8.2.1 Anti-UV Films

For many anti-UV applications, including in particular sun creams and the protection of materials such as cloths, polymers, and coatings, the ideal filter must have maximal absorption in the UV and minimal absorption in the visible part of the spectrum, so as not to affect the natural colour of the skin, substrate, or other underlying material. With its gap at 384 nm, anatase used in nanometric form limits the scattering of light, which would lead to a white colouring[3] (the geisha effect for anti-UV

[2]The species in a supercritical fluid have similar diffusion properties to those in a gas, but with the density of a liquid, which leads to very fast reaction kinetics.

[3]The scattering of light by small particles leading to this white colour is described by Mie's theory. It falls off rapidly with the size of the particle until it is practically imperceptible at around one

creams) and is thus particularly well suited to this application. Note that, in anti-UV filters, nano-TiO_2 is often used in conjunction with other absorbing substances such as zinc oxide or organic molecules. An organic or inorganic shell, e.g., made from aluminium hydroxide, is sometimes used to coat the TiO_2 crystallites to make them chemical compatible with the generally hydrophobic organic matrix into which they are incorporated. By coating the nanoparticles, one can also avoid unwanted photocatalytic effects that would tend to destroy the matrix itself (see below).

8.2.2 Application to Photocatalysis

The creation of electron–hole pairs following the absorption of ultraviolet (UV) photons leads to the formation of free radicals at the surface of the TiO_2 and these will tend to oxidise organic compounds in the surrounding medium. This chemical process requires the presence of water. Nano-TiO_2 films illuminated by UV radiation are thus used to eliminate odours, chemical contaminants, pesticides, and micro-organisms in liquid and gaseous media. Self-cleaning windows are made by depositing a nanometric film of TiO_2 on the glass surface.

Concretes, paints, and self-cleaning coatings containing TiO_2 nanoparticles are also available commercially, but the difficulty is sometimes to avoid the long-term destruction of the substrate itself. Ultraviolet radiation in natural light is converted by the TiO_2 into oxidising radicals that can break up the long organic molecules in the dirt. The resulting molecules and dirt are then washed away by rain (see Fig. 8.2). Note that work is currently being done to increase the capacity of nano-TiO_2 to use the part of the solar radiation beyond UV, i.e., in the visible, by doping it with nitrogen, for example, to increase its efficiency in natural environments.

8.2.3 Application to Photovoltaics

Nanoparticles of TiO_2 are associated with a dye able to pick up a large part of the Sun's radiation and used to make a certain kind of photovoltaic cell known as the Grätzel cell,[4] for the production of electricity. Today, the conversion efficiencies of such cells, inspired by the natural process of photosynthesis, are close to 11 %. Like chlorophyll in plants, when it is excited by solar radiation, the dye in the Grätzel cell produces an electron in a higher energy state than the ground state. This electron can then be transferred into the conduction band of adjacent TiO_2 nanoparticles and thereupon directed to the transparent electrode of the cell and the external circuit for

(Footnote 3 continued)
tenth of the wavelength. Below this threshold, the particle size becomes irrelevant, and scattering is then governed by Rayleigh's law, a limiting case of the Mie theory.

[4]Named after its inventor, a professor at the *Ecole polytechnique fédérale de Lausanne*.

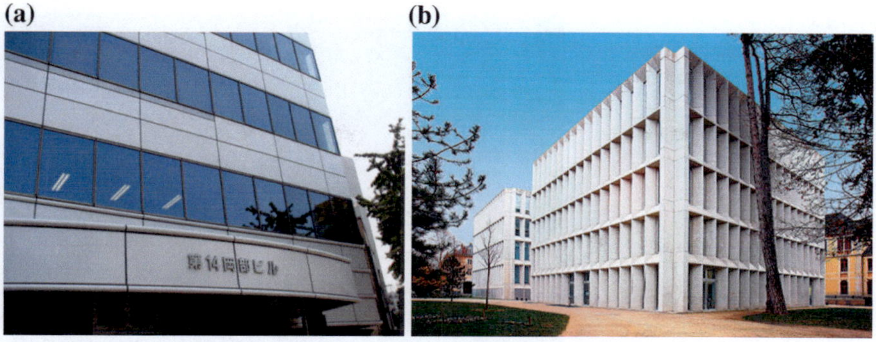

Fig. 8.2 **a** Self-cleaning window coated with a film of photocatalytic titanium dioxide. From [1] ©Elsevier 2010. **b** Building made from self-cleaning concrete at the *Cité des arts et de la musique* in Chambéry, France. Photo gerard-cottet.com

Fig. 8.3 Working principle of the Grätzel photovoltaic cell

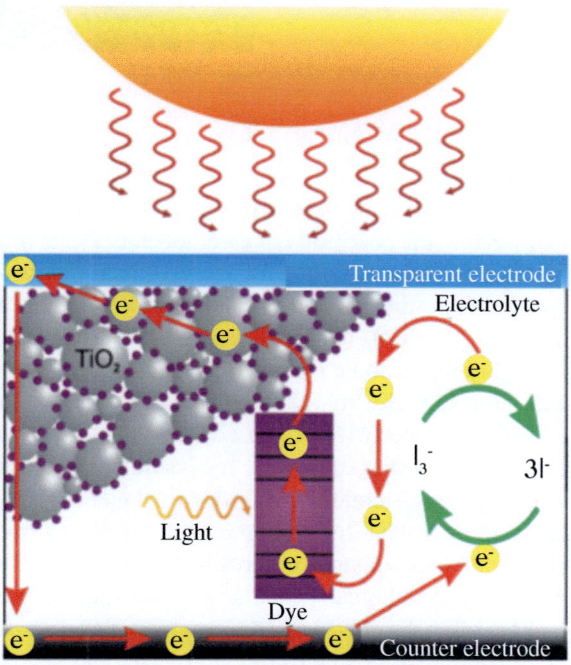

use (see Fig. 8.3). The electron is brought back down to the ground state energy level of the dye by means of a platinum counter electrode.

Grätzel cells constitute an interesting technology because they use compounds that are abundant on Earth and require simpler elaboration techniques than those used in silicon-based photovoltaic technologies, for example. The weak point at the moment is the longevity of the organic part of the cell. Current work aims to replace the dye by an inorganic absorber.

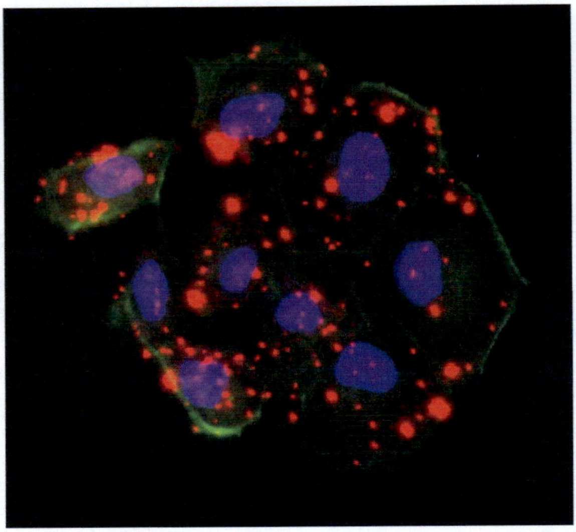

Fig. 8.4 Confocal microscope image on the micrometric scale of TiO_2 nanoparticles marked with rhodamine B (*fluorescent red*) in cell cytoplasm. Image CEA

8.3 Toxicity of Nano-TiO_2

Not so long ago, nanometric TiO_2 was still considered to be a slightly toxic substance. Considering the possibility of carcinogenic effects (not demonstrated), a publication by the US National Office for Occupational Safety and Health (NIOSH)[5] recommended a limiting value in the air of 2.4 mg/m^3 for micrometric TiO_2 particles and 0.3 mg/m^3 for TiO_2 nanoparticles owing to their higher specific surface area.

According to some studies, TiO_2 nanoparticles tend to be absorbed in the lungs and can potentially induce an inflammatory effect and cytotoxicity due to their photocatalytic property (see Chap. 12). On the other hand, a great many studies carried out on sun creams show that, for healthy skin, there is no risk of these particles reaching the systemic route through cutaneous exposure. The TiO_2 nanoparticles get no further than the stratum corneum (the outermost layer of the epidermis) for healthy skin in human skin models. However, some recent studies have shown that TiO_2 nanoparticles can penetrate as far as the epidermis [2].

Regarding the effects of TiO_2 nanoparticles on ecosystems, there have been many often contradictory studies of cytotoxicity and genotoxicity on plants and fish (see Chap. 13). The probability of single nanoparticles ending up in the environment is very low owing to their tendency to cluster together or agglomerate around other particles, something which modifies their behaviour with regard to transport in soils or aquatic environments. Nevertheless, even agglomerated nanoparticles may produce toxic effects.

[5]This is the federal agency of the US Department of Health in charge of investigating whether chemical products are harmful and making recommendations for the prevention of work-related injury and illness.

So for the moment there is no formal proof of the toxicity of TiO_2 nanoparticles in humans or in the environment. However, nanotoxicity studies need to be pursued further before clear conclusions can be drawn. New metrological tools are being developed in research centers, such as micro X-ray analysis and nanotracers (nanoparticles carrying a magnetic, fluorescent, or radioactive tracer) which can be used to localise nanoparticles in living organisms. Figure 8.4 exemplifies the latter with a confocal microscope observation of the distribution of TiO_2 nanotracers marked with fluorescent rhodamine B in cell cytoplasm.

References

1. F.E. Budde, Self-cleaning, odor reducing, and water shedding properties of titanium dioxide (TiO_2) photocatalytic oxidizing coatings. Metal Finish. **108**(1), 34–36 (2010)
2. AFSSET report, Évaluation des risques liés aux nanomatériaux pour la population générale et pour l'environnement, (March 2010) p. 107

Part III
Nanobioscience, Nanomedicine, and Nanotoxicology

Chapter 9
Nanobiosciences: New Ideas and Tools for Investigating and Manipulating Living Systems

Maxime Dahan

Abstract Nanoscience has reached a point today where we can begin to create hybrid objects at the interface between nanomaterials and the life sciences. These objects combine advanced physical properties, viz., optical, mechanical, magnetic, etc., with the remarkable reactivity of biological molecules. In particular, they can be used as probes or micromanipulation tools, from the molecular and cellular scale right up to the scale of organisms. This should bring us new insights into the organisation of living things, and new prospects for diagnosis and therapy in nanomedicine.

The beginning of the twenty-first century was marked by an encounter between two major scientific disciplines: on the one hand, materials science, which studies the elaboration, structure, and properties of matter made up of inert entities, and on the other, the life sciences which seek to understand the function—and indeed the malfunction—of living systems. Materials science traditionally borrows from physics, chemistry, and mechanics, while the life sciences are associated with disciplines like biology, biochemistry, and medicine.

To a large extent, this encounter has been catalysed by the rapid expansion of nanoscience over the past fifteen years. Today, it has become possible to synthesise nanomaterials wherein not only the size (from a few nanometres to a few hundred nanometres), but also the shape and composition can be very tightly controlled. Quantum dots, nanotubes, nanowires, and nanoribbons are but a few examples of nano-objects that can now be prepared down to the finest detail (see Parts I and II of this book). They exhibit unique optical, electric, mechanical, and magnetic properties depending on the materials they are made from. And what is striking is that the nanometric scale is precisely that of the objects making up living systems like proteins, nucleic acids, macromolecular assemblies, viruses, and so on. These biological molecules, while they often manifest less spectacular physical characteristics than nanomaterials, nevertheless exhibit quite remarkable properties of chemical reactivity and specificity. Indeed, these properties, optimised by nature during the long process of evolution, are often impossible to reproduce in synthetic systems.

M. Dahan (✉)
Laboratoire de Physico-chimie, Institut Curie, CNRS UMR 168, 29 Rue d 'Ulm, 75248 Paris Cedex 05, France
e-mail: maxime.dahan@curie.fr

J.-M. Lourtioz et al. (eds.), *Nanosciences and Nanotechnology*,
DOI 10.1007/978-3-319-19360-1_9

Thanks to this convergence of size at the nanometric scale, nanomaterials and biological molecules can be combined to make hybrid objects with quite unique physical, chemical, and biological characteristics. Naturally, this merger between the life sciences and materials science cannot be carried out without raising some difficulties. Building together objects as different as a semiconducting nanowire and a protein involves sophisticated chemical techniques. And exploiting them in a biological or medical context means controlling their reactivity in the living environment. Actually detecting them in complex environments like cells, living tissue, or living organisms is often a considerable challenge, requiring advanced physical methods. Nanobioscience is therefore by its very definition a cross-disciplinary field whose potential can only be realised by bringing together knowhow from a range of backgrounds.

The examples given below, while necessarily a subjective choice in some ways, should serve to illustrate some of the techniques of nanoscience used in biology, and in particular their ability to probe and manipulate living systems with unprecedented sensitivity. However, the contribution of nanoscience is not limited merely to novel techniques. Since they make available new quantitative measurements that were inaccessible to conventional tools, nano-objects can reveal unexpected aspects of the biological world and by doing so call into question some of our preconceptions about the organisation and dynamics of living systems.

9.1 Nano-Objects as Functional Probes for Nanoscale Exploration of Living Systems

One of the most common applications of nano-objects in a living environment is the functional probe. The best example is when the nano-object is used as a sensor in a biological environment. We can thus measure the local characteristics of the environment through changes in the physical properties of the nano-object itself. In many cases, we use changes in the intensity or spectrum of the optical response of a nanoparticle to measure biochemical quantities such as ion concentrations [1], or physical parameters such as the temperature [2] or pressure. Note that these really are local measurements, on the scale of the nano-object, so we can study the biological environment with nanometric resolution. In other cases, a measurement of electrical conductivity is used to detect molecules or biological entities such as viruses through their interaction with the probe [3]. Probes can also be targeted, labelling some particular molecule of interest. Provided they do not perturb the molecules to which they are attached, such probes then serve to locate them within some living sample, to measure their functional activity, or to determine their concentration.

The transformation of a nanomaterial into a functional probe is often a major challenge. To ensure that it has the necessary specificity for chemical detection or targeting, its surface must be modified by grafting reactive biological molecules or chemical groups onto it. Such a functionalisation chemistry is a delicate matter, but it is essential for making this kind of measurement. A detailed description can be found in Sect. 5.4 of Chap. 5, dealing with nanochemistry.

In the case just described where nano-objects are used to probe the environment or to label biomolecules, the role of nanoscience is not just conceptual. However, the physicochemical characteristics of the nano-objects make them choice systems for increasing sensitivity or specificity and thereby carrying out experiments that would be difficult or even impossible using conventional methods. An important and even emblematic example illustrating the impact of nano-objects in the life sciences is provided by semiconductor nanocrystals known as quantum dots [4] (see Sect. 2.3 of Chap. 2 on nanophysics). These are nanoparticles with sizes typically in the range 2–20 nm. Not only can their dimensions be adjusted to within a few atomic layers, but their shape, e.g., spherical or threadlike, can also be very closely specified. Since they are made from a semiconducting material such as cadmium selenide, these nanoparticles have fluorescence properties. On the nanoscale, the optical response is determined by quantum effects. So for example, the fluorescence spectrum will depend on the size of the nanoparticle, shifting from the blue toward the red as the radius increases.

Moreover, it has been known for fifteen or more years how to functionalise quantum dots so as to make them compatible with biological media. They can thus be used as multicolour fluorescent probes, much brighter and more photostable than organic markers. The quantum dots, which act like little nanoscale light bulbs, provide the possibility of ultrasensitive detection and have largely contributed to the development of a new field of study: the imaging of single proteins in living media [5].

9.2 Tracking Single Biomolecules in the Cell

On the molecular scale, objects move in a very different way to what we see in our own macroscopic world. Under the effects of thermal agitation and diffusion, molecules are in a permanent state of motion. In order to understand the functional architecture of the cell, it is essential to be able to describe these motions and analyse their role in molecular interactions. One particularly fruitful approach is to track the motions of individually labelled proteins.

When tracking molecular motions in this way, the signal from the probe attached to the relevant protein must be recorded in a sequence of images. The experimental challenge is to actually detect the signal from a single marker against the background noise produced within a living sample. The use of bright probes like nanoparticles is then often a decisive advantage. There are two important features of such measurements. On the one hand, the position of a molecule can be determined with an accuracy that depends only on the signal-to-noise ratio and can in practice reach about 10 nm. Plotting the trajectory of a protein thus amounts to exploring the cell medium on a nanometric scale. On the other hand, the motion of biological molecules is governed partly by diffusion and partly by their interactions with molecular partners. In many cases, one thus has direct access to the kinetics of these interactions in the context of the living cell.

A particularly striking illustration of these experimental techniques concerns glycine receptors, membrane proteins involved in the transmission of inhibitory signals in neuronal cells (see Fig. 9.1a, b, and c). To track their motion, a quantum dot is coupled with an antibody which recognises an extracellular domain of the protein [6]. When their trajectories are recorded, these proteins are found to be extremely mobile, much more than one would expect. They diffuse in the neuronal membrane and end up in the synapses, where they interact with molecular partners. However, these are transient interactions and the receptors subsequently escape from the synapses to resume their exploration of the membrane. Such observations raise a fundamental question: given that the synapses constitute key structures in the connections between nerve cells, playing an important role in the fundamental processes of learning and memory, how can they maintain their functional stability when the molecular components making them up undergo a permanent process of association and dissociation?

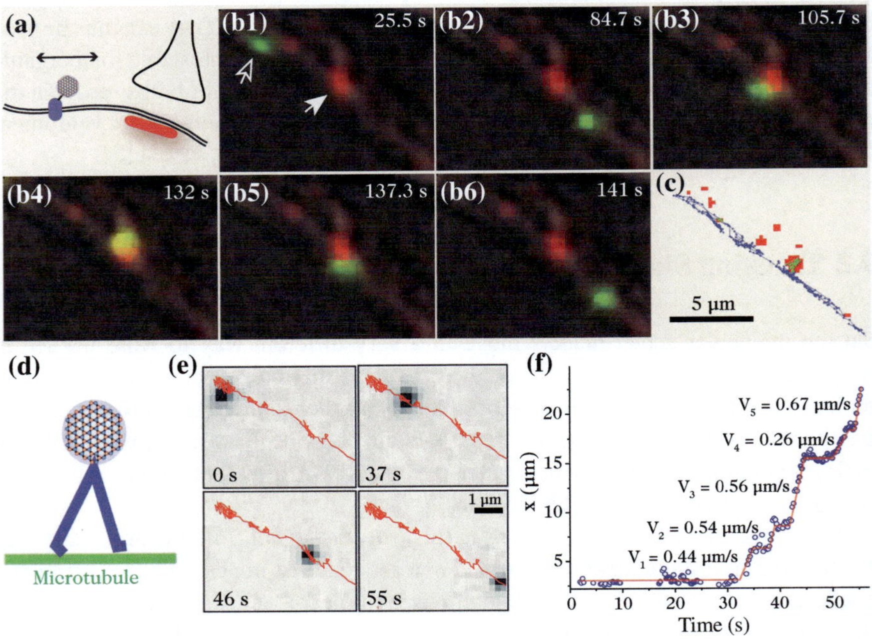

Fig. 9.1 Single-molecule visualisation using nanoprobes. **a** Diagram of a receptor (*blue*) labelled by a quantum dot moving in the neuronal membrane. The receptor can stabilise at the synapse by interacting with scaffold proteins (*red*). **b1–6** Image sequence showing the motion of a single receptor (*green*) relative to a synaptic site (*red*). **c** Reconstruction of the trajectory (*blue*). *Green points* correspond to the part of the trajectory during which the receptor is trapped at the synapse. **d** Diagram of the motor protein kinesin, labelled by a quantum dot and moving along a microtubule filament. **e** Image sequence showing the motion of the motor in the cytoplasm of a living cell. **f** Analysis of the motion along the trajectory, showing alternating phases of directed motion and pauses. **b–d** Adapted from [7], © 2007 Elsevier, **e** and **f** adapted from [8], © 2006 American Chemical Society, reproduced with permission

Tracking measurements are not restricted to membrane proteins. Single molecules can also be monitored in the cytoplasm inside living cells. A particularly fascinating case concerns motor proteins. These are proteins that can transform chemical energy in the form of adenosine triphosphate (ATP) into mechanical energy. Such motors operate with extraordinary efficiency, close to 100 %, well beyond what can be reproduced by chemical synthesis. For example, kinesin, a molecular motor which moves along the microtubule cytoskeleton, is involved in many intracellular transport mechanisms (see Fig. 9.1d). By marking it with a quantum dot, its motion can be tracked within the cell [8]. It is observed to move in a directed manner along a microtubule at a speed of about 0.5 μm/s. However, this displacement phase lasts only a certain time. After a fraction of a second, the motor detaches itself from the microtubule and diffuses freely for a while before reconnecting with the cytoskeleton (see Fig. 9.1e and f).

While fluorescence imaging methods remain the most common, there are other detection modes for imaging single molecules. It is in fact a feature of nanobioscience that the biological systems investigated and the questions raised regularly require the design of novel nanoprobes and the development of ever more elaborate measurement tools. As an example, gold nanoparticles can be used for single-molecule detection. These nanoparticles do not fluoresce but are nevertheless effective light scatterers, with spectral properties that can be modulated by changing their size and shape (see Sect. 5.5.1 of Chap. 5). To increase sensitivity and stand out above the background signal from other cell structures, an elegant approach is to combine the scattering properties of gold nanoparticles with a photoinduced modulation of their temperature. This photothermal method allows a highly specific measurement and accurate tracking of the object to which the gold nanoparticle is attached [9].

Single-molecule experiments provide direct access to the many specificities of biological media, something that would be difficult to apprehend from collective measurements involving molecular ensembles. To begin with, the cell environment turns out to be extremely inhomogeneous and in permanent evolution. Moreover, the interactions between molecular partners are governed by stochastic rather than deterministic processes. This leads to considerable variability in molecular behaviour. It is largely an open question as to how cells can function so effectively despite, or perhaps thanks to, this permanent molecular noise, and this is an active subject of research today.

Quite generally, tracking a molecule in a cell and measuring the kinetics of its interactions with molecular partners is an efficient way to understand the reaction–diffusion processes governing the reactivity of molecular species. Thanks to nano-objects and the sensitivity of the observations they make possible, biochemical measurements can be carried out *in situ* in the cell. This represents a considerable advantage over *in vitro* measurements, where it is often difficult to reproduce the heterogeneity, size constraints, and molecular crowding of cell environments.

9.3 Manipulating Living Systems on the Molecular and Cellular Scales

Nano-objects are not just exceptional tools for observing living systems. They can also be used to manipulate and perturb biological systems on the molecular and cellular scales. To do this, a whole range of micromanipulation tools have been developed over the past twenty years [10]. Exploiting the optical or magnetic properties of nano-objects, forces can be applied on a scale from a fraction of a piconewton to several tens of nanonewtons. In this way, one can probe the relationship between mechanical properties and biological activity at the molecular level, or control the events that trigger cell processes.

The *in vitro* micromanipulation of biological molecules can be illustrated by the example of optical tweezers, introduced in Sect. 2.4 of Chap. 2 on nanophysics. A light intensity gradient is created on the scale of the wavelength (of the order of 1 μm, see Fig. 9.2a) by strongly focusing a laser in an aqueous medium. Particles with sizes in the range 100–1 000 nm can be trapped near the focal point. When these particles are functionalised with biological molecules, we may also apply forces to these molecules in order to study their mechanical response. For example, by pulling on a DNA molecule grafted at the same time onto a glass surface and onto the trapped particle, we may measure the elasticity of DNA (see Fig. 9.2a left).

This is also a useful way of studying interactions between DNA and proteins [11]. For example, RNA polymerase, the enzyme that synthesises RNA from nucleotides, acts locally on its substrate by separating the double helix structure there. In doing so, it modifies the mechanical properties of the DNA and these modifications can be detected through the motion of the particle, which thus acts here as a force transducer (see Fig. 9.2a right). The sensitivity of these measurements is so good that we may now detect effects on the scale of a single base pair, i.e., at distances of barely 0.3 nm.

Micromanipulation *in vitro* is not restricted to DNA–protein interactions. Using optical or magnetic tweezers, or an atomic force microscope, we can study the properties of many systems, including kinesins and myosins, molecular motors which act on microtubules and actin filaments, respectively (see Fig. 9.2b–d). Globally speaking, these experiments reveal the complex relationship between the mechanics of molecular systems and their biochemical reactivity.

The response to mechanical stresses is a central issue in cell biology. Using nanoparticle micromanipulation tools, the effects of mechanical forces can be very precisely measured. Hence, by pulling on a nanoparticle attached to the cell membrane, we can study mechanical transduction mechanisms, i.e., those biological processes whereby the cell detects mechanical perturbation and responds.

Nano-objects can be used to control cell signalling, without necessarily inducing any mechanical perturbation. For example, small paramagnetic nanoparticles ($\sim$ 30 nm) can be coupled specifically to membrane receptors. Under the effect of a constant magnetic field, these nanoparticles acquire a magnetic dipole. The dipole–dipole interaction induces aggregation of the particles (see Fig. 9.2b), and thus also of the receptors to which they are attached, thereby activating signalling channels

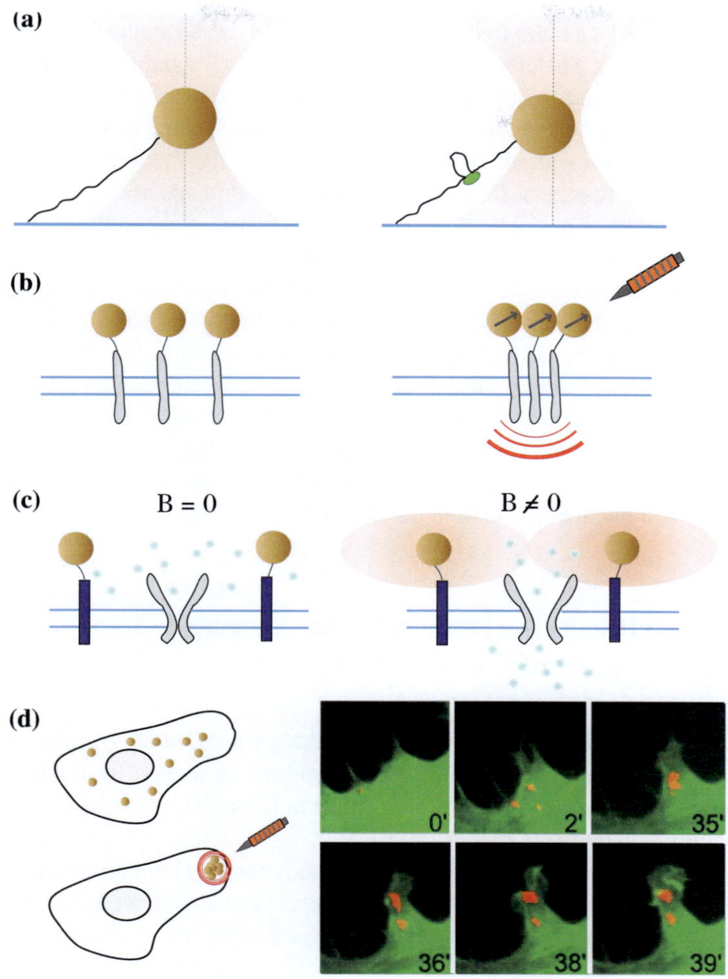

Fig. 9.2 Examples of micromanipulation on the molecular and cellular scale. **a** *Left* A focused laser beam (*pink*) traps a nanoparticle. The latter can be coupled to a DNA molecule which is itself attached to a glass surface and which can thus be put under tension. *Right* When a protein (*green*) interacts mechanically with the DNA, the motion of the nanoparticle can be used to determine the forces generated. **b** *Left* Magnetic nanoparticles are coupled with membrane proteins. *Right* In the presence of a static field, these nanoparticles interact and the aggregation thereby induced triggers a biochemical signal. **c** *Left* Magnetic nanoparticles are coupled with membrane proteins. *Right* A radio frequency magnetic field *B* causes the nanoparticles to heat up, causing the heat-sensitive membrane channels to open. **d** *Left* Nanoparticles functionalised with signalling proteins are injected into a cell. By means of a magnetic force, they are then localised in a particular region of the cell where their accumulation triggers a response from the cell. *Right* Image sequence illustrating this process. Nanobeads (*red*) are carried to the cell membrane where they induce the formation of protrusions. The time is indicated in minutes. Adapted from [12], © 2013 Nature Publishing Group

[13]. Alternatively, a radio frequency magnetic field can be applied instead of a static one (see Fig. 9.2c). The effect is to heat the nanoparticles and this local temperature rise can be exploited to activate heat-sensitive receptors which themselves trigger cascades of intracellular signals [14]. Another approach is to introduce nanoparticles functionalised with signalling proteins directly into the cytoplasm (see Fig. 9.2d). Using magnetic tweezers, they can then be localised in specific regions of the cell, whence it becomes possible to measure the response of the cell to a biochemical perturbation in different local cell environments [12].

Quite generally, whether it be through mechanical effects or direct activation of signalling channels, nano-objects provide a way to apply spatially and temporally controlled perturbations. They are thus invaluable tools for the development of a new discipline, systems biology, which seeks to understand the integrated properties of cell systems. By analogy with the study of electrical circuits, the aim is to measure the cell response (output) to a controlled stimulation (input). Determining the input–output relation is an important step in figuring out the functioning of the molecular circuits which handle information transfer within the cell.

9.4 Nano-Objects for Diagnosis and Therapy: The Prospect of Nanomedicine

Applications of nano-objects in the life sciences are not limited to molecular and cell biology. Indeed, more and more uses are being found for them in the field of medicine. These applications, described in more detail in Chap. 10, are today gathered under the heading of nanomedicine.

The sensitivity of diagnostic imaging methods can thus be improved by exploiting the special physical properties of nanoparticles. For example, we know today how to synthesise quantum dots or carbon nanotubes emitting in the near-infrared [15]. In this spectral range, biological media are relatively transparent, so such objects can be used to image tissues in depth. Another possibility is to use magnetic nanoparticles as contrast agents in magnetic resonance imaging (MRI) [16]. In most cases, nanoparticles are functionalised in such a way that they target particular regions of the organism such as a tumour or inflammation, while they are carefully passivated to ensure that they do not interact with healthy tissues. Achieving high specificity in the recognition mechanisms of the nano-objects, a process that is already complex in cell samples, often represents a much more considerable challenge in the medical context, where the nano-objects have to circulate in the vascular system.

Medical applications of nano-objects concern not only detection and diagnosis, but also therapy. One of the most attractive prospects in nanomedicine is the use of nanosystems as multipurpose platforms, combining a contrast mechanism with targeting and therapeutic functions. This approach, known as theranostics, aims to combine diagnosis and therapy. It forms a natural component in the development of personalised medicine, which seeks to customise treatments. The therapeutic action can be implemented in a targeted way, with localised drug release (see Chap. 11), thereby improving the efficiency of the treatment and reducing the side-effects of the drugs on healthy cells. Physical mechanisms can also be exploited in thermal therapy, where nanoparticles are heated up by irradiating them with light [17] or, as above, by application of a radio frequency magnetic field [18]. The rise in temperature can kill targeted cells, and in particular, tumour cells, in which the nanoparticles have accumulated.

Despite the many positive aspects of nanobioscience, it also raises some questions and a certain apprehension. For the moment, we do not have a good understanding of the mechanisms governing the interactions between nanoparticles and living organisms, nor the degradation of nanoparticles in such a context. Given the growing number of applications and the high economic stakes, it is crucial to assess the potential toxicity of nano-objects and their general effects on health (see Chap. 12). The high chemical reactivity of nanoparticles is also likely to have environmental consequences (see Chap. 13). Whatever the situation, it is essential to be able to precisely specify the dose beyond which exposure becomes a risk. On the other hand, this is often a delicate matter given the sensitive dependence of toxic effects on physico-chemical parameters such as size, shape, composition, and surface characteristics of nanoparticles, not to mention the type of exposure.

9.5 Conclusion and Prospects

Nanoscience has become an important factor in the development of both fundamental and applied biomedical research. The potential of nanoscience in the life sciences was stressed [19] during the launch in the United States of the project known as *Brain Research through Advancing Innovative Neurotechnologies* (BRAIN) in the spring of 2013. This project, announced at the same time as the *Human Brain Project* in the European Community, commits a considerable investment to improving our understanding of neuronal mechanisms in the human brain. To achieve this, a great many tools will be needed to visualise and monitor neuronal activity, either optically or electrically. Eminent research scientists in the field have pointed out that nanoscience will have a key role to play, thanks to its capacity to design systems on the molecular or subcellular scale, interfaced with biological systems while communicating with the macroscopic world [19].

To end, we may quote Sydney Brenner, Nobel Prize for Medicine in 2002 and one of the founding fathers of molecular biology and developmental genetics: "Progress in science depends on new techniques, new discoveries and new ideas, probably in that order." This introduction to nanobioscience together with the next four chapters is an invitation to discover the rich panoply of innovative techniques available today for researchers and engineers in a range of disciplines, describing some of the results already obtained here. It is our hope that the reader will find a source of inspiration for the new ideas which will one day lie at the very heart of many areas of science.

References

1. D. Casanova et al., Single europium-doped nanoparticles measure temporal pattern of reactive oxygen species production inside cells. Nat. Nanotechnol. **4**, 581–585 (2009)
2. G. Kucsko et al., Nanometre-scale thermometry in a living cell. Nature **500**, 54–58 (2013)
3. F. Patolsky et al., Electrical detection of single viruses. Proc. Natl. Acad. Sci. USA **101**, 14017–14022 (2004)
4. A.P. Alivisatos, W. Gu, C. Larabell, Quantum dots as cellular probes. Annu. Rev. Biomed. Engl. **7**, 55–76 (2005)
5. F. Pinaud et al., Probing cellular events, one quantum dot at a time. Nat. Methods **7**, 275–285 (2010)
6. M. Dahan et al., Diffusion dynamics of glycine receptors revealed by single-quantum dot tracking. Science **302**, 442–445 (2003)
7. M.V. Ehrensperger, C. Hanus, C. Vannier, A. Triller, M. Dahan, Multiple association states between glycine receptors and gephyrin identified by SPT analysis. Biophys. J. **92**, 3706–3718 (2007)
8. S. Courty et al., Tracking individual kinesin motors in living cells using single quantum dot imaging. Nano Lett. **6**, 1491–1495 (2006)
9. D. Lasne et al., Single nanoparticle photothermal tracking (SNaPT) of 5-nm gold beads in live cells. Biophys. J. **91**, 4598–4604 (2006)
10. K.C. Neuman, A. Nagy, Single-molecule force spectroscopy: optical tweezers, magnetic tweezers and atomic force microscopy. Nat. Methods **5**, 491–505 (2008)
11. C. Bustamante, W. Cheng, Y.X. Mejia, Revisiting the central dogma one molecule at a time. Cell **144**, 480–497 (2011)
12. F. Etoc et al., Subcellular control of RacGTPase signalling by magnetogenetic manipulation inside living cells. Nat. Nanotechnol. **8**, 193–198 (2013)
13. R.J. Mannix et al., Nanomagnetic actuation of receptor-mediated signal transduction. Nat. Nanotechnol. **3**, 36–40 (2008)
14. H. Huang et al., Remote control of ion channels and neurons through magnetic field heating of nanoparticles. Nat. Nanotechnol. **5**, 602–606 (2010)
15. S. Kim et al., Near-infrared fluorescent type II quantum dots for sentinel lymph node mapping. Nat. Biotechnol. **22**, 93–97 (2004)
16. C. Sun, J.S. Lee, M. Zhang, Magnetic nanoparticles in MR imaging and drug delivery. Adv. Drug Deliv. Rev. **60**, 1252–1265 (2008)
17. L.R. Hirsch et al., Nanoshell-mediated near-infrared thermal therapy of tumors under magnetic resonance guidance. Proc. Natl. Acad. Sci. USA **100**, 13549–13554 (2003)
18. J.H. Lee et al., Exchange-coupled magnetic nanoparticles for efficient heat induction. Nat. Nanotechnol. **6**, 418–422 (2011)
19. A.P. Alivisatos et al., Nanotools for neuroscience and brain activity mapping. ACS Nano **7**, 1850–1856 (2013)

Chapter 10
Nanomedicine and Nanotechnology for Medicine

Patrick Boisseau

Abstract Nanomedicine is a new field of research and industrial development which aims to use nano-objects for therapeutic purposes. Through miniaturisation, labs-on-chips can improve the performance of medical analysis and diagnosis, while biosensors become less invasive, hence better tolerated or implanted. With the development of nanoscale contrast agents, clinical imaging benefits from better resolution. In the field of therapy, nanocarriers can transport medicines right up to the sick tissue within the organism, while metal nanoparticles can be used for localised destruction of tumours. The possibility of combining diagnosis and therapy by means of these nano-objects (theranostics) also offers hope of improving medical practice generally. Likewise, the controlled combination of stem cells and biomaterials should lead to a new form of regenerative medicine, able to regenerate and repair damaged or destroyed tissue *in vivo*. This chapter deals precisely with these developments made possible by nanomedicine, discussing the relevant ethical problems which arise also for conventional medicine. It ends with a discussion of the industrial prospects, describing some of the many clinical trials now under way and the promise of innovative drugs for the benefit of the patient.

Nanomedical research began for real at the beginning of this millennium, riding on the wave of enthusiasm generated by nanotechnology and nanomaterials, even if the first faltering steps can be found as far back as the 1970s under a different name. It has become common practice to cite the lecture entitled *There is plenty of room at the bottom*, given by physicist Richard Feynman to the American Physical Society in Caltech on 29 December 1959, as the origin for the ensuing interest in nanotechnology. The large American university research centres, financed by the National Nanotech Initiative, then the National Cancer Institute, and the European cross-disciplinary research centres, with both national and European funding, thus began to build bridges between physicists, chemists, biologists, and later clinicians,

P. Boisseau (✉)
Commissariat à l'Energie Atomique et aux Energies Alternatives (CEA), Campus MINATEC, 17 rue des Martyrs, 38054 Grenoble Cedex, France
e-mail: patrick.boisseau@cea.fr

J.-M. Lourtioz et al. (eds.), *Nanosciences and Nanotechnology*,
DOI 10.1007/978-3-319-19360-1_10

who would work together to investigate applications of nanotechnology to medicine. And so nanomedicine was born!

There followed a ten-years period of great intellectual creativity, with a considerable increase in research which threw up a whole range of potential applications of nanotechnology in medicine. And as happens with many new technologies, nanomedicine began to look like the answer to all unresolved medical problems! However, scientific reality and industrial constraints soon came to bear, damping the enthusiasm of both researchers and politicians. So we enter the second decade of the millennium with a rationalisation of research which focuses on the more realistic and most promising applications of nanomedicine, those with a reasonable chance of actually being carried through and improving the public health situation. A steady increase in the number of nano-based medicines approved by American and European regulatory bodies has also been observed.

As in many branches of high technology, such as molecular biology, gene therapy, cell therapy, or information and communication technology, the nanotechnologies used in medicine have sometimes raised quite impossible hopes, in line with the expectations of patients both in diagnostics and therapeutics. This tendency was sometimes based on the overambitious and exaggerated claims of researchers themselves, either to attract or to justify research funding. Through its intimate dialogue with matter on the molecular scale, nanotechnology would thus stay one step ahead of disease, cutting it short before it could even get started. Early diagnosis and personalised therapy would be the response of modern medicine.

However, a few years on, the reality of experiments on cell cultures and in animal trials had whittled down the areas where nanotechnology looked likely to bring anything new to diagnosis and therapy. The complexity of phenomena on the molecular scale and the vast multiplicity of molecular interactions *in vivo* have pushed certain applications well into the future. Just mastering the manufacture of nanoparticles in no way solves the problem of controlling their many interactions with biomolecules, especially once they have been injected into a living organism. This is the case, for example, with certain solid inorganic particles known as quantum dots. These have proven extremely useful in imaging thanks to their brightness and colour, which vary with their size (see Fig. 10.1), but their toxicity makes them useless for *in vivo* human applications.

Nanomedicine makes great use of nanometric molecular assemblies, e.g., to transfer therapeutic molecules. These are sometimes referred to as nanoparticles by abuse of language. Their ultrasmall size means they can enter inside cells, because these particles are something like a hundred to a thousand times smaller than a cell. However, just as it is a considerable achievement to carry a drug into the very heart of the cell, so this same ability of the particle to go so deep is a subject of concern for certain groups of people. What would happen if we were no longer able to control this phenomenon? How should we manage nanomaterials that can go so deep within us? In fact, many of the advantages of nanomedicine as advocated by its protagonists can be turned against it and viewed as a source of concern among the population. And the same was surely true with the discovery of bacteria and micro-organisms.

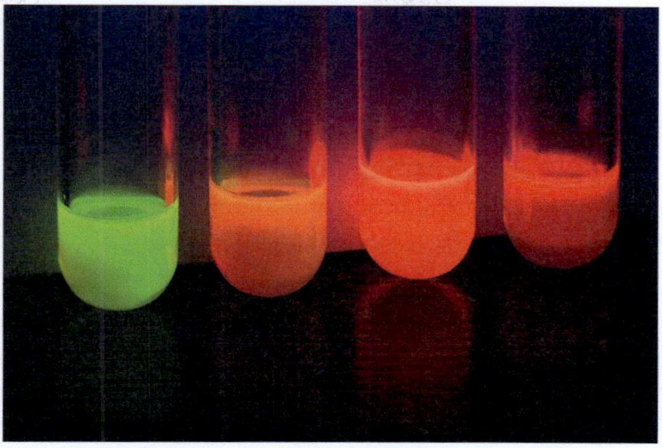

Fig. 10.1 Fluorescent quantum dots. The colour of the different quantum dots depends on their diameter. *Credit Artechnique*/CEA

Training, information, communication, exchanges, participation, and debate must immediately be strengthened in order to identify and understand the advantages and limitations of medical nanotechnology, so that we can ensure responsible development for the benefit of the patient. While they originated in the laboratories of physicists, chemists, and biologists, medical applications of nanotechnology have now entered the public sphere where their future development will be determined by exchanges and interactions between all the stakeholders working for the medicine of tomorrow.

10.1 Nanomedicine or Nanotechnology for Medicine?

The term 'nanomedicine' first appeared at the end of the 1960s at the *Ecole polytechnique fédérale* (ETH) in Zurich. Today, this term refers exclusively to the use of synthetic entities or systems of nanometric size, designed rationally and intentionally for a specific medical application. Contrary to a commonly held opinion, it is obviously not enough merely to exploit interactions on the molecular scale in order to speak of nanomedicine. For instance, an antibody alone, even though it may have nanometric size, is not an example of nanomedicine. On the other hand, an antibody–drug conjugate would provide such an example. Otherwise, one would have to consider the whole of medicine as nanomedicine, since all pathological mechanisms have a molecular origin. Moreover, even if naturally occurring and self-assembled molecular assemblies are of nanometric size and play a role in medicine, these applications still cannot be classified under the heading of nanomedicine. One example among others is provided by the viruses used in gene therapy, which are not considered to constitute instances of nanomedicine. The same can be said when molecules are used as drugs.

As a matter of fact, nanomedicine is not even a field of medicine as such, like diagnostics, pharmacology, or surgery. It would be more fairly defined as *the application of nanotechnology to medicine*, for it does indeed derive from the sometimes novel properties of materials synthesised on the nanometric scale (between one and a few hundred nanometres). From this standpoint, nanotechnology brings genuinely new approaches and new solutions to biological analysis, *in vitro* diagnostics, drug delivery, surgery, cell therapy, therapeutic monitoring, and others.

And neither is there any industrial component to nanomedicine! There are pharmaceutical companies, medical imaging companies, and firms making medical equipment, but there is no nanomedical industry or profession. Indeed, in medicine, the clinical objective remains the priority, whatever technical solution is proposed, provided it is approved and authorised. Hence, nanodrugs are approved by regulatory bodies and put on the market, but they are not listed as such, because although they are medicines formulated in a specific way, they are nevertheless, first and foremost, medicines! Nanodrugs must obey exactly the same regulations as any other therapeutic molecules, and their health benefits will be assessed regardless of their nanometric dimensions. This is why there are still no specific regulations for nanodrugs in France, or even in Europe.

The above remarks might thus make us prefer to speak of *nanotechnologies for medicine*, rather than *nanomedicine*. However, for purely practical reasons, and consistency with common usage in the scientific community, we shall continue to use the term 'nanomedicine' throughout this chapter.

10.2 The Ideal Length Scale in Medicine

The aim of nanomedicine can be roughly defined as the detailed monitoring, control, construction, repair, and defence of any human biological system, starting from the molecular scale and using nanometric molecular assemblies or devices to obtain medical benefit. Nanomedicine thus encompasses molecular assemblies dispersed in a biological medium, but also nanometric components included in a microdevice which is itself macroscopic. This is the concept known as nano-inside, proposed by analogy with the label indicating the presence of certain processors in Intel® computers.

Nanomedicine thus makes use of nanometric entities, i.e., 80,000 times smaller than a human hair, and their interactions with biomolecules in a reaction medium or in a subcellular or cellular medium. Recall that, in the length range up to 100 nm, we find ourselves precisely at the scale of the biological molecules or intracellular structures that control the processes of life itself (see Fig. 10.2). There is thus a certain similarity of size between medical nanostructures or nanosystems and biomolecules such as proteins, sugars, and nucleic acids.

If we consider the example of nanocarriers (see Fig. 10.3), i.e., nanoscopic molecular assemblies designed for the delivery of drugs[1] (see the CNRS film [1]), these

[1]It is common practice to refer to the combined system of the nanocarrier and the drug it transports as the nanodrug.

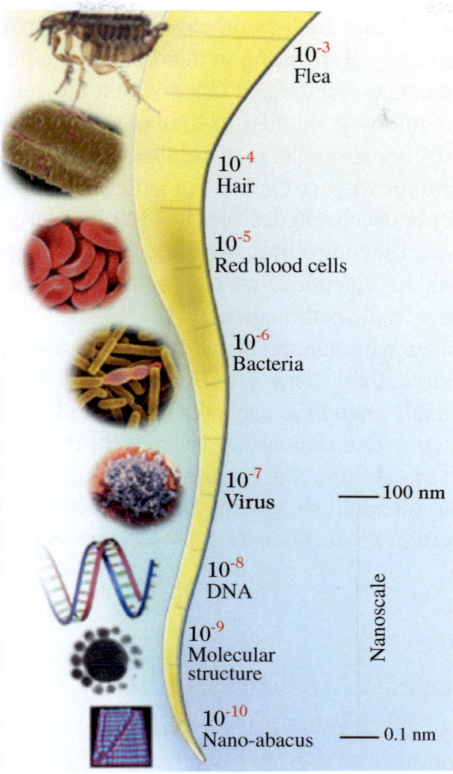

Fig. 10.2 Length scale of biological entities. Adapted from Discover Nano, Northwestern University

Fig. 10.3 Artist's view of a Lipidot® lipid nano-delivery system. *Source* CEA-Leti

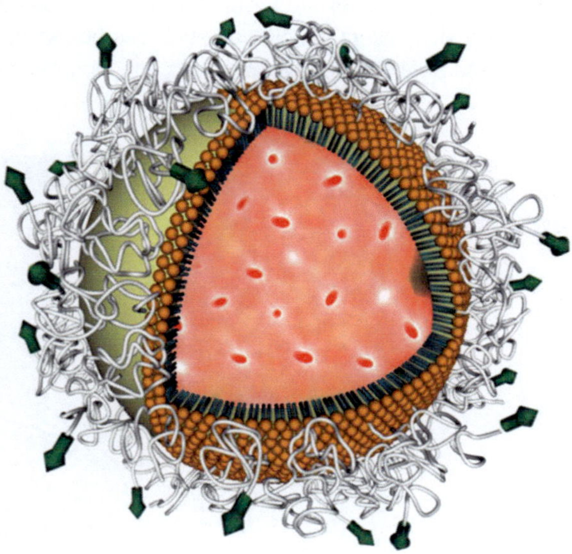

can interact with biomolecules such as proteins or sugars present in the organism, but they can also penetrate cells, bearing in mind that the latter are about a hundred times bigger than such nano-assemblies! Once within the cell, the nanocarrier can grasp, pull, perturb, or otherwise modify the biomolecules involved in a biochemical disorder, these generally being smaller than the nanocarrier itself. This size proximity is crucial when looking for ways to exploit properties specific to the nanoscale.

It was this similarity between length scales that inspired early hopes for nanomedicine, because the prospect of controlling, blocking, or signalling a molecular anomaly in its early stages would in principle allow one to limit the appearance of disease and prevent its development to the pathological stage.

We may thus wonder why nanomedicine has not become a much broader and more ambitious field of activity. There are in fact several explanations. The first is that, while it is reasonable enough to conceive of controlling molecular interactions in a simple culture *in vitro* in the laboratory, when it comes to animal tests or clinical trials on humans, the situation is much more complex. Indeed, there are far more biological entities and interactions, leading to a great deal more interference. The second reason comes from the fact that the nanoscale is not always necessarily ideal in medicine.

10.3 The Nanoscale: A Medical Eldorado?

Why is the nanometric scale so specific? What advantages can be reaped at this tiny length scale? The point is that the surface-to-volume ratio of nanometric entities is very high and thus provides a considerable surface area for exchange and interaction (see the introductory chapter to this book). This is reflected in very fast chemical reaction rates and thus shorter time scales for reaction or analysis.

Nanoscale sensors, or nanosensors, can be made. Indeed, thanks to progress in micro- and nanoelectronics, these can now be produced industrially (see Fig. 10.4).

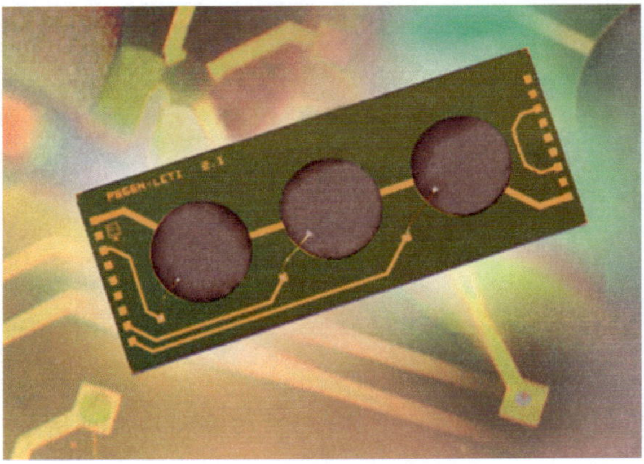

Fig. 10.4 Implantable micropump. Length 2.5 cm. *Source* CEA-Leti

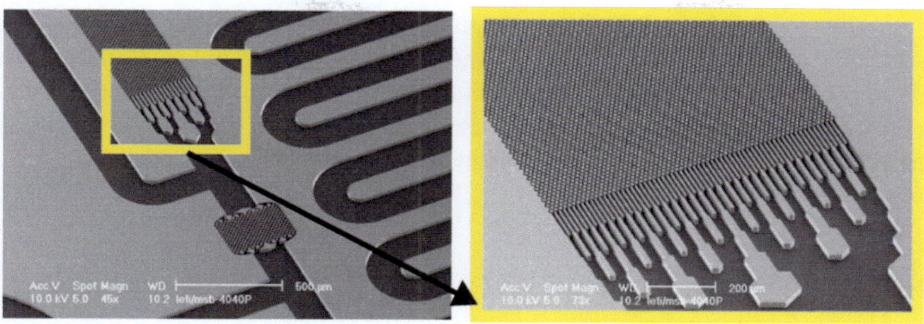

Fig. 10.5 Details of the silicon lab-on-a-chip BioChipLab. *Source* CEA-Leti

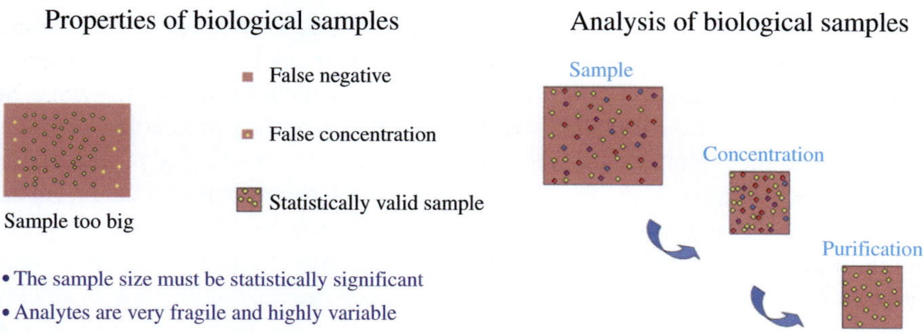

Fig. 10.6 Example of a biological sample, e.g., blood, urine, or saliva, containing many molecules in extremely variable concentrations. In order to isolate a particular category of molecule such as a biomarker, the first step is to increase the molecular concentration of the sample to reduce the analysis volume. The next step is to separate the relevant molecule from the other molecules by processes of purification or molecular sorting which exploit the physical, chemical, or biological characteristics of the molecules

A nanosensor can thus be inserted, even implanted, in the body in a weakly invasive manner, i.e., without significantly affecting the relevant organ. Likewise, nanosystems designed to carry out biological analyses require only very small biological samples and ideally apply to biopsies or manipulation of rare cells (see Fig. 10.5).

Note, however, that the detection of very small amounts of biomolecules in a sample that is itself very small can pose a problem. One example among many concerns early biomarkers for disease. Going by the old saying, we can look for a needle in a haystack, provided that the haystack is big enough to have some chance of finding a needle in it; or alternatively, provided that some kind of filter has been applied, rather as a magnet would attract a needle from the hay. In the case of nanosystems for biological analysis, rather than use a magnet, the idea would be to carry out a preconcentration of the sample. This is what is done to detect a few cells, e.g., bacteria, or a few molecules, e.g., a biomarker, in a blood sample (see Fig. 10.6).

Finally, on the nanometric scale, certain physical laws assume a different form as compared with the micrometric scale. For example, fluids tend to 'stick' to the walls of the capillaries in nanosystems and become more difficult to move around, just as it is easier to drink fruit juice with a thick straw than with a very thin one! This is the emerging field of nanofluidics, a recent extension of microfluidics.

10.4 Medical Diagnostics

Medical diagnostics *in vitro* is a tedious and time-consuming business. At the hospital, blood, urine, or sweat samples are sent to the central laboratory for analysis. The process can take several hours or even days, depending on the parameters to be measured and the techniques used. And all this requires a lot of labour. Difficulties include the risk of sample degradation, cost, delays even in urgent cases, lack of precision for very small samples, the problem of integrating results provided by different methods, and the limited standardisation of sampling protocols.

Thanks to the techniques of microelectronics and the possibility of integrating several parallel functions on the same device, a new generation of smaller, faster, and extremely accurate automatic systems has come into being, operating in push-button mode. These analysis devices accept much smaller samples than conventional systems and can provide more complete and more accurate results on a single sample. Indeed, micro- and nanosystems can simultaneously measure several biochemical parameters on the same sample, in close analogy with the idea of multiplexing.

By processing only very small biological samples, it becomes possible to be much less invasive and hence less traumatising when the samples are taken from the patient. Nanotechnology also makes it possible to design microsystems with very high-throughput analysis, capable of measuring a large number of samples very quickly in parallel, in what is effectively a miniaturised production line.

The techniques of microelectronics have enabled not only extreme miniaturisation of sensors, but also mass production, as we see for microchips in mobile phones. Even though these microsystems often employ a costly material, namely silicon, the unit cost remains reasonable thanks to the economies of scale inherent in mass production. In fact, certain microsystems for biological analysis use microelectronic components from everyday products such as cameras and cell phones [2].

10.4.1 Medical Laboratories

Various operations involved in the preparation and analysis of biological samples can be miniaturised and incorporated in a microsystem known as a lab-on-a-chip (LoC) (see Fig. 10.7). Such integrated devices can mix, process, and separate fluids, and they can analyse and identify samples. With certain components as small as around ten nanometres, they can measure hundreds or even thousands of signals from a

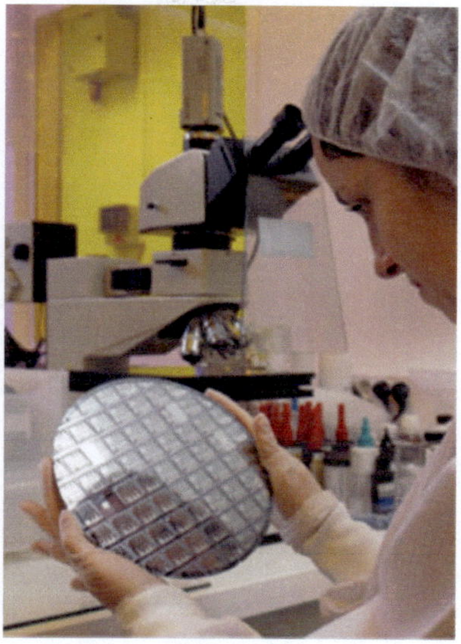

Fig. 10.7 Fabrication of 60 labs-on-chips on a single silicon wafer with diameter 20 cm. *Source* CEA-Leti

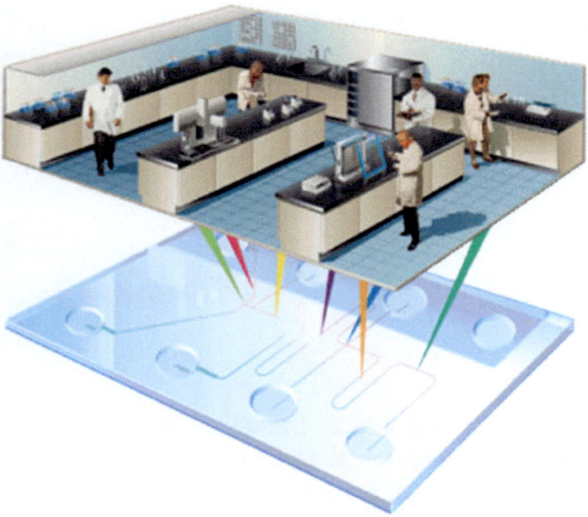

Fig. 10.8 The lab-on-a-chip can carry out the same work as a conventional laboratory for biological analysis, but on an area 1,000,000 times smaller. *Source* From the French biotechnology site www.biorigami.com

single biological sample and thus supply the doctor with vastly more information than was previously possible (see Fig. 10.8).

The best known examples of labs-on-chips were those developed to decode the genome or proteome using DNA fragments or antibodies as recognition biomolecules. The company Affymetrix currently leads this market for research (see Fig. 10.9). Other labs-on-chip can be used for early diagnosis of disease, but also for therapeutic monitoring.

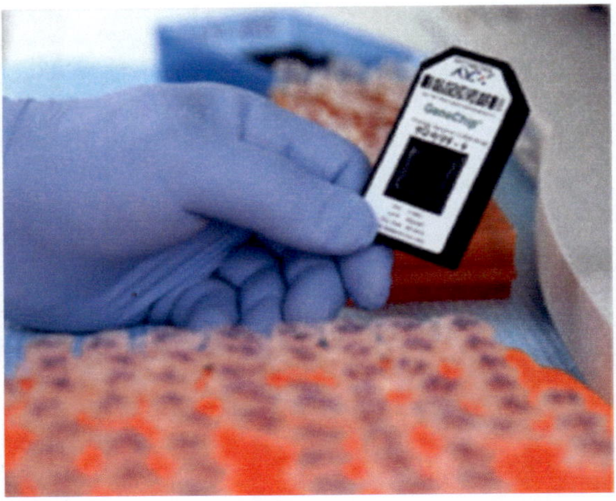

Fig. 10.9 A DNA chip called GeneChip, made by Affymetrix GeneChip. *Source* Rhoda Baer, Wikimedia

Fig. 10.10 Lab-on-a-chip. Integrated analysis system InCheck. Length 20 mm. *Source* STMicroelectronics

Another miniaturised laboratory known as a cell-on-a-chip uses whole cells as systems to detect intoxication or to analyse pathogenic germs. Human cells are used in particular as ultra-miniaturised analysis laboratories. The French company Cytoo has thus developed cells-on-chips for high-throughput screening of pharmacological substances.

All these miniaturised laboratories are the size of a postage stamp (see Fig. 10.10), and today microfluidics can accurately control the displacement of fluids using micropumps, microvalves, micromixers, and so on. Labs-on-chips can be fully automated and provide a full analysis from a single sample. Such systems can be used for the detection of avian flu and swine flu.

Nanotechnology contributes to improving *in vitro* diagnostics in two main points: point-of-care diagnosis, i.e., at the patient's home, at the bedside, at the hospital, or in the GP's surgery, and diagnosis in a central laboratory. In both cases, miniaturisation, integration, and multiplexing make it possible to design and make push-button, automated, and ever smaller systems. The automation of analysis is important in that it eliminates the risk of human error in sample manipulation.

Centralised laboratories require high performance automated analysis systems particularly with regard to throughput, speed, and accuracy. Today, manufacturers focus mainly on equipping these laboratories, which are usually in hospitals, but given the huge numbers of samples and the likely saturation of this kind of centre, we may expect a certain fraction of these analyses to be delocalised to specialist doctors, closer to the patient. We can even envisage some tests being carried out by the GP, or even by the patients themselves in certain simple cases. In that case, we shall need robust automated devices that are easy to use, without the need for any kind of training or special skills on the part of the user, providing a quick and clear response, easily interpretable by the GP or the patient.

10.4.2 Medical Imaging Revisited

Being able to see inside the human body has always been, and remains, a major challenge for medicine. For centuries, we have tried to explore the human body, to understand how it works or cure it of some ill (see Fig. 10.11).

Medical diagnosis *in vivo* usually appeals to imaging methods, but sometimes also to implanted sensors (see Sect. 10.4.3). Molecular imaging visualises molecules or molecular assemblies. The most common methods are optical imaging, various spectroscopic techniques, nuclear imaging with radioactive tracers, magnetic resonance imaging (MRI), ultrasound (echography), and X rays. Some use tracers that must first be injected into the body before the clinical examination.

Over the past 25 years, medical imaging has led to tremendous progress. Previously, only the morphological alterations of tissues could be visualised, and at an advanced stage of the symptoms, by X-ray or echography, for example. Then con-

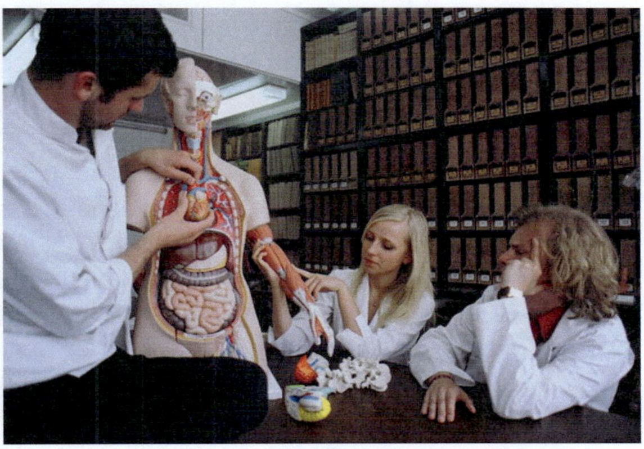

Fig. 10.11 Anatomy lesson at the University of Gdansk. *Source* Courtesy of Jolanta Świerczyńska

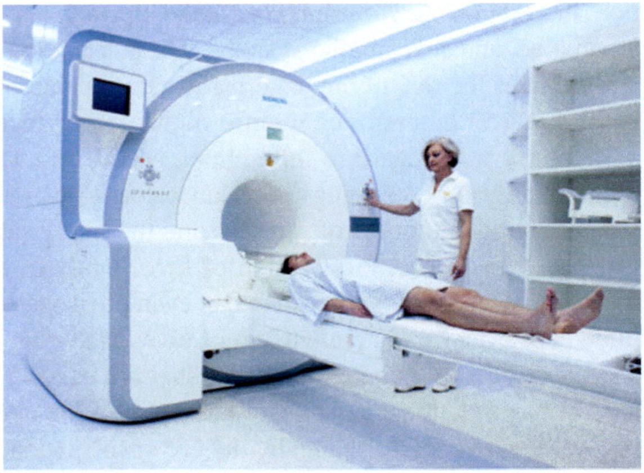

Fig. 10.12 The Biograph mMR scanner made by Siemens can simultaneously carry out MRI and positron emission tomography (PET). It is the first of its kind and was approved in 2010. *Source* Courtesy of Siemens AG

trast agents, or tracers, were introduced to improve identification and localisation of pathologies. Today, progress in the development of nanocarriers means we can envisage even earlier and more accurate localisation of disease, whence it should be possible to implement therapy at a very early, anticipative stage, at the very first signs of cell disorder.

It is also possible to associate different imaging modes in order to enrich image quality, e.g., echography and scanner, or MRI and fluorescence (see Fig. 10.12). We are thus moving toward the doctor's dream of viewing disease on the cellular level and monitoring the effects of therapy. Indeed, we now speak of theranostics, which associates the notions of therapy and imaging diagnosis. We shall return to this new area of research (see Sect. 10.6). Hence the European Sonodrug project (2008–2012), run by Philips Healthcare, developed a therapy in which the release of drugs in a target organ is activated by ultrasound, after combined ultrasound–MRI imaging [3].

Targeted molecular imaging can thus be used to identify inflammation and tumours and their stage of development, as well as visualising vascularised tissues and anatomic regions (see Fig. 10.13). Targeted molecular imaging is also very useful for studying the mechanism, transport, accumulation, and elimination of new drugs.

The localisation and identification of a specific tissue like a tumour, atheromatous plaque, or inflammatory tissue requires special targeting in which the imaging tracer accumulates solely, or almost solely, in the targeted organ (see Fig. 10.14). To achieve this, homing molecules must be grafted onto the nanocarrier, that is, targeting ligands that will specifically recognise the target tissue. Indeed, it is through an interaction and a bond between the homing molecule and the molecules at the surface of the target tissue that the tracer will fix onto the region to be imaged. This idea of active targeting thus involves a ligand and a receptor. The more specific and sensitive this recognition interaction, the better will be the diagnostic. This is the grail sought by the imaging and pharmaceutical industries.

The biomolecules used for targeting can be antibodies, peptides, or sugars. A key advantage of nanoparticles is that several targeting ligands can be grafted onto their

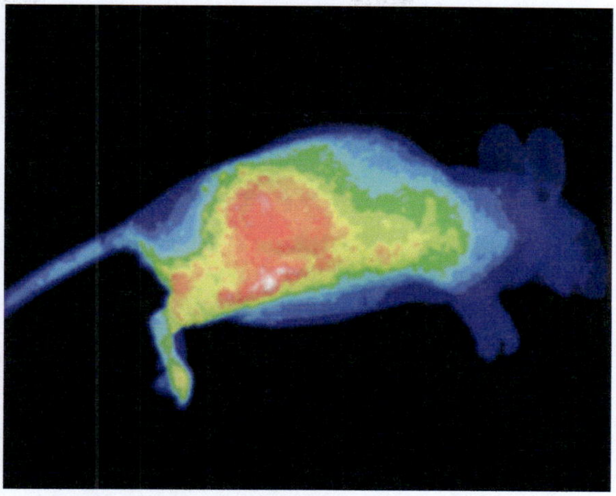

Fig. 10.13 Mouse carrying a xenografted tumour. Fluorescence optical image. *Source* Dr Jean-Luc Coll, *institut Albert-Bonniot*

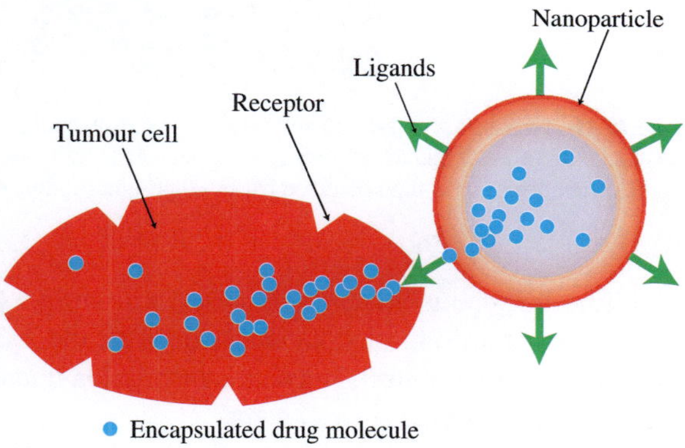

Fig. 10.14 Just as an archer aims at the target, so nanocarriers aim at the sick organ or a tumour, using specific recognition molecules. Adapted from [4]

surface, thereby increasing the probability of recognising the target tissues and fixing onto them.

10.4.3 Smaller and Smaller Implants and Sensors

The ultra-miniaturisation achieved by nanotechnology means that very small and weakly invasive devices can actually be implanted within an organ. Examples are artificial retinas and cochlear implants which can be installed permanently, but also devices that can be swallowed like pills and carry out imaging or analysis of the digestive tube (see Fig. 10.15). The measurement of circulating molecules such as glucose in the case of diabetes or infection biomarkers in the blood or other biological fluids is another promising application (see Fig. 10.16).

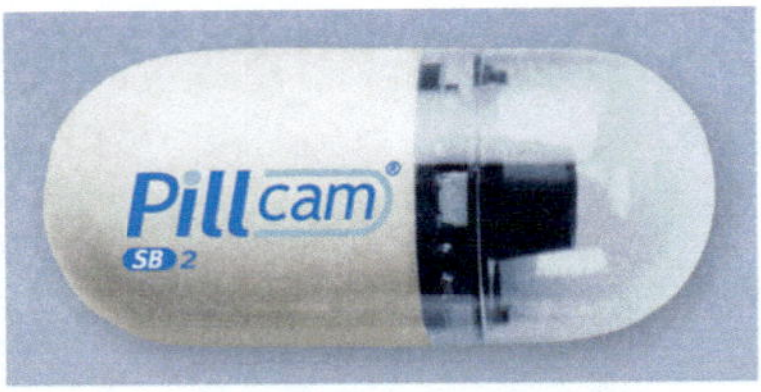

Fig. 10.15 PillCam® COLON capsule, developed by Given Imaging, contains a mini-camera that images the inner wall of the colon. *Source* Given Imaging Ltd

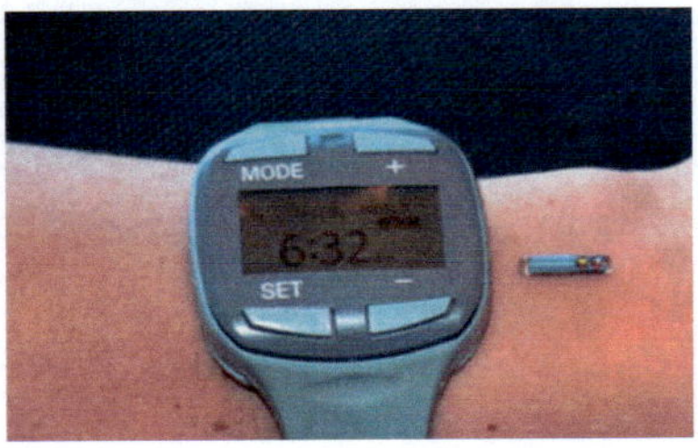

Fig. 10.16 Non-invasive glucometer. Glucose is extracted through the skin by reverse iontophoresis, i.e., extraction of electrically charged molecules from a biological medium under the effects of an electric current. It is then detected by an amperometric biosensor. From [5]

While ultra-miniaturisation combined with surface functionalisation with selected molecules can lead to biocompatible devices, the issues of energy autonomy, self-diagnosis, remote control, and data transmission remain considerable challenges for the implementation of such systems.

In their early stages, certain pathologies are only manifest through small amounts of biomarkers in the blood. These are molecules attesting to the existence of a physiological disorder. If they were more numerous, they could be detected by taking a blood sample, and this is indeed the conventional method for detecting such disorders. However, blood is a highly complex medium, containing huge amounts of different molecules. One only needs to glance at the results of a blood test to see this! Hence, the chances of detecting a low concentration biomarker in a blood sample become hopeless, especially since proteins like albumin and immunoglobulins which are present in very large amounts tend to mask the minority proteins.

Nanoparticles can then provide an interesting alternative way of capturing these markers directly in the blood vessels. This is known as nano-harvesting. Nanobeads wrapped in a biocompatible bag, rather like a tea bag, can be designed with specific molecules grafted to the surface to pick up the relevant biomarkers. The nanobeads are then extracted, e.g., by magnetic sorting, and analysed. This idea is being studied by several laboratories, but is not yet used in a routine manner in the clinic [6].

Tissue sampling, otherwise known as biopsy, is another area where the miniaturisation made possible by nanotechnology has opened new horizons. However, it is impossible, or extremely difficult, to take samples from certain organs like the brain,

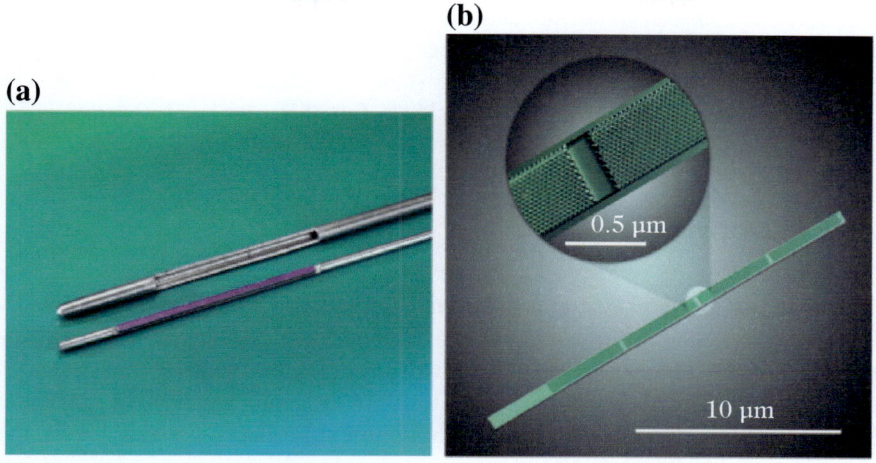

Fig. 10.17 **a** Protool medical device for sampling cells and biomolecules. **b** Enlarged view of the nanostructured capture surface. *Source* CEA-Leti

while it would be useful to establish, for example, the exact localisation and stage of development of a brain tumour. A tool has thus been developed to sample circulating biomolecules which can provide information without incurring damage to the brain (ProTool technology developed by CEA-Leti, Grenoble, see Fig. 10.17).

This tool is in fact a stylus. Its surface is roughened by nanoreliefs in the form of a multitude of tiny pillars on which the circulating molecules or cells of the brain will adhere. These nanopillars can also be coated by molecules for specific recognition of a molecule or range of molecules. The stylus is then analysed by mass spectrometry to identify the harvested molecules. It is only the very small dimensions of the stylus and its very large surface area for exchange and capture that make this kind of sampling possible for the first time.

10.4.4 Personalised Medicine

In an ideal world our doctor would be able to diagnose a possible illness so early on that almost no damage would yet have been incurred. He would then treat the disorder without side effects, and without consequence for the surrounding healthy tissues.

The tracers and detection systems that have brought such spectacular success to *in vivo* diagnosis are destined to become more and more widely used. Today's slogan in medicine is thus: *find, fight, and follow*. This combination of therapy and diagnosis is itself subsumed under the heading of theranostics, a term which covers a wide range of situations: imaging can track the distribution of a drug right up to the target organ, but it can also activate the release of the drug only when it reaches this target.

Some nanocarriers can simultaneously transport imaging tracers and pharmaceutical active principles, whence these two molecules will be co-injected. The ultimate goal of the doctor is still to visualise and understand the reactions occurring within the human body, which remains a world unto itself, full of mystery and surprises.

10.5 Therapy

The traditional aim of pharmacy is to formulate and administer drugs (see Fig. 10.18). It is an approach that can effectively treat a broad range of pathologies. Pharmaceutical active principles are often small chemical molecules, or sometimes biomolecules like proteins, peptides, or nucleic acids. However, systemic injection, i.e., injection into the blood circulation, can produce prohibitive side effects. For this reason, the pharmaceutical industry invests in the search for alternative ways of administering active principles. Encapsulation within carriers is an attractive option because the carrier protects the active molecule while it is circulating in the blood. It also protects healthy tissues from the drug and improves the efficiency of the treatment by delivering the drug solely to the site that needs to be treated.

Over the past thirty years or so, nanotechnology has become more and more attractive in this field thanks to the wide variety of nanocarriers that can be synthesised

Fig. 10.18 The *Musée François-Tillequin* houses the collections of the pharmacognosy laboratory at the University of Paris-Descartes. It currently contains around 25,000 samples of plant and animal extracts, together with objects associated with their production, transport, trade, and use. In pharmacy, these extracts could be any natural substance conserved by drying and stored for the purposes of preparing drugs. Courtesy of Sylvie Michel, *Musée François-Tillequin*, previously *Musée de la matière médicale*

depending on the active principle to be transported and the tissues to be targeted. This is currently one of the most active fields of research. Chemists have demonstrated immense, even limitless creativity. A vast range of nanocarriers has already been designed for applications in pharmacy.

10.5.1 Drug Nanodroplets and Nanocarriers

In the short and mid-term, molecular assemblies will be developed in nanomedicine with a view to targeted delivery of drugs and pharmaceutical active principles. These molecular assemblies are referred to as nanocarriers, nanodroplets, nanodrugs, etc. (see Figs. 10.19 and 10.20). There have been three successive generations of nanocarriers:

- The first and best known are the nanospheres and nanocapsules with solid inorganic core.

Fig. 10.19 Lipidot® lipid nanodroplets for drug encapsulation. The different colours correspond to different encapsulated molecules. *Source* CEA-Leti

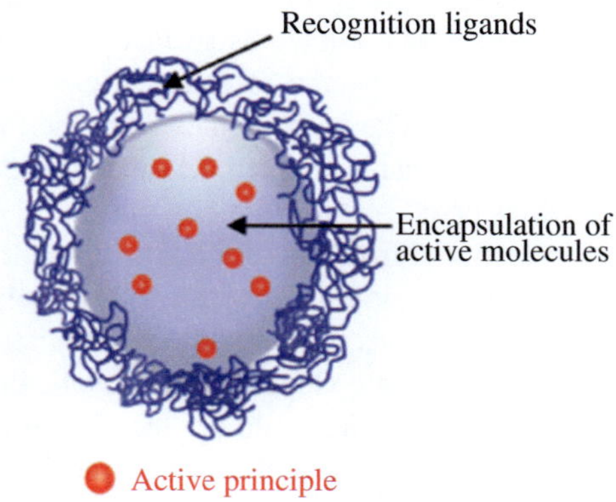

Fig. 10.20 Schematic view of a drug nanocarrier

- The second are particles coated with hydrophilic polymers such as polyethylene glycol (PEG). They are often called PEGylated particles. These particles have the great advantage of being barely visible, or even invisible, to the immune system when injected into the blood circulation. They are then called stealth particles.
- The third concerns PEGylated particles, like the second generation, but now with molecules grafted onto their surface which allow them to target particular pathological tissues such as a tumour. Some of these particles also have a biodegradable lipid, peptide, or polymer core, which facilitate their degradation and elimination by the metabolism of the organism. These particles have mobilised many researchers because they combine stealth, which prevents their immediate elimination, targeting, which concentrates them in the target tissue, and biodegradability, which avoids unnecessary accumulation in the organism.

Today, most academic and industrial research focuses on third generation particles (see Chap. 11).

Chemotherapy, which is routinely used for cancer treatment, administers highly toxic chemical substances that kill cancer cells. Unfortunately, these substances also kill the neighbouring healthy cells, producing unwanted side effects such as nausea, kidney disorders, hair loss, tiredness, and weakened immune response. The great advantage when using nanoparticles to carry these substances in a targeted way directly to the pathological tissue is that healthy tissue remains largely unaffected and unwanted side effects are limited.

In short, the advantages of nanocarriers are:

- to protect the drug from degradation by proteins present in the blood during transport in the blood circulation,
- to improve absorption of the drug in the target cells,
- to improve the duration and distribution of the drugs in the target tissue,
- to prevent the drug from interacting with healthy tissues.

Targeting of a particular tissue like a tumour, an atheromatous plaque, or an infected or inflamed region exploits two sometimes complementary approaches (see Fig. 10.21). The first is active targeting, in which molecules or ligands grafted onto the nanocarrier surface recognise specific molecules that are particularly abundant on the surface of the target tissue, known as surface biomarkers. The second form is called passive targeting. It is specific to nanoparticles, organic or otherwise, but always smaller than 150 nm. The name comes from the fact that only the size of the nanoparticle determines what is targeted, i.e., there is no molecular recognition, in contrast to the previous case.

Indeed, the nanoparticles arrive through the walls of the blood vessels irrigating the tumour, exploiting imperfections or holes in the wall, which allow such leakage. The nanoparticles then slip into the interstitial tissue of the tumour and accumulate there, for there is no longer the lymphatic drainage that would exist in healthy tissue. As the nanoparticles build up, they begin to penetrate the tumour cells. This is the so-called EPR effect (enhanced permeability and retention), raising great hopes for nanoparticle drug delivery.

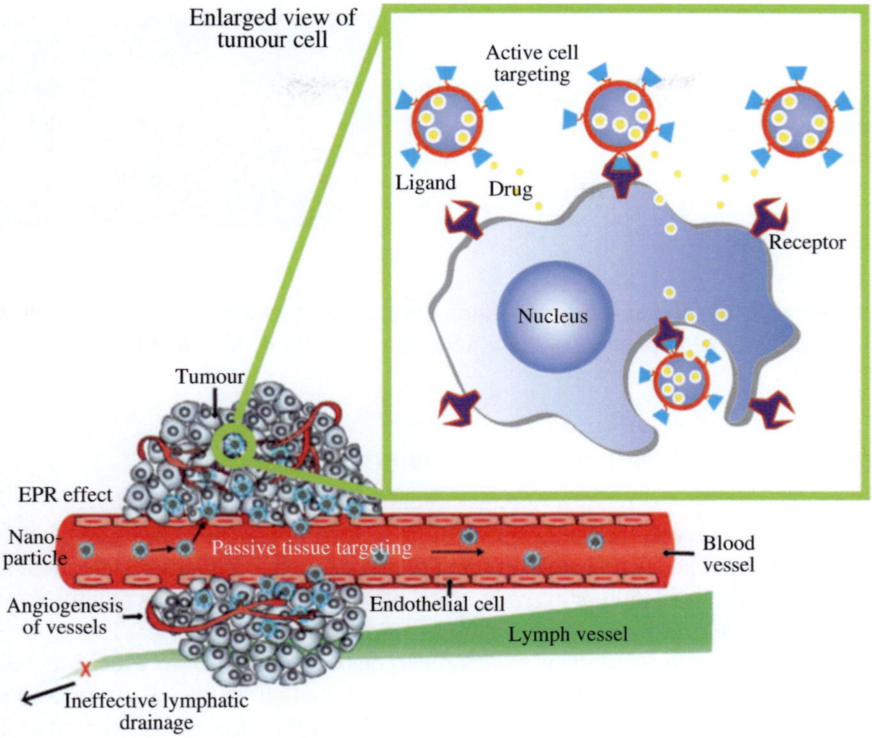

Fig. 10.21 Active and passive tumour targeting. Figure inspired by an illustration in Nature Nanotechnology. *Reference* http://www.nature.com/nnano/journal/v2/n12/images/nnano.2007.387-f1.jpg

10.5.2 Drug Pumps

An alternative to chemotherapy or injection of drug-bearing nanocarriers into the blood circulation is drug delivery by implanted microsystems that have been miniaturised by nanotechnology. An example is provided by the miniature pump known as Nanopump, developed by the Swiss company Debiotech. These are microelectromechanical systems (MEMS), which include one or more mechanical elements and use electricity as energy supply to function as a sensor and/or actuator with at least one structure having micrometric dimensions.

Nanopump is an implantable microsystem, like a pacemaker, designed to deliver finely dosed drugs in an adjustable manner over long periods of time. Such microsystems can be implanted in the brain, for example, to diffuse the drug as close as possible to the region requiring treatment. Nanopump was developed to deliver insulin (see Fig. 10.22). Thanks to the accuracy of the microfabrication and microtechniques, which results from accumulated knowhow in microelectronics, very small systems

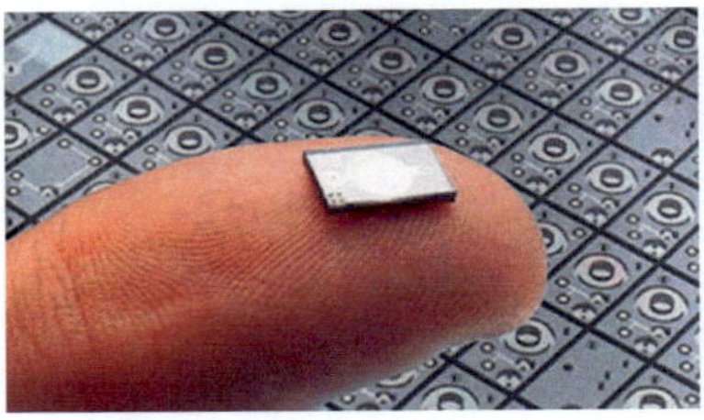

Fig. 10.22 Nanopump. Miniature pump for delivering insulin, under development at Debiotech. © Debiotech SA, Switzerland

can be made containing reservoirs, pumps, and actuators which closely control drug release. Certain components of these microsystems may be nanometric.

10.5.3 Physical Destruction of Tumours

Another alternative to chemotherapy, or more exactly a complementary approach, is physical rather than chemical destruction of the tumour. The conventional method has been surgical ablation or exeresis. Another widely used method is radiotherapy, in which tumours are exposed to ionising radiation which destroys cells. This process can be further improved by using nanoparticles made from heavy ions which increase the ionising effect and amplify the effects of the radiation on the tumour cells (see Fig. 10.23). The nanoparticles are first injected into the tumour, and the latter is subsequently irradiated with conventional radiotherapy equipment, with the same

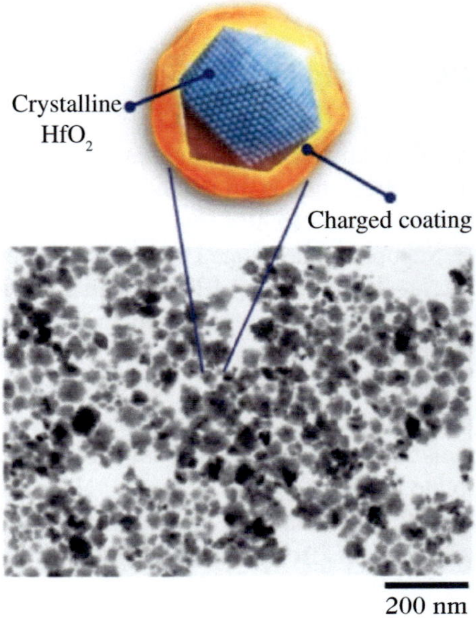

Fig. 10.23 The NanoXray nanoparticle for radiotherapy. These nanoparticules have been undergoing clinical trials since 2011. © Nanobiotix

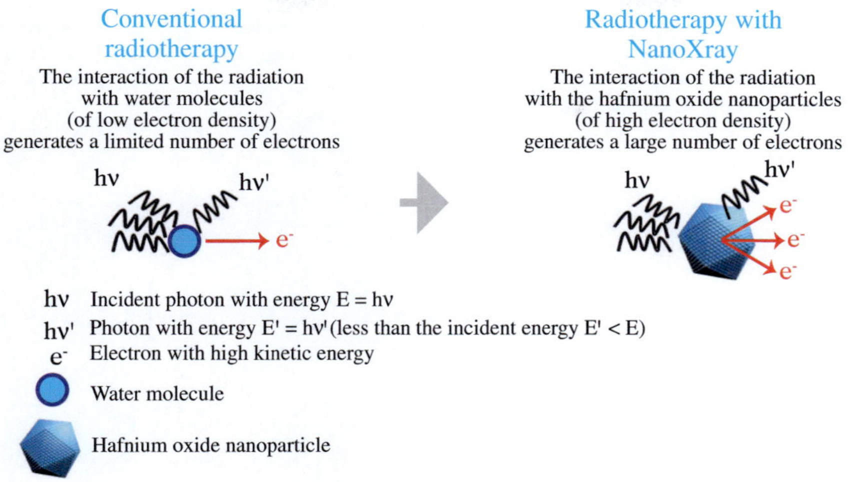

Fig. 10.24 Comparison between conventional radiotherapy and radiotherapy using NanoXray nanoparticles. Figure inspired by an illustration from Nanobiotix

protocol as is currently used in radiotherapy (see Fig. 10.24). The elegance of the process lies in the fact that it barely changes existing clinical radiotherapy protocols, while significantly enhancing their efficiency.

Nanotechnology can also improve another long-standing physical therapy, namely, local hyperthermia. The idea here is to inject magnetic metal nanoparticles into the tumour, then expose the site to a magnetic field like the one provided by a magnetic resonance imaging system (see Fig. 10.25). Under the effect of the magnetic field, the nanoparticles become local heat sources. The German company MagForce has carried out trials to deliver such magnetic particles by the blood system, and they have received authorisation to commercialise their NanoTherm® particles. Nanoparticles responding to infrared (IR) radiation rather than a magnetic field have also been used

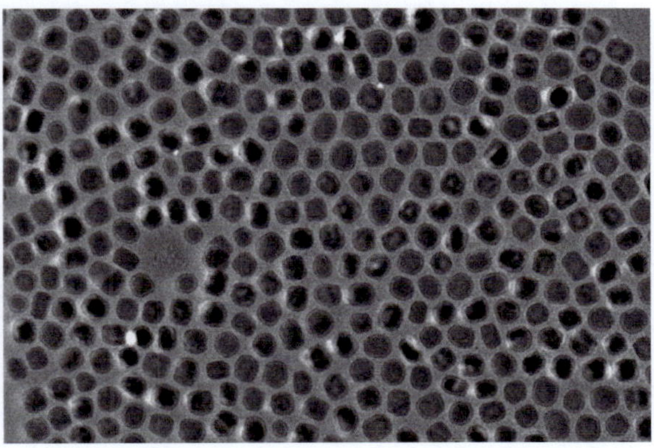

Fig. 10.25 Iron oxide nanoparticles with diameter 10–15 nm, developed by MagForce for radiotherapy. © Boutiq 2014

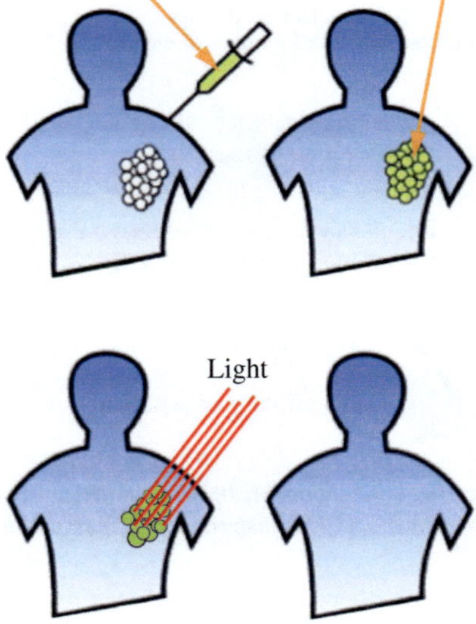

Fig. 10.26 Photodynamic therapy (PDT). Adapted from Belfield Research Group, University of Central Florida

successfully. The nanoparticles are agitated by the IR radiation and once again cause highly localised heating which literally burns the tumour cells. The beauty of local hyperthermia, either infrared or magnetic, lies in the extremely precise localisation within the tumour, thus leaving the surrounding healthy tissues intact. This condition is particularly important when treating brain cancers.

Finally, nanotechnology and the use of chemical nanocarriers, in particular, may provide a way to improve physical treatment of tumours in a method known as photodynamic therapy (PDT). The main steps are summarised in Fig. 10.26.

10.6 Combining Diagnosis and Therapy: Theranostics

The ability of certain nanoparticles to transport simultaneously a contrast agent and a pharmaceutical active principle opens highly innovative prospects in medicine (see Fig. 10.27). This is an example of theranostics, a term coined by contracting 'therapeutics' and 'diagnostics' [7]. This combination can appear in many forms. For example, imaging may be able to track the distribution of a drug in the body and check that it is indeed concentrating effectively in the target. Moreover, laser irradiation, localised heating, or ultrasound can then be used to activate from the outside and release the relevant drugs.

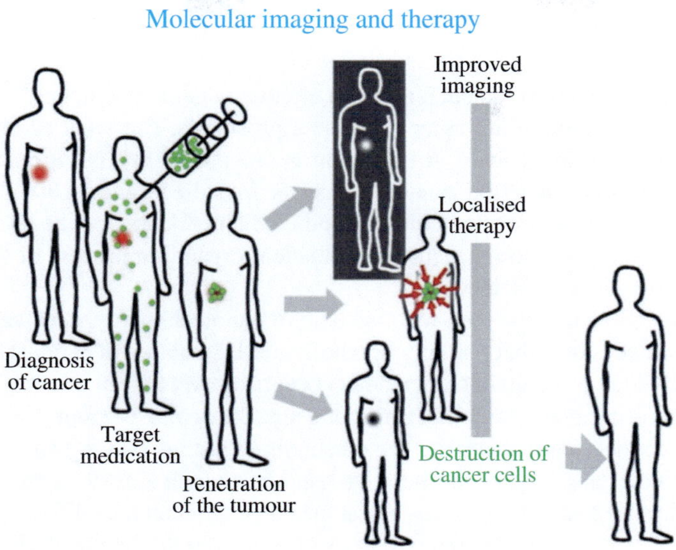

Fig. 10.27 Different clinical stages in which nanoparticles can improve existing practice

Nanotechnology can also be used to design contrast agents to suit specific imaging modes, such as optical imaging (see Fig. 10.28). Moreover, the synergy between *in vitro* diagnostics and *in vivo* imaging can be exploited to target the treatment or ablation of tumours. The tumour is first marked out by nanoparticles which specifically recognise its surface, whereupon it can be identified by imaging. If the nanoparticles can also react to external stimuli such as microwaves, a magnetic field, or laser irradiation, a well localised and non-invasive destruction of the tumour then becomes possible (see Sect. 10.5.3).

Fig. 10.28 Fluobeam® probe for intraoperative optical imaging, using contrast agents encapsulated in nanocarriers. *Source* Fluoptics.com

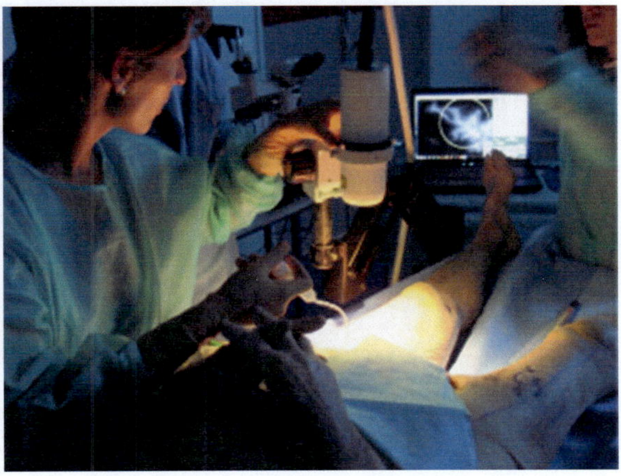

10.7 Regenerative Medicine

How can we make up for a deficient organ after damage or destruction? One rather primitive way to replace an organ is to use a prosthesis. But what can be done if the organ is the heart or liver? A transplant is one possibility, but it raises serious problems of immune reaction in some cases (see Figs. 10.29 and 10.30). A new and rapidly expanding branch of cell biology and medicine focuses on the possibility of helping the body itself to regenerate the deficient organ in a process of self-repair. This is regenerative medicine.

It is reasonable to hope that, with the help of nanotechnology, new ways will be found to compensate either partially or wholly for degradation of certain parts of the human body such as organs or cells, whether occurring over time or present from birth, as for example in patients suffering from congenital muscular dystrophy or myopathy, or indeed cardiac malformations. Nanotechnology is thus playing an ever greater role in the emerging field of regenerative medicine which aims to repair or replace damaged tissues and organs using natural means or the methods of bioengineering.

A first approach to regenerating tissues or organs in the human body, damaged by aging, sickness, accident, or some genetic cause, is to stimulate the tissues in the body so that they actually repair themselves. When the human body is no longer able to repair itself, a second approach is to grow tissues or organs in the laboratory and then implant them in the body. In both cases, regeneration and also fabrication

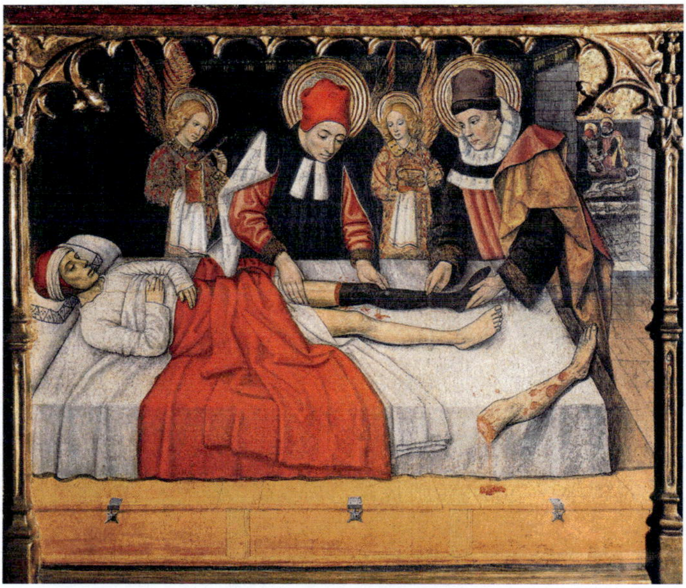

Fig. 10.29 Legendary transplantation of a Moor's leg on Justinian by Saints Cosmas and Damian. Ditzingen, sixteenth century. © Aisa/Leemage

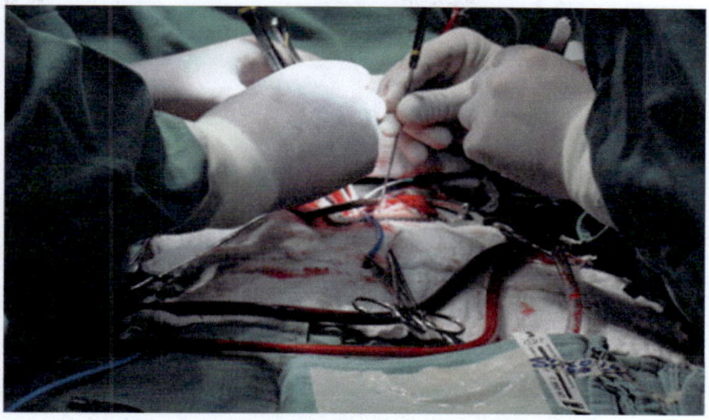

Fig. 10.30 Heart transplant. *Photo credit* Jerry Hecht

of tissues can be achieved by combining living cells, which supply the biological functionality, with biomaterials, which serve as a support for the growing cells.

Regenerative medicine thus raises great therapeutic hopes, but it depends on an extensive knowledge and understanding of cell biology, and in particular, tissue repair processes and stem cell differentiation mechanisms, not to mention a strategic method for instigating, then controlling regenerative processes *in vivo*.

10.7.1 Stem Cells and Cell Therapy

The main aim of current research effort in regenerative medicine is to exploit the huge self-repair potential of adult stem cells. Given the logistic complexity and high costs of current therapies based on autologous re-implantation of differentiated cells multiplied *ex vivo*, it is clear that the next generation of therapies will benefit from progress in tissue engineering in order to use undifferentiated cells such as adult stem cells.

Nanotechnology will contribute to this area of therapeutics. On the one hand, it will provide a way to identify the signalling systems that regulate the self-repair potential of endogenous adult stem cells, and on the other, it will provide systems for targeted delivery of therapies based on these cells.

Recently, damaged neurons have been repaired and regrown by injecting molecules designed to self-assemble in the form of nanofibres in the tissues of the spinal cord [8]. As they form, the nanofibres localise in the tissue areas where they activate certain biological processes able to repair the damaged cells. Samuel Stupp and coworkers at the Northwestern University in the USA have also discovered that these nanofibres can steer the differentiation of stem cells toward production of neurons. These discoveries may have important spin-offs for the treatment of Parkinson's and Alzheimer's diseases in which brains cells cease to function as they should.

Another promising strategy for regenerative medicine uses nanostructured implants impregnated with cells able to send differentiation signals to adult stem cells already present at the implantation site (see Fig. 10.31). Having said this, the reac-

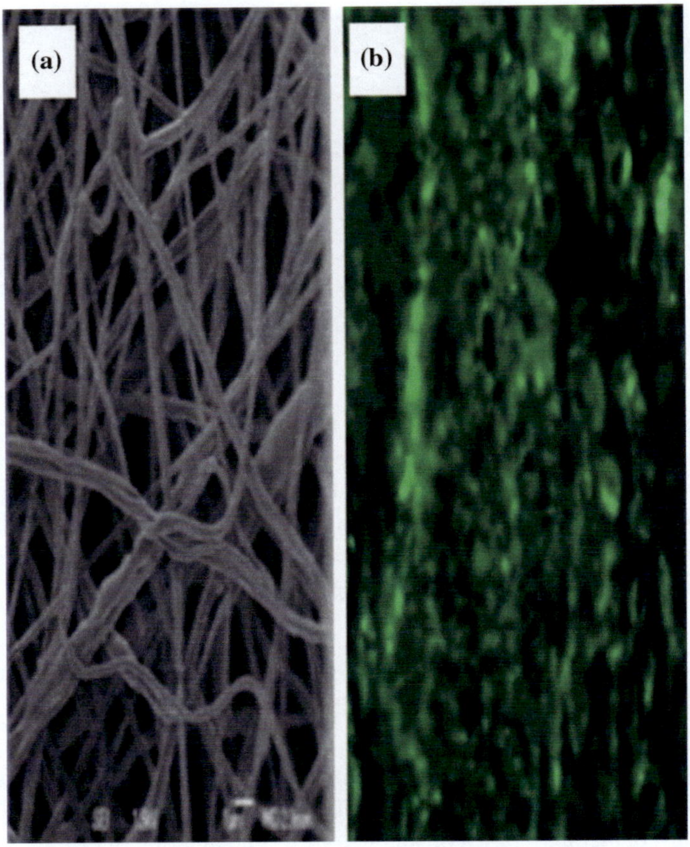

Fig. 10.31 **a** Polycaprolactone (PCL) membrane fabricated by electrospinning. **b** Bone tissue induction *in vivo*. Expression of osteopontin (*green*) on the PCL membrane. Courtesy of Nadia Jessel, INSERM U977, Strasbourg

tion of adult stem cells to a nanomaterial implanted at the regeneration site remains unknown at the present time, and preparatory studies will thus be necessary. Instead of using nanostructured implants impregnated with cells, an alternative would be to implant only an intelligent bioactive material that would itself produce the differentiation signals regulating the self-repair potential of the patient's stem cells.

10.7.2 Biomaterials

As we have just seen, tissue engineering has recently been evolving toward hybrid materials combining nanostructured biocompatible materials, ideally capable of self-assembly (organic polymers or inorganic materials), with living cells to replace defective tissues. The aim is thus to minimise any risk of subsequent rejection by the immune system. The techniques mainly involve implantation of cells, tissues, or molecular scaffolds, reconstructed *in vitro*. At the heart of stem cell research is control of cell differentiation. Indeed, stem cells, which can differentiate into various cell types, will probably prove to be the key to any serious attempt at tissue regeneration, even though this line of research is still in its early stages and there are no tangible results for the time being.

Mammalian cells react *in vivo* to features of their environment, whether they be chemical substances or physical structures such as nanostructures. So today we seek to stimulate the differentiation of adult stem cells directly by means of nanostructured surfaces, i.e., nanoreliefs whose form and periodicity can guide differentiation toward a given cell type (see Fig. 10.32). Reliefs of various kinds, such as spheres, fibres, sponges, or tree-like filaments can be etched on the surface of a biomaterial. In the best cases, the stem cells, which are sensitive to their immediate physical and chemical environment, will indeed differentiate and evolve toward the cells of the organs one would like to regenerate.

Biomaterials implanted in an organ to repair it and assist in the growth of stem cells must therefore reproduce the differentiation signals and if necessary deliver active principles to complete or direct this differentiation. Furthermore, they must simultaneously increase the tissue regeneration capacities, while at the same time reducing the immune response and avoiding infections. Promising trials have already been carried out to regenerate bone, cardiac, cartilage, and neural tissues. Finally, not only must they be able to interact with proteins or cells without disturbing their biological activity, but they must also conserve their physical properties, even after chemical functionalisation of their surface, while remaining biocompatible and non-toxic.

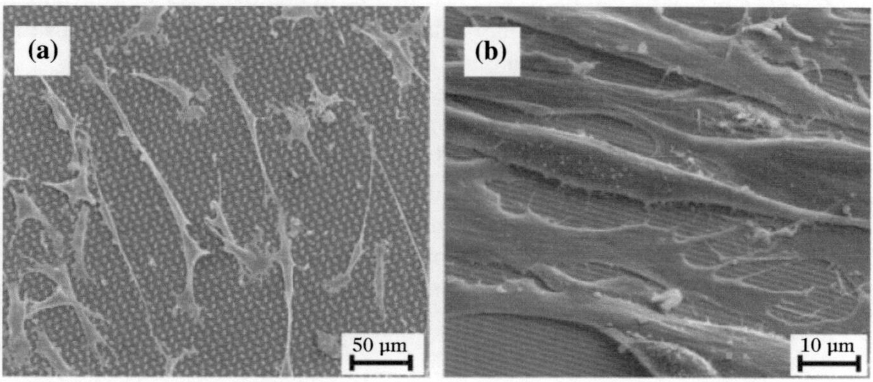

Fig. 10.32 Scanning electron microscope image of rat mesenchymal cells, grown on **a** a poly(methyl methacrylate) (PMMA) surface structured with 2D reliefs with characteristic size 2 μm and **b** a PMMA surface structured with etched grooves of width 2 μm. From [9] © 2009 Elsevier. Reproduced with permission

The controlled fabrication of intelligent biomaterials fulfilling the above conditions provides an interesting opportunity in the context of nanotechnology for the regeneration and differentiation of adult stem cells *in vivo* and the development of regenerative medicine as a whole.

10.8 Is Nanotechnology Essential to Medicine?

Nanotechnology has already brought much progress to medicine, and will continue to do so in the future. Given the similar size of nano-objects and biomolecules, medical nanotechnology may look like a new field of medicine, promising a revolution in diagnosis and therapy. Just as the invention of microscopy revolutionised microbiology, techniques for visualising and manipulating nano-objects open the way to further applications in medicine. At the very heart of diagnosis and therapy lies the ability to see, displace, modify, and transport biological or medicinal molecules. It is therefore quite legitimate to expect nano-objects and nanotechnology to have a real impact in medicine. The previous chapters have reviewed a great many applications, some of which will be developed on a large scale.

However, experience built up in research and development over the past few years requires us to moderate this somewhat optimistic view. Indeed, there are many counterexamples in which nanotechnology does not in fact provide the right response. An example is the detection of rare biological species in very low concentrations. When we analyse a very small sample, the probability of detecting a rare molecule is low, unless there has been a prior concentration phase. The preparation of biological samples, and in particular the step from a macroscopic to a nanoscopic sample, or from a complex to a pure sample, is also crucial. It is a considerable technical feat to concentrate and assay a few molecules in a volume of liquid the size of a swimming pool. The same goes when assaying biomolecules in very low concentrations in a blood sample. Extreme miniaturisation is not always the best solution.

Concerning nanomaterials used *in vivo*, we must clearly demonstrate their innocuousness and determine exactly what becomes of them later on. Such demonstrations obviously depend on the intended clinical applications in a risk–benefit analysis.

Finally, the cost of a technology is a deciding factor in the acceptance of such innovations within a health system, given the strict economic constraints. Silicon-based devices can be mass-produced at low unit cost, as happens for mobile phone components. In contrast, costly production processes, like those required for certain chemical syntheses or in certain cases for assembly of components, may well spell the end for some forms of nanotechnology.

The benefits of medical nanotechnology must therefore be weighed up on a case-by-case basis against alternative techniques. If they nevertheless provide a low cost–benefit ratio while providing a better medical service than the one already in place, they may eventually come to be included in the range of medical technologies available to practitioners. In short, while nanotechnology clearly offers a route to progress, taking us towards new techniques that may complement existing methods, those new techniques must always be compared with the existing methods when assessing the risk–benefit ratio.

10.9 Nanomedicine and Ethical Questions

While nanotechnology undoubtedly opens the way to innovation in medicine, with the promise of many benefits to the patient, most developments are still at the research stage. Certainly, about thirty nanodrugs have already been approved in the United States and several microsystems have now been commercialised, but there is nevertheless a risk of overselling unrealistic medical applications in some cases and exaggerating the risks of nanomedicine in others.

The application of nanotechnology to medicine may raise delicate questions regarding the protection of genetic information, the gulf that separates diagnosis from therapy, or the impact of new methods on health expenditure. For example, should we commercialise a diagnostic test for an incurable disease? What does it mean to detect a disease when there is no apparent symptom? What does it mean to be sick if one feels in perfect health?

Nanotechnology shifts the dividing line between living and non-living through implanted microsystems, for example. This requires a delicate analysis of the ethical consequences of these developments according to the usual rules and standards of the European ethical committees. In order to carry out this analysis, nanotechnologists must first receive an education in ethics, while conversely, ethicists must be introduced to the reality of nanomedical developments, and not left to fall foul of rumour-mongering. This coevolution should ensure that nanomedicine gain acceptance both socially and ethically.

In Europe, the European Group on Ethics in Science and New Technology (EGE) is one of the relevant structures within the European Commission that deals with questions relating to nanomedicine (see Fig. 10.33).

Fig. 10.33 Members of the European Group on Ethics in Science and New Technology (EGE). *Source* European Ethics Committee

10.10 Regulating Nanodrugs

Medical nanotechnologies are used to fabricate what the regulators refer to either as drugs or medical devices. The French authority responsible for authorising clinical trials of new nanoparticles, their commercialisation, and the certification of medical devices is the *Agence nationale de sécurité du médicament et des produits de santé* (ANSM, previously AFSSAPS). In 2009, an AFSSAPS working group made recommendations for toxicological assessment of drugs involving nanoparticles, which serve as a guide for the various partners when designing such drugs. The general principles of conventional toxicology apply equally to nanomaterials, even though certain experimental protocols must be adapted to suit the specific features of these materials. The same is true for cell tests. However, no new regulations specific to the nanomaterials used in medicine have yet been judged necessary by the US Food and Drug Administration and the European Medicines Agency (EMA).

At the present time, 32 drugs and 5 imaging agents involving nanoparticles have been given the go-ahead commercially in the United States [8]. However, the regulatory bodies remain attentive to scientific progress that will lead to a better understanding of the interactions between nanoparticle assemblies, living organisms, and the environment, so that regulation can be adapted in consequence.

10.11 A Newly Emerging Industry

Medical nanotechnology is no longer a purely academic matter. There is also the beginnings of an industry and a market for products and services. However, the still rather vague definition of nanomedicine (see Sect. 10.1) and the realities of the industrial situation mean that few market studies provide clear and consistent estimates. There is no doubt that drug delivery represents the largest sector of application, while biomaterials will show the strongest growth over the period 2006–2015. The commercial nanomedicine field is fragmented, being made up of niche markets.

Examples of companies now active in the field of nanomedicine are Nanospheres Inc., Par Pharmaceutical Companies Inc., Elan Corporation PLC, Abraxis Bioscience, Flamel Technologies SA, Oxonica PLC, MagForce GmbH, Nanobiotix, Arrowhead Research Corporation, and BioAlliance.

The global market was estimated at 53 billion dollars in 2009, with an annual growth forecast of 13.5 %, expected to reach 130 billion dollars in 2016 (source BCC Research). Applications to cancer treatment represent one of the main market sectors at 65 %, i.e., some 20 billion dollars in 2009. The annual growth of this sector is 11 % and should reach 53 billion dollars in 2014. Applications of nanotechnology to diseases of the central nervous system form another major sector, with a value estimated at 1 billion dollars in 2009. With an annual growth of 11.1 %, this sector should represent some 18 billion dollars in 2014.

Even though there would not appear to be many companies engaged in these activities, a great many clinical trials involving nanoparticles are currently under way. Seventy clinical trials were thus listed in the United States for 2011 (NIH database), particularly in phases I and II. In fact there are four clinical phases (see Fig. 10.34). In the first phase, the new drug is tested for the first time on a small group of people to assess its tolerance, to determine acceptable doses, and to identify undesirable side effects. The drug then goes to phase II, where it is prescribed to a larger group of people to test its effectiveness and make a more detailed tolerance assessment. Phase III, also carried out on a large group of people, serves to confirm its effectiveness, to control side effects, to compare with other typical drugs, and to collect information which will improve safety if the drug is eventually certified. In the last phase, once the drug has been commercialised, studies are carried out to gather further information about its effectiveness on various populations and its longer term side effects, if any. A large proportion of clinical trials carried out today in this area prefigure the nanodrugs that will become available tomorrow.

The advent of nanotechnology in medicine can help to reduce certain health costs when it applies to major pathologies, by effectively reducing the cost of treatment, with fewer side effects and greater efficiency, and by reducing the necessary labour costs. On the other hand, nanotechnology may well increase health costs when it applies to pathologies with lower economic impact and when it adds to existing technologies or processes with an unfavourable cost–benefit ratio.

The typical chain of innovation in nanomedicine is of the type A shown in Fig. 10.35, i.e., discoveries and innovation are generally produced in high-tech start-

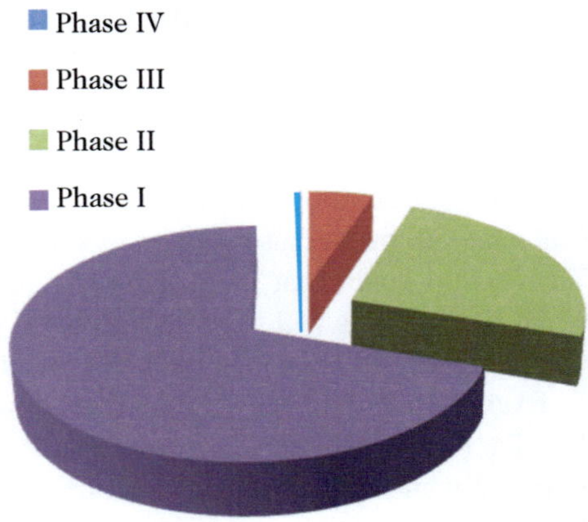

Fig. 10.34 Clinical trials with nanoparticles. Adapted from [10]

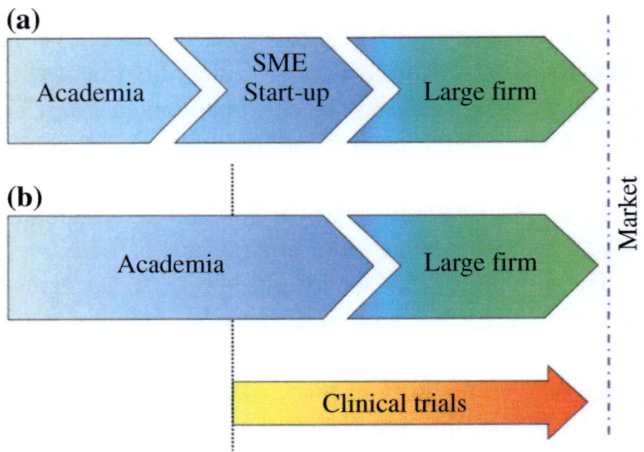

Fig. 10.35 Two chains of innovation in nanomedicine. **a** Through new start-up or SME firms. **b** Through the major industrial groups. Adapted from the European Technology Platform on Nanomedicine, 2013

ups or SMEs which validate their proof of concept and initiate the first clinical trials in a niche market. Thereafter, very large multinational firms acquire the most promising of the validated technologies and introduce them onto the market after clinical trials and certification by the appropriate regulatory bodies.

10.12 The Situation in France Today

Nanomedical research and innovation are being actively pursued in France. Since the beginning of the 2000s, research centres, universities, hospitals, then SMEs and several large firms have been investing and actively participating in various national and European initiatives and programs. Examples are the European program EuroNanoMed and the French program *Investissements d'avenir/Nanobiotechnologies*. Many French academic and industrial teams submit nanomedical projects to these programs and many are coming close to clinical transfer and industrial development.

The French ministries concerned with research and industry have considered nanotechnology to be a strategic priority for several years now. Since 2003, access to technology platforms has been facilitated by a skills network. Some 5 300 research scientists and around 250 research centres are involved in nanotechnological research in France, and the country currently ranks third in Europe after Germany and the United Kingdom in terms of publications, although it remains well placed as far as patent applications are concerned.

In 2009, the French Nano-INNOV initiative provided for investment in nanotechnology in three cross-disciplinary centres: Saclay, Toulouse, and Grenoble.

During the national public debate on nanotechnology held between 2009 and 2010 in 17 French towns, the theme of nanomedicine was the most favourably received, raising far less opposition than nanomaterials in manufactured or food products. Nanomedicine is thus perceived as part of medical research and as such comes under the existing regulatory framework.

The health industries involving nanomedicine are not classified as such. Nanotechnology is used by any industry producing drugs, medical systems, and onboard sensors, or general well-being. The biotechnology database of the Adebiotech committee lists around 50 companies operating in France in activities relating to nanomedicine. These include BioAlliance Pharma, BioMérieux, Carlina Technologies, Cezanne, Cytoo, Diatos, Etypharm, Flamel Technologies, Fluoptics, Genoptics, Guerbet, Imstar, InoDiag, Medsqual, Nano-H, NanoBioTix, and others.

10.13 Nanomedicine: Evolution or Revolution?

As for all the high technologies contributing to the development of health products where the market is very strictly regulated, the average time required to develop a nanomedical product is something like ten years before the first clinical trials. This is particularly true for molecules or systems that are administered *in vivo*. Given the huge research effort made in Europe in the 2000s, and more particularly in certain member states such as France, Germany, the United Kingdom, and Spain, a significant increase in the number of innovative nanomedical products is expected in the period 2014–2020. This is corroborated by the significant rise in the number of clinical trials now in phases I and II in Europe, viz., 25 in 2010 and 85 in 2011 [11]. A similar evolution is observed in the United States.

So all things considered, should we expect nanomedicine to be an evolution or a revolution? The discovery of the scanning tunnelling microscope in 1981 was a genuine revolution in physics, because it suddenly became possible to 'see' atoms, thus opening the way to nanotechnology (see Chaps. 2 and 5 and Fig. 10.36).

Today, nanotechnology is applied to a variety of areas from materials, to electronics, to medicine. As far as medicine is concerned, it brings improved performance as compared with the state of the art, but few applications as yet represent a genuine conceptual breakthrough. For the most part, what we see are incremental improve-

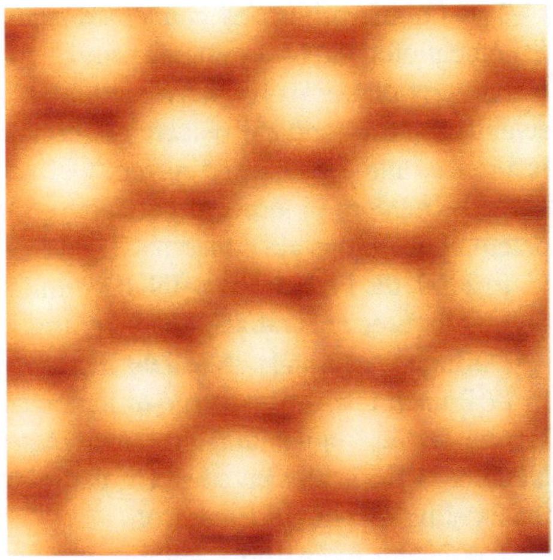

Fig. 10.36 Silicon atoms observed at the surface of a silicon carbide (SiC) crystal using a scanning tunnelling microscope. *Source* Guillaume Baffou, ISMO Orsay

ments, although much appreciated, which should often lead to much better patient care.

Let us hope that the huge potential of nanotechnology in medicine will be harmoniously put to use for the benefit of the patient, in a consensual context where hopes and fears are suitably assessed and weighed up.

References

1. www.cnrs.fr/cw/dossiers/dosnano/cine/film/anim/vectoEntier.htm
2. V. Poher, C.P. Allier, J.G. Coutard, L. Herve, J.M. Dinten, Lens free in-line holographic detection of bacteria, in *Advanced Microscopy Techniques II. Proceedings of SPIE*, vol. 8086 (2011)
3. www.sonodrugs.eu/
4. www.understandingnano.com/nanomedicine-nanoparticles-targeted-chemotherapy.html
5. A.E. Colvin, H. Jiang, Increased *in vivo* stability and functional lifetime of an implantable glucose sensor through platinum catalysis. J. Biomed. Mater. Res. **101A**, 1274–1282 (2013). © 2013, Wiley
6. P.D. Rakowska, M.G. Ryadnov, Nano-enabled biomarker discovery and detection. Biomark. Med. **5**, 387–96 (2011)
7. F.J. Picard, M.G. Bergeron, Rapid molecular theranostics in infectious diseases. Drug Discov. Today **7**, 1092–1101 (2002)
8. M.L. Etheridge, S.A. Campbell, A.G. Erdman, L. Haynes, S.M. Wolf, J. McCullough, The big picture on nanomedicine: the state of investigational and approved nanomedicine products. Nanomed.: Nanotechnol. Biol. Med. **9**, 1–14 (2013)
9. E. Martinez, E. Engel, J.A. Planell, J. Samitier, Annals of Anatomy (Anatomischer Anzeiger) **191**, 126–135 (2009)
10. http://nanocolors.wordpress.com/2009/03/08/nanoparticles-in-clinical-trials-repartition-by-phases
11. http://clinicaltrials.gov/

Chapter 11
Squalenoylation: A Novel Technology for Anticancer and Antibiotic Drugs with Enhanced Activity

Patrick Couvreur

Abstract This chapter describes the 'squalenoylation' technology, a platform for the discovery of new nanomedicines. The design of nanomedicines is generally based on the physical encapsulation, adsorption, or entrapment of a drug in a nanocarrier. This generally results in poor drug loading, and often an uncontrolled fast release of the drug (known as burst release). To overcome those limitations, the squalenoylation concept is based on the chemical (rather than physical) loading of drugs in nanomedicines. The idea is to link a biologically active compound (anticancer, antibiotic, antiviral, MRI imaging agent, etc.) to squalene, a natural and biocompatible lipid. Due to the unique, dynamically folded molecular conformation of squalene, the resulting squalene–drug bioconjugates self-assemble spontaneously in water to form nanoparticles. The resulting nanoassemblies have been shown to have enhanced pharmacological activity, and with reduced toxicity, thus paving the way to a new concept in the field of drug delivery.

Imagine a nanoparticle arriving at the surface of a sick cell to deliver drug molecules or a gene designed to reprogram the cell. This is the ultimate goal of nanometric pharmaceutical carriers, or nanocarriers. The idea was first formulated at the beginning of the twentieth century by the savant Paul Ehrlich, who referred to it as the magic bullet. However, the world had to wait until the 1970s before it was shown that artificial particles measuring a few hundred nanometres could penetrate into the heart of a cell, causing the accumulation of molecules that would not normally diffuse there: so this was the real birth of the nanocarrier.

More generally, the products of nanotechnology are not merely a miniaturisation of larger objects. Because they are nanometric, they have properties that would not be found on longer length scales. And the field of medication is no exception to this rule. Until the beginning of the 1970s, intravenous administration of pharmaceutical suspensions such as a dispersion of solid particles in a liquid was considered

P. Couvreur (✉)
Institut Galien, UMR CNRS 8612, Université Paris-Sud,
5, Rue Jean-Baptiste Clément, 92296 Châtenay-Malabry, France
e-mail: patrick.couvreur@u-psud.fr

J.-M. Lourtioz et al. (eds.), *Nanosciences and Nanotechnology*,
DOI 10.1007/978-3-319-19360-1_11

unthinkable, owing to the obvious risk of possibly fatal embolism. Today, however, nanoparticulate suspensions can be designed to carry drug molecules (nanodrugs) and thereby increase the therapeutic index of many compounds; that is, their activity is enhanced and/or their toxicity is reduced following intravenous administration. The transition from a size of a few tens of microns to a few tens or hundreds of nanometres has thus led to an important technological and medical breakthrough.

The expected advantages of nanodrugs are:

- protection of the active molecule against its own degradation,
- targeting of the active molecule to the relevant tissue, cell, or subcellular compartment (cytoplasm, lysosomes, nucleus),
- the possibility of using a contrast agent to visualise the tissue and/or the distribution of the nanocarrier while it accomplishes its therapeutic activity (theranostics).

By using ‘intelligent’ biomaterials, one can also arrange for these nanosystems to release the active principle they carry in response to some exogenous or endogenous stimulus, e.g., a modification of the pH or ionic strength, a change in temperature, or the application of an extracorporeal magnetic field. Hence, the slightly acidic environment of a tumour tissue can increase the number of protons H^+ in the transport material, making it soluble or changing its conformation, in such a way as to induce release of the active principle within the tumour. Drug release in the tumour can also be triggered by carrying out anti-tumoral hyperthermia, e.g., if the lipid used to carry the drug has a phase transition temperature of the order of 40–42 °C. The application of an extracorporeal magnetic field is another method for achieving targeted drug delivery, provided that the drug in question is carried by a magnetic nanocarrier, e.g., containing maghemite, an iron oxide with chemical formula Fe_2O_3.

Ever more intelligent nanodrugs can thus be prepared thanks to progress in the techniques of supramolecular assembly and self-assembly, but also using new ways of functionalising the surface of nano-objects, including the methods known as click chemistry (see Chap. 5).

The applications of nanotechnology to therapeutics are thus undeniably related to the ability of nanosystems to control the release of active molecules both in time and in space. This is why these technologies apply more particularly to active molecules, whose toxicity for healthy tissues currently restricts the use of large doses. It is not therefore surprising that the therapeutic indications of nanodrugs already on the market concern the treatment of cancer (Caelyx®, Doxil®, DepoCyt®, Abraxane®) and deep mycoses (Ambisome®). Despite the considerable progress made in the field of materials chemistry and pharmaceutical formulation (galenic pharmacy), there is no escaping the fact that the nanodrugs available today remain limited to the areas just mentioned. Indeed, there are several important technological bottlenecks:

1. The low level of encapsulation of nanoparticulate systems, the weight percentage of the active principle relative to the carrier material rarely exceeding 5 %.
2. The fast release, or burst release, of the encapsulated drug which may occur before the therapeutic target is reached. This uncontrolled release is due to the fraction

of drug molecules that are simply adsorbed onto the surface of the nanocarrier and not actually encapsulated within it.
3. The difficulty in obtaining low toxicity, biodegradable synthetic materials which do not induce cell or tissue thesaurismosis.

There is thus a broad consensus that it has become urgent to propose and validate new concepts that will introduce breakthrough technologies in the field of drug delivery. This must be achieved by introducing new and more efficient biocompatible materials. The coupling of squalene with drugs currently used to treat cancer and viral pathologies is a perfect example of such an innovative technology, known as squalenoylation. It leads to more efficient nanodrugs for the treatment of these serious diseases.

11.1 Origin of the Concept of Squalenoylation

Squalene is a lipid in the terpene family. It is a naturally occurring cyclic molecule, found in large quantities in the plant kingdom, e.g., in olive oil, wheat germ, rice, and so on, and also in the animal kingdom, e.g., in shark liver oil. It is also ubiquitous in human tissues, and in particular in the skin. It can thus be considered as completely lacking in toxicity.

Squalene has the unique property of adopting a cyclic structure in lanosterol, itself a precursor of cholesterol, by spontaneously rolling up into a highly compact molecular conformation in an aqueous medium. Indeed, thanks to its molecular flexibility and compact nature, squalene can adopt various conformations which allow it to enter the hydrophobic pocket of the enzyme oxidosqualene cyclase, where the cyclisation reaction takes place. Note that, quite remarkably, this reaction occurs without the need for coenzymes and without the supply of biochemical energy in the form of adenosine triphosphate (ATP).

This unique property of squalene to adopt a highly compact molecular conformation has been exploited to obtain particulate nanosystems by coupling this lipid with anticancer or anti-infection molecules, and in particular with nucleoside analogues [1]. Squalenoylated nanodrugs obtained in this way have proved to be much more effective than their parent molecules. Indeed, these nanosystems:

1. can be administered orally or intravenously,
2. favour intracellular penetration,
3. facilitate passage through membranes,
4. protect the active molecule from degradation and/or metabolisation.

As we shall see later, the nature of the chemical bond between squalene and the drug can allow selective release of the latter at the biological target.

11.2 Coupling Squalene with Drug Molecules

Figure 11.1 is a schematic view of the chemical coupling of a squalene molecule (the carrier) with different kinds of nucleoside analogues with anticancer or antiviral activity (the drug molecules or active principles). Squalene (the chemical radical R here) is first oxidised in acid, before coupling with a given group of the nucleoside. The chemical bonding groups can be of different kinds (sugar: OH–, amine : N–, group sensitive to the surrounding pH, etc.). The derivative molecules thus obtained assemble almost miraculously into nanoparticles, either crystallised or otherwise and with sizes in the range 100–300 nm, thereby constituting the nanodrug. If the nanoparticle surface is then decorated by chains of the hydrophilic polymer polyethylene glycol (PEG), the nanocarrier can avoid otherwise rapid capture by macrophages, while at the same time being functionalised by chemical ligands able to recognise tumoral or infected cells (see Chap. 5).

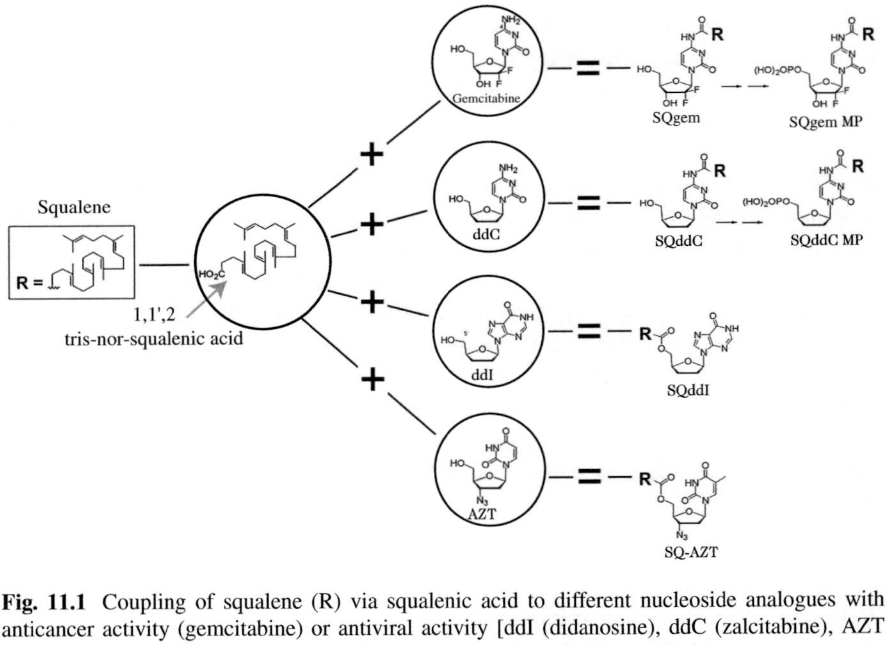

Fig. 11.1 Coupling of squalene (R) via squalenic acid to different nucleoside analogues with anticancer activity (gemcitabine) or antiviral activity [ddI (didanosine), ddC (zalcitabine), AZT (zidovudine)]. Adapted from [2], © 2012 Elsevier. Reproduced with permission

11.3 Characteristics and Morphology of the Main Squalene Derivatives

Table 11.1 lists the main drug molecules that have so far been coupled with squalene, together with the therapeutic treatments they are intended for (cancers, infectious diseases, etc.), the type of molecular coupling, and the size and organisation of the nanoparticles obtained by clustering of the squalenoylated molecules.

The examples in Table 11.1 display the extraordinarily universal and flexible nature of squalenoylation technology, which can be adapted to many molecules with therapeutic activity, be it anticancer, antiviral, or antibiotic, with a variety of physico-chemical characteristics. There are hydrophilic molecules (nucleoside analogues like gemcitabine), lipophilic molecules (paclitaxel), small molecules (penicillin, etc.), and macromolecules (siRNA). In every case, the squalene conjugates self-assemble spontaneously in an aqueous medium to form nanoparticles, by virtue of molecular mechanisms that are still poorly understood. In some cases, the supramolecular organisation leads to hexagonal phases, viz., squalene–gemcitabine nanoparticles (see Fig. 11.2) or cubic phases, viz., ddC–squalene. In others, a total absence of organisation is observed, viz., penicillin–squalene or paclitaxel–squalene nanoparticles (see Fig. 11.3).

11.4 Applications to Cancer Treatment

11.4.1 Nanoparticles of Gemcitabine Coupled with Squalene

In the field of cancer treatment, squalenoylation was first applied to gemcitabine (Gemzar®) to produce SQgem, used clinically to treat solid tumours such as bronchial cancer (not small cells), either locally advanced or metastatic, pancreatic cancer, bladder cancer, and breast cancer. However, gemcitabine is quickly metabolised by the action of deoxycytidine deaminase, mainly located in the blood, the liver, and the kidney, thereby forming the uracil derivative which turns out to be totally inactive. Administered intravenously, gemcitabine thus has a non-optimal anticancer activity. In addition, despite a slightly increased lipophilicity due to the presence of two fluorine atoms, this molecule is still highly hydrophilic, and this limits its passive diffusion through the plasma membrane. This means that the intracellular penetration of gemcitabine can only be achieved by active transport using a protein, the nucleoside transporter hENT1. Inhibited expression of this transporter then often implies resistance to treatment. This happens with pancreatic cancer, for example.

When gemcitabine couples with squalenic acid, the resulting molecular derivative has the quite exceptional property that, in an aqueous medium, it can spontaneously form uniformly sized nanoparticles (130 nm) that can be administered intravenously. Preclinical trials have thus been carried out to compare the squalene–gemcitabine (SQgem) nanoparticles with free gemcitabine (gem) solutions.

Table 11.1 Main characteristics of squalenoylated nanodrugs so far reported

Drug molecule (nucleoside, antibiotic, etc.)	Expected therapeutic effect	Squalene–drug molecule coupling	Organisation resulting from coupling after precipitation
Gemcitabine (dFdC)—gem	Anticancer—cancers of the pancreas, bronchi, bladder, breast, etc.	Amine group N–	Formation of organised 100–300 nm nanoparticles. Examples: inverted Hexagonal phase for SQgem [3] (Fig. 11.2); cubic phase for SQddC [4]; several crystalline phases for doxorubicin nanoparticles, with elongated shape (nanospaghetti), inducing a remarkable vascular remanence [5]
Zalcitabine (ddC)	Antiretroviral—AIDS	Amine group N–	
Didanosine (ddI)	Antiretroviral—AIDS	Sugar group OH–	
Zidovudine (AZT)	Antiretroviral—AIDS	Sugar group OH–	
Cytarabine	Anticancer—leukemia	Amine group N–	
Acyclovir	Antiviral—herpes	Amine group N–	
Adenosine	Various biological processes	Amine group N–	
Thymidine	Various biological processes	Sugar group OH–	
Doxorubicin	Anticancer	OH– group	
Paclitaxel (mitosis inhibitor)	Anticancer—cancers of the lung, ovary, breast, head and neck, etc.	2′–OH group [6]	Formation of amorphous 100–300 nm nanoparticles
Penicillin G	Antibiotic—various bacterial strains	Ester group with bond sensitive or insensitive to pH [7]	Formation of amorphous 150 nm nanoparticles (Fig. 11.3)
siRNA (small interfering RNA)	Hydrophilic anticancer macromolecule—cancer of the thyroid	Coupling of squalene maleimide to the RNA coding strand [8]	Formation of amorphous 160 nm nanoparticles

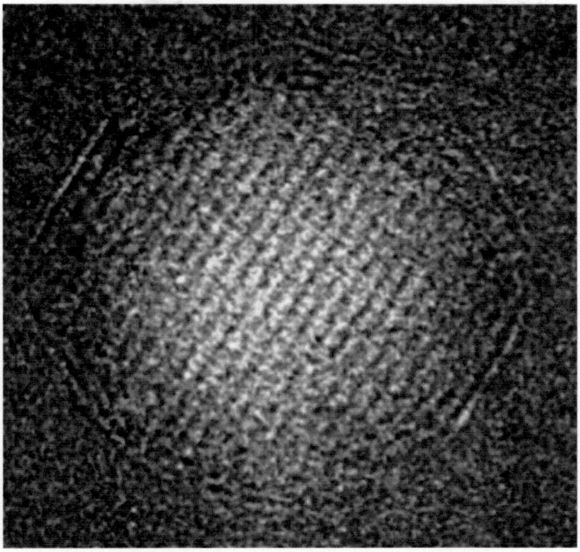

Fig. 11.2 Morphological appearance of a squalene–gemcitabine (SQgem) nanoparticle. Image obtained by transmission electron cryomicroscopy (cryoTEM). The hexagonal structure is visible. Adapted from [3], © 2008 Wiley. Reproduced with permission

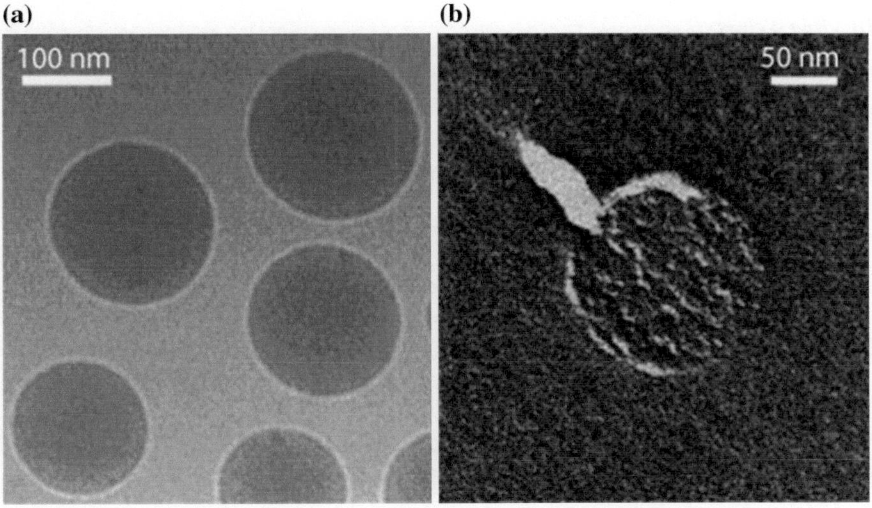

Fig. 11.3 Morphological appearance of penicillin–squalene nanoparticles (either pH-insensitive or pH-sensitive). **a** Image obtained by transmission electron cryomicroscopy (cryoTEM), not showing the internal organisation of the nanoparticles. **b** Image obtained with cryofracture for better visualisation of the internal relief of the nanoparticle. The absence of organisation is confirmed. Adapted from [7], © 2012 American Physical Society

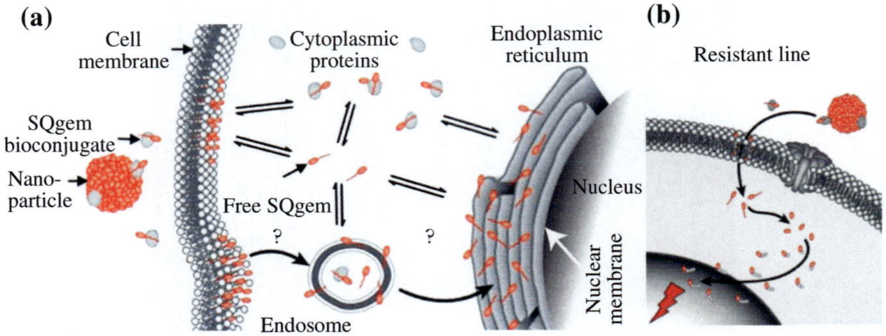

Fig. 11.4 **a** Interaction mechanism between squalene–gemcitabine (SQgem) nanoparticles and a cancer cell. After crossing the cell membrane, the SQgem bioconjugate distributes itself between the intracellular lipid membranes and the more hydrophilic cytoplasmic medium (*double arrows*). The gemcitabine is released by the action of cell enzymes and, after phosphorylation, can then reach the DNA in the cell nucleus. One question, indicated by the *question mark*, today answered in the affirmative, was whether this phosphorylation did actually occur in the cytoplasm. **b** Schematic view of the pathway followed by the gemcitabine molecules from the nanoparticle to the cell nucleus, without the need for the nucleoside transporter hENT1. The *grey-shaded* appendages of the gemcitabine molecules symbolise their phosphorylation (to gem-triphosphate). Adapted from [10], © 2010 Elsevier. Reproduced with permission

In Vitro Trials

These trials have shown that, in contrast to non-squalenoylated gemcitabine, SQgem nanoparticles are stable in the presence of blood plasma, since there is no metabolisation. They release the gemcitabine under the action of intracellular enzymes [1], according to the following mechanism (see Fig. 11.4a):

1. In the extracellular medium, the nanoparticles release SQgem in molecular form [9]. Released individually in this way, the molecules associate with extracellular proteins, favouring their diffusion toward the cell membrane where they accumulate in large quantities.
2. The SQgem molecules are then distributed between the cell membrane and the intracellular membranes (endolysosomal and endoplasmic reticulum membranes) [10].
3. The SQgem is subsequently metabolised by intracellular enzymes (cathepsins B and D), and this releases the gemcitabine in the cell cytoplasm.
4. Finally, phosphorylation of the gemcitabine by deoxycytidine kinase, then by pyridine kinases, serves to integrate the molecule into the DNA, leading to its cytotoxic effect.

When the cells are resistant to gemcitabine owing to reduced expression of the transporter hENT1, penetration of the SQgem within the cancer cell is not therefore inhibited, in contrast to what happens with free gemcitabine. This therefore provides a way round the gemcitabine resistance of the cancer cells (see Fig. 11.4b) [10].

In Vivo Trials

Pharmacokinetic trials have been carried out after intravenous administration of gemcitabine in free form or in the form of squalenoylated nanoparticles at a dose of 15 mg/kg. It was observed that the SQgem nanoparticles induced much higher plasma concentrations of gemcitabine than when the gemcitabine was administered in free form. This shows that the nanoparticulate form of gemcitabine protects the drug molecule from metabolisation and degradation. Furthermore, a biodistribution study of the labelled product showed that gemcitabine concentrations in the main organs are always higher when the anticancer agent is administered in the form of nanoparticles.

Testing the Efficacy of Squalenoylated Gemcitabine on Mouse Leukemia

The anticancer activity of SQgem nanoparticles was first tested on a mouse leukemia model (L1210) grafted intravenously on laboratory animals [11]. This is a highly aggressive metastatic model which induces fast weight loss. The animals die 20 days after the tumour cells are grafted. Treatment with free gemcitabine extends survival up to 40 days, but there are no long term survivors. No animal is cured. In contrast, treatment by SQgem nanoparticles cures 75 % of the animals, i.e., some 75 % of them survive in the long term. Moreover, tumour cell samples showed that the SQgem nanoparticles induce the arrest of cell division in phase S and a much higher level of apoptosis than after treatment by free gemcitabine. These results are explained by the better distribution of the anticancer agent in the deep organs where the metastasis takes place. Preliminary toxicological trials specified the maximal tolerated dose, and at this dose, no major toxicity was observed [11]. At higher doses, the toxicological profile (hematopoietic toxicity) is comparable with that of free gemcitabine.

Since squalene is a lipid that is particularly well absorbed orally, the efficacy of SQgem nanoparticles was also tested on a metastatic model of rats carrying the lymphocytic leukemia RNK-16 LGL. In this model, and following treatment by SQgem nanoparticles, 60 % of the animals survived in the long term, whereas no animal was cured by treatment with gemcitabine in the free form, administered orally at the same dose [1]. This result is explained by an increase in the concentration of gemcitabine in the lymphoid organs.

After intravenous administration, the SQgem nanoparticles are also much more active than gemcitabine on the mouse leukemia model P388 grafted subcutaneously (solid tumour). Following treatment by SQgem nanoparticles, the mice exhibited no visible tumour nodule, whereas a five times higher dose of free gemcitabine did not reduce the tumour mass compared with non-treated animals [12].

Testing the Efficacy of Squalenoylated Gemcitabine on Pancreatic Cancer

With a survival rate of only 3 % at five years and a median survival period of just six months, pancreatic cancer is one of the tumour pathologies with the worst prognosis. Gemcitabine is the primary treatment. This is why we tested the efficacy of SQgem nanoparticles on an orthotopic mouse model of a human pancreatic cancer (panc-1)

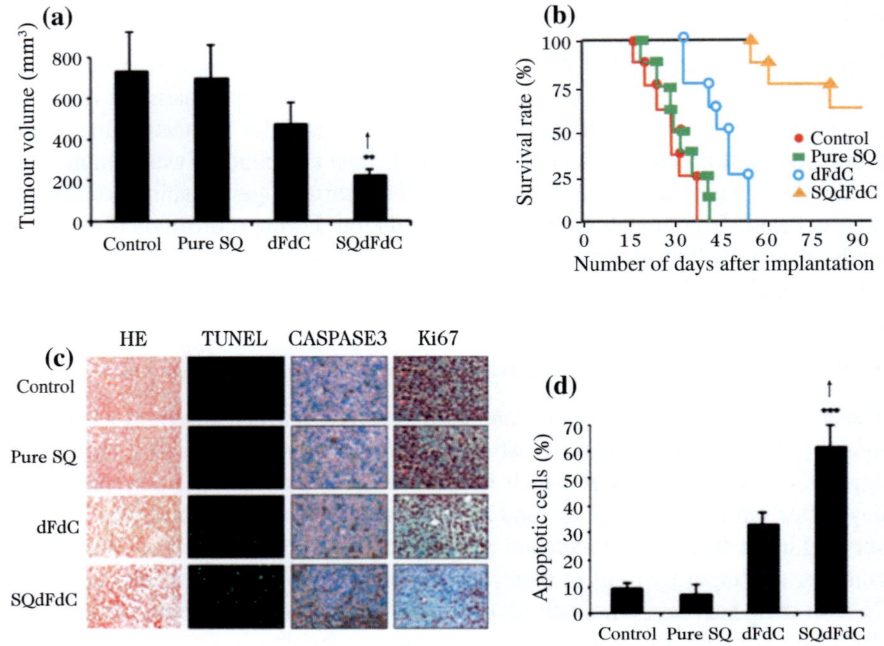

Fig. 11.5 SQgem nanoparticles (denoted here by SQdFdC) exhibit greater antitumour activity than free gemcitabine (denoted by dFdC) and pure squalene nanoparticles (pure SQ) on the orthotopic model of human pancreatic cancer panc-1. **a** Size of tumour nodules for the different treatments. **b** Survival rate of the animals for the different treatments. **c** Histology and immunohistochemistry of tumour biopsies after treatment. **d** Cells in apoptose after treatment. Adapted from [13], © 2011 Elsevier. Reproduced with permission

[13]. In this case, the tumour cells are grafted in the pancreatic head. The size of the tumour (see Fig. 11.5a) and the survival of the animals (see Fig. 11.5b) were assessed as a function of the treatment. For example, SQgem nanoparticle treatment produces a spectacular reduction in the size of the pancreatic tumour nodule, leading to a 65 % long term survival rate, whereas all the other treatments (nanoparticles of pure squalene or free gemcitabine) leave no survivors in the long term. These results are confirmed by immunohistochemistry (see Fig. 11.5c) and measurement of cancer cell apoptosis for the different treatments (see Fig. 11.5d). Indeed, the nanoparticle treatment induces an increase in the number of cells in apoptosis, viz., 60 % as compared with 30 % for the free gemcitabine treatment, while at the same time reducing the number of cells in proliferation.

Hence the bright red spots produced by staining tumour biopsies with hematoxylin and eosine (HE, see Fig. 11.5c) indicate the presence of large numbers of dead cells following treatment by SQgem nanoparticles. The Tunel labelling immunohistochemical test reveals the programmed apoptosis of tumour cells by green fluorescence (rupture of the DNA). The caspase-3 labelling test confirms the increased number of cell deaths by a dark brown colour. On the other hand, labelling with the Ki67 antigen attests, by a bright stain (here, blue), to reduced proliferation of tumour cells due to the treatment.

It has been shown recently that it is also possible to associate two drugs in the same nanoparticle: one of them, SQgem, induces tumour cell death, while the other iso-Combretastatin, reduces tumour vascularisation. This double action has proved to be extremely effective on an experimental model of colon cancer. This is the principle of the multidrug nanoparticle [14].

11.4.2 Nanoparticles of Doxorubicin Coupled with Squalene

Doxorubicin-loaded nanocarriers have gained increasing interest as they can improve the treatment of tumours and reduce drug-mediated cardiotoxicity. However, the need to surface-functionalise conventional nanocarriers such as liposomes with PEG raises toxicological issues because PEG is not biodegradable. In this context, it has been discovered that, through a simple manufacturing procedure, the linkage of doxorubicin to squalene allows the synthesis of 130 nm non-PEGylated nanoparticles (SQ-Dox) with impressively high drug loading (i.e., 57 %), slow drug release, and a novel 'loop-tail' elongated structure never observed before [15]. Further physico-chemical and morphological investigations will certainly be needed to explain these surprising results, but the elongated morphology of SQ-Dox nanoparticles is likely to be the explanation for their ability to circulate for long periods in the bloodstream after intravenous injection. The fact that 'loop-tail' SQ-Dox can align with the bloodstream makes them invisible for liver macrophage recognition.

Cell culture viability tests and apoptosis assays showed that SQ-Dox displayed about the same antiproliferative and cytotoxic effects as native doxorubicin. *In vivo* experiments showed that SQ-Dox dramatically improved anticancer treatments as compared with free doxorubicin. In particular, M109 lung tumours that did not respond to doxorubicin treatment were found to be inhibited by 90 % when treated with SQ-Dox. Similarly, SQ-Dox-treated MiaPaCa-2 pancreatic human tumours xenografted in mice were found to decrease by 95 % instead of the 29 % reduction achieved with native doxorubicin [15]. It was also shown that the maximum tolerated dose for SQ-Dox nanoassemblies was five times higher than for the free drug and that SQ-Dox did not cause any myocardial lesion such as those induced by a free doxorubicin treatment. All these results demonstrate that nanoparticles of doxorubicin coupled with squalene make tumour cells more sensitive to doxorubicin and reduce the cardiac toxicity, thus providing a remarkable improvement in the drug's therapeutic index.

11.4.3 Nanoparticles of SiRNA Coupled with Squalene (SQsiRNA)

The nucleotide sequences of fusion oncogenes are unique to the cancer cells to which they belong and represent specific targets for the development of new anticancer molecules. Hence, for papillary thyroid cancer, the most frequent endocrine tumour of the thyroid, inhibition of the fusion oncogene (RET/PTC1) should provide a way to halt the tumour process. The development of a small interfering RNA (siRNA)[1] has the advantage of precisely targeting the fusion oncogene and hence avoiding side effects due to inhibition of normal genes in non-tumour cells, as happens with non-specific treatments. Indeed, these siRNA molecules are highly promising therapeutic agents because of their specific action: they are active at low dosage and exhibit low toxicity. On the other hand, they are extremely hydrophilic macromolecules that degrade very quickly in the plasma and do not spontaneously penetrate cells.

An siRNA molecule with sequence complementary to the fusion oncogene messenger RNA was coupled to squalene maleimide (see Table 11.1). This is a novel approach because most nucleic acid nanocarriers use cationic lipids or polymers owing to the affinity of positive charges for cell membranes. However, these materials are not without a certain level of toxicity, and in particular the destruction of red blood cells (hemolysis). By coupling siRNA to a lipid like squalene, one can avoid the toxicity inherent in polycations, while taking advantage of the strong affinity of squalene for most cell membranes.

11.5 Application to Treatment of Infectious Diseases

11.5.1 Antiretroviral Nucleoside Analogues

The idea of squalenoylation has been applied to nucleoside analogues with antiviral activity, such as zalcitabine (ddC) and didanosine (ddI). However they bind to the squalene, on the sugar or the heterocycle (see Fig. 11.1), all the squalenoylated molecules self-assemble in water to form 100–300 nm nanoparticles. Their derivatives have been tested *in vitro* for their anti-HIV activity on infected lymphocytes from three different donors. The two squalenoylated derivatives proved to be two to three times more active than the corresponding parent molecules with regard to their ability to inhibit viral multiplication by 50 %.

The antiviral activity of SQddI and SQddC nanoparticles was then tested on human lymphocytes infected by certain resistant viral strains of the AIDS virus (HIV-1-144 and HIV-1-146). The antiviral activity of the squalenoylated derivatives was much greater than that of ddl or ddC on these same resistant strains. For example, SQddl nanoparticles proved to be ten times more active than ddl on the HIV-1-146 strain.

[1] Small interfering RNA (siRNA) is a short ribonucleic acid with 20 or 25 base pairs.

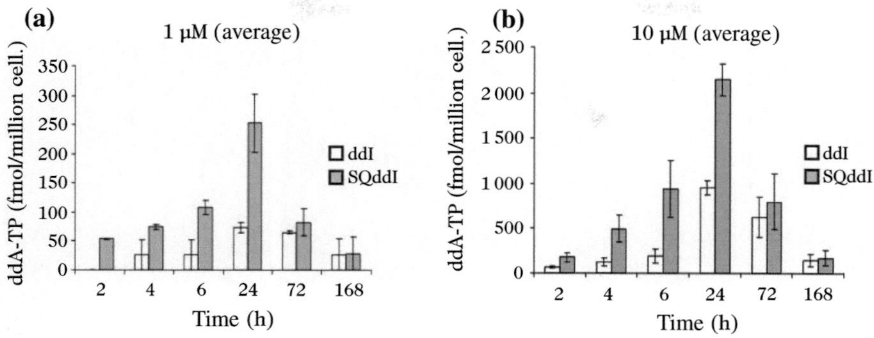

Fig. 11.6 Intralymphocyte concentrations of ddA-TP after incubation of lymphocytes with ddl (*white*) or with SQddI (*grey*) at a concentration of 1 μM (**a**) or 10 μM (**b**). Adapted from [15], © 2008 Elsevier. Reproduced with permission

It is also found that SQddI nanoparticles induce much higher intracellular concentrations of the active form of the ddI molecule, viz., ddA-TP (2′, 3′-dideoxyadenosine-5′-triphosphate), and this whatever the concentration of the product incubated with the cells. This result suggests that phosphorylation is much more efficient with SQddI nanoparticles than with ddI (see Fig. 11.6).

11.5.2 Intracellular Antibiotherapy

The treatment of intracellular infections is another promising field of application of squalenoylated nanodrugs. When they arrive in the blood compartment, the bacteria are opsonised to facilitate capture by macrophages in the body's defence system (the liver and the spleen) where they are destroyed, in particular by the lysosomal enzymes of the cells. In certain situations, such as immunodepression, opportunistic infections, and others, the lysosomes of these macrophages actually provide a kind of haven for the multiplication of intracellular bacteria. Many antibiotics are largely inactive against these germs located within the cell, either because they are degraded in the intracellular medium, or because they do not diffuse easily inside the cells, or again because they are rapidly washed out of the cells and never reach the infected intracellular compartments (endosomes or lysosomes) in sufficient concentrations. In other words, the infected macrophages end up as reservoirs of bacteria that are resistant to most conventional antibiotic treatments because these only eradicate extracellular germs. There is thus a real need to develop antibiotic nanocarriers targeting the intracellular environment.

This has been done by coupling penicillin to squalene by both pH-sensitive and pH-insensitive links. When the link is pH sensitive, the release of the antibiotic is triggered by the acidic pH of the infected lysosomes [7]. These pH-sensitive or pH-insensitive penicillin–squalene nanoparticles penetrate the macrophages (J777) by

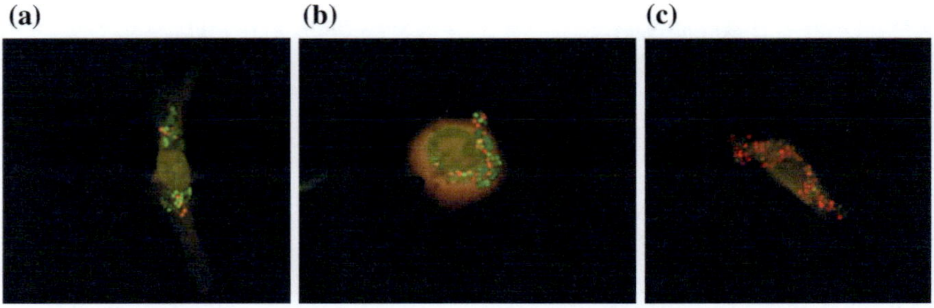

Fig. 11.7 Fluorescence confocal microscopy showing living (*green*) and dead (*red*) intracellular bacteria. **a** Non-treated macrophages. **b** Macrophages treated with free penicillin. **c** Macrophages treated with penicillin–squalene nanoparticles. Adapted from [7], © 2012 American Physical Society

endocytosis and end up in the cell lysosomes. The antibacterial activity then proves to be much more effective in the case of pH-sensitive penicillin nanoparticles [7]. Indeed, on a model of macrophages (J774) infected by *Staphylococcus aureus*, these nanoparticles can very quickly kill the intracellular bacteria (see Fig. 11.7), proving that nanotechnology can provide effective tools for treating resistant intracellular infections.

11.6 Application to the Treatment of Neurological Disorders

Drug delivery remains the main challenge in central nervous system (CNS) drug development, due to the rapid metabolisation and/or rapid blood clearance of most CNS drugs, and sometimes poor diffusion through the blood–brain barrier (BBB) and the blood–spinal cord barrier (BSCB). A typical example of the problems that can arise for drug delivery in neurological pathologies is adenosine. Indeed, this molecule represents a class of potential therapeutic agents with significant beneficial activity in severe neurological disorders such as strokes, spinal cord injury, or multiple sclerosis. However, adenosine exhibits serious limitations due to its short plasma half-life (a few seconds) following rapid metabolisation, the advent of moderate side-effects, and its inability to cross the BBB and BSCB, and for these reasons, it has never been used for the treatment of neurological diseases.

Notwithstanding, a very simple and easy way to use the currently unusable adenosine as a neuroprotective drug with intravenous injection is the conjugation of adenosine with squalene (SQAd) and the subsequent formation of nanoassemblies allowing prolonged circulation of this nucleoside. This approach has been shown to provide neuroprotection in a mouse stroke model and a rat spinal cord injury model [21]. The animals receiving systemic administration of SQAd nanoassemblies showed a significant improvement of neurological deficit score in the case of cerebral ischemia, and an early motor recovery of the hind limbs in the case of spinal cord injury

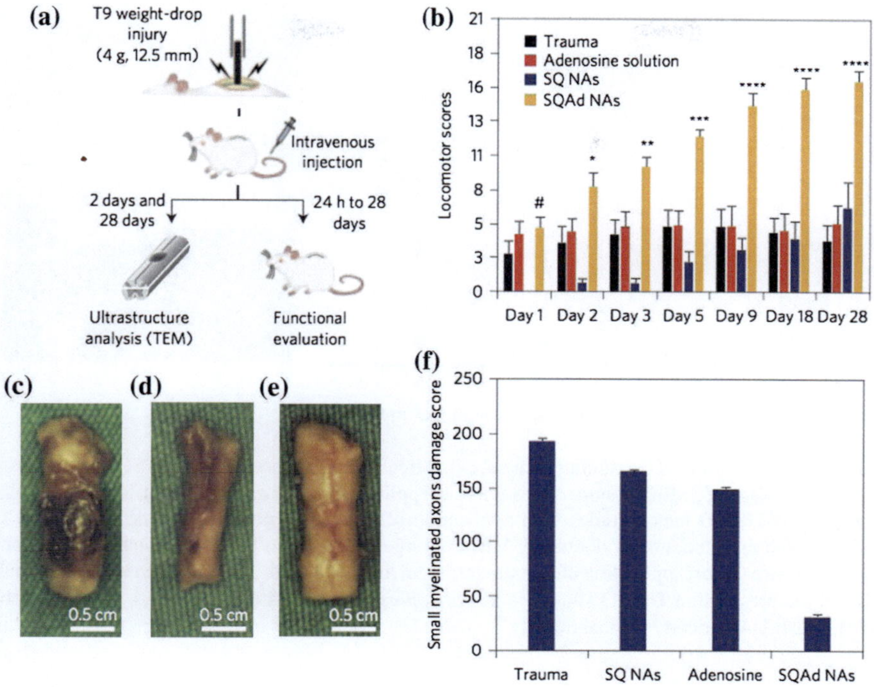

Fig. 11.8 Pharmacological efficiency of the SQAd nanoassemblies in a model of spinal cord injury in rats. The pharmacological efficiency of the squalene–adenosine nanoassemblies (SQAd NA) was assessed in a T9 contusion spinal cord injury model. Within 5 min, the animals were intravenously injected with either dextrose 5 %, SQAd NAs, or free adenosine (**a**). After 24, 48, and 72 h, and up to 28 days post-trauma, the animals were functionally graded using the Basso, Beattie, and Bresnahan grading (**b**). After 72 h, the SQAd NA injected animals showed a complete recovery of their hind limbs, in accordance with the absence of any visible traumatic area on the cord (**e**) compared to trauma group (**c**) and the adenosine treated group (**d**). Quantification of the damage to the small myelinated axons showed that the SQAd NAs dramatically reduced the damage score (**f**) compared to all other groups. Reprinted from [21] by permission from Macmillan Publishers Ltd

(see Fig. 11.8). Moreover, in FRET experiments (see Sect. 2.7.2), it has been demonstrated for the first time that nanoparticles can provide neurological protection despite the fact they are unable to translocate the blood–brain barrier. The reason is that these nanoparticles can favour a peripheral vascular mechanism, leading in turn to secondary parenchymal neuroprotection [21].

11.7 Magnet Guidance, Imaging, and Theranostics

Iron oxide particles (magnetite) can be included in the core of the nanocarrier in such a way that it will respond to an extracorporeal magnetic field. This idea has been used to guide gemcitabine in a transplanted mouse tumour model. In practice,

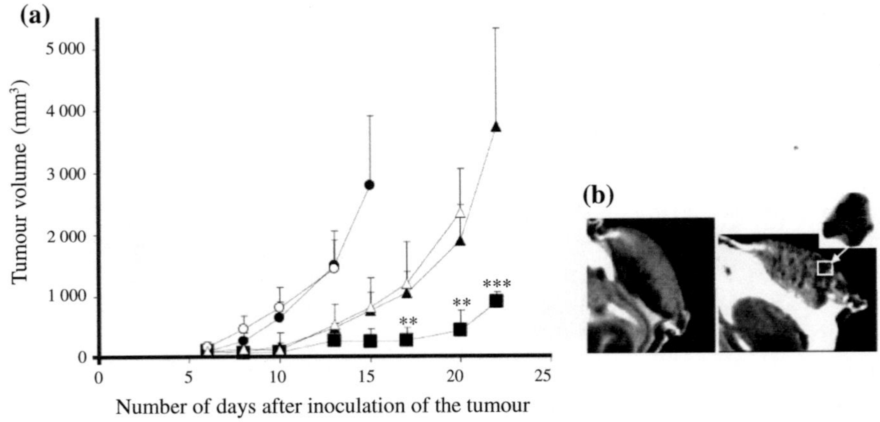

Fig. 11.9 **a** Evolution of the tumour volume: (•) no treatment, (○) gemcitabine, (▲) SQgem nanoparticles, (△) SQgem/USPIO nanoparticles without application of an extracorporeal magnetic field, (■) SQgem/USPIO nanoparticles with application of an extracorporeal magnetic field of 1.1 T. **b** Visualisation of the tumour nodule by MRI after injecting SQgem/USPIO nanoparticles without (*left*) and with (*right*) application of an extracorporeal magnetic field. The reduction in the T2 signal due to the presence of USPIO shows up as *dark spots* within the tumour (*insert*). Adapted from [17], © 2011 American Physical Society

ultrasmall magnetic nanoparticles (ultrasmall superparamagnetic iron oxide USPIO) measuring only a few nanometres, are trapped within a nanomatrix made from SQgem molecules and having dimensions of the order of a hundred or so nanometres [16]. These nanosystems are then injected intravenously while a magnetic is placed close to the tumour. The inhibition of tumour growth is spectacular and can be monitored by magnetic resonance imaging (MRI) thanks to the presence of the iron oxide particles which, within the lipid nanomatrix, induce a reduction in the MRI T2 signal [17] (see Fig. 11.9).

The core of the nanocarrier thus plays both a therapeutic and a diagnostic role (nanotheranostics). This still highly experimental idea has been extended to other molecules with cytotoxic activity, such as paclitaxel, doxorubicin, and cisplatin, and also to other contrast agents, in particular squalene coupled with gadolinium. The technique can thus be adapted to many active principles and contrast agents. The association of a drug with an imaging tool within the same nanocarrier provides an ideal route to personalised medicine. Indeed, being able to monitor both the nanocarrier distribution by MRI and at the same time also its therapeutic efficacy by visualising the tumour size means that it will be possible to make a clinical decision about stopping or pursuing the treatment on a case-by-case basis.

11.8 From Squalenoylation to Terpenoylation

As mentioned in Sect. 11.2, squalene belongs to the terpene (or isoprene) family, which is an extremely diverse group of compounds, both with regard to chemistry and from the structural and functional point of view [18]. Their lipophilic nature and their permanent presence in living organisms mean that they are likely to play a role in the elaboration of the most primitive biological membranes [19]. Surprisingly, although most terpenes, both natural and synthetic, are easy to handle and biocompatible polymers, and despite the fact that their physicochemical characteristics allow them to adapt to a wide range of active principles, these compounds have never been used to transport and deliver drugs, with the exception of squalene, whose self-assembly into nanoparticles was discovered by researchers at the *institut Galien* of the Paris-Sud university.

The aim now is therefore to extend the idea of squalenoylation to terpenoylation by chemically coupling a biologically active molecule to different kinds of terpenes in order to identify which bioconjugates will or will not be able to self-assemble to form nanoparticles in an aqueous medium. For example, gemcitabine has been coupled to several terpenes containing one to six isoprene units. Figure 11.10 shows the chemical structure of one isoprene unit. To get an idea, squalene contains 6 isoprene units and generally speaking there are many possible isoprene chains for any given number of units.

In the first *in vitro* experiments, it was observed that the anticancer activity of the nanoparticles was closely linked to the size of the polyisoprene chain. Nanoparticles containing three isoprene units (farnesyl) proved to be the most effective on different lines of tumour cells. Trials carried out *in vivo* on an experimental xenograft model of human pancreatic cancer confirmed the anticancer activity of certain nanoparticles of gemcitabine–farnesyl derivatives (leading to 76 % tumour inhibition) and gemcitabine–squalene nanoparticles (leading to 41 % tumour inhibition), whereas free gemcitabine proved to have no antitumour effect on this model. The level of anticancer activity was in fact correlated to the gemcitabine release kinetics. These observations show that the cancer cells can be made more sensitive to the gemcitabine treatment by modulating the length of the polyisoprene chain, and this without conspicuous toxicity.

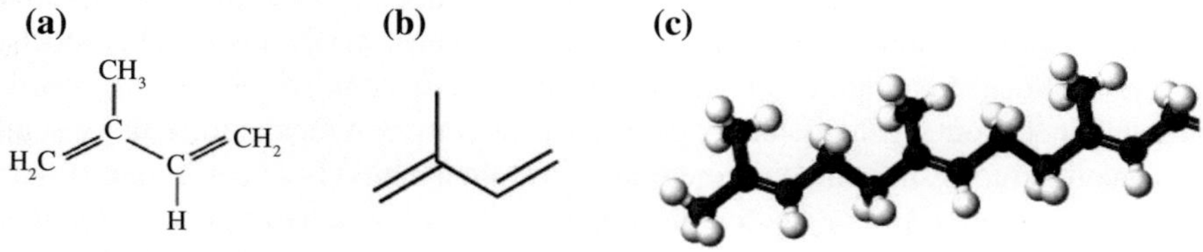

Fig. 11.10 **a** Chemical structure of an isoprene unit. **b** Simplified representation of one unit. **c** Three-dimensional representation of farnesyl containing three isoprene units

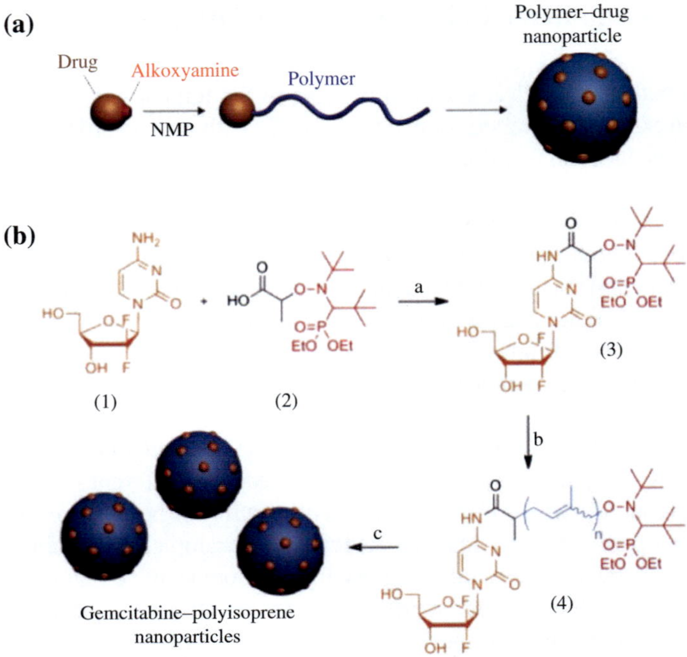

Fig. 11.11 Construction of gemcitabine–polyisoprene nanoparticles. **a** General principle of controlled radical polymerisation from a drug serving as a trigger for the polymerisation reaction. This produces a conjugate in which the end of each polyisoprene chain is coupled to a drug molecule. The macromolecule thus obtained forms stable nanoparticles in water. **b** Application to gemcitabine carrying a macro-alkoxyamine group. The reaction (*1*) + (*2*) $\longrightarrow$ (*3*) followed by polymerisation (*4*) leads to a gemcitabine–polyisoprene conjugate which forms stable nanoparticles in water. Adapted from [20], © 2013 Wiley. Reproduced with permission

Recently, with a view to increasing the number of isoprene units beyond six, researchers at the *institut Galien* have developed a novel way of building nanoparticles using longer polyterpene sequences [20]. This approach is based on chemical modification of the anticancer agent using an alkoxyamine function to initiate controlled isoprene polymerisation (see Fig. 11.11b). This so called nitroxide mediated polymerization (NMP) is a radical polymerisation reaction providing good control over the macromolecular characteristics of the polymer. For example, the length of the polyisoprene chain attached to the gemcitabine can be easily adjusted to produce the most effective nanoparticles. The excellent anticancer activity of the nanoparticles obtained from a gemcitabine–polyterpene derivative of mean molecular weight 2 510 dalton (corresponding to more than 35 isoprene units) has been demonstrated [20]. This coupling strategy based on controlled growth of a hydrophobic oligomer from an anticancer drug carrying a macro-alkoxyamine group leads to an anticancer agent at each end of the synthesised polyterpene chain (see Fig. 11.11a). This discovery, which could be applied to other active principles than gemcitabine, looks likely to extend the available pharmacological applications.

11.9 Conclusion

The idea of squalenoylation, and its extension to terpenoylation, provide an example of a highly original nanotechnology platform for the design of new anticancer, neuroprotective, and anti-infection nanodrugs. Constituting as it does a completely innovative approach, this highly flexible technique may well meet major societal needs by opening the way to new prospects for the treatment of cancers, neurological disorders, or infections that resist conventional practice. Researchers at the *institut Galien* where this work was initiated have already developed a first type of nanodrug for the treatment of resistant forms of hepatocarcinoma, and it is currently undergoing phase III clinical trials, the last phase prior to industrialisation, by the company BioAlliance, now ONXEO. This innovation has just been awarded the European Inventor Award 2013, thus illustrating the dynamical relationship between fundamental research and the pharmaceutical industry in the field of nanodrug development. As in many similar cases, the transition from research to pre-industrialisation can only be made through company start-ups, while continuing to maintain the work of the research group, with the support of both public and private funding, and taking care to patent all inventions before publishing research results in the scientific journals.

References

1. P. Couvreur, B. Stella, L.H. Reddy, H. Hillaireau, C. Dubernet, D. Desmaële, S. Lepêtre-Mouelhi, F. Rocco, N. Dereuddre-Bosquet, P. Clayette, V. Rosilio, V. Marsaud, M. Renoir, L. Cattel, Squalenoyl nanomedicines as potential therapeutics. Nano Lett. **6**, 2544–2548 (2006)
2. D. Desmaële, R. Gref, P. Couvreur, Squalenoylation: a generic platform for nanoparticular drug delivery. J. Control. Release **161**, 609–618 (2012)
3. L.H. Reddy, S. Mangenot, J.H. Poupaert, D. Desmaele, S. Lepetre-Mouelhi, B. Pili, C. Bourgaux, H. Amenitsch, M. Ollivon, P. Couvreur, Discovery of new hexagonal supramolecular nanostructures formed by squalenoylation of an anticancer nucleoside analogue. Small **4**, 247–253 (2008)
4. F. Bekkara-Aounallah, R. Gref, M. Othman, L.H. Reddy, B. Pili, V. Allain, C. Bourgaux, H. Hillaireau, S. Lepêtre-Mouelhi, D. Desmaële, J. Nicolas, N. Chafi, P. Couvreur, Novel PEGylated nanoassemblies made of self-assembled squalenoyl nucleoside analogues. Adv. Funct. Mater. **18**, 3715–3725 (2008)
5. A. Maksimenko, F. Dosio, J. Mougin, A. Ferrero, S. Wack, R.L. Harivardhan, A.A. Weyn, E. Lepeltier, C. Bourgaux, B. Stella, L. Cattel, P. Couvreur, Squalenoylated doxorubicin: a new long circulating and non pegylated anticancer nanomedicine. Proc. Natl. Acad. Sci. USA **2**, E217–E226 (2014)
6. J. Caron, A. Maksimenko, S. Wack, E. Lepeltier, C. Bourgaux, E. Morvan, K. Leblanc, P. Couvreur, D. Desmaele, Improving the antitumor activity of squalenoyl-paclitaxel conjugate nanoassemblies by manipulating the linker between paclitaxel and squalene. Adv. Healthc. Mater. **2**, 172–185 (2012)
7. N. Semiramoth, C. Di Meo, F. Zouhiri, F. Saïd-Hassane, S. Valetti, R. Gorges, V. Nicolas, J. Poupaert, S. Chollet-Martin, D. Desmaele, R. Gref, P. Couvreur, Self-assembled squalenoylated penicillin bioconjugates: an original approach for the treatment of intracellular infections. ACS Nano **6**, 3820–3831 (2012)

8. M. Raouane, D. Desmaele, M. Gilbert-Sirieix, C. Gueutin, F. Zouhiri, C. Bourgaux, E. Lepeltier, R. Gref, R. Ben Salah, G. Clayman, L. Massaad-Massade, P. Couvreur, Synthesis, characterization, and *in vivo* delivery of siRNA-squalene nanoparticles targeting fusion oncogene in papillary thyroid carcinoma. J. Med. Chem. **54**, 4067–4076 (2011)
9. L. Bildstein, V. Marsaud, H. Chacun, S. Lepêtre-Mouelhi, D. Desmaële, P. Couvreur, C. Dubernet, Extracellular-protein-enhanced cellular uptake of squalenoyl gemcitabine from nanoassemblies. Soft Matter **21**, 5570–5580 (2010)
10. L. Bildstein, C. Dubernet, V. Marsaud, H. Chacun, V. Nicolas, C. Gueutin, A. Sarasin, H. Bénech, S. Lepêtre-Mouelhi, D. Desmaële, P. Couvreur, Transmembrane diffusion of gemcitabine by a nanoparticulate squalenoyl prodrug: an original drug delivery pathway. J. Control. Release **147**, 163–170 (2010)
11. L.H. Reddy, P.E. Marque, C. Dubernet, S. Lepetre-Mouelhi, D. Desmaele, P. Couvreur, Preclinical toxicology (sub-acute and acute) and efficacy of a new squalenoyl gemcitabine anticancer nanomedicine. J. Pharmacol. Exp. Therapeut. **325**, 484–490 (2008)
12. L.H. Reddy, J.M. Renoir, V. Marsaud, S. Lepetre-Mouelhi, D. Desmaele, P. Couvreur, Anticancer efficacy of squalenoyl gemcitabine nanomedicine on 60 human tumor cell panel and on experimental tumor. Mol. Pharm. **6**, 1526–1535 (2009)
13. S. Réjiba, L.H. Reddy, C. Bigand, C. Parmentier, P. Couvreur, A. Hajiri, Squalenoyl gemcitabine nanomedicine overcomes the low efficacy of gemcitabine therapy in pancreatic cancer. Nanomedicine **7**, 841–849 (2011)
14. A. Maksimenko, M. Alami, F. Zouhiri, J.D. Brion, A. Pruvost, J. Mougin, A. Hamze, T. Boissenot, O. Provot, D. Desmaele, P. Couvreur, Therapeutic modalities of squalenoyl nanocomposites in colon cancer: an ongoing search for improved efficacy. ACS Nano **8**, 2018–2032 (2014)
15. H. Hillaireau, N. Dereuddre-Bosquet, R. Skanji, F. Bekkara-Aounallah, J. Caron, S. Lepêtre, S. Argote, L. Bauduin, R. Yousfi, C. Rogez-Kreuz, D. Desmaële, B. Rousseau, R. Gref, K. Andrieux, P. Clayette, P. Couvreur, Anti-HIV efficacy and biodistribution of nucleoside reverse transcriptase inhibitors delivered as squalenoylated prodrug nanoassemblies. Biomaterials **34**, 4831–4838 (2013)
16. J.L. Arias, L.H. Reddy, P. Couvreur, Magnetoresponsive squalenoyl gemcitabine composite nanoparticles for cancer active targeting. Langmuir **24**, 7512–7519 (2008)
17. J.L. Arias, L. Harivardhan Reddy, M. Othman, B. Gillet, D. Desmaële, F. Zouhiri, F. Dosio, R. Gref, P. Couvreur, Squalene based nanocomposites: a new platform for the design of multifunctional pharmaceutical theragnostics. ACS Nano **5**, 1513–1521 (2011)
18. J. Gershenzon, N. Dudareva, The function of terpene natural products in the natural world. Nat. Chem. Biol. **3**, 408–414 (2007)
19. G. Ourisson, The evolution of terpenes to sterols. Pure Appl. Chem. **61**, 345–348 (1989)
20. S. Harrisson, J. Nicolas, A. Maksimenko, D. Trung Bui, J. Mougin, P. Couvreur, Nanoparticles with *in vivo* anticancer activity from polymer prodrug amphiphiles prepared by living radical polymerization. Angew. Chem. Int. Ed. Engl. **52**, 1678–1682 (2013)
21. A. Gaudin, M. Yemisci, H. Eroglu, S. Lepêtre-Mouelhi, O.F. Turkoglu, B. Dönmez-Demir, S. Caban, M.F. Sargon, S. Garcia-Argote, G. Pieters, O. Loreau, B. Rousseau, O. Tagit, N. Hildebrandt, Y. Le Dantec, J. Mougin, S. Valetti, H. Chacun, V. Nicolas, D. Desmaële, K. Andrieux, Y. Capan, T. Dalkara, P. Couvreur, Squalenoyl adenosine nanoparticles provide neuroprotection after stroke and spinal cord injury. Nat. Nanotechnol. **9**, 1054–1063 (2014)

Chapter 12
Health Impacts of Nanomaterials

Francelyne Marano and Rina Guadagnini

Abstract Applications of nanotechnology can now be found in many everyday products, from cosmetics to personal hygiene products, foods, clothing, and others. The consumer is exposed to nanomaterials without realising it, owing to lack of adequate regulations concerning labelling. It has therefore become essential to better assess the health impacts of nanoparticles. From the firsts experimental studies, we know that nanoparticles can cross the biological barriers in small amounts (less than 1 %) whenever they enter the organism, the lungs by the respiratory route or the digestive system by the oral route. They are then distributed throughout the organism by the blood system and can, depending on what kind of particles they are, accumulate in organs like the liver, the lungs, or the kidneys. Certain metal nanoparticles, if they persist in these organs, will engender an oxidative stress that may cause inflammation and lead eventually to more serious pathologies. Epidemiological studies are nevertheless necessary to establish whether these sometimes worrying consequences can be extrapolated to humans.

In the previous chapters, it has been shown how the properties of matter on the nanometric scale can lead to an extremely varied range of applications, and the astonishing development of nanoscience has not gone unnoticed by the world of industry. At the present time, developments in nanotechnology cover almost every sector of industry, especially those relating to materials and electronics, but also those producing everyday products. Although not always very well informed, owing to the lack of regulation concerning labelling, consumers are more and more often coming into contact with products containing nanoparticles. The Nanotechnology Consumer Product Inventory is drawn up regularly by the Woodrow Wilson International Center

F. Marano (✉) · R. Guadagnini
Laboratoire des réponses moléculaires et cellulaires aux xénobiotiques (RMCX),
Unité de biologie fonctionnelle et adaptative (BFA), CNRS EAC 4413 case
7073 Bâtiment Buffon, Université Paris Diderot-Paris 7, 4 rue Marie-Andrée Lagroua,
Weill-Hallé, 75205 Paris Cedex 13, France
e-mail: marano@univ-paris-diderot.fr

R. Guadagnini
e-mail: rina.guadagnini@gmail.com

J.-M. Lourtioz et al. (eds.), *Nanosciences and Nanotechnology*,
DOI 10.1007/978-3-319-19360-1_12

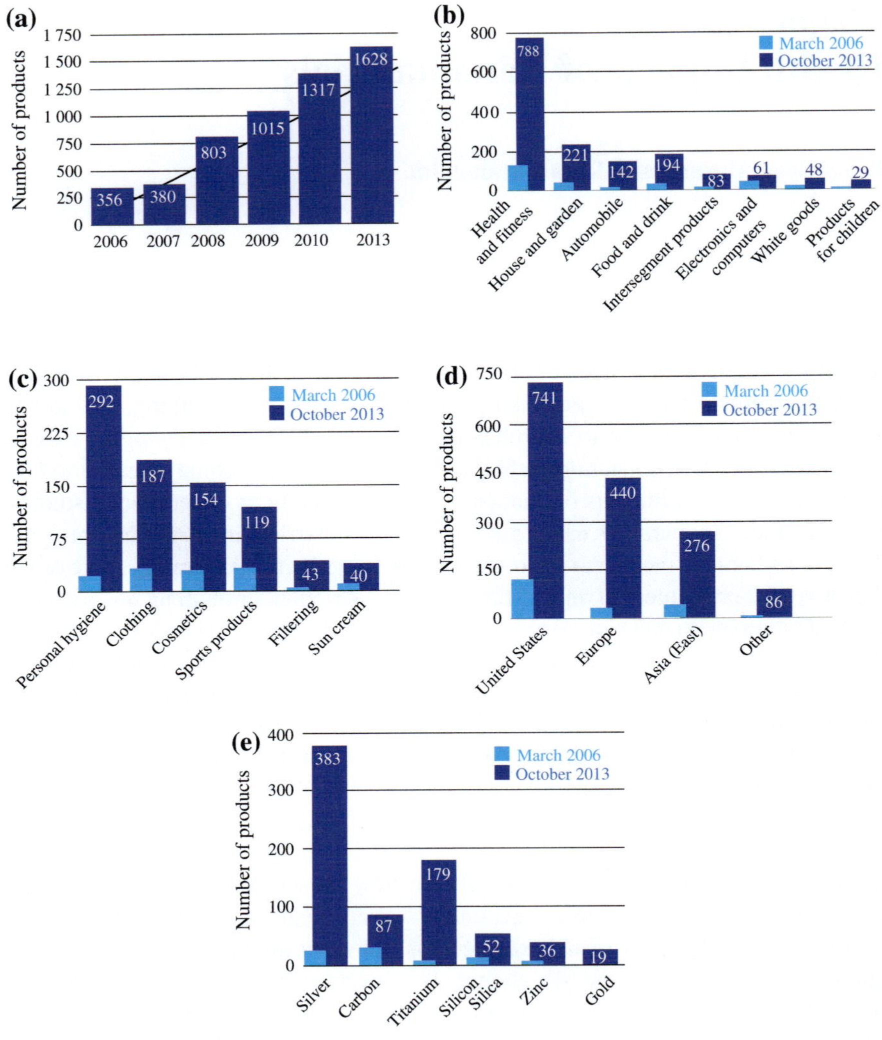

Fig. 12.1 Document produced by the Project on Emerging Nanotechnologies and showing: **a** the increase in consumer products containing nanomaterials over the past five years, **b** the main categories of products, **c** a detailed analysis of the category containing personal hygiene products and cosmetics, **d** the continents where the products are made, and **e** the nanomaterials used in these products. From [1]

for Scholars (Washington DC, 2013) on the basis of industrial declarations. Since these are not obligatory, we should consider that the number of such products is underestimated. Even so, this number has increased by more than 500 % over the last five years (see Fig. 12.1a)! One finds in particular products for everyday use, e.g., for personal hygiene, cosmetics, the home, and the garden (see Fig. 12.1b, c).

One question causing some concern at the moment is the use of nanoparticles in the food industry, where there is unfortunately all too little information. If we consider the origin of these products (see Fig. 12.1d), the United States is the biggest producer, followed by Europe, then South-East Asia, with a huge increase from 2006 to 2013. The nanomaterial occurring most often is silver (see Fig. 12.1e). This turns up in all kinds of preparations by virtue of its bactericidal properties, which have been known for several decades. Apart from its traditional use in medicine, it is also found in treatments for clothing like shirts and socks, on the surface of food packaging, fridge compartments, and computer keyboards, and in personal hygiene products. Titanium oxide nanoparticles are very widely used in the cosmetics industry, in various creams like sun creams and in makeup products. The materials industry has also shown interest, for example by incorporating them in anti-soil wall paints and cements, whence this nanomaterial is now produced in tens of thousands of tonnes. Products containing carbon black nanoparticles have been part of our everyday lives for a long time now, e.g., in the inks of computer printers. Colloidal silica nanoparticles are widely used in the food industry to prevent aggregation effects in salt, sugar, and flour.

These few examples of the most widely used nanomaterials show that we are more and more exposed to nanoparticles. This exposure may occur in the workplace for workers in production units or subcontractors, and such products are now manufactured in almost 600 companies in 30 countries. Generally speaking, it is seriously taken into consideration during the fabrication of nanomaterials, and strict safety regulations have been set up in the United States and Europe. One important problem is the exposure of workers in factories where finished products are made using these nanomaterials, but without always setting up the appropriate safety measures, often through lack of awareness of the potential risks. However, the most difficult problem is nevertheless to protect consumers. Indeed, the sketchy descriptions of nanoparticle use clearly show that there are many kinds of exposure, that they are difficult to assess, and that they often occur on a daily basis.

The question of the impact on nanomaterials on health must therefore be given due consideration if we wish to benefit from the promising developments of nanotechnology while respecting human health and the environment. Since 2005 and the first publications raising the question of their potential toxicity, the toxicology community has been extremely active in trying to understand the interactions that can exist between nanoparticles and living systems. The industrial developments here are too recent for epidemiological studies to have been carried out on professional or environmental exposure, whence health studies remain experimental: exposure of laboratory animals, mainly rats and mice, cell cultures, including human cells from different organs considered to be potential targets, and macromolecules such as proteins and nucleic acids. The number of publications has increased exponentially over the past few years. However, much uncertainty remains regarding the ever increasing numbers of nanoparticles with extremely varied chemical compositions. In particular, it is still hard to suggest to regulatory bodies clear strategies for evaluating the risks to workers or consumers.

12.1 Comparing the Health Risks of Ultrafine Atmospheric Particles and Nanoparticles

By their size and shape, spherical nanoparticles are comparable with ultrafine atmospheric particles, particularly those produced by combustion processes, like the particles rejected by Diesel engines whose harmful health effects are now well established. Likewise, by their length and diameter, nanotubes are comparable with asbestos fibres, even though the chemical composition of these manufactured products is very different. In 2005, inspired by these observations, toxicologists specialising in the health impacts of atmospheric particles in the professional and environmental context alerted us to the possibility of similar impacts from nanoparticles [2, 3]. Indeed, it was at this point that nanotoxicology came into existence.

Questions concerning the potential dangers of nanoparticles are thus ultimately inspired by what is known about fine and ultrafine atmospheric particles (airborne FPs and UFPs, also known as aerosols) and particles rejected by Diesel engines [4]. They also relate to experimental studies comparing the biological and toxicological effects of various manufactured FPs and UFPs, and in particular carbon, silica, TiO_2, and ZnO nanoparticles, among others. The results of recent epidemiological studies which relate the amounts of UFPs in the atmosphere with increased cardiorespiratory morbidity and mortality, and the classification of Diesel emissions in group 1, carcinogenic for humans, by the World Health Organisation (WHO) in 2012, show that these concerns are justified.

Many epidemiological and toxicological studies carried out over the past few years have shown that fine and ultrafine atmospheric particles, after being inhaled, may have systemic effects on organs like the heart, even though they are not the direct target. It is thus important to understand the relevant mechanisms and to determine whether or not these particles can get past the biological barriers, and in particular the respiratory barrier, to distribute themselves throughout the organism. The questions about the health effects of atmospheric particles were raised again in the debate about the toxicity of nanoparticles and the introduction of strategies for assessing the associated risks.

Regarding nanoparticles and nanotubes, the only toxicological data currently available remains somewhat piecemeal and is usually derived from studies carried out on cell cultures, and occasional animal studies. Most of this work yields rather surprising results. Indeed, if we compare particles with the same chemical composition but different sizes (micrometric or nanometric), then at the same concentrations the biological consequences are generally very different, in the sense that nanoparticles may exhibit toxicity where the corresponding microparticles do not. This toxicity seems to be due to the increase in specific surface area, and hence in the number of atoms at the surface of insoluble particles, which favours their interaction with biological molecules. Increased surface reactivity is therefore one of the criteria to be taken into account when assessing the hazards of nanoparticles. Furthermore, for a constant mass of a particle ensemble, the smaller the size of the particles, the more of them there will be, in such a way that it becomes crucial to quantify exposure in terms of particle mass, number, and even surface area. Other important factors to take into account are the solubility of the particles and their ability to cluster.

12.2 Nanoparticle Exposure Routes and Potential Health Effects

Exposure to nanoparticles occurs predominantly by the respiratory route, especially for workers who may be exposed to powders while manipulating nanomaterials during the production process. As far as consumers are concerned, exposure also occurs through the skin, in particular for cosmetics and personal hygiene products. It may also occur in the digestive system through foodstuffs. Whatever the type of exposure, there is currently very little information, and indeed very few possibilities, for making a realistic assessment of such exposure for humans, except in the case of professional exposure [5]. The first evaluations of professional exposure indicate that very low mass concentrations (50 $\mu g/m^3$) are detected in the workplace [6]. These low concentrations must nevertheless be treated with great caution, because even if the masses are low, given the nanometric size of the particles, they represent a very large number of particles.

Current understanding of translocation, accumulation, and persistence in organs, and also the fundamental transport mechanisms, is based on experiments carried out with model nanoparticles on laboratory animals. These nanoparticles have generally been labelled by radioactive tracers in order to track them within the organism, detect potential accumulation sites, and estimate excretion rates. Attempts have been made to assess qualitative changes in the biokinetics of nanoparticles in sick organisms or during specific exposure periods during the lifetime (foetus, childhood, old age) using animal models intended to represent these various situations. In brief then, exposure to nanoparticles can occur through three main channels to be discussed below: the respiratory route, the cutaneous route, and the oral route (see Fig. 12.2).

12.2.1 Exposure by the Respiratory Route

Nanoparticle uptake by inhalation occurs through the respiratory system, whereupon the particles are deposited at different points of the respiratory tract, depending on their size. Indeed, the penetration and deposition of inhaled particles in the lungs involves different kinds of processes for different particle sizes, including sedimentation, inertial impaction, interception (particle–surface contact), diffusion, and electrostatic attraction. One must also consider the geometry and ramifications of the airways, the respiratory rate, and the respiratory mode (nasal or buccal) (see Fig. 12.3).

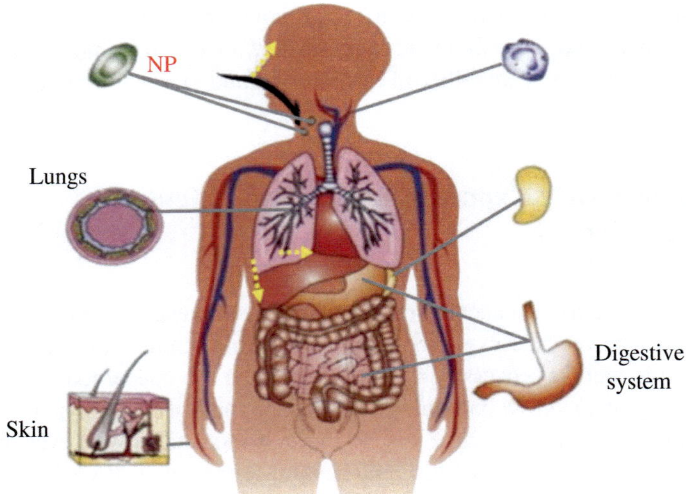

Fig. 12.2 Nanoparticle exposure routes. The main route is respiratory, and potentially through the olfactory nerve from the nose to the brain. The other exposure routes are the skin and the digestive system

It is generally accepted that the smallest particles, with diameters less than or equal to 1 μm, can reach the pulmonary alveoli. This is where gas exchange takes place, and the proximity of the alveolar epithelium and the endothelium of the blood capillaries suggests that nanoparticles might be able to transfer, leading to a systemic exposure to inhaled nanoparticles.

Having said this, it is now considered that nanoparticles with sizes smaller than 20 nm are essentially deposited by diffusion mechanisms and are thus mainly located in the nasopharynx. More quantitatively, it is estimated that 90 % of 1 nm nanoparticles are deposited in the nasopharynx and the remaining 10 % in the tracheobronchial tree. In the bronchial region, deposition may be heterogeneous and the main points for deposition have been identified as being in the bronchial bifurcations, especially in the case of particles measuring more than 100 nm [7]. Nanoparticles inhaled in the form of aggregates or agglomerates, with sizes up to 1 μm and more, will end up in the alveoli.

Inhaled nanoparticles can be eliminated by physiological clearance mechanisms: mucociliary clearance, when the nanoparticles are deposited in the airways, and macrophage clearance, when they accumulate in the alveoli. This elimination will depend on various factors, but in particular, the size, number, solubility, and chemical composition of the nanoparticles, as well as the deposition site and the state of the respiratory tract. Pulmonary retention of nanoparticles will be all the more significant if the patient suffers from an obstructive pathology such as asthma or chronic obstructive pulmonary disease (COPD) [8]. Soluble nanoparticles are dissolved in the layer of mucus produced by cells of the bronchial epithelium which covers the airways. They are then eliminated by mucociliary transport, a kind of conveyor belt of

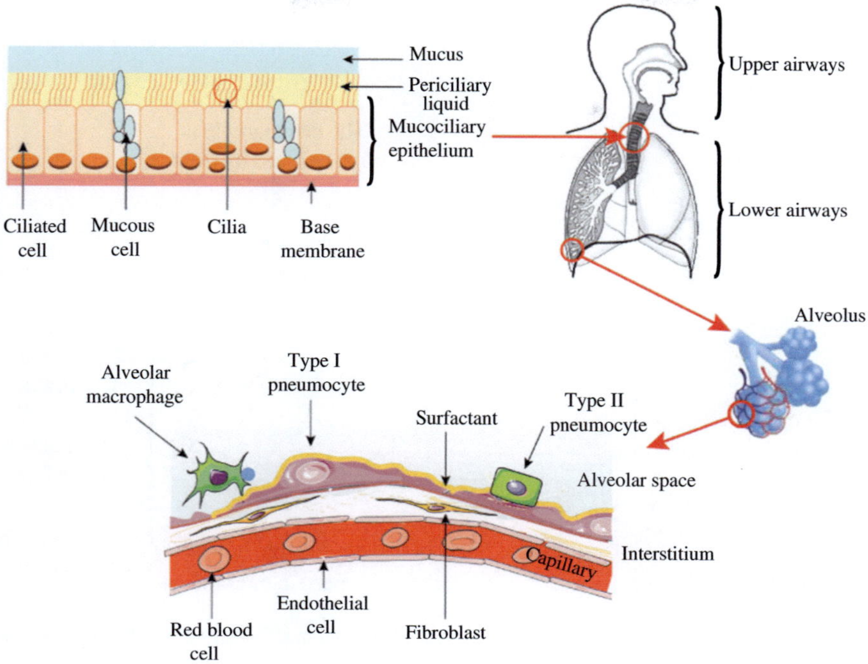

Fig. 12.3 Respiratory system. The mucociliary epithelium, made of ciliated cells and mucous cells and covered with a layer of mucus, lines the upper and lower airways. The alveolar epithelium, made up of type I and type II pneumocytes and coated with surfactant, lines the alveolus. The blood capillaries bounded by the endothelium are very close to the alveolar epithelium. The alveolar macrophages also reside in the pulmonary alveoli

mucus which continually ascends the deep airways toward the nasopharynx, thanks to the beating motion of the cilia on the ciliated cells. Those nanoparticles that are insoluble, or only poorly soluble, can also get trapped in the mucus where they are deposited and then get expectorated or swallowed with the help of mucociliary transport. Indeed, mucociliary clearance is extremely effective. It is faster in the bronchi, but slows down in the bronchioli (see Fig. 12.4).

However, if the nanoparticles are not correctly eliminated, and in particular if a surfeit builds up, they can enter the mucous layer and accumulate in the epithelial cells (see Fig. 12.4). This particle retention in the lungs can produce an inflammatory reaction and lead to pulmonary infections [8, 9]. This has indeed been demonstrated with fine and ultrafine atmospheric particles which trigger attacks in asthma sufferers. It seems likely that a chronic accumulation in highly polluted atmospheres, as has been revealed in workers exposed to exhaust fumes from Diesel engines, may be a factor in the development of lung cancers, particularly in sensitive individuals.

Nanoparticles reaching the alveoli are phagocytosed by alveolar macrophages, depending on their size (see Fig. 12.5). Once loaded with particles, the macrophages can leave the alveoli and migrate into the alveolar interstitium, approaching the

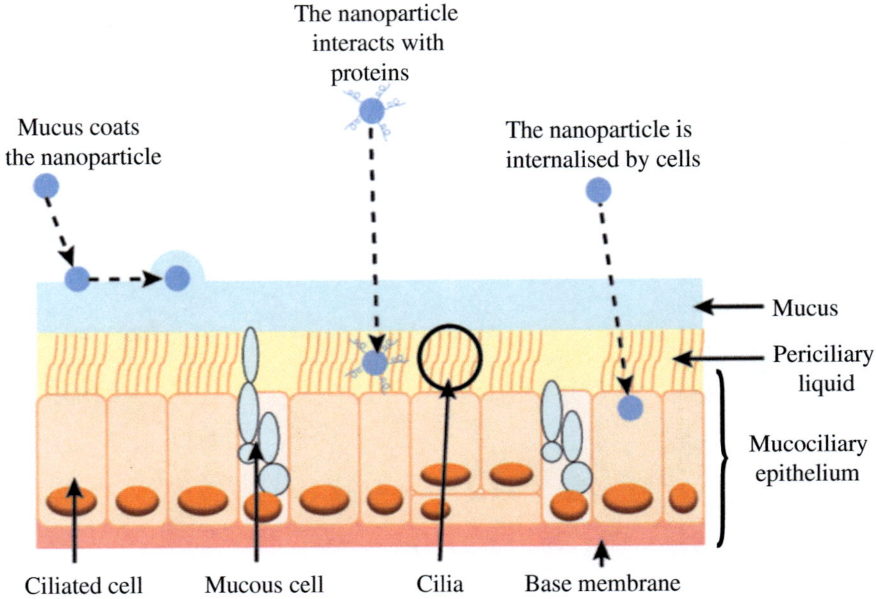

Fig. 12.4 Mucociliary clearance of nanoparticles. When clearance is incomplete, the nanoparticles can come into contact with cells and be ingested

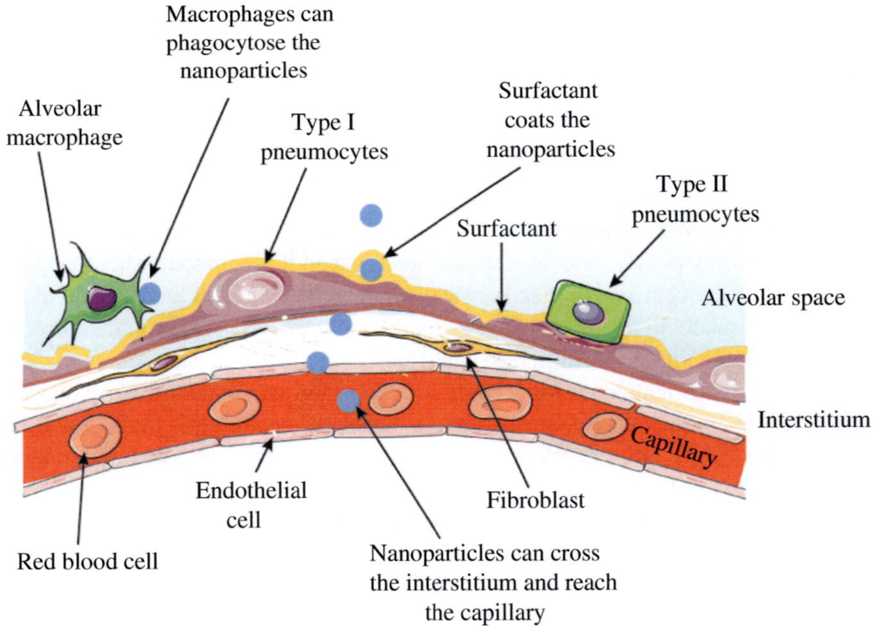

Fig. 12.5 The alveolar–capillary (or blood–air) barrier and transfer of nanoparticles into the circulatory system. The nanoparticles can be phagocytosed by the alveolar macrophages. They can also be coated in pulmonary surfactant and cross the alveolar–capillary barrier to enter the blood

lymphatic ganglions, where cells of the immune system proliferate and differentiate, and entering the blood circulation. A fraction estimated at 10–20 % of insoluble particles is never eliminated from the human lung under physiological conditions due to the very slow kinetics of alveolar particle clearance, implying very long retention times in humans. This fraction might lead to fibrosis in cases of high particle exposure.

In the alveoli, the distance between the air and the blood is very short (around 1 μm), so the alveoli are likely to be a favourable entry point into the systemic circulation, i.e., the general circulatory system (see Fig. 12.5). Studies carried out on rodents, i.e., rats or mice, to estimate the translocation of inhaled particles across the air–blood barrier give results that depend on both the kind of nanoparticle and the administration route (instillation or inhalation).

The most striking results were reported on rats made to inhale radio-labelled iridium nanoparticles (15–80 nm) for 1 h. One percent of the radioactivity was observed seven days later in secondary organs such as the liver, the spleen, the heart, and the brain [10]. These results have been confirmed recently after inhalation by rats of aerosols comprising different 20 nm nanoparticles, viz., iridium, carbon, titanium dioxide, and gold. Small fractions of the nanoparticles were subsequently found in all the secondary organs examined, including the brain and the heart. Having said this, no definitive conclusion about systemic translocation of inhaled nanoparticles can yet be drawn owing to the lack of studies on humans [11].

There is no incontrovertible proof of respiratory pathologies induced in humans by the inhalation of manufactured nanoparticles. All current data has been obtained in studies on animals or cell cultures. They show that certain kinds of nanoparticles are likely to cause inflammation and oxidative stress, factors which could lead to respiratory pathologies.

A publication from 2009 [12] focused on observations of pleural effusion leading to death, pulmonary fibrosis, and granuloma in workers exposed professionally to polyacrylate nanoparticles. A controversy ensued because the causal connection with the nanoparticles found in the biopsies and highlighted by the authors was not at all clear. Indeed, the existence of other risk factors had not been considered in the absence of accurate measurements of the aerosol to which the workers had been exposed.

Concerning the possible carcinogenic effects of nanoparticles, studies carried out so far do not provide enough evidence. For example, titanium oxide (TiO_2) seems to be able to induce pulmonary tumours in rats after repeated instillation of nanoparticles. However, no definitive conclusion can be drawn regarding the carcinogenic effect of TiO_2 nanoparticles on humans. We can only say that, on the basis of animal data, TiO_2 is suspected of being potentially carcinogenic for humans by the International Agency for Research on Cancer (IARC), which has classified these nanoparticles in class 2B (possible human carcinogenicity).

Recently, several publications have raised the alert over the possible dangers of carbon nanotubes, analogous to the dangers posed by asbestos fibres, and this from studies on rats and mice. We must therefore assume the possibility of health risks following exposure to carbon nanotubes, and this has led to proposals to provide maximal protection for workers in factories producing these nanotubes.

12.2.2 Exposure by the Cutaneous Route

The skin represents more than 10 % of our body weight and plays a key role as a barrier with respect to our external environment, serving both to protect and to maintain homeostasis. It comprises three main parts: the epidermis, the dermis, and the hypodermis. The question of exposure of the epidermis to nanoparticles has become critical in the past few years, because the cosmetics industry now incorporates titanium oxide and zinc oxide nanoparticles in 'total block' sun creams. This is supposed to provide more effective sun protection thanks to the particular absorption properties of nanoparticles for UV radiation, while these remain transparent to visible radiation. At the same time therefore, these creams become transparent and hence more aesthetic than others containing very white microparticles!

Studies carried out on skins reconstituted in the laboratory or on animals, in particular pigs, conclude that nanoparticles do not cross to the dermis for healthy skin. Nanoparticles can penetrate the stratum corneum of the epidermis and the hair follicles or sudoriferous (sweat) glands. The nanoparticles remain on the surface of the epidermis or in the stratum corneum, the layer at the surface of the skin, but are unable to penetrate or cross the living layers of the epidermis, even if they can accumulate and persist in the hair follicles (see Fig. 12.6). However, this situation may change when the integrity of the cutaneous barrier is compromised, or in the case of skin diseases. Allergic contact dermatitis (ACD), atopic eczema, and psoriasis,

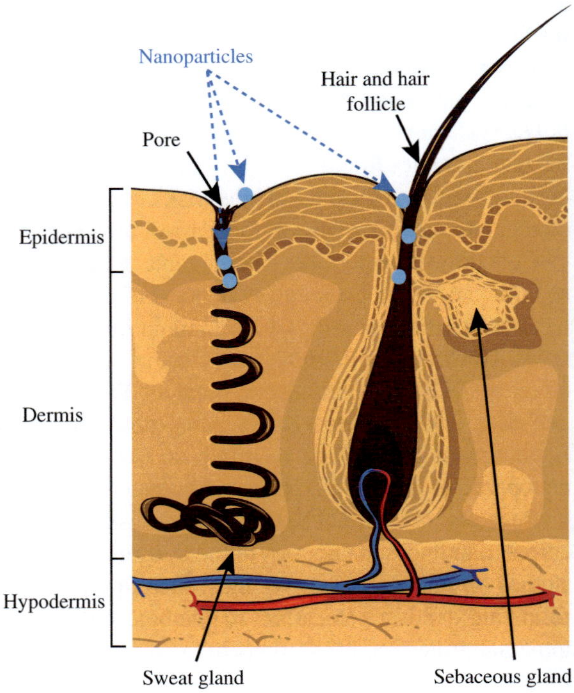

Fig. 12.6 Structure of human skin with its three layers: the epidermis, dermis, and hypodermis. Nanoparticles can penetrate the epidermis and accumulate in the hair follicles, and also in the sudoriferous (sweat) glands

but also mechanical flexion and the use of irritant detergents and chemical products can increase cutaneous absorption [13].

12.2.3 Exposure by the Oral Route

Nanotechnology is gradually appearing in the food sector, essentially in techniques for encapsulating nutrients and vitamins, but also in products for enhancing flavour and taste. These capsules are generally based on liposomes and are thus made from biological molecules which, in principle, raise no health problems. However, food packaging may contain nanoclays and silver nanofilms to prevent deterioration and oxidation of the foodstuff. We should therefore raise the issue of the stability of these films and their possible transfer into foods. In addition, silica nanoparticles are widely used to fluidify salt, sugar, and flour. Finally, we should add to this list all the links in the food chain which, intentionally or otherwise, are impacted by the

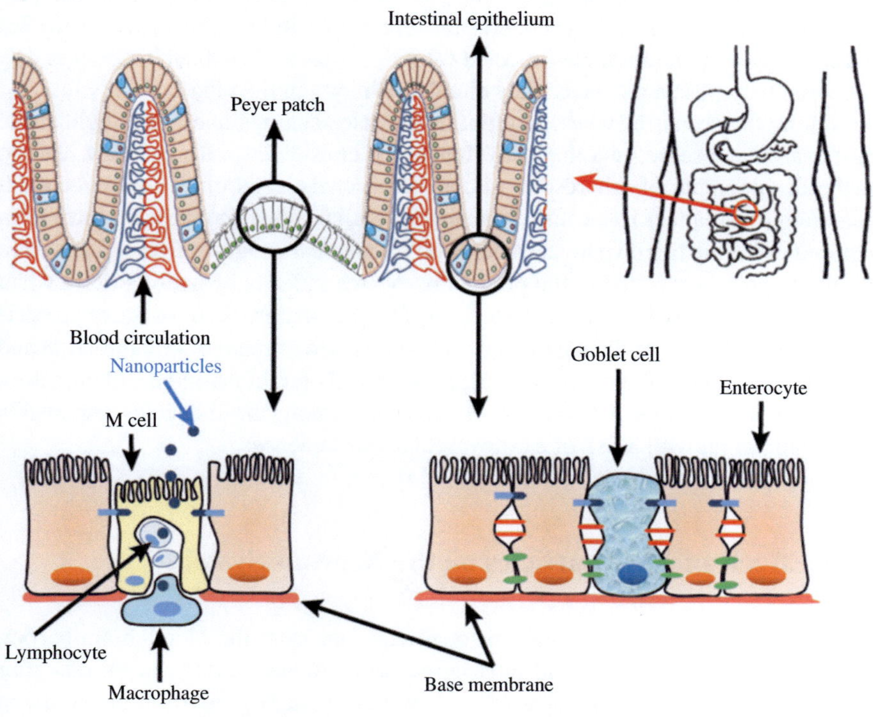

Fig. 12.7 Intestinal mucus comprises a layer of enterocytes and goblet cells. The M cells and Peyer patches are within this epithelium. Nanoparticles might get through preferentially here, given their role in the immune system and their proximity to macrophages in the blood circulation

presence of nanoparticles and nanomaterials in the environment, and in particularly in contact with water (see Chap. 13).

The inner wall of the small intestine is lined with a single-layered epithelium containing mucous cells and enterocytes, which absorb nutrients, and covered with many microvilli, small protrusions which increase the absorption area. At the present time, we have only a very little information about this nanoparticle absorption route. However, results obtained by introducing nanoparticles into food for rodents show that a small fraction of these particles is subsequently found in the internal organs. This means that they are able to go through the intestinal barrier. The most likely mechanism is through absorption by M cells and Peyer patches (lymphoid aggregates in the small intestine) [14]. The nanoparticles might then be absorbed in different ways depending on their quantity and lipid solubility (see Fig. 12.7).

12.3 Effects of Nanoparticles on the Cardiovascular System

There is still little information about the possible cardiovascular effects of nanoparticles. The only available data concerns the effects of ultrafine atmospheric particles (UFP). However, nanoparticles injected for the purposes of medical imaging or as a nanodrug may pose a risk since they circulate directly in the blood.

Among the scenarios whereby ultrafine particles and nanoparticles might cause cardiovascular disease, endothelial dysfunction is considered as a possibility, leading to the development of atherosclerosis, acute coronary syndrome, and myocardial infarction (heart attack). Recently, human endothelial cells from the aorta have been exposed in culture to iron nanoparticles used for medical imaging. Toxic effects were observed, possibly related to the production of free radicals by nanoparticles when they interact directly with the endothelium. The production of an oxidative stress is considered to be one of the significant risk factors downstream of inflammation and the initial stages of atherogenesis [15]. Free radicals might cause an inflammatory response with secretion of cytokines, which feature among the molecules responsible for inflammation with a risk of cardiovascular consequences.

12.4 Effects of Nanoparticles on the Nervous System

There is still debate over whether nanoparticles can cross the blood–brain barrier, even though biodistribution and bioaccumulation studies carried out on rats after inhalation of radioactive nanoparticles show that a small proportion of them end up in the brain [16]. These studies have been contested because the radioactivity detected in the brain might have been due to the radioactive label detaching itself from the nanoparticle without the latter having actually crossed the blood–brain barrier. However, it has been suspected that aerosols may cross this barrier in humans in very polluted areas like Mexico City. The nanoparticles would then be able to induce the formation of free radicals and engender oxidative stress, known to underlie brain pathologies. Free radicals have been linked to several neurodegenerative disorders, such as Parkinson's disease, Alzheimer's disease, and Huntington's disease. The brain is particularly vulnerable to oxidative stress damage owing to its high unsaturated fatty acid content, its high oxygen consumption, and its relatively low antioxidant enzyme content compared with other organs.

12.5 Conclusion

Although research into the health risks of nanoparticles has developed enormously over the past few years, much remains to be done in order to understand their biological effects and assess the risks for our health. Most existing data, obtained experimentally on animals and cell cultures, cannot be directly extrapolated to humans. However, this information has considerably improved our understanding of the interactions between nanoparticles and living organisms. Discoveries concerning the ability of nanoparticles to interact with biological fluids such as mucus, surfactants, and serum proteins, will help us to assess their chances of crossing the various biological barriers, viz., the alveolar–capillary, intestinal, and blood–brain barriers, and eventually also the placental barrier. It is already clear that the surface properties of nanoparticles play a key role in such translocation. Another key factor that would surely be of decisive importance when assessing health risks is the accumulation of nanoparticles in target organs and their biopersistence, even if the levels of transfer and accumulation are very low [10].

Since the health crises associated with silica in coal mines and asbestos, we know that this accumulation can in the long term give rise to pathologies as serious as pulmonary fibrosis and cancers of the lungs and pleura. This must certainly be taken into account most carefully, given that exposure to nanoparticles, even at very low levels, occurs on a daily basis for workers involved in industrial production and also for many consumers. We still do not know anything like enough to be able to draw conclusions, all the more so in that the many publications over the past five years show that, while we can establish the main mechanisms occurring in cells, such as phagocytosis or the induction of oxidative stress, the biological response may vary considerably. It depends not only on many physicochemical factors intrinsic to the nanoparticle, but also on factors relating to the environment which must necessarily be taken into account to understand the origin of these biological responses and to be able to make proper comparisons. This means that research into the relationships between nanoparticles and health must necessarily be cross-disciplinary, bringing together metrologists, chemists, and toxicologists to develop our fundamental understanding of the biological properties of these nanoparticles, but also to formulate appropriate regulatory strategies to protect humans and their environment.

References

1. www.nanotechproject.org/inventories/consumer/
2. G. Oberdörster, E. Oberdörster, J. Oberdörster, Nanotoxicology: an emerging discipline evolving from studies of ultrafine particles. Environ. Health Perspect. **113**, 823–839 (2005)
3. K. Donaldson, V. Stone, C.L. Tran, W. Kreyling, P.J. Borm, Nanotoxicology. Occup. Environ. Med. **61**, 727–728 (2004)
4. F. Marano, Les particules atmosphériques fines et ultrafines: l'apport de la toxicologie dans l'évaluation des risques pour la santé. Pollution Atmosphérique 29–35 (2010)
5. R.J. Aitken, M.Q. Chaudhry, A.B. Boxall, M. Hull, Manufacture and use of nanomaterials: current status in the UK and global trends. Occup. Med. (Lond.) **56**, 300–306 (2006)
6. A.D. Maynard, P.A. Baron, M. Foley, A.A. Shvedova, E.R. Kisin, V. Castranova, Exposure to carbon nanotube material: aerosol release during the handling of unrefined single-walled carbon nanotube material. J. Toxicol. Environ. Health A **67**, 87–107 (2004)
7. P. Borm, F.C. Klaessig, T.D. Landry, B. Moudgil, J. Pauluhn, K. Thomas, R. Trottier, S. Wood, Research strategies for safety evaluation of nanomaterials. Part V: role of dissolution in biological fate and effects of nanoscale particles. Toxicol. Sci. **90**, 23–32 (2006)
8. P. Andujar, S. Lanone, P. Brochard, J. Boczkowski, Effets respiratoires des nanoparticules manufacturées. Rev. Mal. Respir. **26**, 625–637 (2009)
9. M. Geiser, W.G. Kreyling, Deposition and biokinetics of inhaled nanoparticles. Part. Fibre Toxicol. **7**, 2 (2010)
10. W.G. Kreyling, M. Semmler, F. Erbe, P. Mayer, S. Takenaka, H. Schulz, G. Oberdörster, A. Ziesenis, Translocation of ultrafine insoluble iridium particles from lung epithelium to extrapulmonary organs is size dependent but very low. J. Toxicol. Environ. Health A **65**, 1513–1530 (2002)
11. W. Yang, J.I. Peters, R.O. Williams, Inhaled nanoparticles—a current review. Int. J. Pharm. **22**, 239–247 (2008)
12. Y. Song, X. Li, X. Du, Exposure to nanoparticles is related to pleural effusion, pulmonary fibrosis and granuloma. Eur. Respir. J. **34**, 559–567 (2009)
13. M. Crosera, M. Bovenzi, G. Maina, G. Adami, C. Zanette, C. Florio, F. Filon, Larese: nanoparticle dermal absorption and toxicity: a review of the literature. Int. Arch. Occup. Environ. Health **82**, 1043–1055 (2009)
14. A. Cordeiro-da-Silva, S.G. Romeijn, M. Amidi, A. de Sousa, G. Borchard, H.E. Junginger, Uptake studies in rat Peyer's patches, cytotoxicity and release studies of alginate coated chitosan nanoparticles for mucosal vaccination. J. Control. Release **12**, 348–358 (2006)
15. M.T. Zhu, B. Wang, Y. Wang, L. Yuan, H.J. Wang, M. Wang, H. Ouyang, Z.F. Chai, W.Y. Feng, Y.L. Zhao, Endothelial dysfunction and inflammation induced by iron oxide nanoparticle exposure: risk factors for early atherosclerosis. Toxicol. Lett. **10**, 162–171 (2011)
16. G. Oberdörster, Z. Sharp, V. Atudorei, A. Elder, R. Gelein, W. Kreyling, C. Cox, Translocation of inhaled ultrafine particles to the brain. Inhal. Toxicol. **16**, 437–445 (2004)

Chapter 13
Environmental Risks of Nanotechnology: A New Challenge?

Jean-Yves Bottero

Abstract A major research effort is underway around the world to understand the impacts of nanomaterials on our environment. Some occur naturally or are produced by human activity, while others are manufactured on an industrial scale, and their presence in the environment cannot be disputed. The questions raised here concern their transformations in contact with water, the mechanisms and kinetic aspects of the exposure of biological organisms in water, sediments, or soils, and their possible harmful effects on living beings. What characterises some of these nanomaterials is the fact that they dissolve in water, then reprecipitate in the form of different minerals that are less toxic because they lose their initial properties. The effects on living beings, particularly within a trophic cascade, are still poorly understood, especially since the concentrations are very low. The challenge here is to obtain a sufficient understanding of the exposure and transformation phenomena, together with the effects on organisms within the relevant trophic cascades, to be able to model the risks and hence arrive at a predictive assessment when the nanomaterials are both chemically and structurally complex. Research in this area can only be cross-disciplinary.

Toxicological research on fine and ultrafine particles (UFP) is a highly complex matter given their many possible forms and the wide range of mechanisms brought into play when they come into contact with different parts of the biological or natural environment (atmosphere, sea, rivers, etc.). The aim in this chapter is to present research on the reaction mechanisms between nanoparticles and our environment, and in particular water, and their possible effects on different ecosystems and hence on humans. With this understanding, it should be possible to determine an effective strategy for the treatment of domestic drinking water. We shall also describe an attempt to model the interaction processes between nanoparticles and water, together with the potential risks for the environment.

According to the investigations of the Woodrow Wilson Center in Washington in 2010 [1], there are more than a thousand products on the market containing nanopar-

J.-Y. Bottero (✉)
Centre Européen de recherche et d'enseignements des géosciences de l'environnement (CEREGE) UMR 7330, CNRS/Université d'Aix-Marseille, Europole de l'Arbois BP 80,13545 Aix-en-Provence Cedex 4, France
e-mail: bottero@cerege.fr

J.-M. Lourtioz et al. (eds.), *Nanosciences and Nanotechnology*,
DOI 10.1007/978-3-319-19360-1_13

Fig. 13.1 Factory at the water's edge. Environmental pollution long predates nanotechnology

ticles, of which a significant proportion concerns general well-being, i.e., cosmetics, drinks, etc. (see Fig. 12.1b). Nanoparticles are also incorporated into many everyday objects, e.g., paints, tyres, cements, and glasses, to improve performance. The first toxicological studies of the potential risks of nanoparticles for the environment were already being carried out at the end of the 1980s [2], following the very real problems caused by asbestos and the inhalation of fine and ultrafine particles so prevalent in certain urban environments. These studies were undertaken mainly in laboratories that had already participated in work on asbestos and the toxicology of micrometric contaminant particles with proven toxicity such as arsenic, polycyclic aromatic hydrocarbons (PAH), and pesticides (see Fig. 13.1).

13.1 From Naturally Occurring Nanoparticles to Manufactured Nanoparticles

For our purposes, we consider nanoparticles to fall into three categories: naturally occurring nanoparticles, those produced as a by-product of human and industrial activities, and manufactured nanoparticles.

13.1.1 Naturally Occurring Nanoparticles

Nanoparticles are continually being produced in nature without any need for human intervention. Consider two examples:

- Volcanic soils in La Réunion, Hawaï, and elsewhere synthesize clays known as imogolites which regularly diffuse into the upper atmosphere. They take the form of nanotubes about 2 nm in diameter and 100 nm long (see Fig. 13.2) and are made up of aluminium and silicon atoms in the molar ratio [Al]/[Si] $=$ 2. They have

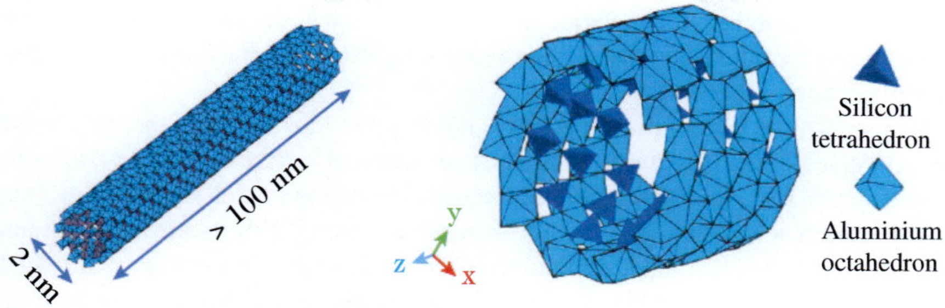

Fig. 13.2 Nanotubes containing aluminium and silicon (imogolites) found in some volcanic soils

a remarkable capacity to concentrate metals like nickel and copper coming either from volcanic activity, or from recycled agricultural waste such as fertilisers. They are also used as coatings for photographic papers.

- Enormous amounts of silica nanoparticles (SiO_2) are carried along by rivers and winds. They are poorly crystallised or amorphous. They are commonly used in industry in materials like tyres to enhance their resistance to wear, and in certain special cements to reduce the amount of water needed for the cement to set (see Chap. 7).

13.1.2 Nanoparticle By-Products of Human and Industrial Activities

These are the by-products of industrial and domestic heating systems, car exhaust pipes, smokers, and dusts emitted in most professional activities, whatever the size of the company (paint, metallurgy, etc.). These nanoparticles are much closer to us. They are very carefully monitored by health watch institutions, as attested by campaigns regarding the harmfulness of Diesel exhaust emissions.

13.1.3 Manufactured Nanoparticles

This last category includes nanoparticles that have been made intentionally for their industrial applications. By controlling parameters such as size, composition, and morphology, one can obtain particles with quite different chemical reactivity and physical properties (electronic, optical, mechanical, biological, etc.) as compared with larger objects. These nanoparticles can be used as such or else incorporated into other materials referred to by extension as nanomaterials.

Regarding the quantities involved here, we may cite the significant industrial production of silica, cerium and titanium oxides, and carbon and aluminate nanotubes, which are used in sectors as varied as drugs, cosmetics, the food industry, the car industry, the aeronautics industry, the chemical industry, and the fabrication of components, among others.

Among all the nanoparticles, gold is certainly the most emblematic. Of course, gold is valued for its brightness, stability, and chemical inertness. However, at nanometric dimensions, it behaves quite differently. For nanoparticles smaller than 3 nm, gold becomes a catalyst, i.e., it can exchange electrons with its environment and regenerate [3]. Another singular feature is the sudden drop in melting point which occurs for gold nanoparticles of around 20 nm, from 1 336 K for the bulk material to just 300 K! Finally, as a nanometric powder, gold naturally changes colour (see Chaps. 2 and 5).

Nanoparticles are often designed for their optical properties. For example, quantum dots[1] are used to image biological reactions or localise objects in living organisms or in cells (see Chap. 9). Metal nanoparticles for their part serve to colour materials like glasses, to make them more attractive to the consumer.

Nanoparticles are also used for their catalytic properties. Cerium oxide (CeO_2) nanoparticles are used in diesel fuel to reduce the combustion temperature and oxidation of carbon monoxide. Below a size of a few nanometres, the existence of vacant oxygen sites in its structure endows it with anti-oxidant properties. Beyond a few nanometres, it reacts with oxygen to transform it into reactive oxygen species (ROS) [4], and among these one finds the free radicals which are sometimes very powerful oxidising agents and extremely harmful to living organisms.

In nanoparticulate form, titanium dioxide (TiO_2) has photocatalytic properties whose efficacity depends on the size and crystallographic nature of the particles (anatase or rutile) (see Chap. 8). Optimal efficacity is obtained when the particles are smaller than 15 nm. These can oxidise organic pollutants. For example, trichloroethylene is destroyed by 7 nm TiO_2 particles, and chloroform by 11 nm TiO_2 particles.

Nanoparticles can also have size-dependent thermal properties. This happens with indium and tin nanoparticles, whose melting points are reduced by 120 and 80 °C, respectively, when their sizes are decreased from 100 to 10 nm. Their latent heat of fusion actually decreases exponentially with their size once the latter goes below 20 nm.

Considering the direct exposure of factory workers producing nanoparticles, health institutions issue recommendations about the means of protection to be used when manipulating them, and also about ventilation systems, etc. Since 1 January 2013, it has become obligatory to mention their presence and their level of dangerousness.

But the point to bear in mind is that all the nanoparticles just discussed will eventually end up being rejected into the environment, and it would be illusory to expect to predict a priori all the problems this might raise. Their presence in the air

---

[1] These are metal or semiconductor nanocrystals measuring 6–8 nm and which can emit light by fluorescence under certain thermal or lighting conditions. The wavelength of this emission, i.e., its colour, depends on the kind of material used and also the size of the dots (see Sect. 2.3 of Chap. 2).

and toxic effects caused by their inhalation or ingestion have already been discussed in Chap. 12. Here we shall focus on studies characterising some aspects of their behaviour in contact with water or with organisms present in aquatic environments.

13.2 The Importance of Contact with Water

The most common interaction that nanomaterials are likely to undergo when they come into contact with the natural environment is undoubtedly their interaction with water in all its various forms, viz., in the form of vapour in the air, in liquid form in rivers or in biological organisms, or even in solid form. Their affinity with water will sometimes modify certain of their properties, such as:

- faster or less energetic solubilisation, the nanoparticles transforming into solvated ions,
- transition from a hydrophobic to a hydrophilic state,
- increased surface reactivity, transforming the nanoparticle into a Trojan horse for proven pollutants.

13.2.1 Do Nanoparticles Solubilise According to the Same Laws as Microparticles?

The dissolution of a solid compound in a solvent is characterised by an equilibrium constant, the solubility product K_{sp}, which is a thermodynamic quantity. The solubility K_b of this compound, that is, the maximal concentration in moles per litre that can be dissolved or dissociated in the solvent at a given temperature, is thus necessarily related to K_{sp}. In the case of a nanoparticle crystal, the solubility also depends on the radius r of the nanoparticle and its surface energy γ, which characterises its interaction with the surrounding medium, here the solvent. Recall, for example, that the spreading of a water droplet at the surface of a solid depends on this interface energy between the solid and the water. It turns out that the solubility of a nanoparticle crystal is related to the above defined quantities according to

$$\ln K_b = \ln K_{sp} + C(\gamma/r) ,$$

where C is a constant for a given material.

The solubility therefore increases as the radius of the nanoparticle decreases, a relationship that has been well confirmed for nanoparticles measuring more than 25 nm. For smaller nanoparticles, their morphology begins to play an important role and the dependence of γ on the size can generally no longer be neglected. However, there are exceptions: recent work has shown that, for metallic silver nanoparticles (Ag^0) with sizes from a few nanometres to fifty nanometres, the above equation remains valid [5]. Each nanoparticle must therefore be considered on a case by case basis.

13.2.2 Transition from Hydrophobic to Hydrophilic Properties

The fullerenes (C_{60}, C_{70}, C_{80}), which have a more or less spherical shape and are made up of 60, 70, or 80 carbon atoms, are the nanoparticles whose interactions with water have been most widely studied. These materials are initially hydrophobic, and when they first come into contact with water, they tend to remain at the interface with the air. However, when agitated, some of them will disperse in the aqueous phase and form stable suspensions. These nanoparticles thus become hydrophilic. The effective affinity of fullerenes for water has been demonstrated by adsorption/desorption of water vapour on previously dehydrated C_{60} powder (see Fig. 13.3).

Depending on the partial vapour pressure P_{H_2O}/P_0 at the surface, the adsorption/desorption cycle exhibits a first adsorption phase characteristic of a hydrophobic material (increasing curve). The ensuing desorption phase (decreasing curve) returns to zero water vapour pressure ($P_{H_2O}/P_0 = 0$) and gives rise to the deposition of a water monolayer at the surface of the nanoparticle. The initially hydrophobic C_{60} nanoparticle then becomes hydrophilic [6, 7] through the presence of hydroxyl groups (OH) bound to the carbon atoms and through their dissociation, viz., C–OH $\rightarrow$ $C\text{–}O^- + H^+$, which bestows a negative electric charge carried by the oxygen atoms.

13.2.3 A Possible Trojan Horse Effect

As the particle size becomes nanometric, the percentage of atoms at their surface increases almost exponentially (see Fig. 13.4). Consider the case of maghemite, a

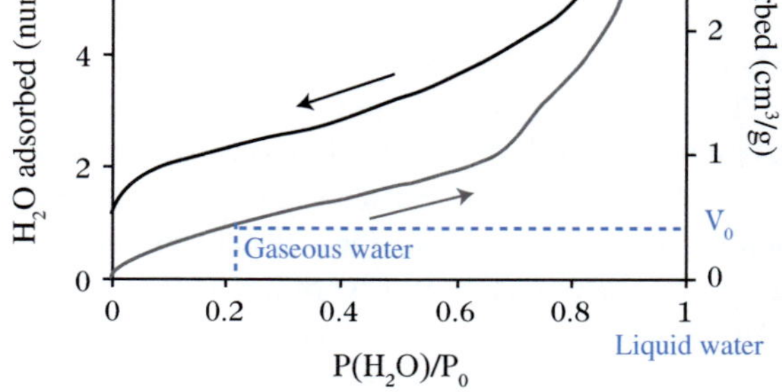

Fig. 13.3 Adsorption of water vapour as a function of the partial vapour pressure at the surface of a nanoparticle, e.g., a fullerene C_{60} particle. The water vapour adsorption curve is indicated by an *upward arrow*, the desorption curve by a *downward arrow*. The vertical increase for $P_{H_2O}/P_0 \sim 0.9$ corresponds to liquefaction

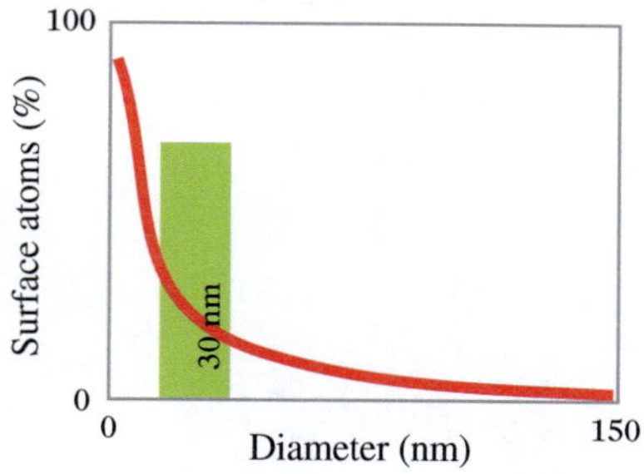

Fig. 13.4 Percentage of surface atoms for different particle sizes

superparamagnetic oxide[2] of iron (III) or of iron (II) and iron (III), used in particular in magnetic imaging. Below 30 nm, the percentage of surface atoms exceeds 30 % and the interatomic potentials are such that atomic sites become vacant, thereby modifying the surface chemistry.

Figure 13.5 shows the increase in the number of vacant sites for constant surface area when going from maghemite particles measuring 100 nm to ones of less than 10 nm [8]. This results in the appearance of adsorption sites with a strong affinity for certain molecules. This is the case for arsenic hydroxide $As(OH)_3$, which is a very powerful contaminant for humans and has a particularly strong affinity for maghemite nanoparticles with diameter less than 10 nm.

While there have been many studies concerning the capacity of iron oxide nanoparticles to adsorb metals and metalloids [9], far fewer have focused on carbon-bearing nanoparticles, fullerenes C_{60} or carbon nanotubes, and their ability to adsorb hydrophobic or weakly hydrophilic organic pollutants. However, the first such studies have shown that fullerene C_{60} nanoparticles have a much greater capacity to retain organic pollutants than activated carbon, used today to eliminate pollutants in the treatment of drinking water and humic substances in river water. There is more and more evidence that nanoparticles can trap contaminants. At the moment, it is quite impossible to predict the environmental consequences, because there is inadequate experimental data concerning the transport role played by these objects with regard to the organisms and plants in different ecosystems.

13.2.4 Aggregation Properties Associated with Reactivity at the Water or Solvent Interface

Particles contained in surface waters often occur in an aggregated form, including both biological organic matter and inorganics. These aggregates will either remain in

[2]Superparamagnetism is a typical behaviour of ferromagnetic and ferrimagnetic materials when they occur as very small grains or nanoparticles. Their magnetisation reverses spontaneously, in such a way that it seems on average to be zero. An external field can then magnetise them like a paramagnetic material, except that the magnetic susceptibility of the grains is much higher.

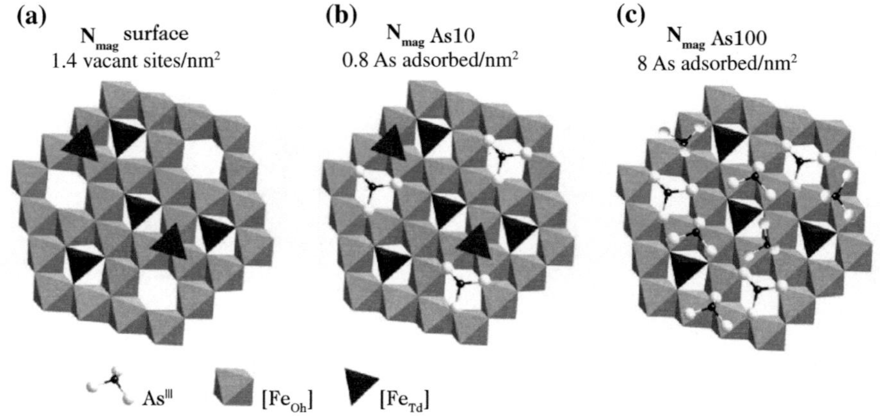

Fig. 13.5 Vacant sites at the surface of maghemite nanoparticles Nmag (**a**), and adsorption of $As(OH)_3$ molecules on 10 nm maghemite particles: NmagAs10 (**b**), and NmagAs100 (**c**) correspond to 10 % and 100 % of the surface layer, respectively. Many of the vacant sites occupied by the $As(OH)_3$ molecules correspond initially to tetrahedral iron atoms (**a**) and (**b**). The molecules are then associated with six iron atoms. The other molecules of $As(OH)_3$ are adsorbed on sites by complexing with two or three iron atoms. The number of molecules adsorbed per unit area increases by a factor of around 10 when the size of the maghemite grains goes below 20 nm (**b**) → (**c**)

suspension for particles with sizes around the micron, or decant to form the sediment at the bottom of rivers in the case of larger particles.

Nanoparticles in suspension can in some cases repel one another and in others aggregate either together or with other particles present in the medium. Naturally, their interactions in aqueous media depend on their chemical nature and the nature of the surrounding medium. In biological media, or in aqueous media in general (rivers, waterlogged soils), the main components other than water are salts of ions like Na^+, Ca^{2+}, K^+, NO_3^-, SO_4^{2-}, PO_4^{3-}, etc., and organic molecules ranging from proteins through humic acids, to compounds of industrial origin like herbicides, pesticides, and solvents. The affinity of the latter molecules for nanoparticles depends on the pH, their hydrophobicity, and their transport conditions.

All these features must be taken into account in analytic studies of aqueous media and water treatment. In the following, we shall focus in particular on the case of fullerenes and also of metal oxide particles, for which several studies have now been published.

13.2.4.1 Aggregation of Fullerenes in Water and Solvents

Fullerenes have variable size distributions depending on whether they are dispersed in water or in different solvents, with or without agitation [6]. Figure 13.6 shows that the size of the nanoparticles is bigger in water than in solvents, even after agitation.

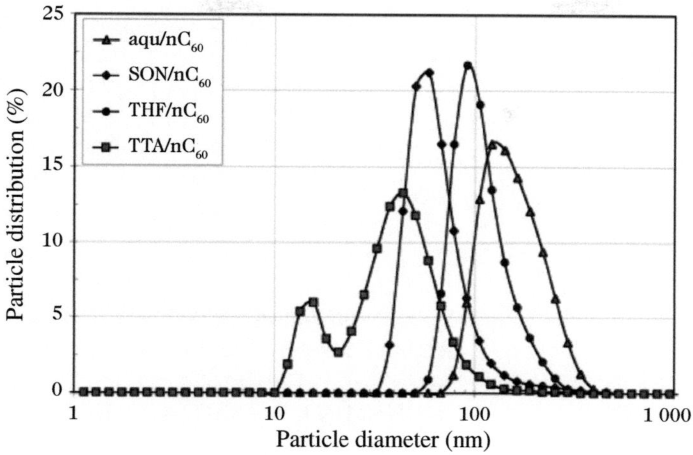

Fig. 13.6 Granulometric distribution of C_{60} nanoparticles in water with agitation by rotation (aqu/nC_{60}), in water after ultrasound treatment (SON/nC_{60}), in tetrahydrofuran (THF/nC_{60}), and in toluene (TTA/nC_{60}) [7]

Fullerene C_{60} particles have a natural tendency to aggregate in water. The carbon atoms in the C_{60} bind easily to hydroxyl OH groups (see Sect. 13.2.2), which dissociate by the reaction C–OH $\rightarrow$ C–O^- + H^+ and thereby generate a negative charge on each oxygen atom bound to a carbon. The hydroxylated fullerenes, known as fullerols, will therefore be attracted by cations like Na^+ or Ca^{2+} present in the water, in such a way that the ions become a kind of bridge between neighbouring fullerene particles [6].

In the presence of light, hydroxylated fullerenes behave differently to non-hydroxylated fullerenes because of their enhanced aggregation in water. Indeed, we observe the production of reactive oxygen species (ROS), which are toxic for living beings, whereas non-hydroxylated fullerenes do not produce these [10]. The non-hydroxylated particles can also form aggregates, but these have a very close-packed face-centered cubic (cfc) structure, while the fullerol aggregates have fractal structures, more open and porous, where each nanoparticle is in contact with many neighbours but interacts more with the surrounding water.

13.2.4.2 Aggregation of Oxide Particles: Charge Effect

A significant fraction of industrially produced nanoparticles are made from metal oxides. Today, the tendency of these oxides to disperse or aggregate in aqueous media depending on their hydrophilic nature is well established. Indeed, the aggregation of metal oxide particles is a phenomenon that has received a great deal of interest since the discovery of the surface charge which results from the pH-dependent dissociation of hydroxyl groups.

Table 13.1 Values of the zero point charge (ZPC) for different metal oxides

Metal oxide	Zero point charge (ZPC)
TiO_2	5.8
CeO_2	8.1
SiO_2	2
Fe_3O_4	6.5
ZnO	9

As an example, the following equation shows the action of a proton H^+ on a surface atom of sulfur (S)—a metalloid or semi-metal—associated with an oxygen atom:

$$SO^- + H^+ \longrightarrow S\text{–}OH + H_2O \longrightarrow S\text{–}OH_2^+ + HO^- .$$

This exchange reaction changes the electric charge at the surface of the particles depending on the number of protons associating with the S atom through the oxygen atom. The way this charge changes thus depends on the pH of the medium, i.e., its acidity, and the concentrations of H_3O^+ or H^+ ions. The negative charge decreases with the concentration of H_3O^+ or H^+ ions, eventually reaching zero. This situation corresponds to a pH called the zero point charge (ZPC).[3] Table 13.1 gives the values of the zero point charge for different metal oxide nanoparticles.

The presence in water of cations like Na^+, Ca^{2+}, etc., or anions like Cl^-, NO_3^-, SO_4^{2-}, and PO_4^{3-}, will modify the dispersion of the particles [11]. For example, Ca^{2+} cations will be attracted by negative particles and thereby form bridges with other particles having the same surface charge. The particles, whose surface charge has the same sign, will thus tend to aggregate together, whereas in the absence of cations, they will of course repel one another (see Fig. 13.7).

Among the metal oxides, titanium dioxide TiO_2 nanoparticles are efficient photocatalysts in aqueous media. Work in this area has shown that the photocatalytic production of hydrogen in anaerobic media is significantly accelerated at the pH value corresponding to the zero point charge, i.e., when there is a high level of aggregation of the TiO_2 nanoparticles, in contrast to what happens for an acidic pH (=2), where the nanoparticles were dispersed. This shows why it is useful to understand the role of particle aggregation in the photocatalysis process, given that this process plays an important role in the toxicity of nanoparticles [12, 13].

[3]The zero point charge is the pH at which the surface of an oxide is neutral. Below this point, the surface is positively charged and above, it is negatively charged. It is essential to determine the ZPC in order to control the adsorption of anions or cations on the oxide support. The pH of a solution can be measured in various ways, notably by electrochemistry using a device called a pH metre, or with a pH indicator substance whose colour changes at a given value of the pH.

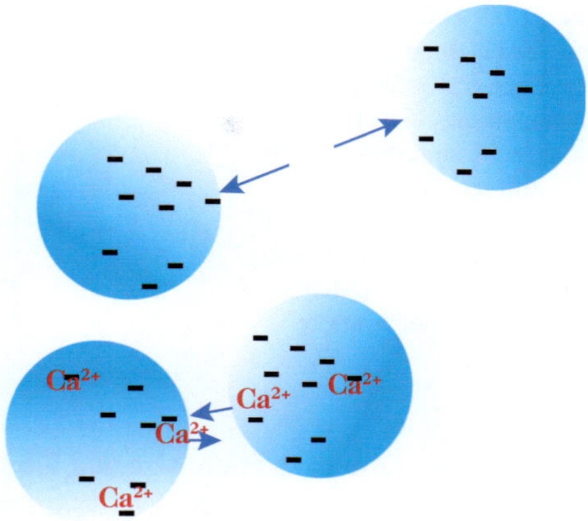

Fig. 13.7 Interactions between negative nanoparticles with and without the presence of Ca^{2+} cations. The nanoparticles have surface charges with the same sign and hence repel one another. If there are Ca^{2+} ions on the surface, this creates points with positive charge which can attract neighbouring particles with the same negative charge

13.3 Nanoparticle Transport in Porous Media

Many natural environments fall into the category of porous media, comprising a fixed solid phase entangled with the aqueous phase, e.g. soils, sediments, most rocks, and certain living materials. Such media are selective natural barriers to the migration of nanoparticles. Depending on whether the supply of nanoparticles is perfectly dispersed in the aqueous phase or trapped on a porous collector, the effects on living organisms, access to the food chain, and environmental impact on the life cycle will be different. Figure 13.8 illustrates the parameters associated with nanoparticle transfer in porous media. The fate of nanoparticles depends both on the efficacity of transport, which is governed by the water flow, and on the attachment process at the collector surface.

The transport of nanoparticles, carried along by water, occurs essentially by diffusion. Interception phenomena and gravity are generally negligible [15]. How efficient it is depends on the porosity of the medium (i.e., the ratio of pore volume to total

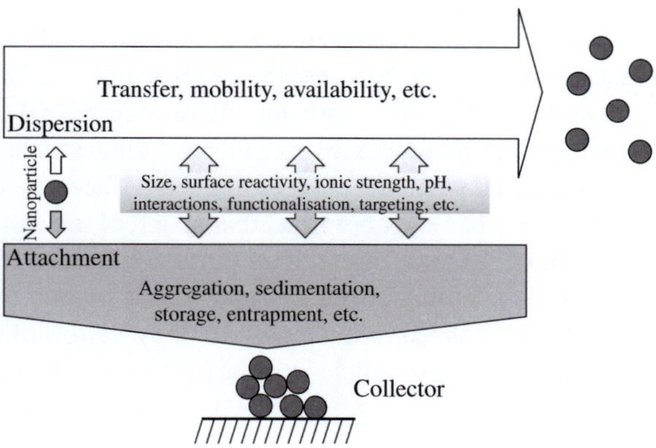

Fig. 13.8 Parameters controlling the transfer, fate, and bioavailability of nanoparticles in an aqueous environment [14]

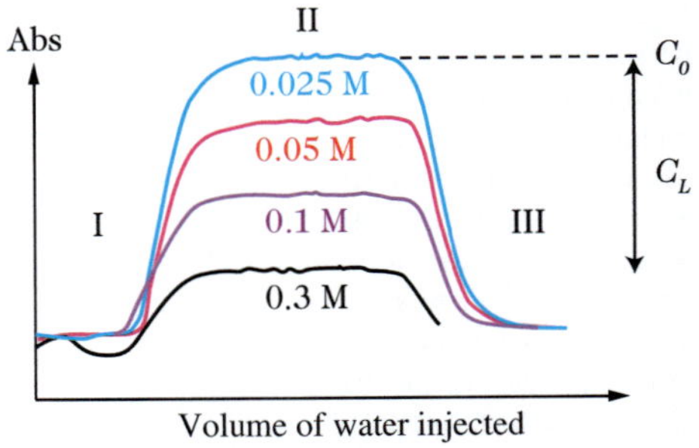

Fig. 13.9 Breakthrough curves obtained experimentally for C_{60} nanoparticles in a column filled with silica microbeads for different NaCl concentrations. The number of nanoparticles is quantified by the absorbance (Abs) in UV–visible spectrometry

volume), the ratio of the pore size and the size of the moving nanoparticles, and the speed of approach of the particles toward the collector surface.

The attachment efficiency is zero when the interaction between the nanoparticle and the collector surface is totally repulsive. It obviously increases with the proportion of attractive interactions. The forces exerted at the interface between nanoparticle and collector are in fact attractive forces of chemical origin at short ranges and repulsive forces of electrostatic origin at longer range. However, the range of these forces, whatever they may be, is here of the same order as the pore size. These interface forces are thus likely to significantly constrain the diffusive transport of the nanoparticles.

The attachment efficiency is measured in the laboratory using a column filled with the porous medium, saturated with water, and by injecting a given volume of nanoparticles in suspension upstream of the column. The nanoparticle concentration in this volume is C_0, while that measured at the column outlet at various times is $C_L(t)$. The resulting variation has the form of a breakthrough curve,[4] typically box-shaped, with a plateau that corresponds to the maximal concentration of nanoparticles that do not attach to the pore surfaces (see Fig. 13.9). Once the whole volume of nanoparticles has been injected, the concentration C_L measured at the outlet begins to go back down until it returns to zero. By modifying the many parameters of the mobile phase, e.g., pH, ionic strength, hydrodynamic flux, or those of the stationary phase, e.g., porosity, it thus becomes possible to study their influence on the nanoparticle attachment.

The first results published in the literature tend to show the importance of electrostatic interactions between the nanoparticles and the solid surface of the porous collector [16, 17]. The dominant influence of the pH and the ionic strength between the charges at the interfaces corroborates the fact that the mobility of the nanoparticles is reduced when the charges at the interfaces are opposite or screened, whereas it increases when they are exacerbated and of the same sign. Figure 13.9 shows the influence of the ionic strength on the attachment of fullerene nC_{60} nanoparticles in a column containing an ideally constituted porous medium of silica beads. When the NaCl salt concentration increases, the electrostatic repulsion between the (negatively charged) nanoparticles and the (negatively charged) silica beads is screened, the

[4]The nanoparticles cross the porous medium formed by solid grains.

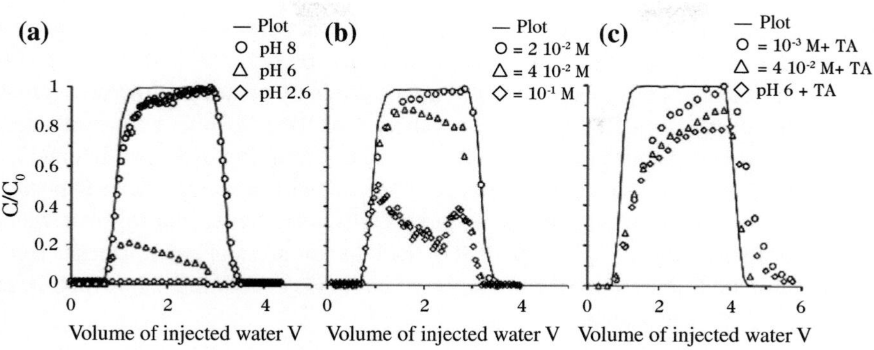

Fig. 13.10 Breakthrough curves measured for TiO_2 nanoparticles in a porous medium, made from 300 μm sand grains, as a function of the pH (**a**), at constant pH = 8 but varying ionic strength (i.e., varying molar concentration of NaCl) (**b**), and in the presence of tannic acid (**c**). *Continuous curves* are ideal breakthrough curves before reaching a critical pH or a critical concentration of salt or tannic acid

attractive interactions become predominant, the proportion of nanoparticles leaving the column drops sharply, and the attachment coefficient rises.

It should be noted that the critical salt concentration above which the plateau of the breakthrough curve begins to fall (the attachment coefficient increases) is less than that corresponding to the aggregation of nanoparticles in a dilute medium (see Sect. 13.2.4). What drives attachment is thus the electrostatic interaction between the nanoparticles and the silica beads, rather than mechanical entrapment in a putative network of nanoparticle aggregates.

Through its affinity for inorganic surfaces, the presence of organic matter is also likely to affect the mobility of nanoparticles in porous media. When organic molecules coat the surface of the nanoparticles or the collector, the surface properties are modified. Figure 13.10 uses breakthrough curves to give a quantitative illustration of the influence of pH, ionic strength, and the presence of tannic acid on the transport of TiO_2 nanoparticles in a porous medium made up of sand grains. Nanoparticle attachment increases with the concentration of salt (NaCl) or tannic acid, whereas it decreases with the pH. All the particles attach themselves to the porous collector for an acidic pH of 3.6.

For a given porous medium, it is clear that the attachment will also depend on the chemical nature of the nanoparticles. In waterlogged sand, fullerols and carbon nanotubes thus have much weaker attachments than inorganic nanoparticles such as silica or aluminium oxide. Recall that when fullerenes come into contact with water they become hydrophilic and negatively charged (see Sect. 13.2.2), which favours their mobility.

Finally, many factors may come together in natural systems to change and often increase the mobility of nanoparticles, i.e., reduce their attachment. It is found that the adsorption of natural polyelectrolytes like bacterial polysaccharides or humic and fulvic acids facilitates the transport of nanoparticles toward the water supply. The functionalisation of carbon nanotubes often makes them more mobile. The existence of fractures and other kinds of heterogeneity in natural porous media also improves the transfer of nanoparticles. On the other hand, the presence of iron oxide or aluminium hydroxide tends to slow down these transfers by creating attractive regions due to the electrical neutrality of these particles or their positive surface charges under typical pH conditions. Such mobility studies are in fact only just beginning. We must clearly get a better understanding of ways to collect nanoparticles in porous media if we are to prevent dispersion and thereby have better control over water treatment. In this context, it is important to track the progress of nanomaterials from a production unit, or monitor the alteration of materials containing nanoparticles, in a real water catchment area.

13.4 Main Nanoparticle Toxicity Mechanisms: Experiments on Living Organisms

Several recent papers [4] have tried to classify nanoparticles according to the harmful effects they may produce in living organisms. Three main toxicity mechanisms stand out:

1. The production of oxygen radicals at the interface between the nanoparticles and water, producing oxidative stress in living organisms and hence alteration of proteins, or even DNA, lipids, and biological membranes (see Fig. 13.11).
2. Nanoparticle dissolution by oxidation and release of toxic cations (e.g., Ag^0 or metallic Ag which oxidises by releasing highly toxic Ag^+ cations) or by

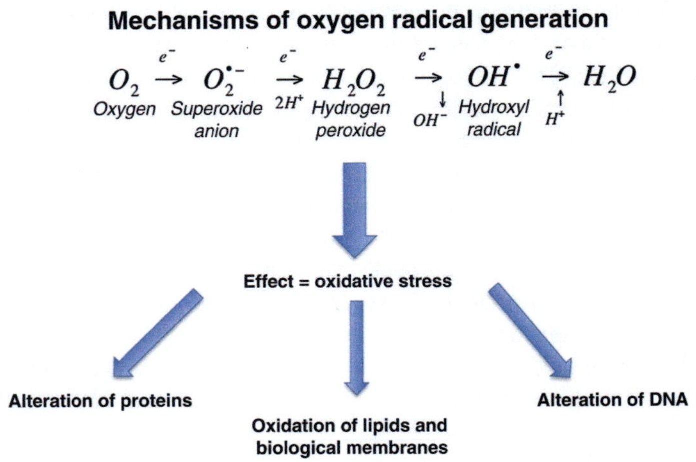

Fig. 13.11 Nanoparticle toxicity due to production of oxygen radicals

reduction with release and production of oxidising species. The following describes a coupled set of chemical reactions involving iron nanoparticles:

$$Fe^0 + O_2 + 2H^+ \longrightarrow Fe^{2+} + H_2O_2 \ ,$$
$$Fe^0 + H_2O_2 \longrightarrow Fe^{2+} + 2OH^- \ ,$$
$$Fe^{2+} + H_2O_2 \longrightarrow Fe^{3+} + OH^* + OH^- \ .$$

Here the toxicity comes from the formation of radicals like OH^*.

3. Changes in the conformation of proteins that normally serve as antioxidants.

Table 13.2 summarises our present state of understanding with regard to toxicity mechanisms for certain nanoparticles [4, 18]. In fact, for each nanoparticle, toxicity studies have been based largely on experiments concerning the interaction between the particle and a given type of cell or biological species, usually bacteria [19].

Table 13.2 Known properties of certain nanoparticles and their toxic consequences [4, 18]

Nanoparticle	Mechanisms associated with cytotoxicity
TiO_2	Production of oxygen free radicals in the presence of UV light
	Oxidative stress, cell death
ZnO	Dissolution associated with the production of free radicals
	Inflammation
Ag	Oxidative dissolution to Ag^+
	Inhibition of respiration enzyme activity and adenosine
	Triphosphate (ATP) production due to free radical production
	Rupture of membrane integrity and transport processes
Au	Modified protein conformation
CdSe	Dissolution and release of Cd^{2+} and Se^{2+} cations
SiO_2	Production of free radicals due to surface defects and impurities
	Modified protein conformation
Fe_3O_4	Production of free radicals
	Oxidative stress
	Perturbation of cation equilibrium and transport
CeO_2	Reductive dissolution and free radical production
	Modified protein conformation
	Alteration of DNA and chromosomes
	Genotoxicity at low concentrations
C nanotubes	Frustrated phagocytosis
	Chronic inflammation
	Generation of free radicals due to the oxidation–reduction of the metal catalysts present within the CNT
Fullerenes	Production of photoactivated free radicals
Cu/CuO	Oxidative stress
	Alteration of DNA
MoO_3	Membrane rupture

Publications concerning algae, protozoa, and invertebrates are more recent, and indeed only began in 2006. Research on bactericidal effects were then extended to biofilms and bacterial communities including fungi, algae, and protozoa, but there is still a lack of research on marine and freshwater environments [20].

Studies of interactions between species are more interesting and more promising for understanding long term effects, in particular on biodiversity. We shall illustrate the complexity of these interactions with several examples.

It turns out in particular that the results on biofilms[5] do not always coincide with those obtained *in vitro* on a single species. For example, TiO_2 nanoparticles have little effect on the marine bacterium *Vibrio fischeri* [20], but the addition of 5.3 mg/L of TiO_2 to a bacterial microcosm under UV radiation damages the cells of planktonic micro-organisms owing to the accumulation of TiO_2 in benthic biofilms (i.e., living near the bottom of seas, lakes, or rivers), and favours the production of oxygen free radicals [21].

Metallic silver nanoparticles induce desquamation of biofilms of the soil bacteria *Pseudomonas putida*. The presence of fulvic acid reduces this desquamation, but increases the bio-accumulation of silver [22]. In the activated sludges of waste water treatment stations, metallic silver nanoparticles at concentrations well below ppm inhibit the proliferation of nitrifying micro-organisms by boosting intracellular oxygen free radicals [23]. In an arctic ecosystem, metallic silver nanoparticles inhibit species fixing the nitrogen essential for plant nutrition.

The exposure of soil bacterial communities to ZnO nanoparticles at levels of 0.05–0.5 mg per gram of soil for 60 days alters the structure and reduces the bacterial biomass [24]. On the other hand, the exposure of a marine sediment bacterial community to 1 mg/L doses of 58 nm silver nanoparticles for 20 days does not produce any change in structure. This difference is due to a change in the speciation[6] of the Ag^+ ion, which is poorly soluble in the presence of Cl^- ions. Recent studies on simplified trophic chains comprising only daphnia and their predator the zebra fish (*Danio rerio*) did not reveal any firm evidence of toxicity of TiO_2 nanoparticles when these were injected into the medium (see Fig. 13.12) [25].

Having said this, such studies do not take into account a true assessment of what is ingested by the organisms. Indeed, the destabilisation of the nanoparticles and their varying degrees of sedimentation are important factors in the long term exposure of living organisms. One study taking into account the exposure parameters [26], almost the only one of its kind, has shown that positively charged gold nanoparticles (65 nm $\times$ 15 nm) in contact with estuarine mesocosms (sediments, biofilms, primary producers, filter organisms, omnivores, etc.) concentrated mainly in the filter organisms and biofilms. The biofilms represented only 0.5 % of the mass of the mesocosm but concentrated some 60 % of the nanoparticles. Likewise, the filter organisms

[5]Biofilms are multicellular communities of micro-organisms (bacteria, fungi, algae, or protozoa) with varying degrees of complexity, often symbiotic, sticking together on a surface and distinguished by the secretion of an adhesive protective matrix. They generally form in water or in aqueous media.

[6]In chemistry, speciation refers to the quantitative distribution of the various states, e.g., anionic, cationic, neutral, reduced, oxidised, adsorbed, etc., of an element in the medium. In biology, speciation is the evolutionary process whereby new living species arise.

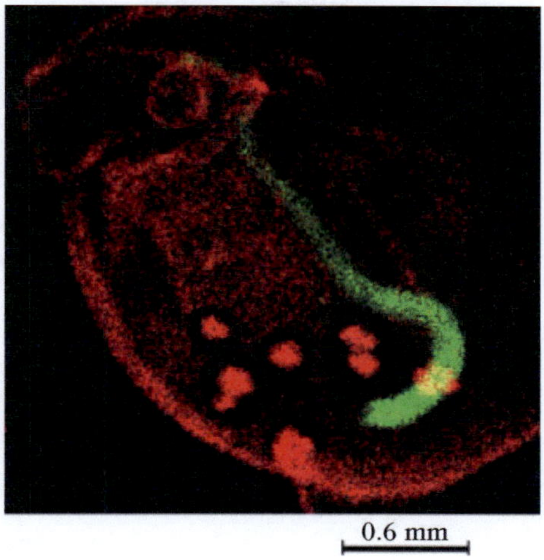

Fig. 13.12 Image obtained by X-ray microfluorescence of a daphnia whose digestive tube is filled with TiO_2 nanoparticles (*green*). The *red colour* indicates calcium. From [25], with the kind permission of Elsevier

(*Mercenaria mercenaria*), representing only 0.01 % of the mass of the system, concentrated 5 % of the nanoparticles. It should be remembered that these biofilms and filter organisms are consumed by humans and thus constitute a non-negligible route for nanoparticles to transfer to humans [27].

13.5 Assessing the Risks of Nanomaterials: Toward Predictive Models?

13.5.1 Difficulties Inherent in Risk Assessment

The notion of risk associated with a nanomaterial is not a simple one, for it must include the exposure to the nanomaterial and all the biological effects correlated with this exposure. For many nanomaterials, the prediction of risk is thus made difficult due to uncertainties in the amounts produced, inadequate understanding of physicochemical characteristics, poorly determined causes of exposure or uses, and above all lack of data concerning effects on organisms or ecosystems. A full assessment of the risks associated with the emergence of manufactured nanomaterials is therefore impossible using conventional methods based on proven experimental facts and/or chemical analysis. In particular, the factors controlling environmental exposure to nanomaterials must be taken into account.

A strategy for risk management accounting for all possible exposure routes with a fundamental understanding of the associated mechanisms leads to a very broad range of options depending on the kind of products used to elaborate the nanomaterial, the systems used to protect workers, the existing standards for eliminating or

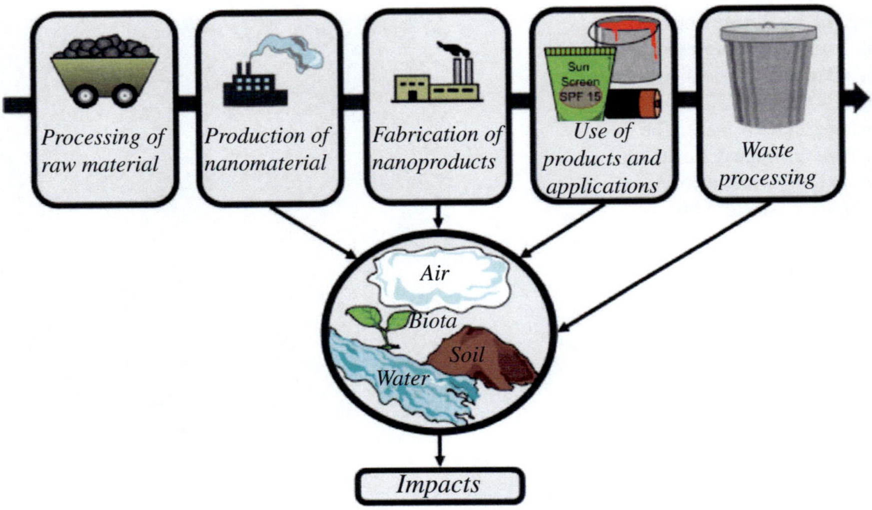

Fig. 13.13 Risks of exposure during the lifetime of a manufactured nanomaterial

recycling products, the implementation of systems for reducing pollution, and above all, changes in human behaviour.

Figure 13.13 illustrates the risks of exposure during the life cycle of a manufactured nanomaterial, from the moment of its fabrication right through to its elimination. This cycle can be considered schematically as a series of steps leading from the initial processing of the basic resources used to synthesise the nanomaterial, through all the production stages, fabrication of derivative products and utilisation, to waste management when the material reaches the end of its life. Note that in principle risk assessment requires one to know for each step the mechanisms whereby the nanomaterial is released into the environment, as well as the probabilities associated with these mechanisms. However, data concerning these aspects is often rather incomplete. Furthermore, the same nanomaterials may be produced unintentionally by the natural environment or human activity, and one must then take into account these other sources.

Figure 13.14 represents the production of manufactured nanomaterials and their incorporation into various products along a value chain. The source at the beginning of the chain corresponds to the fabrication of the nanomaterials themselves. These are then incorporated into products which can be qualified as intermediate, and which are themselves incorporated into other products. The chain can of course extend much further and, at each point of the diagram, the nanomaterial can be released into different sectors of the environment. The total fraction of the original nanomaterial that turns up in the water treatment station could be calculated, for example, by adding up all the fractions of nanomaterial released at the different production points. In practice, this total fraction can be measured by comparing the amount of nanomaterial in the water treatment plant with the amount originally

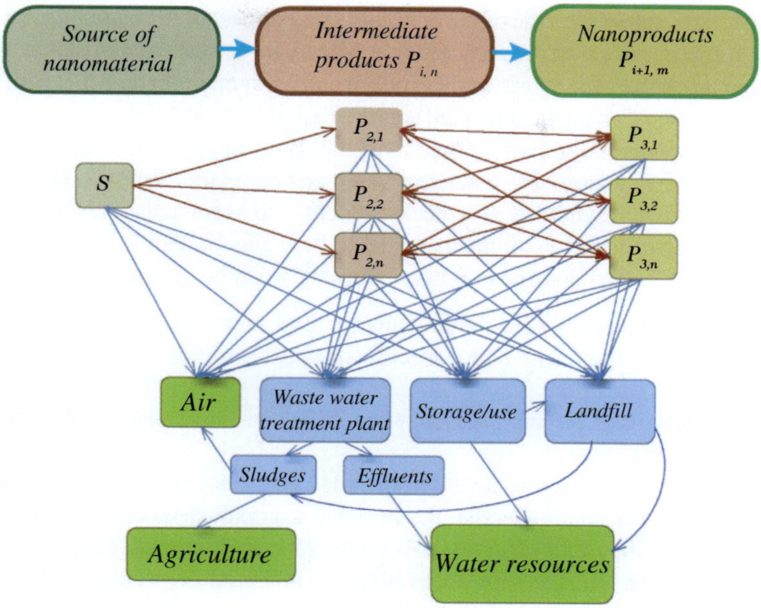

Fig. 13.14 Simplified diagram showing the fate of nanomaterials from the moment they are produced right through to the various sectors of the environment (air, soil, water) and after processing of effluents and sludges from water treatment stations. S represents nanoparticles made in a factory, $P_{2,i}$ are intermediate nanomaterials formulated in a second factory, and $P_{3,i}$ are products made with the nanomaterials in a third company. For example, S = nanoTiO_2, $P_{2,i}$ = nanoTiO_2 + a layer of organic and/or inorganic material on the surface, and $P_{3,i}$ = cosmetic. *Arrows* indicate the connections between the various steps in the synthesis, but also the fate of the materials during their life cycle: losses at each stage and ultimate fate

produced. The fraction of nanomaterial arising from a particular product can be estimated by looking at the way the product is used. What is striking here is the enormous number of weighting factors arising in the diagram of Fig. 13.14. But even if we knew all the parameters in the diagram, in order to assess the risk to humans, we would still have to estimate the chances of deterioration or modification of the nanomaterial in different environmental sectors, such as water treatment, but also in the ecosystem as a whole. A few examples of significant environmental modification were given in Sect. 13.4.

13.5.2 Experimental Approach to Risk Assessment Integrating Collateral Damage

Another thing to be considered is the risk due to collateral damage caused by nanomaterials, and in particular the environmental impacts of the production and use of nanomaterials, rather than the direct environmental impact of the nanomaterial itself. The high energy consumption sometimes necessary for the production of materials or components on the nanometric scale is one example among others. An experimental approach used in Canada in 2005 [28] was the first of its kind to take into account the indirect environmental impacts. The aim of the study was to assess the risks due to the industrial fabrication of five nanomaterials and compare them with the potential risks involved in the synthesis of typical products such as solvents. This work reveals the complexity of the evaluations that need to be made, as well as the considerable number of parameters coming into play. This study can be summarised as follows:

- The five nanomaterials were single-walled carbon nanotubes, fullerenes C_{60}, ZnSe quantum dots, alumoxane nanoparticles, and titanium dioxide (TiO_2) nanoparticles.
- The solvents used for comparison were thiophene, benzene, ethylenediamine, xylene, toluene, etc.
- The products and processes analysed included all the chemical substances needed to synthesise the nanomaterials, the intrinsic dangerousness of these substances when known, the fabrication processes and work environment, conditions of use and manipulation of the substances with the potential or proven environmental risks due to the nanomaterials themselves.
- A qualitative method was devised for analysing risk exposure and a proposal made for a toxicity scale from 1 to 4 with reference to the recommendations of the Canadian Centre for Occupational Health and Safety (CCOHS).
- Ranking of risks by level of dangerousness.
- A first database was set up with reference to criteria of toxicity and dangerousness.

These studies showed, for example, that the production of carbon nanotubes [29] could in turn produce non-nano waste products which themselves generate other risks. According to the authors, carbon nanotubes will end up in the environment in the coming years, and problems might arise due to the following:

- The presence of metals (catalysts) used to synthesise the nanotubes.
- Adsorption of polycyclic aromatic hydrocarbons (PAH) which are known to be toxic.

Another important conclusion to be drawn from this work is that it should be possible to devise greener ways of fabricating nanomaterials, e.g., using less toxic solvents or less energy-costly procedures. The hopes for such a trend do of course depend on an increased awareness of the risks of pollution.

13.5.3 Risk Forecasting Models for Nanomaterials

The amount and quality of data concerning the environment, health, and safety with respect to nanoparticles are rapidly improving, and this information is freely available. For example, since the 2005 study described above and up until 2011, the number of scientific publications dealing with risks relating to nanomaterials has gone from 200 to more than 1 500 (source ISI Web).

The creation of a specific database which is kept up to date with progress in our knowledge of the latest nanomaterials, the associated risks, and their interactions with the environment could then be used to set up a robust model for estimating the risks associated with other nanomaterials in the future. Such a model must be able to predict the biological activity of the nanomaterials, their impact on the ecosystem, and related exposure factors given their physicochemical properties.

A first kind of model would be based on quantitative functional relations formulated mathematically, like those commonly used between chemical structure and biological activity (quantitative structure–activity relationship, QSAR), except that the functional relations would be more general, more complex, and more numerous in the present case. Indeed, nanomaterials can produce second order effects due to their redox reactivity and their capacity to adsorb contaminants like arsenic, themselves correlated with their structural properties (morphology, electronic structure, crystal structure, surface defects, surface functionality, etc.) [18, 30, 31]. A model generalising the QSAR approach would very likely require advanced computer techniques like those used for shape recognition, whose future capacity is still hard to predict.

Another possibility here would be a Bayesian probabilistic model,[7] constructed graphically as a probability network. The network is represented by a set of nodes (the system attributes) connected together by causal links. Each node is characterised by probabilities or probabilistic mathematical expressions representing our knowledge of the given attributes. These mathematical expressions can be:

- a mechanistic description such as would be given by chemical reaction kinetics,
- empirical relations such as linear regression models,
- relations following from expert assessments, depending on the information available about the relationships characterising a particular node.

The possible results at each node are expressed probabilistically. The possibility of incorporating mechanistic, empirical, and expert assessments makes the Bayesian approach extremely flexible and facilitates extension to non-traditional areas of great public interest. In a sense, this approach leads to an expert system that can be updated when new knowledge is acquired.

Fortunately, the causal relations just mentioned are not completely opaque in a Bayesian risk assessment model for nanomaterials. For example, the influence of nanoparticle size and charge parameters, and also the interactions with macromolecules, have been under investigation for more than a decade in studies of nanomaterial transport and transformation [32]. In many cases, the role of the ionic strength, i.e., the amount of ions, and macromolecules has already been well established in the scientific literature on colloids. We also have a great deal of information about the role of attractive and repulsive forces between surfaces, and also about the structure of the solid–water interface. In many cases, the mechanisms are well understood and

[7] Thomas Bayes was a British mathematician and Presbyterian minister, born in London in 1701.

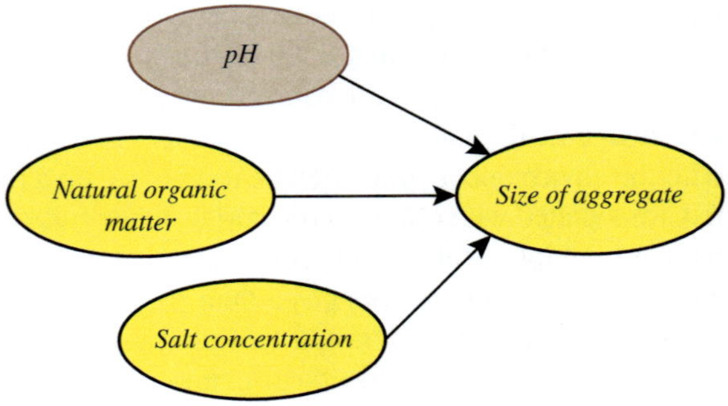

Fig. 13.15 Fragment of a Bayesian probability network for nanoparticle aggregation. The pH (*shaded*) is assumed constant here (pH = 7.2)

an expert system or a statistical approach for describing the relations is not actually necessary.

Figure 13.15 shows a fragment of a Bayesian network that could be conceived to assess the risks associated with nanomaterials. This fragment concerns the size of aggregates, e.g., of hydroxylated fullerenes (fullerols), depending on two parameters: the presence of natural organic matter such as tannic acid in the medium and the salt (NaCl) concentration. The pH is assumed constant here (=7.2), but it could also be treated as a variable parameter in the more general case. The presence or otherwise of natural organic matter is associated with a binary variable (0 or 1, yes or no). The molar salt concentration is classified according to several registers corresponding to commonly encountered ranges of values.

The size of the aggregate in the given medium can be measured at any given time by light scattering and the values obtained are also classified according to several registers. Table 13.3 summarises the probabilistic data associated with this fragment of the network. As an example, the value 0.875 given in the top left of the table means that there is an 87.5 % chance of measuring a fullerol aggregate size smaller than 300 nm for an NaCl molar concentration less than 0.1 M and with tannic acid present in the medium. Any other, more accurate information obtained as time goes by about the size of the aggregates could be used to supplement and update the table.

Table 13.3 Conditional probabilities giving the size of fullerol aggregates depending on the presence or otherwise of tannic acid (Y = yes, N = no) and the molar concentration of NaCl in the given medium

NaCl concentration	Size (nm)		Size (nm)		Size (nm)		Size (nm)	
	$0 < x < 300$		$300 < x < 600$		$600 < x < 900$		$900 < x < 1\,200$	
	Tannic acid							
	Y	N	Y	N	Y	N	Y	N
0.00–0.10	0.875	0.3	0.125	0.6	0	0.08	0	0.025
0.10–0.20	1	0	0	0.25	0	0.4	0	0.35
0.20–0.40	0.3	0	0.7	0	0	0	0	1

Courtesy of C. Hendren and E. Money, Duke University

Quite generally, an evolutive risk assessment model will have six key features:

1. The ability to make predictions associated with different levels of uncertainty for short term questions.
2. The ability to assess relevant sources of nanomaterials.
3. A systemic approach to the impacts of the production of nanomaterials and their use during their life cycle, going beyond toxicological approaches.
4. The possibility of updating predictions as soon as new information becomes available.
5. The possibility of identifying the main areas where understanding must be improved in order to increase the accuracy of the model.
6. The ability to draw conclusions that will help to reduce the impact of nanomaterials by improving production processes.

13.6 Concluding Remarks

In this chapter, we have presented certain reaction mechanisms that are important for nanoparticles in the environment, and particularly in water, together with their possible effects on ecosystems and on humans. Clearly, it is hard to apprehend the full complexity of the relevant phenomena, given the great variety of situations encountered. More recent studies nevertheless show that, in aqueous media [33], nanoparticles exhibit specific kinds of behaviour that differ from those of larger particles. Further studies of this kind should allow us to develop better strategies for trapping them, lessening their toxic effects, and optimising the treatment of drinking water, so vital for humanity.

Today there are more and more toxicity studies and thanks to the increasing volume of data and rigorous documentation, it is now possible to develop ever more accurate predictive models concerning the risks associated with nanomaterials and nanoparticles. The approach to adopt here is to study the potential toxicity of nanoparticles at the moment of synthesis, taking into account their environment and the reaction mechanisms they generate. As this progress is made, it will also be important to make the results available to the general public.

References

1. Woodrow Wilson Database (2010). www.nanotechproject.org
2. The big down, from genomes to atoms. ETC Group (January 2003)
3. M. Turner, V.B. Golovko, O.P.H. Vaughan, P. Abdulkin, A. Berenguer-Murcia, M.S. Tikhov, B.F.G. Johnson, R.M. Lambert, Selective oxidation with dioxygen by gold nanoparticle catalysts derived from 55-atom clusters. Nature **454**, 981–983 (2008)
4. M. Auffan, J. Rose, J.-Y. Bottero, G.V. Lowry, J.P. Jolivet, M.R. Wiesner, Towards a definition of inorganic nanoparticles from an environmental, health and safety perspective. Nat. Nanotechnol. **4**, 634–641 (2009)

5. R. Ma, C. Levard, S.T. Makarinos, Y. Chen, L. Liu, F.M. Michel, G.E. Brown, G.V. Lowry, Size-controlled dissolution of organic-coated silver nanoparticles. Environ. Sci. Technol. **46**, 752–759 (2012)
6. J. Labille, A. Masion, F. Ziarelli, J. Rose, J. Brant, F. Villiéras, M. Pelletier, M. Wiesner, M. Giuliano, J.-Y. Bottero, Evidence of fullerene C_{60} hydrophilic character. Hydration and dispersion in aqueous system. Langmuir **25**, 11232–5 (2009)
7. J. Brant, J. Labille, J.-Y. Bottero, M. Wiesner, Characterizing the impact of preparation method on fullerene cluster structure and chemistry. Langmuir **22**, 3878–3885 (2006)
8. M. Auffan, J. Rose, O. Proux, D. Borschneck, A. Masion, P. Chaurand, J.L. Hazemann, C. Chanéac, J.P. Jolivet, M.R. Wiesner, A. VanGeen, J.-Y. Bottero, Enhanced adsorption of arsenic onto nano-maghemites: As(III) as a probe of the surface structure and heterogeneity. Langmuir **24**, 3215–3222 (2008)
9. J.-Y. Bottero, M. Auffan, J. Rose, C. Mouneyrac, C. Botta, J. Labille, A. Masion, A. Thill, C. Chanéac, Manufactured metal and metal-oxide nanoparticles: properties and perturbing mechanisms of their biological activity in ecosystems. C. R. Geosci. **343**, 168–176 (2011)
10. J. Lee, J.D. Fortner, J.B. Hughes, J.H. Kim, Photochemical production of reactive oxygen species by C_{60} in the aqueous phase during UV irradiation. Environ. Sci. Technol. **41**, 2529–2535 (2007)
11. N. Solovitch, J. Labille, J. Rose, P. Chaurand, D. Borschneck, M.R. Wiesner, J.-Y. Bottero, Concurrent aggregation and deposition of TiO_2 nanoparticles in a sandy porous media. Environ. Sci. Technol. **44**, 4897–4902 (2010)
12. N. Lakshminarasimhan, W. Kim, W. Choi, Effect of the agglomerated state on the photocatalytic hydrogen production with in situ agglomeration of colloidal TiO_2 nanoparticles. J. Phys. Chem. C **112**, 20451–20457 (2008)
13. M. Auffan, J. Rose, M.R. Wiesner, J.-Y. Bottero, Chemical stability of metallic nanoparticles: a parameter controlling their potential cellular toxicity in vitro. Environ. Pollut. **157**, 1127–1133 (2009)
14. J. Labille, J.-Y. Bottero, Fate of nanoparticles in aqueous media, in *Nanoethics and Nanotoxicology*, ed. by M. Lahmani, F. Marano, P. Houdy (Springer, Heidelberg, 2010), pp. 291–324
15. B. Espinasse, E.M. Hotze, M.R. Wiesner, Transport and retention of colloidal aggregates of C_{60} in porous media: effects of organic macromolecules, ionic composition, and preparation method. Environ. Sci. Technol. **41**, 7396–7402 (2007)
16. N. Solovitch, J. Labille, A. Mansion, J.-Y. Bottero, Transfert of TiO_2 nanoparticles in a sandy porous medium, driven by sorbing natural organic matter. Submitted to Environ. Pollut. (2013)
17. S. Lin, M.R. Wiesner, *Influence of Fe_2 on the Transfer of Nanoparticles in a Sandy Porous Medium* (Séminaire I-CEINT, Porquerolles, 2010)
18. M. Auffan, D. Bertin, P. Chaurand, C. Pailles, C. Dominici, J. Rose, J.-Y. Bottero, A. Thiery, Role of molting on biodistribution of CeO_2 nanoparticles within Daphnia pulex. Water Res. **47**, 3921–30 (2013)
19. N.C. Mueller, B. Nowack, Exposure modeling of engineered nanoparticles in the environment. Environ. Sci. Technol. **42**, 4447–4453 (2008)
20. M. Auffan, E. Flahaut, A. Thill, F. Mouchet, M. Carriere, L. Gauthier, W. Achouak, J. Rose, M.R. Wiesner, J.-Y. Bottero, Ecotoxicology: nanoparticle reactivity and living organisms, in *Nanoethics and Nanotoxicology*, ed. by M. Lahmani, F. Marano, P. Houdy (Springer, Heidelberg, 2010), pp. 325–358
21. M. Auffan, C. Santaella, A. Thiéry, C. Pailles, J. Rose, W. Achouak, A. Thill, A. Masion, M.R. Wiesner, J.-Y. Bottero, Ecotoxicity of inorganic nanoparticles: from unicellular organisms to invertebrates, in: *Encyclopedia of Nanotechnology*, (Bhusan, Springer 2012)
22. T.J. Battin, F.V.D. Kammer, A. Weilhartner, S. Ottofuelling, T. Hofmann, Nanostructured TiO_2: transport behavior and effects on aquatic microbial communities under environmental conditions. Environ. Sci. Technol. **43**, 8098–8104 (2009)
23. J. Fabrega, S.R. Fawcett, J.C. Renshaw, J.R. Lead, Silver nanoparticle impact on bacterial growth: effect of pH, concentration, and organic matter. Environ. Sci. Technol. **43**, 7285–7290 (2009)

24. Y. Ge, J.P. Schimel, P.A. Holden, Evidence for negative effects of TiO_2 and ZnO nanoparticles on soil bacterial communities. Environ. Sci. Technol. **45**, 1659–1664 (2011)
25. X. Zhu, J. Wang, X. Zhang, Y. Chang, Y. Chen, Trophic transfer of TiO_2 nanoparticles from daphnia to zebrafish in a simplified freshwater food chain. Chemosphere **79**, 928–933 (2010)
26. J.L. Ferry, P. Craig, C. Hexel, P. Sisco, R. Frey, P.L. Pennington, M.H. Fulton, I.G. Scott, A.W. Decho, S. Kashiwada, C.J. Murphy, T.J. Shaw, Transfer of gold nanoparticles from the water column to the estuarine food. Nat. Nanotechnol. **4**, 441–444 (2009)
27. J.D. Judy, J.M. Unrine, P.M. Bertsch, Evidence for biomagnification of gold nanoparticles within a terrestrial food chain. Environ. Sci. Technol. **45**, 776–781 (2010)
28. C.O. Robichaud, D. Tanzil, U. Weilenmann, M.R. Wiesner, A relative risk analysis of several manufactured nanomaterials: an insurance industry context. Environ. Sci. Technol. **39**, 8985–8994 (2005)
29. D.L. Plata, P.M. Gschwend, C.M. Reddy, Industrially synthesized single-walled carbon nanotubes: compositional data for users, environmental risk assessments, and source apportionment. Nanotechnology **19**, 14 (2008)
30. A.E. Nel, L. Mädler, D. Velegol, T. Xia, E.M.V. Hoek, P. Somasundaran, F. Klaessig, V. Castranova, M. Thompson, Understanding biophysicochemical interactions at the nano-bio interface. Nat. Mater. **8**, 543–557 (2009)
31. M. Auffan et al., Nanomaterials as adsorbants in environmental nanotechnology—applications and impacts of nano-materials, in *Environmental Nanotechnology*, ed. by M.R. Wiesner, J.-Y. Bottero (MacGraw-Hill, New York, 2007)
32. J.A. Brant et al., Nanoparticle transport, aggregation, and deposition, in *Environmental Nanotechnology*, ed. by M.R. Wiesner, J.-Y. Bottero (MacGraw-Hill, New York, 2007)
33. E.M. Hotze et al., Mechanisms of photochemistry and reactive oxygen production by fullerene suspensions in water. Environ. Sci. Technol. **42**, 4175–4180 (2008)

Part IV
Nanotechnology and Society

Chapter 14
Research in Nanoscience and Nanotechnology: The French Research System

Patrice Hesto and Jean-Michel Lourtioz

Abstract This chapter presents the public institutions and research laboratories, as well as the main industrial R&D centres in France, describing also the national research programs and their funding. More briefly, it also presents the European and worldwide research programs in the nano field.

The aim in this part of the book is to provide some information about French academic research institutes and industrial R&D in the nano field,[1] but also the public awareness measures set up by the French authorities. We also outline the national research programs and funding. Finally, on a broad scale, we shall discuss the corresponding programs throughout Europe and the rest of the world.

It should be stressed that the nano field is currently one of the most dynamic and productive areas of science. As an example, the number of scientific publications has been growing very fast over the past few years, in fact by about 25 % per year. Hence in 2009, 22 000 articles were published by European research groups, 17 000 in the US, 17 000 in China, and 6 000 in Japan.

14.1 Upstream Research in France

The French nano research scene represents about ten thousand researchers, teacher-researchers, engineers, technicians, and doctoral and post-doctoral students.[2] These depend for the main part on universities and public organisations like the *Centre*

[1] In the following chapters, we shall use the generic term 'nano' to refer to all the fields covered in this book, from the exact sciences to the human and social sciences.

[2] Note that, in the nano field, the proportion of women, although still too low, is nevertheless significant. Doctoral students prepare a thesis in a research institute. Post-doctoral students are young researchers supplementing and perfecting their knowledge in their special field of interest by doing a training course in a different research centre to the one where they did their thesis.

P. Hesto (✉)
31 rue des grands champs, 75020 Paris, France
e-mail: patrice.hesto@laposte.net

J.-M. Lourtioz
Institut d'Electronique Fondamentale Bât 220, Université Paris-Sud, 91405 Orsay Cedex, France
e-mail: jean-michel.lourtioz@u-psud.fr

J.-M. Lourtioz et al. (eds.), *Nanosciences and Nanotechnology*,
DOI 10.1007/978-3-319-19360-1_14

national de la recherche scientifique (CNRS), the *Commissariat à l'énergie atomique et aux énergies alternatives* (CEA), and the *Institut national de la santé et de la recherche médicale* (INSERM). But the world of nano research also comprises resources and programs for coordinating the actions of the various stakeholders.

The vast majority of laboratories in French universities are associated with the CNRS, and in some cases with the CEA or INSERM. The research orientation of such joint research labs thus follows the scientific policy determined in common by universities and national research organisations. Joint research labs receive basic funding from both their university and the associated research organisation. There are also research labs fully managed by one or more national research organisations, which then provide the totality of the basic funding. For laboratories in the nano field, basic funding is typically 20–30 % of the laboratory budget excluding salaries, and laboratories must compete for national, European, and worldwide offers.

At the CNRS, research is organised into ten institutes, several of which are concerned with the nano field: the *Institut des sciences de l'ingénierie et des systèmes* (INSIS) for nanotechnologies, the *Institut de physique* (INP) for nanophysics, the *Institut de chimie* (INC) for nanochemistry, the *Institut des sciences biologiques* (INSB) for nanobiology and nanomedicine, and the *Institut des sciences humaines et sociales* (INSHS) for all the various societal aspects. These institutes specify the main national research orientations in the research labs of the CNRS itself or those associated with it. The CNRS steers the RENATECH network bringing together the main technology facilities financed by the state, the universities, and the regions of France. This network, set up in 2003, comprises six facilities located in Besançon, Grenoble, Lille, Toulouse, and in the south of the Île-de-France (see Fig. 14.1a), each associated with specific research facilities. It makes high performance micro- and nanofabrication installations available for all the French academic research labs. In addition to these major RENATECH facilities, there are also those of CEA-LETI, which combine to form the Basic Technological Research (BTR) network[3] (see Fig. 14.1a). The French ministry of research and the CNRS also finance C'Nano, a network of centres of competence spread across the whole country, which encourage innovative research within the community by favouring, among other things, access to technology facilities.

At the French atomic energy commission (CEA), upstream research is piloted by the *Direction des sciences de la matière* (DSM), while more finalised aspects are covered by the *Direction de la recherche technologique* (DRT). Activities in the nano field, which depend on the DSM, concern physics, chemistry, biology, and nanofabrication, and are carried out at the *Institut rayonnement matière de Saclay* (Iramis) and at the *Institut nanosciences et cryogénie* (Inac) in Grenoble. Within the DRT, there are three relevant research centres:

---

[3] All the BTR facilities have been awarded the title 'major research infrastructures' (*très grandes infrastructures de recherche*, TGIR) by the French research ministry in 2009, at the same level, for example, as the European Synchrotron Radiation Facility (ESRF).

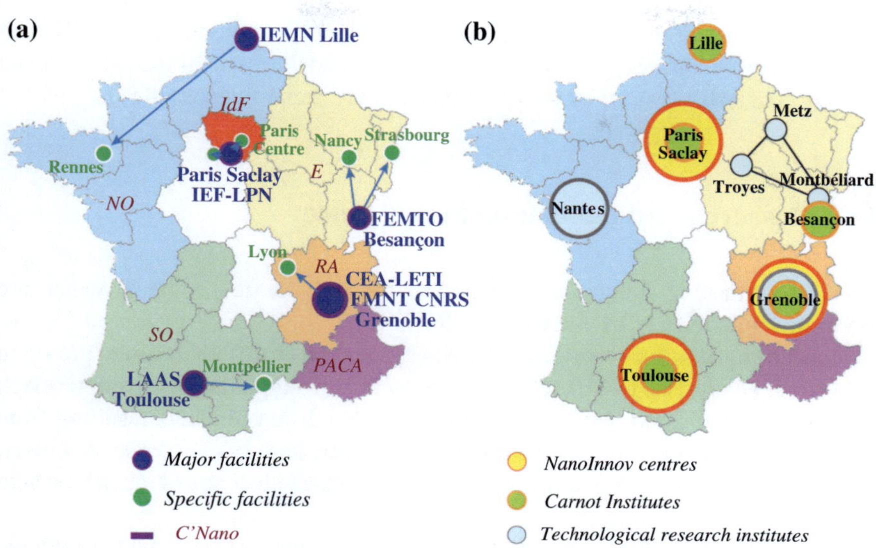

Fig. 14.1 **a** Major and specific facilities in the French Basic Technological Research (BTR) network. The community as a whole is directed by six C'Nano competence centres spread across the country. **b** Main research and innovation initiatives in the French nano field: Carnot institutes, NanoInnov integration centres, and institutes of technology research (see text). *Source* CNRS/CEA/ANR

- The *Laboratoire d'électronique et des technologies de l'information* (LETI) for nanoelectronics and fabrication processes.
- The *Laboratoire d'innovation pour les technologies des énergies nouvelles* (LITEN) for applications of nanos to photovoltaics and the hydrogen fuel cell and energy storage sector.
- The *Laboratoire d'intégration des systèmes et de technologies* (LIST) for the design of systems exploiting nanotechnology.

The CEA is heavily involved in the MINATEC (MIcro-NAno TEChnologies) campus and also the GIANT (Grenoble Innovation on Advanced New Technologies) campus, both in Grenoble. On these sites, it has succeeded in developing an R&D and transfer cluster of European consequence, in close synergy with STMicroelectronics (STM). CEA personnel working on nano themes represent about 20 % of the national effort in this field.

As the name suggests, INSERM is concerned not only with public health problems, but also with medical research. It is organised into ten institutes, one of which is particularly concerned with nano issues, viz., the 'Technologies for Health' institute (*Institut 'Technologies pour la santé'*). One of its most promising themes is the development of nanodrugs. Another important mission for the INSERM research groups is the study and quantification of the health impacts of nanoparticles. The INSERM research groups are often associated with university, CNRS, and CEA research centres.

To get an idea of the way things play out, a third of the subjects treated in French research centres concern physics, a third chemistry, and the rest engineering, systems, biology, and a still very small contribution from the social sciences (4 %).

14.2 Research and Development in France

The large industrial groups have their own research centres, some of which are associated with the CNRS, the CEA, or INSERM.

Concerning start-up companies, the situation is rather different. These are often very small enterprises (VSE) set up by researchers from public or private research centres to capitalise on their research by commercialising products resulting from their work. These companies generally engage in a very high level of research activity, often maintaining a close synergy with the original research centre of the person who set them up.

In parallel with French state aid for cooperative research programs between the public and private sectors which will be discussed below, there is a specific financial tool designed to back research activities developed by companies using their own finances, namely, a research tax credit. This is public aid for the development of in-company research and innovation which includes a specific component for funding innovation in small and medium-sized enterprises (SME).

14.2.1 Organising the Nano Community

The six C'Nano competence centres set up by the French research ministry and the CNRS cover the whole of France and federate French academic research groups working in the nano field, from the natural sciences to the human and social sciences. They organise research in the different regions, animating the relevant scientific community and encouraging cross-disciplinary exchanges. The interactive website can be used to identify the research centres active in a specific field. It can also be used to obtain information about the various training establishments in the nano field (see Chap. 15). Finally, a partnership between the French National Laboratory of Metrology and Testing (*Laboratoire national d'essais* LNE) and the C'Nano centres helped to set up a society devoted to nanoscale metrology, viz., the *Club NanoMétrologie*.

The CNRS and the CEA also set up the *Observatoire des micro-nanotechnologies* (OMNT) dedicated to strategic monitoring of nanoscience and nanotechnology. It incorporates a network of more than 200 experts.

14.2.2 French Programs and Financing

It is sometimes difficult to make sense of the various national research programs and their funding, given the many different ventures. These ventures may also overlap, or a program rendered obsolete by new initiatives may keep going rather than be discontinued. In the following presentation, we shall nevertheless attempt to summarise the main support programs for nano research initiated in France since the 2000s.

To begin with, following specific actions instigated by the CNRS or the ministry of research, the first large scale program, launched in 2003, aimed to set up a network of major technology facilities. Originally financed by the ministry of research, funding was then taken over by the *Agence nationale de la recherche* (ANR), set up in 2006.

Since then, the ANR has financed many other research projects in the nano field in response to tenders from different programs:

- non-thematic programs from exploratory research,
- the thematic *Programme Nanotechnologies Nanosystèmes* (P2N) for more applied projects.

More than half of the projects retained now involve partnerships between academic labs and companies. Since 2012, such partnerships have even become obligatory for P2N projects. In 2011, the overall ANR funding for nano projects has been of the order of 100 M€ including 22 M€ for P2N and 10 M€ for RENATECH.

More regionally biased and designed to promote upstream research, the advanced research networks known as *Réseaux thématiques de recherche avancée* (RTRA) were set up in 2007. Three of these focus more particularly on the nano field: the *Triangle de la physique* in Orsay, *Nanosciences à la limite de la nanoélectronique* in Grenoble, and the *Centre international de recherche aux limites de la chimie* in Strasbourg.

Also regionally biased but designed to back R&D and innovation projects, competitiveness clusters were set up in 2005. In the nano field or neighbouring areas we may cite *Minalogic* (in the Rhône-Alpes region), *Microtechniques* (in the Franche-Comté region), *Solutions Communicantes Sécurisées* (in the Provence Alpes Côte d'Azur or PACA region), *Photonique* (also in PACA), and *Sciences et Systèmes de l'Énergie Électrique* (in the Centre region).

In the same period, in 2006, an excellence label by the name of Carnot was launched, to be awarded by the French Ministry of Research and Higher Education to establishments carrying out research in partnership with socio-economic actors. Five Carnot institutes with their supporting research centres—le CEA-LETI (Grenoble), FEMTO-Innovation (Besançon), the IEMN (Lille), the IOGS (Paris Saclay), and the LAAS (Toulouse)—are heavily involved in nanoscience and nanotechnology (see Fig. 14.1b).

More recently, the aim of the NanoInnov plan, launched in 2009, was to help French industry make the transition to nanotechnology by setting up nanotechnology integration facilities in Grenoble, Paris-Saclay, and Toulouse to develop a synergy between upstream research and industrial R&D. A total of 70 M€ has been set aside for this plan, including 46 M€ for the Saclay integration facility alone.

Also in 2009, the French cabinet adopted a national research and innovation strategy (*Stratégie nationale de recherche et d'innovation*, SNRI). Under the heading 'Information, communication, and nanotechnology' is a series of challenges including the objective of carrying through the nanotechnological revolution in electronics, materials, health technologies, and renewable energies.

The systems previously put in place were consolidated by the program known as *Investissements d'avenir* (investment in the future, IA) in the framework of the *Grand Emprunt* (a government lending scheme) in 2011. For example, it was in this framework that nanotechnology projects were selected as part of the Equipex and Labex programs (facilities and laboratories of excellence), e.g., Labex NanoSaclay in Saclay. The ANR's call for nanobiotechnology projects focusing on nanomedicine, medical diagnosis, and medical imaging also benefited from these IA investments. Three technology research institutes (*instituts de recherche technologique* IRT) were also set up with IA investment to work in the nano area: the IRT NanoElec in Grenoble for nanoelectronics, the IRT M2P in Metz, Troyes, and Montbéliard for materials, metallurgy, and processes, and the IRT Jules-Verne in Nantes for composite materials (see Fig. 14.1b).

To end this chronological listing, we may also cite the Nano2012 program, which associates IBM R&D centres with those of STMicroelectronics and the CEA. The global R&D budget was 2.3 billion euros and the tangible investment over five years was 1.25 billion euros. The French state together with local and regional authorities put up almost 0.5 billion euros. Today, this program has been extended by Nano2017 with a projected budget of 3.5 billion euros (see Chap. 16).

In parallel with these programs, there is a possibility for doctoral students working in private research centres to obtain a grant as part of the *Convention industrielle de formation par la recherche* (CIFRE). This is a three-party convention between the academic research lab, a private company, and the doctoral student who subsequently receives a fixed term or permanent contract.

14.3 International Programs

It is not always easy to obtain information regarding the policies of different states. It is thus difficult to compare data from one country to another, since it does not necessarily cover the same items, depending on whether figures include salaries or infrastructure, whether programs are upstream or more finalised, etc. However, we may still sketch an outline of the nano programs in different parts of the world. Statistics about the publications produced by the main actors in the nano field (United States, Europe, and Asia) are given in Fig. 14.2.

On the European level, calls for tender are made within the Research and Development Framework Programme (RDFP). In the 7th RDFP (2007–2013), nano research appeared in the program 'Nanosciences, nanotechnologies, materials and new production technologies' (NMP). In 2008, the European investment amounted to almost

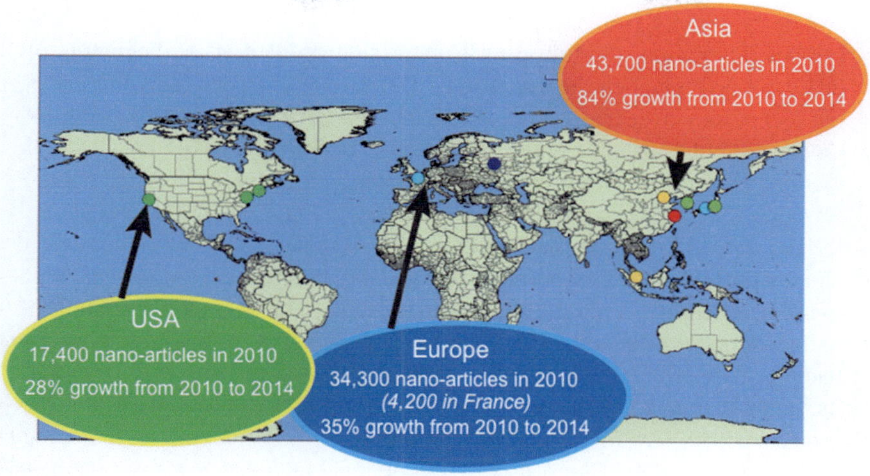

Fig. 14.2 Upstream research in the nano field for the main actors around the world, viz., the United States, Europe, and Asia (http://statnano.com/report/s29). R&D and patents are discussed in Chap. 16 (see for instance Table 16.1)

2 billion dollars. For the period 2010–2015, the NanoAction Plan 2010–2015 marked a trend toward innovation and downstream actions, with a more careful consideration of social expectations.

Graphene: A New Flagship Project for Europe

As emphasised in Chaps. 2 and 5, graphene has quite extraordinary physical and chemical properties: it is the first 2D atomic crystal, in other words the thinnest material in the world, conducting electricity much better than copper, and having a mechanical strength two to three hundred times greater than steel, not to mention a thermal conductivity on a par with its electrical conductivity. In addition, due to its thinness, it is optically transparent, and due to its electronic structure, it has a broad spectral response. Thanks to the European scientists who contributed to its discovery and revealed its potential for different purposes, graphene seems likely to become one of the emblematic materials of the twenty-first century, like steel and plastics in the twentieth century. In particular, there is talk of using it in the place of silicon for information and communication technologies (ICT) in certain cases.

For this reason, the European Commission decided at the beginning of 2013 to make graphene research one of its two flagship projects in emerging technology over the coming ten years. The other flagship is the study of the human brain, with the prospect of treating neurological disorders. Each of these flagship projects receives regular funding, originally in the context of the European FP7 program (2007–2013), which ended in 2013, and now as part of the Horizon 2020 program (2014–2020), which began in 2014 after discussion and finalisation at the European Council and the European Parliament. In order to promote research and innovation, and hopefully create growth and employment as a consequence, the European Commisssion put forward an ambitious budget of 80 billion euros over seven years for the Horizon 2020 program, including the financing of the two flagship projects themselves. Each of these projects will receive 1 billion euros (around 1.14 billion dollars) over ten years, half coming from the European Commission and the other half from the participants themselves, viz., member states of the European Union, universities, research institutions, industrial companies, and so on. The first funding for graphene research amounted to 54 million euros in 2013, in the framework of the specific program for information and communication technologies (ICT).

Graphene research is a perfect example of 'translational' nanotechnology,[4] where the discoveries of academic research centres are rapidly transferred to applications and commercial products. Indeed, graphene and associated materials may have profound impacts on the information and communication technologies in both the short and the long term, especially when we consider the integration of graphene components into silicon-based electronics and the partial replacement of silicon in new applications. But beyond ICT, graphene research is just as promising in the fields of energy, transport, and health. To illustrate this, the research areas and industrial sectors listed for the Graphene Flagship project include fundamental research in physics and chemistry, spintronics, high-frequency electronics, flexible electronics, optoelectronics, the production of materials and nanocomposites, sensors, and applications to health, the environment, and energy.

The Graphene Flagship project is led by professor Jari Kinaret at the University of Chalmers in Sweden. It involves more than 700 research groups, spread over almost 140 different institutions, and includes four Nobel prizewinners. Almost 70 of these groups belong to industrial companies. The current distribution is shown country by country in Fig. 14.3. Note in particular the significant investment from countries like Spain, the United Kingdom, Germany, and France.

[4]The term 'translational' is commonly used in association with medicine. In fact, translational medicine reflects the ability of certain research results to get very quickly transferred from laboratory to bedside.

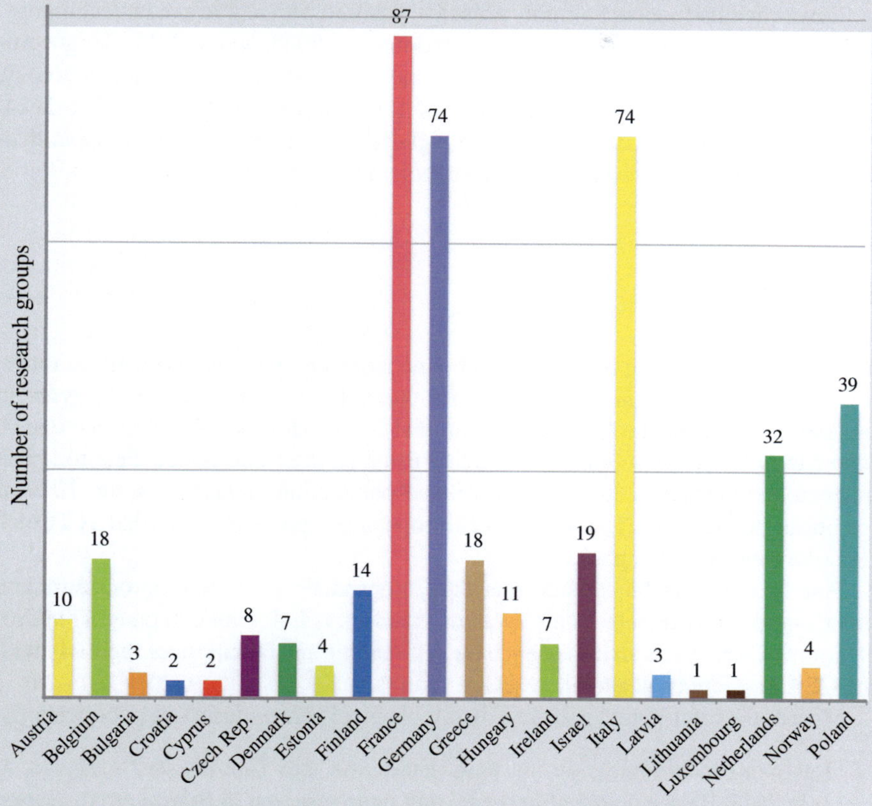

Fig. 14.3 European research groups involved in Europe's *graphene flagship* project

In the United States, programs and coordination between the various federal departments are laid down by the National Nanotechnology Initiative (NNI). Federal investment is typically around 2 billion dollars a year (1.8 in 2010). Several federal departments contribute to the NNI, including the Department of Defence (DoD), but also bodies like the National Aeronautics and Space Administration (NASA), the National Institute of Health (NIH), and the Food and Drug Administration (FDA). Finally, the National Institute of Standards and Technology (NIST) and the Environmental Protection Agency (EPA) are building a huge materials database.

Japan has always been and remains an important contributor to the field of nanotechnology. It has established a national network of five main institutes in the universities of Hokkaido, Tohoku, Tokyo, Osaka, and Kyushu. Annual government funding can be estimated at around 1 billion dollars, with a multiplicative factor of ten if one includes the universities and the various institutions and companies.

China has launched a long term national plan to focus on scientific and technological development for the period 2006–2020, and nanotechnology constitutes a major theme, with the development of new materials and devices for nanoelectronics. Public research funding reached 0.6 billion dollars in 2007.

Taiwan has also adopted a highly proactive position with regard to nanotechnology. Its R&D investments exceeded 0.3 billion dollars in 2008, and in 2011, the government launched a plan for sectors using nanotechnologies such as biotechnology, imaging, and monitoring, with a budget of 1.5 billion euros. From 2010 to 2014, countries in Asia have demonstrated the highest rate of growth in scientific publications in the field of nanoscience and nanotechnology (see Fig. 14.2).

14.4 Conclusion

Even though the French organisation of nano research may seem complex at times, it is important to recognise its efficacy, for this field of activity, extremely dynamic in France, stands at the highest international level, with two Nobel prizewinners, Jean-Marie Lehn for supramolecular chemistry in 1987 and Albert Fert for giant magnetoresistance in 2007. French science publications represent some 10 % of European publications, just as 15 % of European patents have been filed by French research centres or companies.

Finally, it should be stressed that the European leader in microprocessors and other micro- and nanoelectronic systems is the French–Italian company STMicroelectronics, which has an important base in Grenoble and collaborates regularly with the CEA and French academic research centres.

Most of the information presented in this chapter is drawn from two public reports:

1. Les nanotechnologies: un nouveau paradigme. *Les Cahiers de l'ANR*, no. 5, July 2012. Those involved in organising nano research in France are discussed on pp. 17–25 with information about 'academic' research, but also industrial R&D programs. It can be obtained from www.agence-nationale-recherche.fr/magazine/actualites/detail/les-nanotechnologies-un-nouveau-paradigme-cahier-anr-n5-juillet-2012/
2. European nanotechnology landscape report. A publication by the Observatory-NANO Work Package 3. It can be obtained from http://bwcv.es/assets/2011/11/22/European_Nanotechnology_Landscape_Report.pdf

For more information about the French system of research in the nano field, it will be helpful to visit the following websites: CNRS:www.cnrs.fr/; INSERM: www.inserm.fr/; CEA:www.cea.fr/; C'Nano:www.cnano.fr/; and French Ministry of Research: www.enseignementsup-recherche.gouv.fr/.

Chapter 15
Training in Nanoscience and Nanotechnology

Sylvie Retailleau, Claire Dupas-Haeberlin, Jean-Michel Lourtioz and Arnaud Bournel

Abstract In this chapter, we present new training programs set up at all the different levels from first degree to master's and doctoral degrees in French higher education, in universities, engineering schools, and *Grandes Ecoles*, but also in high school or even in primary school, to prepare young people for the new and highly cross-disciplinary field of nanoscience and nanotechnology, together with their many applications. Maps show the availability of nano master's degrees in France and across Europe. We also discuss measures adopted to better inform the general public.

15.1 Opportunities for Cross-Disciplinary Training

As a consequence of the huge efforts made today around the world to encourage research in nanoscience and nanotechnology, there has been a significant renewal in the training courses open to students in higher education and research institutions. Indeed, research continually feeds and restructures the teaching that goes on in such institutions. It is also through their research potential that they can respond to the demands of the economic world and society in general, in particular by preparing students for new professions and training them in new skills.

S. Retailleau (✉) · J.-M. Lourtioz · A. Bournel
Institut d'Electronique Fondamentale, Université Paris-Sud,
Bât 220, 91405 Orsay Cedex, France
e-mail: sylvie.retailleau@u-psud.fr

J.-M. Lourtioz
e-mail: jean-michel.lourtioz@u-psud.fr

A. Bournel
e-mail: arnaud.bournel@u-psud.fr

C. Dupas-Haeberlin
ENS Cachan, I61 Avenue du Pdt Wilson, 94235 Cachan Cedex, France
e-mail: claire.dupas@sfr.fr

J.-M. Lourtioz et al. (eds.), *Nanosciences and Nanotechnology*,
DOI 10.1007/978-3-319-19360-1_15

The first training courses affected by the rapid development of nanoscience and nanotechnology, and also the first didactic introductions to these subjects in French [1], subsequently published in English [2], aimed at the highest levels in the education system, that is, master's and doctoral degrees, leading eventually to the professions of research, teacher-researcher, and R&D engineer, in areas as varied as electronics, optics, energy, chemistry, materials, biotechnology, health, the food industry, the environment, and others. Thanks to the cross-disciplinary nature of these courses, which now exist in France or abroad, students can choose from a broad spectrum of research centres to prepare their doctorate, both academic and industrial, or indeed an association between academic and industrial research units, for example, in the context of a *Convention industrielle de formation par la recherche* (CIFRE). At the present time, it can be estimated that in France some 20 % of the theses presented each year in mathematics, physics, and chemistry are connected in some way with the nano sector.

In fact, the opportunities represented by the nano field in terms of cross-disciplinary training can bring benefits at every level of education, without necessarily having to sacrifice the fundamentals of the traditional disciplines. As has been shown throughout this book, many examples taken from the world of nanoscience can be used to illustrate the basic laws of physics and chemistry and the properties of matter in general, because they are both fascinating and easy to demonstrate. The laws of optical diffraction and interference seem so much more natural when used to explain the iridescent gleam of opal, the wings of certain butterflies (Fig. 15.1), or CDs (compact discs) and DVDs (digital versatile discs) when they are held up to the light. The very same effects are put to use in the development of photonic crystals (see Chap. 2). Redox reactions in chemistry are brilliantly illustrated by the synthesis of gold nanoparticles whose apparent colour depends on their size and hence the extent to which the reaction has proceeded. The rapid degradation of pollutants when illuminated by UV light demonstrates the efficacy of photocatalysis in a nanoparticle medium.

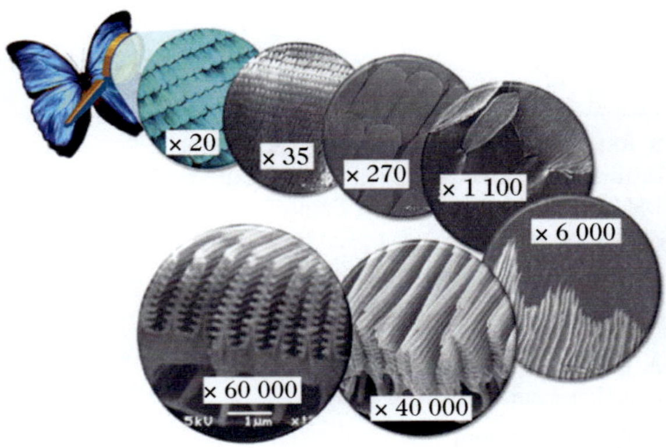

Fig. 15.1 Successive zooms showing the diffractive structures giving rise to the *blue colour* of Morpho butterfly wings. Taken from [3]

These few examples show that it is easy to introduce nano concepts into first degree courses through examples illustrating the basic ideas, to give the student a better grasp of the specific properties of matter on this scale. This can be done either as an illustration for physics and chemistry courses, or through specific introductory modules. This approach is implemented today in most universities, both in France and abroad.

Furthermore, nano ideas can be introduced even at high school, integrating them into the traditional school courses in such a way as to meet the needs of the new programs. In France, this has been done with the 'Create and innovate' program of the scientific stream for penultimate year pupils, which aims to build an understanding of the interactions between science and society through a series of examples. Pupils gather and exploit information about current topics in science and technology, or about scientific and technical professions and training in relation with local resources or in partnership with a research institute.

The new programs for final year pupils in the scientific stream aim "to further the pupil's appreciation of the scientific approach, encouraging observation, understanding, and action, by making use of modern and promising opportunities raised by questioning". For their speciality in the final year, these pupils must be placed in a research situation where they are encouraged to develop the three essential activities:

- experimental practice,
- analysis and synopsis of scientific documents,
- solution of scientific problems.

Nanoscience has been the subject of much public debate over the past decade. It thus provides a way of associating social issues with scientific teaching, the aim being to give the citizens of tomorrow a good foundation for making technological choices and thereby strengthening the new and necessary dialogue between science and society. Very successful measures have opened the way to this new introduction to science, such as the European Nano@school program [4] in Grenoble and the activities of *Nano-École* [5] in the Paris, Toulouse, Lille, and Grenoble centres. We shall return to this point below.

The initiatives developed in Toulouse in the *Nano-École* framework have shown that the quality of teaching and the interest manifested by high school pupils for the scientific aspects of the nano field were considerably increased when they were involved in a socio-scientific debate. The idea is to present scientific progress hand in hand with the concomitant controversy and debate occurring even within the scientific community itself. This is rather different to the usual way of teaching science, where knowledge is often disseminated as indisputable truths, with very little discussion of the downsides, while controversy naturally plays a central role in the scientific process. In this way, pupils can be induced to take into account ethical, legislative, and social aspects of nanotechnology, so that these future citizens should be better able to measure the progress and impacts it may bring about. Such issues can be tackled with teachers of civil, legal, and social education (*éducation civique, juridique et sociale*, ECJS) in the French curriculum.

15.2 BMD and the Role of Nano Teaching in France and Worldwide

The standards and qualifications of higher education have been harmonised in Europe through the reference structure specified by the bachelor–master–doctorate (BMD) system. In this context, the French university curriculum now offers three degree diplomas:

- a three-year bachelor's degree, leading to[1] BAC + 3,
- a two-year master's degree, leading to BAC + 5,
- a three-year doctoral degree, leading to BAC + 8.

A master's degree is characterised by levels M1 and M2 referring to the two years of training, usually with several possible courses in the second year. For example, one can sign up for a physics master's degree and follow courses on nanoscience and nanotechnology at level M2. One can also sign up directly for a master's degree in nanoscience and nanotechnology, if it happens to exist in one's particular choice of higher education institution.

Clearly, it is mainly through university master's degrees, and in particular the second year of such a degree, that students can obtain training in nanoscience and nanotechnology in the French education system. On the other hand, there are still very few engineering courses offering specialisation in this field. We should nevertheless mention:

- The biotechnology and health engineering degree at the *Université de technologie de Compiègne* (UTC), which includes a course on micro- and nanotechnology.
- The micro- and nanotechnology course available in the *Instituts nationaux des sciences appliquées* (INSA).
- The nanophysics course at the *Institut d'optique Graduate School* in Palaiseau (IOGS).
- The physics–nanoscience course at the *École nationale supérieure de physique, électronique et matériaux* in Grenoble (INP Phelma).

It should also be noted that most engineering schools and *Grandes Ecoles* invite their students to follow university master's degree courses which they sometimes jointly supervise with universities. We may cite the example of the M2 nanoscience course in Paris-Saclay, jointly supervised by seven establishments, the universities of Paris-Sud and Versailles-Saint-Quentin-en-Yvelines, the *École Centrale Paris*, the *École normale supérieure* (ENS) in Cachan, the *École Polytechnique*, the IOGS, and SUPELEC. The current distribution across France of master's degrees in nanoscience and nanotechnology (see Fig. 15.2 and Table 15.1) amounted to almost 35 courses in 2013, covering part or all of nanophysics, nanochemistry, nanobiology, nanomaterials, and nanosystems. More precisely, there are 7 master's degrees (M1 and M2) and 28 second year options (M2).

---

[1]BAC is the abbreviation for the French baccalaureat, the high school leaver's certificate.

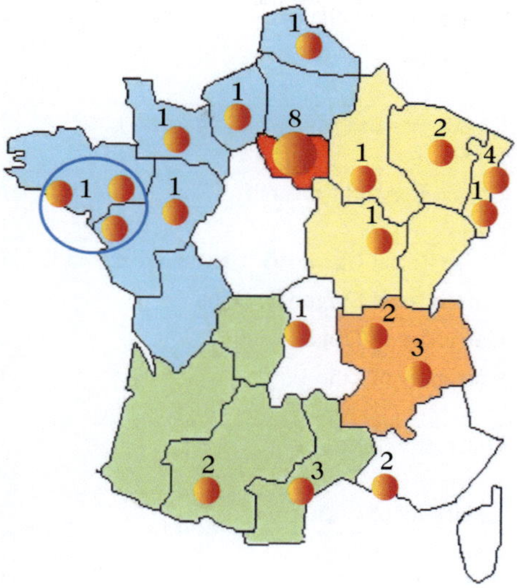

Fig. 15.2 Possibilities for master's degrees in nanoscience and nanotechnology in France [6]

Table 15.1 Master's degrees or specialisations relating to nanoscience or nanotechnology in France

Institution	Training course: title of master's degree (M1+M2) or M2
Universities of Paris-Sud and Versailles-Saint-Quentin-en-Yvelines, Ecole Polytechnique, ENS Cachan, IOGS, Ecole Centrale Paris, Supelec	M2 Nanoscience
Universities of Paris 6, Paris 7, ENS Ulm and Cachan, Ecole Polytechnique, ESPCI, ENSCP	M2 Materials science and nano-objects
University of Paris 7	M2 Nanochemistry, materials, surfaces
University of Paris 7	M2 Surfaces, thin films, nanomaterials
University of Paris 13, CNAM	M2 Physics and nanotechnology
University of Evry	M2 Nanomaterials
University of Evry	M2 Surfaces and nanomaterials
ENS Cachan	M1+M2: Erasmus Mundus: Molecular nano- and bio-photonics for telecommunications and biotechnologies (Monabiphot)
Joseph Fourier University, Grenoble	M1+M2: Nanoscience, nanotechnology
Joseph Fourier University, Grenoble	M1+M2: Erasmus Mundus: Nanophysics, nanostructures
INP Grenoble, EPFL, Politechnico Turin	M1+M2: Nanotech
Paul Sabatier University, Toulouse, INP, INSA, Sup'Aero	M2 Nanomaterials and multimaterials
Paul Sabatier University, Toulouse, INP, INSA	M2 Nanoscience, nanocomponents, nanomeasurements

(continued)

Table 15.1 (continued)

Institution	Training course: title of master's degree (M1+M2) or M2
Claude Bernard University, Lyon, ECL, INSA, ENSL	M1+M2: Micro-nano
Claude Bernard University, Lyon	M2 Instrument development for micro- and nanotechnology
Blaise Pascal University, Clermont Ferrand II	M2 Nanostructure and nanophotonics
University of Burgundy, Dijon	M2 Nanotechnology and nanobioscience
University of Troyes	M2 Optics and nanotechnology
University of Lille 1	M1+M2: Microelectronics, nanotechnology and telecom
University of Rouen	M2 Science of matter: Materials, nanoscience, and energy
University of Caen	M2 Materials, nanoscience, and energy
University of Angers	M2 Fine chemistry, functional materials, and nanoscience
University of Rennes, Lorient and Nantes	M2 Nanoscience, nanomaterials, and nanotechnology
University of Montpellier	M2 Nanophysics
University of Montpellier	M2 Physics and engineering of materials for microelectronics and nanotechnology
University of Montpellier	M2 Electronics, photonics, nanotechnology
University of Aix-Marseille	M1+M2: Micro- and nanoelectronics
University of Aix-Marseille	M2 Advanced materials for nanoscience and energy
University of Strasbourg	M2 Micro- and nanoelectronics
University of Strasbourg	M2 Condensed matter and nanophysics
University of Strasbourg	M2 Vision, automation, nanophotonics
University of Strasbourg	M2 Materials engineering (chemistry and physics of nanostructured systems)
University of Lorraine	M2 Condensed matter and nanophysics (CMN)
University of Lorraine	M2 Photonics and optics for materials
University of Mulhouse	M2 Materials engineering (chemistry and physics of nanostructured systems)

Source C'Nano [6]

Three of the master's degrees can be found in Grenoble, another at the ENS in Cachan, and the last three are spread between Lille, Lyon, and Aix-Marseille. Two of these degrees are associated with the European Erasmus Mundus program,[2] one

[2]Erasmus Mundus is a program designed to improve the quality of European higher education and strengthen cross-cultural understanding. It encourages and supports the mobility of individuals and cooperation between both European and non-European establishments. The aim is to promote the European Union as an area of academic excellence on a worldwide scale, to contribute to the

in Grenoble and the other in Cachan. Concerning the M2 options, 8 of them can be found in the Île-de-France and the other 20 in the provinces.

Across Europe, almost 30 master's degrees were taught in English in 2013. These are listed by country in Table 15.2. The map in Fig. 15.3 shows their geographical distribution. These are mainly two-year master's degrees in nanoscience and nanotechnology. A European student can thus follow one year in the nano field, either the first year or the second, in another country.

Quite generally, the internationalisation of the master's degree is strengthened by the possibility of following courses in English and also by European programs like Erasmus Mundus, which allow the student to obtain at the very minimum a double diploma with a foreign country. The condition is that the diploma should be delivered in at least three countries within the European Union. The Erasmus Mundus program also provides grants for its students. As mentioned before, two of the nano master's degrees in France are associated with and financed by this program.

As a matter of fact, the list in Table 15.2 is not exhaustive, since in many countries like Germany, Italy, and the countries of central Europe, the nano theme is not often advertised as a master's course, but is taught as part of a master's degree in physics or chemistry. It is in fact a strategic choice as to whether a master's course or M2 option is clearly advertised as dealing with the nano field. It depends on the policy of the country or the establishment with respect to their training offer. In many European countries, cross-disciplinary training can be built up by taking several modules from different master's courses.

In Germany, the Fraunhofer institutes openly advertise the teaching of nanoscience and nanotechnology (see the website of the Fraunhofer Nanotechnology Alliance [8]). There are eight Fraunhofer institutes specialising in nanochemistry and nanomaterials, three in nano-optics and nanoelectronics, three in nanoprocesses and nanomanipulation, and two in nanobiotechnology. These institutes, among the most prestigious applied research centres in Europe, stand at the interface between the universities and the world of business, and maintain very strong links with the universities.

The great challenge for cross-disciplinary training lies in the need to cover a broad spectrum of knowledge without neglecting the fundamentals in at least one of the relevant disciplines. These courses must also meet the requirement for a high scientific level with both theoretical and technological skills. It is crucial to have access to high-tech platforms like those developed for research, giving students the possibility of state-of-the-art practical work and projects during their master's and doctoral years.

In this context, the *Coordination nationale de la formation en microélectronique et en nanotechnologies* (CNFM), which comprises twelve centres, organises shared actions to pool together the skills and resources of all the French establishments involved in the teaching of micro- and nanoelectronics in higher education. Thanks

(Footnote 2 continued)
sustainable development of higher education in non-member countries, and to offer students the best possible career prospects. The program had a budget of some 950 million euros for the period 2009–2013.

Table 15.2 English language master's degrees in the field of nanoscience and nanotechnology in Europe as of 1 May 2013

University	Master's degree	Country
University of Dresden	Master's in nanoscience and nanotechnology—Erasmus Mundus (Grenoble)	Germany
University of Freiburg	Master's in materials and nanoscience	Germany
University of Bâle	Master's in nanoscience	Switzerland
ETH Zurich	Master of Science ETH in micro and nanosystems	Switzerland
University of Louvain	Complementary master's in nanotechnology	Belgium
University of Anvers	Master's in physics and nanophysics	Belgium
University of London	MSc in nanotechnology	UK
University of Nottingham	Nanotechnology and nanoscience centre	UK
University of Glasgow	Master of nanoscience and nanotechnology	UK
University of Southampton	MSc Bionanotechnology	UK
University of Dublin	MSc in nanobioscience	Eire
University of Delft	Master of nanoscience	Netherlands
University of Twente	MSc in nanotechnology	Netherlands
University of Groningen	Top master programme in nanoscience	Netherlands
University of Barcelona	Master of nanoscience and nanotechnology	Spain
Autonomous University of Madrid	Master's in nanoscience and molecular nanotechnology	Spain
University of Valencia	Master's in molecular nanoscience and nanotechnology	Spain
University of Saragossa	Master of nanostructured materials for nanotechnology applications	Spain
University of Chalmers	Master of nanotechnology	Sweden
Royal Institute of Technology	Master's programme in nanotechnology	Sweden
University of Lund	Master of nanoscience	Sweden
University of Linköping	Msc in physics of materials and nanotechnology	Sweden
University of Copenhagen	Master's in nanoscience	Denmark
University of Aalborg	Master's in nanophysics and materials	Denmark
University of Aalborg	Master's in nanobiotechnology	Denmark

(continued)

Table 15.2 (continued)

University	Master's degree	Country
University of Silesia	MSc in physics: materials, nanophysics and mesoscopic physics	Poland
University of Jyväskylä	Master's in nanoscience	Finland
University of Aalto	Master's in micro- and nanotechnology	Finland

Source C'Nano [7]

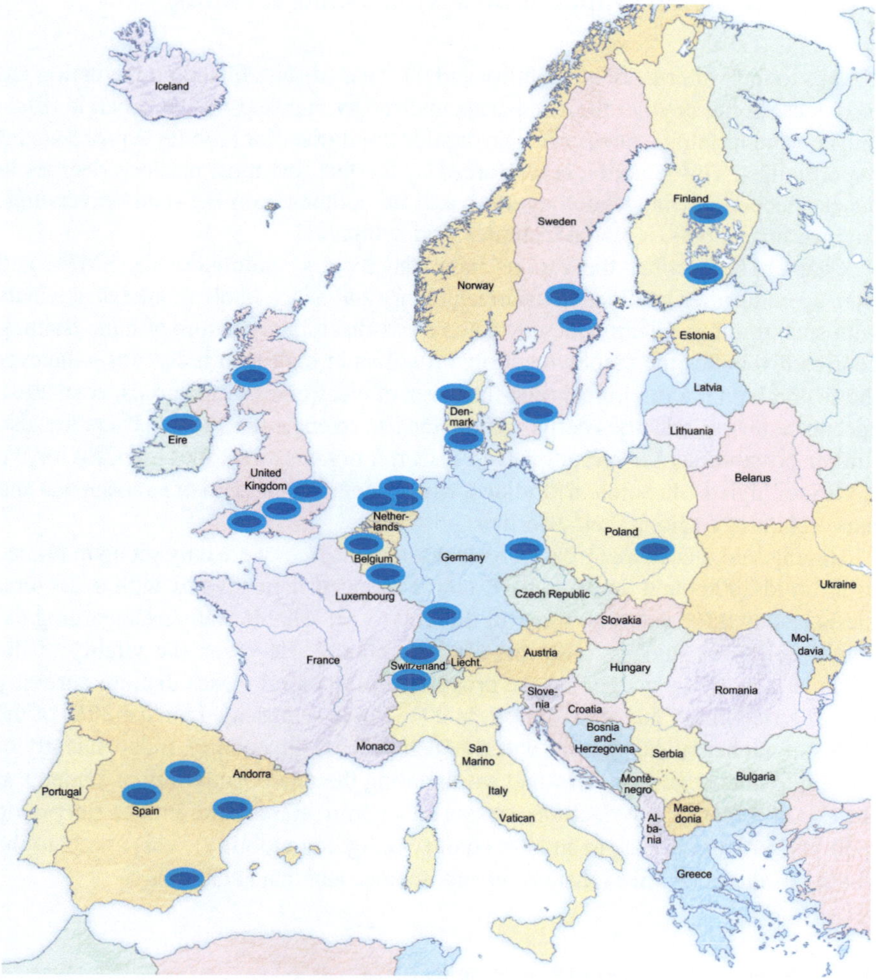

Fig. 15.3 Possibilities for master's degrees in nanoscience and nanotechnology in Europe

to this pooling of resources, the CNFM centres provide operational advantages and high level knowhow, especially in terms of clean room fabrication and characterisation. It should also be mentioned that the *Programme des investissements d'avenir* (PIA) contributes to the experimental training of students and doctoral students using the state-of-the-art facilities acquired in the Equipex framework (*Équipement d'excellence* projects). Furthermore, the Labex research centres have already begun to provide master's grants, thus boosting the attractiveness of these courses.

15.3 Professional Opportunites After Nano Training

Thanks to their international visibility and their use of state-of-the-art technological resources, going beyond the compartmentalisation imposed by the separate disciplines, nano training courses offer favourable conditions for high level professional opportunities. This situation is reinforced by the fact that most master's degrees in nanoscience and nanotechnology are based on collaboration between universities, engineering schools, research institutes, and companies.

Chapter 16 describes the various industrial sectors, multinationals, SMEs, and start-up companies involved in nanotechnology and hence likely to take on students with such training. Clearly, the novel and cross-disciplinary nature of nano training courses diversifies the employment opportunities in high-tech industries, whatever the sector, be it the traditional high-tech area of electronics or materials, cosmetics, special cements, industries deriving from the life sciences, and so on. There are also further possibilities for recruitment in research organisations like the CEA or the CNRS, or higher education institutions, which advertise the field of nanoscience and nanotechnology among their priorities.

Having said this, in many cases, nano master's degrees were only set up in France in the mid-2000s and students have then continued their studies with a doctoral thesis lasting three years, so it is still difficult to get reliable statistics regarding the employability of students with this kind of training. However, the vitality of the field can already be gauged by the proportion of doctoral theses that are currently being carried out in this area. Indeed, in 2013, these represented almost 20 % of the theses in mathematics, physics, and chemistry across the country. Independently of the employment question, it is also worth noting the relative mobility of students at the end of a master's or doctoral degree, either from abroad into France (incoming mobility) or from France to another country (outgoing mobility). This attests to the vitality of international exchanges in nanoscience and nanotechnology.

15.4 Outreach: Nanoscience at School

In a context of increasing disaffection for scientific and technical studies among young people today, and especially for long series of studies, it is important to set up innovative initiatives to make science more attractive. Nanoscience and nanotechnology, ‘new sciences’ touching upon a great many fields of application and likely to witness a very high level of development in the future, are particularly well placed to provide renewed motivation.

For a few years now, a certain number of countries, and in particular the United States, have been investigating ways to introduce nanoscience into secondary school education [9]. There are several reasons for this. Indeed, the aims are to train future specialists in the field, to inform the public of tomorrow about these supposedly sensitive new topics, and to exploit the cross-disciplinary nature of the subject to motivate the younger generation and give science a more concrete image. Hence in the United States, the National Science Foundation (NSF) devotes a sizeable budget through the Nanotechnology Center for Learning and Teaching (NCLT) to introduce nanoscience into the school curriculum, train teachers, and put forward didactic methods for tackling the subject.

In a similar way, original initiatives have also been developed in France. It was in the context of the 2009 economic recovery plan and the national operation known as NanoInnov that the *Nano-École* initiative was launched [10, 11]. Its aim was to experiment with the introduction of nanoscience and nanotechnology in schools, in four different areas, namely, Grenoble, Lille, Paris-Saclay, and Toulouse, by associating science teaching and socio-scientific debate.

The *Nano-École* project was built in collaboration with the respective educational authorities and in partnership with local research groups. The accent is firmly placed on cross-disciplinarity, and interventions in secondary schools encourage active participation on the part of the pupils. Furthermore, pupils are also given an opportunity to visit major research facilities and discuss directly with researchers who are often perceived as hard to understand and inaccessible, living an isolated existence in their laboratories! Societal issues are stressed, a key point for a sector that is still the subject of some controversy. The pedagogical content is elaborated in collaboration with high school teachers, and this serves to identify points of contact between the new ideas brought by nanoscience and the notions put forward in the official programs of the French educational authorities.

The efforts made in the Toulouse area benefit from the longest hands-on experience, notably with the development of a final year science stream specifically dedicated to the nano field at the *lycée St-Sernin*. Pupils have lessons on nanotechnology given by doctoral students from the *Laboratoire d’analyse et d’architecture des systèmes* (LAAS, CNRS) and carry out experimental work on a nano-object. This apprenticeship is followed by a debate on a controversial question relating to some potential application of nanotechnology.

At the Paris-Saclay and Lille sites, interventions with high school classes take place both within the school and in a research centre (Fig. 15.4). Pupils carry out manipulations as sophisticated as realising a device in the clean room. Moreover, in order to be able to do experiments in school, the Paris-Saclay centre has developed teaching packages, distributed by a publisher of educational materials. Also proposed are conferences followed by debates in collaboration with teachers of civil, legal, and social education (*éducation civique, juridique et sociale*, ECJS) and with the *Laboratoire Sciences-Techniques-Éducation-Formation* (STEF) of ENS Cachan.

Fig. 15.4 Some of the themes developed in Île-de-France in the form of teaching packages: colours of the nanoworld, nanoparticles and the Tyndall effect, TiO_2 nanoparticles and depollution, photonic crystals

15.5 Outreach: From Pupils to the General Public

The idea of presenting and explaining the discoveries of nanoscience and nanotechnology should not be reserved solely for the younger generations. Initiatives have been taken to involve the general public, in particular by the C'Nano competence centres distributed across the country (see Chap. 14) and by the national program NanoInnov in 2009–2010. Visitor centres were created, didactic scientific modules and demonstrations were developed, and events were organised [12]. These activities are continuing under the auspices of C'Nano and various non-profit-making organisations.

Figure 15.5 shows a force-feedback nanomanipulator designed by a research group in Grenoble to reveal the forces coming into play during an interaction between an atomic force microscope tip and a surface [13]. This is an absolutely unique pedagogical tool. It is accessible at the Nanomonde platform of the *Centre interuniversitaire de microélectronique et de nanotechnologies* in Grenoble. A version has been built as an outreach facility for the general public in a travelling exhibition entitled *Nanotechnologie—infiniment petit-maxi défis* (Nanotechnology—infinitely small—maximum challenge), and it has been exposed at the *Cité des sciences et de l'industrie* in Paris. Other nano demonstration tools have been set up in visitor centres and entrance halls, like the one at the MINATEC site in Grenoble, which has more than 10 000 visitors a year. Demonstration modules and models of objects designed on the nanoscale are also produced for symposia bringing together the general public and professionals from different areas. Meetings are held regularly to inform the public about all the various aspects of Nanomonde and debate the real or imagined impacts with as broad a public as possible (*Nuit des chercheurs*, *Fête de la science*, etc.).

Fig. 15.5 Force-feedback nanomanipulator developed by a group at the Néel Institute in Grenoble. This device can feel the interaction between an atomic force microscope tip and a surface on which it is displaced virtually via an image on the computer screen. Courtesy of Florence Marchi, Néel Institute, Grenoble

15.6 Prospects

After an initially slow rise in nanoscience and nanotechnology training courses followed by strong growth in the 2000s, the situation in France today consists in a consolidation of the higher education opportunities which need to be better publicised. While many nano training courses were originally just rejuvenated versions of former specialisations in microelectronics or the physics of materials, cross-disciplinarity gradually began to make its way, providing more and more room for the life sciences and nanotechnology proper. This is similar to the evolution on the international scene, where this area of science remains a priority.

The attraction for the cross-disciplinarity of the nano field is not only illustrated by the opportunities for students when they complete such an education, but also by the scientific education of the younger generations, from the primary level to high school. The *Nano-École* project in France was extremely well received by both pupils and teachers. The pursuit of these operations beyond the four test areas and their generalisation on a national level have now been set in motion by the C'Nano network, which will coordinate the relevant regional partners. Furthermore, with the recent introduction of nanotechnology in the high school curriculum, an academic training program (*plan académique de formation*, PAF) is available to secondary school teachers to provide them with a clear overview of the recent progress in nanoscience and nanotechnology research, but also the related social issues.

References

1. M. Lahmani et al., *Les Nanosciences*. Editions Belin, Collection Echelles. Vol. 1: *Les nanotechnologies et la nanophysique*, 3rd edn. (2009); Vol. 2: *Nanomatériaux et nanochimie*, 2nd edn.; Vol. 3: *Nanobiotechnologies et nanobiologies*, 1st edn. (2007); Vol. 4: *Nanotoxicologie et nanoéthique*, 1st edn. (2010)
2. C. Dupas, P. Houdy, M. Lahmani (eds.), *Nanoscience. Nanotechnology and Nanophysics* (Springer, Berlin, 2007); M. Lahmani, C. Bréchignac, P. Houdy (eds.), *Nanoscience. Nanomaterials and Nanochemistry* (Springe, Berlin, 2008); M. Lahmani, P. Boisseau, P. Houdy (eds.), *Nanoscience. Nanobiotechnology and Nanobiology* (Springer, Berlin, 2010); M. Lahmani, F. Marano, P. Houdy (eds.), *Nanotoxicology and Nanoethics* (Springer, Berlin, 2011)
3. S. Berthier, *Photonique des morphos* (Editions Springer, Paris, 2010)
4. www.nanoatschool.org/
5. www.nano-ecole.fr/index.php
6. www.cnano.fr/IMG/swf/carte_formation.swf
7. www.masteretudes.fr/Master/Physique/Master-en-Nanotechnologie/Europe/
8. www.nano.fraunhofer.de/english/competences_instituts.html#ikts
9. B. Hingant, V. Albe, Nanosciences and nanotechnologies learning and teaching in secondary education: a review of literature. Stud. Sci. Educ. **46**, 2 (2010)
10. N. Panissal, J.C. Cau et al., Nanotechnology training before university: a new approach combining scientific and social issues. J. Mater. Educ. **33**, 1–2 (2011)
11. www.nano-ecole.fr
12. A. Sauvageot et al., Colloque Images, mirages et nanosciences. Toulouse, 9–10 December 2010
13. F. Marchi et al., Prix de l'innovation pédagogique 2010 au colloque sur l'enseignement des technologies et des sciences de l'information et des systèmes

Chapter 16
Nanotechnology and Industry

Patrice Hesto, Jean-Michel Lourtioz, Claire Dupas-Haeberlin, Marcel Lahmani and Thomas Dubouchet

Abstract This chapter describes the vast industrial panorama that has been profoundly influenced by the advent of nanoscience and nanotechnology. The nano field has contributed to industrial innovation and development from electronics, through materials and energy, to biomedicine. Here we present an overview of the industrial actors in France, considering the large corporations, SMEs, and start-ups, and giving a brief description of the potential markets and investment on the international level.

Nanotechnology is not in itself an industrial sector, since it applies across many industrial sectors, contributing in varying degrees to their development. Of course, it is a necessary transition for industries like electronics, micromechanics, the associated instrumentation, and materials. It is also a source of innovation for energy production, the sensor industry, medicine, and many other sectors. It has also made its way into many existing products, improving both quality and performance. What we have here is thus a vast and continually expanding industrial panorama which we shall outline in particular through the situation in France.

Having said this, not all the companies exploiting nanotechnology have any real knowhow in this field, and neither do they necessarily invest in it. The list contains companies that incorporate devices developed using nanotechnology and

P. Hesto (✉)
31 rue des grands champs, 75020 Paris, France
e-mail: patrice.hesto@laposte.net

J.-M. Lourtioz
Institut d'Electronique Fondamentale, Bât 220, Université Paris-Sud, 91405 Orsay Cedex, France
e-mail: jean-michel.lourtioz@u-psud.fr

C. Dupas-Haeberlin
ENS Cachan, 61 Avenue du Pdt Wilson, 94235 Cachan Cedex, France
e-mail: claire.dupas@sfr.fr

M. Lahmani
Club NanoMicroTechnologie, 8 Rue de L'Avenir, 91140 Villebon-sur-Yvette, France
e-mail: marcel.lahmani@wanadoo.fr

T. Dubouchet
Nanothinking, 40 Avenue Parmentier, 75011 Paris, France
e-mail: thomas.dubouchet@nanothinking.com

J.-M. Lourtioz et al. (eds.), *Nanosciences and Nanotechnology*,
DOI 10.1007/978-3-319-19360-1_16

others implementing materials whose performance has been improved in a different sector like the chemical industry. It also contains industries that have long exploited nanoscale processes or materials for tyres, cosmetics, catalysis, surface treatments, and others, well before the nano field really took off toward the end of the twentieth century. For these reasons, it is no simple matter to give a precise description of the world nanotechnology market, something we shall attempt to do in the second part of the chapter, while growth in these activities leaves no-one in doubt.

To end our overview, we shall also discuss financing and investment in nanotechnology, considering in particular the systems that have been set up in France.

16.1 The French Industrial Landscape

16.1.1 Multisector Applications

The industrial development of nanotechnology benefits not only from the latest scientific progress in physics, surface science, chemistry, or molecular biology, but also and above all from developments in the methods of micro- and nanofabrication. Even today, these developments on the scale of the nanometre promise a whole range of applications in a variety of industrial sectors. Indeed, practically all sectors are concerned here.

More than 500 applications of nanotechnology have been identified in products developed and commercialised by companies operating in France. This concerns first and foremost the sectors of electronics, optics, biotechnology for health, cosmetics, and transport, and it includes both the production of nanomaterials and the fabrication of nanodevices. Depending on the case, nanotechnology can be used to obtain a new property for a given product, or indeed optimise one or more properties of an existing product.

Among the main orientations and applications, we may cite:

- the production of materials that are both strong and light,
- continued miniaturisation, diversification, and integration of electronic components,
- the development of new, better targeted, more effective, and cheaper therapies,
- new properties for textiles (bactericidal, fire resistant, luminescent, etc.), paints (self-repairing, self-cleaning, anti-scratch, etc.), building materials (steels, glass, and 'intelligent' cements—see, for example, Chap. 7).

Figure 16.1 illustrates the generic nature of nanotechnology, here the property of hydrophobicity obtained by nanostructuring the surface of different kinds of materials. Cloths can be made impermeable and cements water-repellent, while kitchen and bathroom surfaces, or indeed car windscreens, can be protected against infiltration of water. Figure 16.2 is a schematic view of nanotechnology applications for each of the main industrial sectors.

Fig. 16.1 Water droplet on a hydrophobic textured substrate. The droplet maintains its *spherical* shape. This is the lotus effect, obtained for superhydrophobic surfaces. *Insert*: Detail of the specially structured surface. There are applications in impermeable cloths, water-repellent cements, coatings for kitchen and bathroom surfaces, or car windscreens. Source © CNRS Photothèque, David Quéré, Mathilde Callies, *Laboratoire de physique des fluides organisés*, Paris

The broad presence of nanotechnology across all industrial sectors can be explained not only by its cross-disciplinarity, but also by the fact that *the same basic nano building block can be used for many different applications* in different sectors. This feature can be illustrated by two examples, one in materials, the other in fabrication technology:

- The case of carbon nanotubes, which have been the subject of a great deal of research into their exceptional mechanical, electrical, chemical, and optical properties over the past twenty years, is without doubt emblematic (see Chaps. 2 and 5). Applications have been proposed in all the major industrial sectors, in particular to obtain lighter and stronger materials with optimised optical properties, or to produce ultra-miniaturised transistors and high-emissivity nanocathodes, or again to store various gases. The first large scale industrial applications are just beginning and the interest is such that carbon nanotubes are now being overproduced in the world.
- Another example which shows how nanotechnology can be used to fabricate stacks of ultrathin films with sufficient efficiency on an industrial scale is the transfer from research centres to industry of techniques for depositing materials atom by atom to form films containing only a few atomic planes (see Fig. 16.3). These multilayers are put to use in the extreme miniaturisation of microprocessors, but also in the development of magnetic sensors for reading and writing hard disks, the fabrication of semiconductor lasers for optical disks, the production of micro- and nanosystems for position and acceleration sensors (see Chaps. 2 and 4), or the design of very high efficiency photovoltaic cells (see Chap. 3). These films can also be applied to surface treatments (see Chap. 5), particularly for the passivation of electronic components or for the deposition of anti-reflective coatings.

Fig. 16.2 The main industrial sectors using nanotechnology. This figure has been produced using an interactive tool available online [1]

Fig. 16.3 Molecular beam epitaxy (MBE) system able to deposit layers atom by atom. The example in the *insert* is a stack of layers designed for making giant magnetoresistance (GMR) sensors for magnetic read heads (see Chap. 4). Each point of light in the image corresponds to one atom. Source © CNRS Photothèque/THALES, Hubert Raguet/Jean-Luc Maurice, *Unité mixte de physique CNRS/Thalès*, Orsay

16.1.2 Some Key Industrial Sectors

The industrial sectors in which nanotechnology has been most present are electronics, materials, and medicine.

Micro- and Nanoelectronics

Two companies based in France can illustrate the impact of nanotechnology on industrial development in the field of micro- and nanoelectronics.

The first is a company which, following the technology roadmap for semiconductors, has gradually made the transition from microtechnology to nanotechnology. This is STMicroelectronics, set up in 1987 on the initiative of the French and Italian governments by regrouping two existing companies, viz., Thomson Semiconducteurs and SGS Microelettronica. The two country's governments contributed significantly to its development and, during its early years, the fabrication processes of STMicroelectronics owed much to two research centres, namely, LETI in Grenoble (CEA) and CNET in Meylan (France Telecom). In 2013, the company employed almost 50,000 people around the world, with a turnover of some 8 G$ (throughout the book, 1 G$ represents a billion dollars, and 1 G€, a billion euros). Its R&D effort represents 28 % of its turnover. It manufactures and commercialises a whole range of semiconductor components, and in particular, microprocessors. Its research and production activities in France are shared over eight sites, including the one in Crolles, near Grenoble, which has two production lines and R&D centres working closely with academic research institutes and those of the CEA. The cost of a production unit is 7.5 G$ for circuits realised on substrates of diameter 300 mm, and 10 G$ for the most recent factory, which makes them on 450 mm substrates (*Les Échos*, 2012).

The second example is the company PICOGIGA (as in picometre and gigahertz), set up in 1983 by the director of the molecular beam epitaxy facility at the *Laboratoire central de recherches* (LCR) of Thomson CSF, located in Corbeville near Orsay. This company specialises in the deposition of thin films of III–V semiconductors for ultrahigh frequency (UHF) applications, including in particular reception and emission circuits in mobile phones. This company quickly became one of the best suppliers worldwide of epitaxial films used in making heterojunction bipolar transistors (see Fig. 16.4). In 2003, the company joined up with the SOITEC group, which specialises in silicon-on-insulator (SOI) technology, described in more detail in Chap. 2.

In France, apart from the above-mentioned companies, many others are active in the fabrication, characterisation, and instrumentation of micro- and nanoelectronic systems.

Nanomaterials

Here we distinguish two types of industrial activity: production of nanomaterials such as nanoparticles, carbon nanotubes, etc., and integration of these nanomaterials in solid matrices like plastics and metals, or liquid matrices, like paints, inks, varnishes, and so on.

Fig. 16.4 Scanning electron microscope (SEM) view of a heterojunction bipolar transistor during fabrication in the laboratory. The layers are, from *bottom* to *top*, the emitter, the base, and the collector. The films here are just a few atomic planes thick. Courtesy of Jean-Luc Pelouard, *Laboratoire de photonique et nanostructures*, Marcoussis

Companies Producing Nanomaterials

In contrast to the top–down approach used in micro- and nanoelectronics, nanomaterial production is based rather on the bottom–up approach, which consists in assembling matter molecule by molecule to obtain functionalised nanoparticles (see Chaps. 2 and 5). Today, the industrial realities of this approach concern mainly the chemical and/or mechanical methods used to synthesise these nanomaterials.

The nanomaterials produced in the largest amounts are silver, titanium dioxide, and silica nanoparticles. These are followed by carbon nanotubes and zinc oxide and cerium oxide nanoparticles. To get some idea, in 2013, the world production capacity for multiwalled carbon nanotubes reached a few hundred tonnes a year for the main producers. This corresponds to an average price of around 100 € per kilogram. In 2013, the global annual production of silver nanoparticles is estimated at around 320 tonnes [2]. Estimates suggest that world annual production will reach some 1120 tonnes by 2015 [3]. Titanium dioxide (see Chap. 8) and silica are materials that have been produced for several decades now in the form of an ultrafine powder. The size distribution of the particles present in these powders extends from the nanoscale to the microscale. This makes it impossible to estimate the actual production of nanometric particles. However, several hundred thousand tonnes of these two types of powder are produced worldwide each year.

Companies Integrating Nanomaterials

Once the nanomaterials have been produced, the next step is integration or formulation in solid or liquid matrices. The homogeneous dispersion of nanomaterials in these matrices is essential to prevent the formation of aggregates and hence conserve the specific properties arising due to their nanometric size (see Chap. 7 for instance).

Among the many industrial products involving a nanomaterial integration stage, we may cite:

- Construction products like steels with good enamel adherence, cements, glasses, coatings, etc.

- Products for the transport industry, such as Michelin's green tyres or composite structures for aeronautics, and in particular for Airbus, etc.
- Cosmetic products in which different kinds of nanoparticles are integrated, e.g., cream, mascara, varnishes, etc., along with other ingredients to obtain well controlled textures or colours, for example.

Medicine

There are a great many applications of nanotechnology in the medical sector. Here we give brief mention only of therapy and diagnosis.

Therapy

An important field of application of nanotechnology for therapy is drug delivery. Here therapeutic molecules are delivered specifically to the sick cell or tissue (see Chap. 11). A second major objective is to increase the efficiency of active principles, e.g., by improving their solubility, stability, or biocompatibility. It should become possible to administer neurological drugs by means of nanoparticles able to cross the blood–brain barrier (BBB).

Diagnosis

It is now possible to make miniaturised sensors that are both sensitive and selective to ensure permanent prevention of certain pathologies at a much reduced cost. Two examples available today concern the measurement of glycemia, using nanosensors which change colour depending on the blood sugar level, and heart rate measurements, using nano strain gauges (see Chap. 10).

Another important objective is the early diagnosis of cancer, cardiovascular disease, and neurodegenerative disorders. An example is the diagnosis of lung cancer by analysis of chemical markers present in the patient's breath or the early detection of tumours using contrast agents which increase medical imaging resolution.

16.1.3 Industrial Actors in France

The maps in Fig. 16.5 show the distribution of nanotechnology industries in France, with small or large companies in Fig. 16.5a and start-up companies in Fig. 16.5b. Altogether around a hundred companies and 150 start-up companies are identified. The companies are located across the whole country, but with a very high concentration around Paris and Grenoble, while the start-up companies, for their part, are located rather in the vicinity of the main technological research centres (see Chap. 14). It should be noted, however, that the number of industrial actors is probably underestimated here, since some applications of nanotechnology will be concealed as a matter of industrial secrecy. Moreover, the proliferation of intermediaries in many industrial value chains makes it difficult to identify all the industrial actors with nanotechnology knowhow.

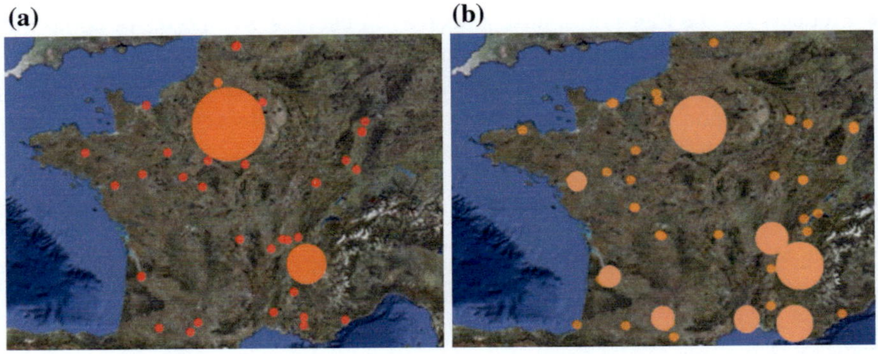

Fig. 16.5 Companies with nanotechnology knowhow operating in France. **a** The main groups and companies of intermediate size. **b** Start-up companies. The area of each disk is proportional to the number of companies, while *small circles* correspond to a single enterprise. These figures were produced using an interactive tool available online [4]

16.1.4 Zoom on Start-Up Companies

The 150 start-up companies identified here are situated mainly around Paris, Grenoble, Lyon, and Marseille, and to a slightly lesser degree around Bordeaux, Toulouse, Montpellier, and Nantes. These companies are extremely dynamic, and indeed many are born, grow, and disappear each year. They also form an extremely heterogeneous group, since the very term 'start-up' covers many different situations, from very young companies with only one or two employees to others with several dozen, which have already been through the fund-raising process several times. The vast majority of technological start-ups are set up by researchers coming from public or private research centres, or by engineers from the main industrial groups, whence their geographical distribution as mentioned above. They are sometimes qualified as spin-offs, when the staff and skills are for a large part derived from an existing company. Often directed by the researchers or engineers who set them up, they benefit from technical, administrative, and financial support from the original structure.

The life cycles of start-ups differ considerably depending on the field of application. For example, the biotechnology sector is characterised by the long delays between the development of products and their commercialisation (more than ten years for a drug). Generally speaking, the creation of a start-up requires a sufficiently important innovation with a significant improvement in performance or some novel use to make up for the additional costs due to the development of the new technology. Otherwise, the continued improvement of existing products or processes will always remain more competitive.

Note that start-up companies are present across all industrial sectors, with activities ranging from the production of nanomaterials to the fabrication of devices. The activities of these small dynamic enterprises can be classified into four categories:

- The production of basic nanomaterials, such as ceramic or TiO_2 nanoparticles.
- The formulation of nanomaterials in a mixture, e.g., a dispersion of carbon nanotubes in a liquid phase to make resins, paints, inks, and so on.
- The fabrication of nanodevices to analyse the composition of fluids, in particular, air and water, or indeed process them.
- The development of instrumentation for nanofabrication and nanocharacterisation with systems integrating optical components.

The activities of start-ups and the use they make of nanotechnology can be illustrated by two examples, one concerned with medicine, the other with the environment.

- The flagship product of the company Nanobiotix is NanoXray, designed for the treatment of tumours. The product consists of hafnium oxide nanoparticles of diameter around 50 nm which are injected into cancer cells where they accumulate by virtue of their specific size and an appropriate choice of chemical ligands grafted onto their surface. The point about these hafnium oxide nanoparticles is that they emit very large amounts of electrons when exposed to ionising X-ray radiation, thereby amplifying the lethal energy dose released within the tumour. This enormously increases the efficacy of the radiotherapy, even for a very low dose of X-rays passing through the healthy tissues.
- The company Ethera produces and commercialises the Profil'air® kits, which can provide a fast and accurate analysis of indoor air pollution thanks to their reliable detection of very small amounts of chemical pollutants, and in particular certain volatile organic compounds such as formaldehyde or benzene. Different kinds of sensor (the consumables here) are offered. The basic Ethera technology consists of nanoporous materials, synthesised using a sol–gel process (see Chap. 5), which trap the target pollutants through the specific reagents incorporated within them. Initially transparent, they change colour depending on the concentration of the contaminant. The result is read visually or using an optical device. This technology can thus provide sensitive, selective, and fast sensors.

16.1.5 But What About the Risks?

At the present time, the possible risks in the use of nanomaterials in industrial products remain rather poorly understood, and often poorly quantified (see Chaps. 12 and 13). Faced with the risks that nanomaterials may represent in terms of toxicity and ecotoxicity, and to meet the requirements of new laws introduced by the *Grenelle de l'environnement* 1 and 2 (round table held in 2009 and 2010), the French ministry responsible for ecology, sustainable development, and energy adopted new regulations which came into force on 1 January 2013, making it compulsory for industrial actors to declare the products containing the nanomaterials they produce. On 30 June 2013, more than 900 companies made 3500 declarations. This is a world first. It remains rather complex to implement but should provide a way to register and trace nanomaterials across the various industrial sectors and react in the best possible way whenever a risk is identified [5] (see also Chap. 17).

16.2 The Industrial Situation Worldwide

16.2.1 A Growth Market

It is no easy matter to evaluate the world nanotechnology market. A first difficulty is to dissociate those companies directly concerned with nanotechnology from those that merely integrate nanomaterials (nanoparticles, carbon nanotubes, etc.) or nanodevices (microprocessors, semiconductor lasers, etc.) in finished products without requiring any strictly nanotechnological competence. Industrial actors are usually categorised according to whether they produce:

1. nanotools for fabrication, characterisation, or modelling,
2. nanomaterials,
3. intermediate products with nano characteristics (coatings, textiles, optical components, etc.),
4. finished products using nano ingredients, such as microprocessors, for example.

Note, however, that there is no clear dividing line between categories (3) and (4). Another difficulty is to distinguish innovative products drawing in some way upon nanotechnology, referred to as nano-enabled products, from those for which the nano element is not the key ingredient. It may also be that, depending on the attitude of the consumer, the nano element may be emphasised or, on the contrary, played down. From this angle, the French regulation initiative (see Sect. 16.1.5) should improve matters in the long term.

Figures Most Often Cited

Three series of figures concerning the nanotechnology market are among the most often cited on Internet.[1] One of these speaks of a market amounting to 500 G$ in 2010 with a forecast of over 900 G$ in 2015. The second also estimates the market as around 500 G$ in 2010, but forecasts 2500 G$ in 2015. The last predicts a growth from 20 G$ in 2011 to reach 50 G$ in 2017. All these studies thus predict considerable growth in the nanotechnology market, but it is clear that they are not referring to the same things. The first two take finished products into account, while the last would appear to exclude them. The difference between the forecast evolutions in the first two studies can probably be explained by the numbers of finished products they consider. To get round these apparent contradictions, the problem can be approached differently.

Figures to Be Retained

There are two well defined industrial sectors concerned with nanotechnology: microelectronics and optoelectronics, in which fabricated components have at least one nanometric dimension (see Chap. 2). The market for logic or analogue integrated

[1]The figures given in the following are drawn from general studies, one by the Helmut Kaiser Consultancy in Tübingen (2009), another by BBC Research 2012, and a third by Lux Research, Inc.

circuits, memories, and microprocessors oscillates around 250 G$, of which 20 % for the latter, with a slight progression of a few percent between 2011 and 2014. The optoelectronics market amounted to 26 G$ in 2012 and should reach 29.5 G$ in 2014–2015.[2] Asia, excluding Japan, covers 56 % of the market, and the United States, Japan, and Europe, 18, 14, and 12 %, respectively.

The figures for the other sectors are lower and less certain. Concerning nanomaterials, the figures cited for 2009 range from 1 to 1.5 G$, with strong growth to reach 3–5 G$ by 2015–2016. Among the nanomaterials, carbon nanotubes represented a market share of 0.2 G$ in 2011 (not to be confused with the tyre market, evaluated at around 126 G$!). Figures of 2 G$, certainly overestimated, have been announced for 2020. Concerning nanochemistry in general, it is very difficult to put forward any figures. In a sense, all chemistry is bottom–up, hence nano! Furthermore, it is sometimes impossible to make the difference between nanometric and micrometric sizes in the production of powders (see Chap. 8).

Regarding nanomedicine, figures of the order of 1 G$ have been put forward for drug nanocarriers (not to be confused with the global nanomedicine market, evaluated at 53 G$ in 2009—see Chap. 10). The annual growth of the global market is estimated at more than 13.5 %, implying a figure of 130 G$ for 2016 (certainly exaggerated).

The market shares corresponding to the instrumentation sector are hard to evaluate, because materials are not always specifically 'nano'. For example, should one include all the thin layer deposition systems, or all the ion implanters (see Chap. 2). However, for information, the world microscope market was slightly more than 1 G$ in 2010.

In fact, most markets involving nano elements are now closely linked to the exceptionally fast development of information and communication technologies, even though the progress in elementary components and instrumentation is driven for the main part by nanophysics. In the coming years, considerable growth is expected in the medical sector and materials in relation with nanochemistry and the nanobiosciences.

16.2.2 State Involvement

Nanotechnology is closely linked to R&D activities in many sectors and benefits from significant government funding, sometimes in the form of subsidies.[3] These subsidies themselves are often the result of agreements between countries within the framework of the World Trade Organisation (WTO), which negotiates compensatory measures.

---

[2]The figures for integrated circuits and optoelectronics were communicated at the European Microelectronics Summit organised in Paris by SITELESC in 2012. The other figures are to be found on more specialised sites, accessible on Internet.

[3]The data in this section comes from a presentation at the OECD/NNI symposium in Washington in March 2012, for which the text is accessible at the National Nanotechnology Initiative (NNI) website [6], and a presentation by Lux Research at the EuroNanoForum 2001, for which the slides can be consulted at the website [7].

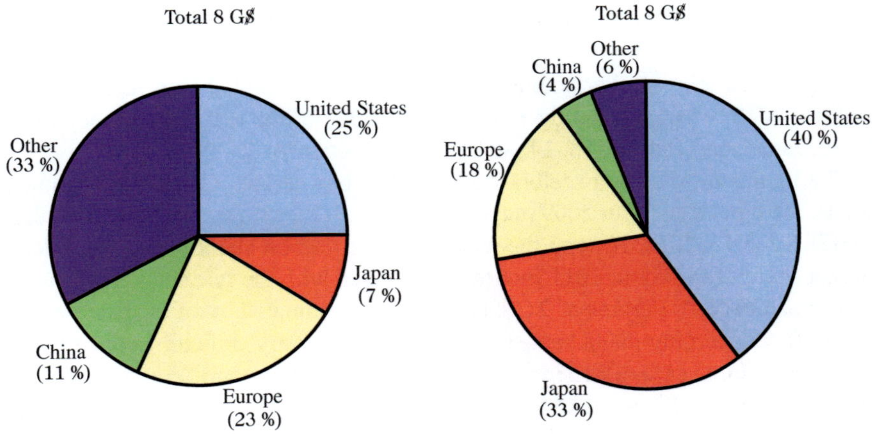

Fig. 16.6 Distribution by country and/or region of government funding (*left*) and self-financing by private entreprise (*right*). Adapted from Lux Research B.V., EuroNanoForum 2011

There are three main types of financing: public, private, and funds raised by risk capital. The distribution of government funding and financing through the equity capital of private enterprise is shown in Fig. 16.6 for different countries and/or different regions of the world. Russia (we shall say more later) and South Korea come under the heading 'Other' here. As we can see, the total of government funding, of the order of 8 G$ a year, is equivalent to the total investment by private enterprise. For its part, risk capital comprises a much smaller share, less than 1 G$, or about 6 % of all investment, which would suggest that most industrial developments involving nanotechnology spring from well established companies. Note also that the share of risk capital is tending to fall and concerns primarily the United States. Regarding government funding, the United States and Europe make the biggest effort, each putting up around 2 G$ a year.

In the United States, it is the National Nanotechnology Initiative (NNI) which coordinates the actions of the main US bodies,[4] with a rather constant budget of 1.8 G$ in 2012, 1.65 G$ in 2013, and in 1.7 G$ in 2014. The NNI also organises public outreach (see Chap. 17).

In Europe, part of the financing will now depend on the pluri-annual Horizon 2020 program which takes over from the 7th Framework Programme for Research and Technological Development for the period 2014–2020. The new feature of Horizon 2020 is that industrial innovation is one of the three main priorities of the European program.[5] The importance of industrial development is thus clearly advertised with a total budget of 20 G€ over 7 years, split in three directions:

---

[4]It would be tedious to list all the US agencies, but the main ones are cited in Chap. 14. For more information, the reader may consult the website [6].

[5]The two other priorities are scientific excellence (28 G€) and the social challenge (36 G€).

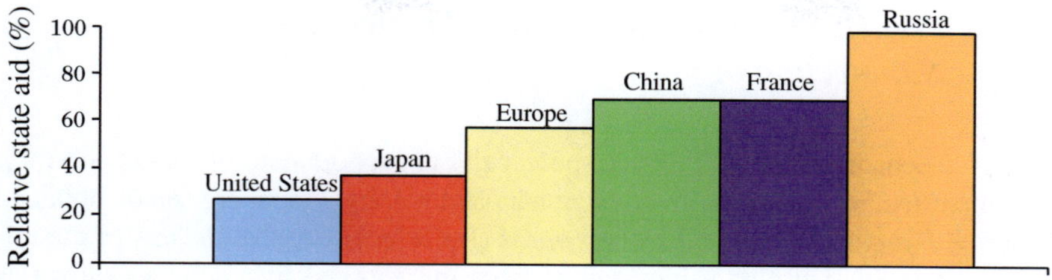

Fig. 16.7 Percentage ratio of governmental and private financing. The world average is 50 %. Adapted from Lux Research B.V., EuroNanoForum 2011

1. Strengthening leadership in generic and industrial technologies, with a budget of 6.6 G€. The generic technologies are identified as micro- and nanoelectronics, advanced materials, photonics, and nanotechnology.[6]
2. Financing of 'risk projects' to a total of 4 G€ across all technologies.
3. Support for innovation in SMEs (0.7 G€).

In addition, the European Institute of Innovation and Technology (EIT),[7] which has become an integral part of Horizon 2020, will bring together higher education institutes, research centres, and excellence enterprises with the aim of producing the entrepreneurs of tomorrow. It will have a budget of 2.8 G€ for the period 2014–2020, compared with only 0.1 G€ per year since its creation.

For other countries or regions around the world, the data is less precise. Some figures were given in Chap. 14.

Concerning private enterprise, it is in the United States and Japan that there is the most self-financing, while in Europe, and even more so in China and the other countries, this self-financing still remains very low or unknown.

The percentage of state aid compared with self-financing by firms is a good indicator for quantifying governmental effort in the field of nanotechnology. Figure 16.7 presents this percentage for the countries and regions of the world considered above, but with the addition of Russia. Note on the one hand the relatively important state effort in China and France, and on the other, the paradoxical situation in Russia, where the government contributes to the development of 1 G$ nanotechnology without apparently stimulating much interest on the part of industry.

[6]There are two other generic technologies, viz., industrial biotechnology and advanced fabrication systems.

[7]This institute was set up in 2008 to facilitate innovation in Europe. One aim is to make the link between scientific research and commercial prospects, encouraging a culture of enterprise in the fields of research and education and helping SMEs to reach a critical mass of users/consumers.

16.2.3 French Provisions for the Development of Nanotechnology

Several mechanisms exist in France specifically to support state-of-the-art industry, and in particular companies involved in nanotechnology. Financing can be obtained from the *Banque publique d'investissement* (Bpifrance) and the budgets of the relevant ministries, but also using interest from the national loan scheme known as the *Grand emprunt*, launched in 2011 (*Investissements d'avenir*). This financial aid is in fact a consequence of actions set up in the framework of the *Investissements d'avenir*, recovery plans (NanoInnov), and the *Agence nationale de la recherche* (ANR). There are two further mechanisms designed to encourage R&D activities in companies: the *Conventions industrielles de formation par la recherche* (CIFRE) and a research tax credit. Finally, in addition to all these mechanisms, there are two particular programs which go beyond the national framework, but for which financial support for the French components are exclusively national, viz., the programs CATRENE and Nano2017.

Financing by Bpifrance

In 2012, three organisations, Oséo, the *Caisse des dépôts et consignation*, and the *Fonds stratégique d'investissement* were merged to form the *Banque publique d'investissement* (Bpifrance), which has thus become a French public enterprise whose role is to support small and medium-sized enterprises (SME), intermediate-sized enterprises (known as ETI in France) and large groups involved in innovation with the backing of national and regional public policy. Bpifrance had a capital of 21 G€ in 2013. Among other things it provides a way to finance innovation, to guarantee company loans, to invest directly in capital equity, and to support exportation. The *Agence nationale pour la valorisation de la recherche* (ANVAR) is now encompassed within this organisation, and a label *Anvar entreprise innovante* has been created [8].

Investment for the Future

In 2011, the French government launched a national loan scheme known as the *Grand emprunt*, the interest from which is used to invest in higher education, professional training schemes, and research, not to mention industry and SMEs. This is the program known as *Investissements d'avenir* (investment for the future). Companies operating in the field of nanotechnology are in fact concerned by the whole of this setup, because they are largely based on progress made in research and R&D centres. More specifically, there are two initiatives to strengthen partnerships between training, research, and industry: the *Instituts de recherche technologiques* (IRT) and the Carnot institutes.

The two main IRTs concerned with nanotechnology are IRT NanoElec, located in Grenoble in the Rhône-Alpes region, and IRT M2P (materials, metallurgy, processes, distributed over three sites, one in Metz in the Lorraine region, another in Belfort/Montbéliard in the Franche-Comté region, and the third in Troyes in the

Champagne-Ardenne region. The partners of these IRTs are industrial companies (more than ten per IRT), higher education institutions such as universities, French *Grandes Ecoles*, engineering schools, and research centres in the relevant regions. Two competitiveness clusters are also partners, viz., Minalogic for IRT NanoElec and Microtechniques for IRT M2P. The competitiveness clusters were set up in 2004 to build ties on the regional level between firms, training centres, and research units. The overall budget for all the IRTs (around ten at the moment) is 2 G€, financed by Bpifrance.

The Carnot excellence label, created in 2006, applies to research structures involved in research partnerships. The geographical distribution of the resulting Carnot institutes is shown in Fig. 14.1b (see Chap. 14). Recently, the Carnot-TIC-MNT alliance was set up to bring together the Carnot institutes with an activity in the fields of information and communication technology (ITC) and micro- and nanotechnology (MNT). This alliance covers all areas of science in the two fields, e.g., CMOS technologies, thin films, optoelectronics, etc., and rallies unique resources in terms of research potential at the European level and worldwide. In particular, it provides for strategic partnerships with other large European centres like Fraunhofter, mentioned earlier, TNO, VTT, and others.[8]

NanoInnov Program

The main aim of the French national program known as NanoInnov, launched in the context of the 2009 recovery plan, is to develop three nanotechnology integration centres in Grenoble, Saclay, and Toulouse. This program has been allotted a budget of 70 M€ of which 46 M€ went to the centre in Saclay. These centres, split into specific platforms, integrate everything from training to research and industrial considerations, with about twenty companies in the Grenoble area, a strong component in power electronics (Freescale) and transport in Toulouse, and the R&D teams of ALCATEL-THALES and HORIBA in Saclay.

ANR Funding

The activities of the *Agence nationale de la recherche* (ANR) were described in Chap. 14. Here we note simply that the projects coming under the thematic P2N program (*Programme Nanotechnologies Nanosystèmes*), active until 2013, were necessarily based on partnerships between research centres and private enterprises. The backing for this program was 22 M€ in 2011. The ANR action plan for 2014–2015, focusing on major social challenges as laid down by the ministry of higher education and research, once again mentions nanotechnology among several of these challenges, especially those relating to energy, industrial revival, health, and the information and knowledge society.

CIFRE

[8]TNO: Toegepast Natuurwetenschappelijk Onderzoek (Dutch organisation for applied science research; VTT: Valtion Teknillinen Tutkimuskeskus (National centre for technical research, a contract research organisation which depends on the Finnish ministry of trade and industry).

Two mechanisms are directly aimed at developing R&D activities in companies and thus promoting innovation: CIFRE and research tax credits. The *Conventions industrielle de formation par la recherche* (CIFRE) already mentioned in the discussion on research and training, provide a way for doctoral students working in private research centres and R&D to benefit from a grant when they prepare their thesis. The CIFRE convention is in fact a three-way agreement between the firm, a research centre or laboratory, and the PhD student, in which the latter obtains a fixed term or permanent contract.

Research Tax Credit

Research tax credit is a tax reduction based on R&D expenditure in a company. The calculation is somewhat complicated, but roughly speaking, this reduction is of the order of 30 % of R&D spending when the latter is relatively low (up to 100 M€). It is capped at 5 % beyond that. The total cost of this tax relief across all branches varies around 3–5 G€ depending on the year. In Europe, only Spain and the United Kingdom have set up similar systems.

CATRENE and Nano2017 Programs

The CATRENE program (Cluster for Application and Technology Research in Europe on NanoElectronics), which will continue until the end of 2015, is part of an intergovernmental initiative by the name of EUREKA, set up in 1985 to support European cooperation in the field of technology. Piloted by the European nanoelectronics industry, it is designed to promote coordinated projects in this sector. There are partners in 19 countries, with 34 % large corporate groups, 39 % SMEs, and 27 % institutes and universities. Its annual global budget is of the order of 4 G€, and it involves 2500 people. For this type of program, financing is done country by country, and in fact through Bpifrance for French partners.

The R&D program Nano2017, which is an extension of Nano2012, associates STMicroelectronics, CEA-LETI, IBM, and a host of industrial partners of the group. The R&D budget is around 3.5 G€. It includes an investment of 1.3 G€ from STMicroelectronics over five years, 600 M€ from the French government, and contributions from local and regional authorities.

16.3 By Way of Conclusion

The above panorama of nanotechnology in France has shown just how many different branches of industry are concerned here. Nanotechnology thus occupies an important position upstream of industrial development in general. But the diversity of applications also means that it can be difficult to specify clear boundaries, or indeed predict the evolution of nanotechnology sector by sector. In contrast to what happens for nanoelectronics, there is no roadmap for nanotechnology in the fields of materials, optics, biology, and medicine, while all the time they become more and more important.

Table 16.1 Number of patents filed in the United States in 2012 in the nano field and the industrial sector in general

Country/region	Nano patents	Total industrial patents	Percentage nano patents
United States	8100	120,000	7
Japan	1950	46,000	4
Europe	1600	34,000	5
South Korea/Taiwan/China	1350	28,000	5

Also indicated is the percentage of nano patents among the total of all patents. Source Nanomagazine.co.uk 2012, based on USPTO figures

While all the indicators suggest a growing world nanotechnology market, it is also difficult at times to find published figures which agree with one another, simply because they do not cover the same revenues or the same products. Indeed it is surprising that, for a global market evaluated in 2013 at around 500 G$, the corresponding global investment should be only 16 G$, including all public aid and funding by private entreprise, a figure which represents less than 4 %. It seems more likely that the R&D investment of innovative companies should be 15–20 %, or even more, so one should remain critical about the published figures. In fact, companies probably do not divulge the most recent data about their financing, in order to avoid speculation on the stock exchange, an apparent connection with governmental aid policy, or possible political agreements which might be reached without their knowledge, as witnessed for example by the problems arising in 2013 between Europe and China regarding the sale of solar panels or top-of-the-range vehicles (*Les Échos*, 2013).

Independently of financial considerations, the vitality of industries dealing in nanotechnology can be assessed by the number of patents filed. While published figures are far from agreeing with one another when we compare different sources, the main trends can nevertheless by identified.[9] As can be seen from Table 16.1 concerning patents filed in the United States in 2012, there is a clear predominance of patents filed by the US itself, with more than 8000 nano patents, followed by a rather uniform distribution among Japan (1950), Europe (1600), and South East Asia (1350). Nanotechnology thus represents an important fraction, between 4 and 7 %, of the patents filed across the whole range of industrial activities. In the European context, France is well placed with 300 nano patents in 2012 for 5000 industrial patents in all, i.e., a ratio of 6 %. France thus holds second place, just behind Germany.

The clear difference between the number of patents filed *in* the United States *by* the United States on the one hand, and *by* European countries or Asian countries on the other hand, clearly attests to a less well developed practice of filing patents, but

[9]The difficulty in actually counting the patents comes from the fact that, in contrast to a publication, a patent is not necessarily valid worldwide. It is filed in the country where one would like it to apply. For example, a patent filed at the European Patent Office (EPO) [9] is only valid in the US if it is also filed at the United States Patent and Trademark Office (USPTO) [10].

also to the greater benefits of filing a patent at the USPTO[10] rather than in Europe. Note that, over the past few years, much progress has been made in France through the application of appropriate research and R&D policies. On the other hand, the level of industrialisation of products, and also the position occupied by France in the European and world markets, does not yet reflect this progression.

References

1. www.nanothinking.com/nanofinder.html
2. B. Nowack et al., 120 years of nanosilver history: implications for policy makers. Environ. Sci. Technol. **45**, 1177–1183 (2011)
3. M.C. Stensberg et al., Toxicological studies on silver nanoparticles: challenges and opportunities in assessment, monitoring and imaging. Nanomedicine **6**, 879–889 (2011)
4. www.nanothinking.com/nanotechmap.html
5. www.r-nano.fr
6. www.nano.gov
7. www.euronanoforum2011.eu/wp-content/uploads/2011/09/enf2011_support-commercialisation_raje_fin.pdf
8. www.bpifrance.fr/aides-entreprise
9. www.epo.org
10. www.uspto.gov

[10]The difference between the US and the rest of the world is much less marked if we consider EPO figures for patents filed *in* Europe where, for 2012 and in the nano field, 850 patents were filed *by* the United States, 950 *by* Japan, and 1050 *by* European countries.

Chapter 17
Societal Approach to Nanoscience and Nanotechnology: When Technology Reflects and Shapes Society

Françoise Roure

Abstract Through its convergence with a whole range of technologies developed in information science, biology, and cognitive science, nanotechnology reflects and shapes society. Because it has a tremendous capacity to transform our perceptions, practices, and representations of the future, it has political as well as economic and social impacts. Making the nanoworld more accessible by informing the public and introducing appropriate laws is an ethical imperative. Apart from education and incorporation in industrial processes and professions, this broadened access is a precondition for a proper public debate on the uses and also the limits of innovation, whenever the latter does not correspond to the general principles of responsibility and precaution. One aim here is to reflect upon the potential of nanotechnology in the context of the transhumanist agenda. It is essential to go beyond the traditional cost–benefit approach to an assessment based on the systemic nature of the risks induced by nanotechnology. The main goal of this chapter is to introduce the reader to this perspective.

The emergence of the term 'nanotechnology' in the 2000s stirred up much social debate of a kind the scientific community was not familiar with. There are many reasons for this discussion around people's perception of the changes brought about by nanotechnology, which may engineer innovation in an incremental way or lead to technological breakthroughs. Some concern the dimension of nanomanufactured materials, others real problems of pollution and public health, with reference to the conditions of production and life cycle of asbestos, for example, and still others an economic competition whose financial promise seem unquestionable.

The need to raise public awareness of these problems has led governments to create regulatory frameworks and research programs and back them up with considerable human and financial resources. However, the various measures still need to be adapted to the realities of the industries they were intended for, and this by making scientific and technical information available to the public and by implementing a

F. Roure (✉)
Ministère de l'Economie, de l'Industrie et du Numérique, 139 rue de Bercy,
75012 Paris, France
e-mail: francoise.roure@finances.gouv.fr

J.-M. Lourtioz et al. (eds.), *Nanosciences and Nanotechnology*,
DOI 10.1007/978-3-319-19360-1_17

dialogue open to all comers through international consensus meetings (ICM), citizens' conferences, and local initiatives like the *Bars des sciences* in France (Cafés scientifiques in the UK and Science cafés in the US).

These initiatives have certainly been useful in many ways:

- They have made scientists aware of the potential impacts of their work, and in particular the risks, when their discoveries are given practical application, and also the perception of these risks by the general public.
- They have led to the creation of registers for declaring product data, as well as bibliographies and webographies.
- They have encouraged the world of industry to introduce some elements of traceability, user guides, and work station classifications dealing with ways to avoid the dissemination of materials in forms that might involve danger or exposure, i.e., a risk.
- Finally, they have built bridges between the scientific community and civil society.

But has this been enough? It seems that much remains to be done. Since the publication of certain works of fiction, sometimes apocalyptic, e.g., *Prey* by M. Crichton in 2002 [1], and until quite recently, newspapers with very wide circulation alternate between extolling the exceptional virtues of new materials like graphene, to which the European Union has recently allocated a billion euros, and detailing the hazards they would represent for the human species if ever they were deployed beyond the scope of socially responsible investment.

It is not easy to settle between the two extremes, but it is crucial to be clear about the important areas in which nanotechnology plays a role that can (and must) be beneficial, and those for which care and control—in the strict sense—must be ensured because effectively they have, both directly and indirectly, worldwide consequences and a systemic long term impact on the economic, ecological, and environmental levels.

The aim in this chapter is to reflect upon the interactions between emerging, converging, and enabling technologies on the one hand, and the multidimensional society on the other, together with their effects on these important subjects, by highlighting the following issues:

- The citizen's desire for information.
- Efforts to set up programs associating research and industrial applications in a spirit of shared responsibility.
- Doubts about the industrial deployment and the use of private capital in this area owing to the uncertainties over the return on investment, given the relative immaturity of nanotechnology and hence the high risks involved.
- Certain futuristic scenarios concerning the possibility that nanotechnology, if it can provide ways to engineer living systems, may produce a bifurcation in the human species.

The conclusion will detail recent developments regarding these different points and present an overview of the international negotiations likely to occur in this field, which involves risks but also significant benefits.

Philosophers, sociologists, and politicians have debated these questions at great length and we shall attempt to outline their various points of view. Some may seem excessive to the world of the exact sciences from which they appropriate the language, translating it into terms accessible to all. Others comment upon the supposed dangers of nanoscience and nanotechnology.

However, it seems important to approach this through a rational dialogue between intellectual communities with different backgrounds and training and the various components of society.

"Man is the only being we know able to have responsibility. And because he is *able* to have it, he has it."[1]

Hans Jonas, *Pour une éthique du futur*, 1993 [2]

17.1 Making the Nanoworld Accessible, Informing, and Regulating: An Ethical, Legal, and Societal Imperative

Modern society demands to be informed about the applications of new technologies, fearing its impacts on the environment, and even more so, on health. This is a perfectly respectable demand and every possible effort must be made by scientists and media to honour it. However, if there are risks, society must be protected by regulations and legal measures. It should be remembered though that even deeper fears, not to be underestimated, are expressed with regard to engineering of living systems, and these must certainly be taken into account. Efforts made by governments, scientists, and associations have attempted to resolve these problems of information, but while commendable results have been obtained, there remains much to do, and it is important to continue to mobilise scientists capable of popularising their work.

There is no point in trying to list all the resources implemented to channel the tremendous amount of work that has been carried out. Here we merely cite a few examples:

- Scientific communication in specialised books and reviews or in the popular media, not forgetting the *Bars des sciences* (Cafés scientifiques) and open discussion areas.
- The main national and international programs.
- The role of consumer protection associations and the issues they raise.

This is what we shall try to demonstrate throughout the rest of this chapter.

[1] *L'homme est le seul être connu de nous qui puisse avoir une responsabilité. En pouvant l'avoir, il l'a.*

17.1.1 Informing

The means of communication available today are certainly responsible for the major revolution in the dissemination of information in the broad sense, i.e., for absolutely everyone, from specialists to laypeople, and from those with good intentions to those with bad. So what are the consequences of this revolution in the area that concerns us here? How should one clearly and honestly disseminate information in such a way as to avoid any risk of its being exploited in an abusive or malicious way? What questions does the layperson ask, far from the world of science, and how can we inform such a person? What are the responsibilities of research scientists? What should be the role of the powers that be?

How Should One Inform the General Public?

What does the nanoworld look like to the lay public, and how can we help them to get their bearings in this infinitely small space? This field, lying as it does between the atom and the micron, is physical very close, even if it remains imperceptible to us: it is on the scale of the cells that make up our own bodies. Originally exclusively natural (volcanic smoke, natural aerosols, dusts, etc.), it is becoming ever richer today with nanoparticles generated by industry (combustion products from the increase in world transport or the return to intensive use of coal) and, recently, artificial objects, nanosystems and machines designed and produced by humans who seek to use nanotechnology as a kind of toolbox.

Naturally, it would be impossible to perceive, represent, or in any sense access this infinitely small space without a certain scientific maturity, and indeed an understanding of the technological and cognitive prostheses currently in use, which some would claim actually extend and increase our human aptitudes. These developments, which would not be without consequence when fixing priorities for responsible research, development, and innovation, raise real problems for society, and the coverage of these areas by sensation-seeking media is clearly causing many citizens great alarm.

But how can one honestly inform the public? There exist specialised publications, magazines that popularise science—but unfortunately too few and with too small a readership when it comes to nanotechnology and its offshoots—, the more general media, and even the social networks. How can one broaden access to scientific and technical culture in these fields, given that members of the public generally have little interest in devoting time to a basic self-training that would turn them into well informed consumers?

This broadening of awareness, this access to understanding of nanoscience and the associated technologies, will require a culture shift and also deliberate action on the part of the powers that be, with significant educational and training efforts. This drive should include the media, which are not necessarily specialised in scientific and technical matters, and which should aim to disseminate information using a genuine scientific approach, rather than a largely emotional one.

We shall see, however, that deep concerns persist over the impact of nanotechnology on humans, and in particular that some excesses are possible, which can be exploited to the detriment of scientific progress.

Making Information Available to a Specialised Public: Programs Devoted to this Goal

Since the beginning of the 2000s, the United States and the European Union have been among the instigators of worldwide nanoscience and nanotechnology programs, backed by considerable human and financial resources, directly under the responsibility of the highest political authorities in the relevant countries. These federative programs are:

1. The National Nanotechnology Initiative (NNI) in the United States.
2. The strategic action plan for nanoscience in the European Union.

They are still running, or extended by the European framework program for research, and assessments and progress reports for the various projects are published regularly. Anybody can consult these. In fact, they are easy to consult on the web at any moment. They cover many scientific and industrial projects, with step by step progress reports. Here is an outline for each, together with a few recent updates.

The NNI Program

Since it was set up in 2000, this has received funding of 15 billion dollars. It brings together the 25 state departments in the US, viz., education, science and technology, health, defence, etc., scientific and technical institutes like MIT, Caltech, etc., through their supervisory ministries, and various public and private organisations. The whole entity reports directly to the President of the United States each year.

It covers the seven themes listed below (themes also appearing in the European action plan):

- Fundamental research on nanoscale mechanisms.
- Systems relating to nanoelectronics.
- Instrumentation.
- Nanofabrication (methodology).
- Industrialisation.
- Environment, health, and safety.
- Education and social aspects.

The 2012 report, entitled Nano2, illustrates recent progress in all these areas. Given the different interests people have for different domains, it is very hard to select one or other among them. Anyway this site can be consulted at any time [3], and it is highly recommended for its incredible wealth and variety, as well as the results obtained, particularly in science and education. What stands out here is the short term aim of transferring nanoscale technologies to the benefit of trade and in the public interest. There is thus a clearly stated desire to profit from this area of science, but without forgetting to recognise the needs of society and the other emerging technologies. However, the latter objective concerns the 2020 horizon, as can be seen from Fig. 17.1.

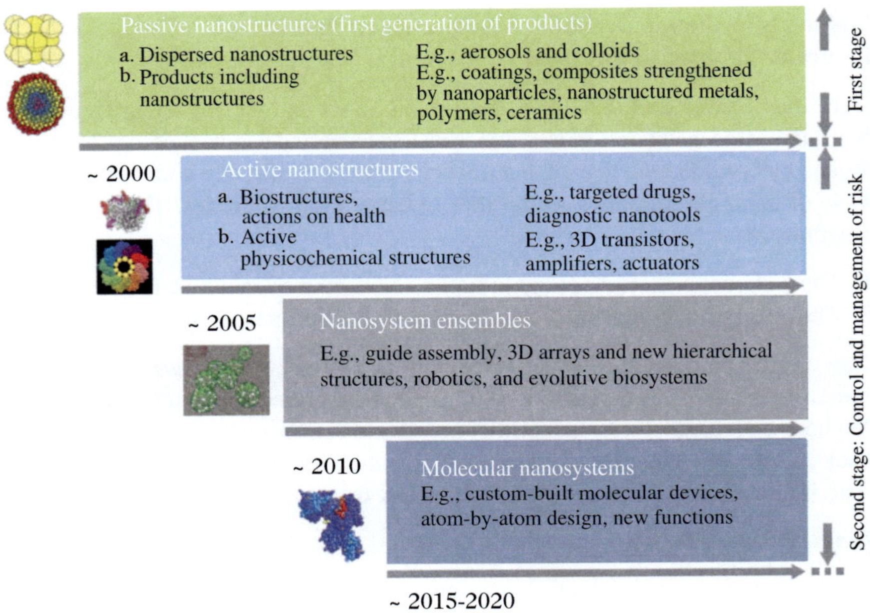

Fig. 17.1 The four generations of nanotechnology products and production processes [4]. Risks and deadlines for potential harm factors. Taken from the NNI program

Observatory Nano

Among the many projects belonging to the European framework program, it is worth mentioning the Observatory Nano, although unfortunately its activities have now been suspended. The directions chosen by the Observatory Nano in 2009 deliberately lean toward more social considerations. There are five themes:

- The ageing population.
- The need to save energy.
- Food security and sustainable development.
- Healthy and intelligent international relations.
- Best use of resources and efficient industrial production.

These are indeed noble objectives, and indeed, much more universal than if they referred only to the nanoworld. But where do things stand today? To promote transparency, information concerning the contribution of nanotechnology applications to each of these societal issues is available and can always be consulted at the website of the Observatory Nano [5] in the recent and extremely detailed progress report published in 2012 under the title *European Landscape Report 2012*. It contains a list of the participating countries, public institutions, private companies, programs now running, scientific and industrial databases, scientific papers, and patents. The conclusions of this report—as in the American report—observe progress in all areas and in particular those relating to energy and the life sciences.

To conclude regarding this work in progress, there is a clearly stated desire on both continents to introduce regulation that will guarantee industrial property rights for inventors and safety for the population.

French Programs

As has been shown by the public debate in France since the beginning of the 2000s, those solicited by the *Commission particulière du débat public sur les nanotechnologies* (CPDP) did everything to persuade the relevant government bodies that they should accept to debate the ultimate goals of nanoscience and nanotechnology. Indeed they considered that, quite wrongly, the only subjects actually debated had been the modalities of their development and incidentally the question of the governance of those modalities. The report made by the *Commission nationale du débat public* (CNDP) to the government on the basis of the work done by the CPDP brought out the need to design and implement a reflexive governance, adapted to what is at stake and easy to see, easy to understand, in a word, comprehensible and useful to all concerned.

The debates organised in the context of the *Grenelle de l'environnement* (a round table on environmental issues) finally led in 2007 to regulatory proposals in an area where the modalities for applying the law are a very long way from being operational. This regulatory framework deserves at least a European approach, with reference to the international consensus within standard development organisations (SDO) and, beyond that, with reference to a cross-disciplinary scientific approach to nanomaterials which remains to be established. This is indeed the price to pay to overcome in the mid-term the commercial, legal, and institutional disagreements over the emergence of non-tariff trade barriers. Without this, they are likely, even inevitable in the short term, as a consequence of the lack of international dialogue and anticipation with regard to the question of governance.

Since 2008, the visibility and comprehensibility of French state action have been focused on the NanoInnov project, with its three geographical sites in the Rhône-Alpes, Midi-Pyrénées, and Île-de-France regions, and with the *Commissariat à l'énergie atomique* (CEA) at the heart of the regional initiatives in Grenoble and Saclay.

French institutional expertise has not been idle in these areas and we may mention the action of the *Agence nationale de sécurité sanitaire pour l'évaluation des risques sanitaires* (ANSES), which coordinates the European project NanoGenotox, involving 18 organisations and 13 countries, whose goal is to identify the toxicity of 14 materials for genes and DNA. Or again, the federative project NANORIS run by INERIS, which aims to assess the risks associated with the production and use of nanomaterials. The role of ANSES will be discussed further below.

Government has long been advised and informed about the technological revolution currently under way and the potential societal and systemic changes that may be induced by human capacity to act on matter, and in particular living matter, with atomic precision. Represented by the *Conseil national de la consommation* (national consumer association), whose administration is entrusted to the *Direction générale de la concurrence, de la consommation et de la répression des fraudes* (directorate general for competition, consumer affairs, and fraud control, DGCCRF), consumers finally agreed to recognise the crucial importance of high quality national, European, and international governance in the emerging field of nanotechnology, in response to a growing awareness of the potential changes they are likely to bring about in the world of consumerism and trade.

It is crucial for the fate of scientific and technical progress in the twenty-first century to create and maintain informed consumer confidence in products resulting from convergent technologies at the nanoscale.[2] Because this progress is linked to the very possibility of progress in general, within the limits of the essence and dignity of the human being, who is as fragile and vulnerable as the civilisation he has shaped, it is not only legitimate, but even eminently desirable that consumers and citizens should take into account ethical, legal, and social aspects of nanotechnology. Since these issues go well beyond the governance of private enterprise and national authorities, both now and for some time to come, it must be recognised that international institutional dynamics provide a relevant and indeed indispensable space for the exercise of public responsibility in these areas, while respecting the principle of subsidiarity.

Other information programs have been implemented on the national level (see Chaps. 14–16).

17.1.2 Regulating

In a world where competition between humans still predominates over cooperation and tends to be transformed into a profitable market, the gradual determination of rules for authorised and/or prohibited use is currently under way at both the national and international levels. This is attested by the disparity between the situations in different European countries when it comes to the implementation of laws regulating the fabrication and use of materials and objects arising from nanotechnology, as can be seen from Fig. 17.2, taken from the website [6] of the Observatory Nano as of 20 May 2013.

We shall not describe in this section the countless debates over the many requests for regulation formulated by consumer associations, because it seems that decisions have finally been made.

Regarding regulation, we shall mention two documents published in January 2013. The first (number 1 below) was produced by the United States Environmental Protection Agency (EPA) [7], and the second (number 2 below) by the Council of Europe (CoE) [8]. They review the steps taken on the two continents to regulate the use of nanomaterials and nanotechnology:

[2]Convergent technologies at the nanoscale provide ways to combine components involving at least one nanoscale dimension in the observation, simulation, manipulation, and production of nano-objects, nanomaterials, or nanosystems, whether these components are specified by their physical, chemical, or biological properties, or by a combination of them.

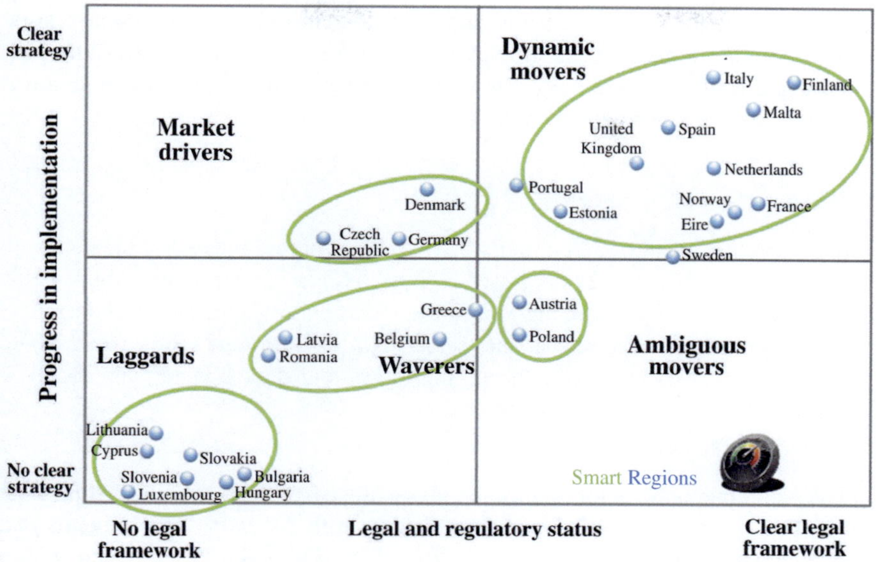

Fig. 17.2 Disparity in the determination to regulate nanotechnology within the European Community

1. The American document stipulates that any person who intends to produce, import, or distribute nanomaterials or products containing them in the United States must inform the EPA 90 days prior to any such action. The information that must be supplied concerns production volumes, the processes used, and any available data concerning health and safety. These elements should allow the EPA to reduce risk to individuals and the environment by authorising or prohibiting their use. The regulations have been applicable in the United States since July 2013 (publication of the EPA bulletin).

It is important to note that, in this same context, France has established since 1 January 2013 the obligation for all stakeholders (producers, exporters, importers) to declare all substances with dimensions from 1 to 100 nanometres to provide better information about commercialised substances in the nanoparticle state and their uses, and also to provide traceability within the sectors where they are used and a better knowledge of the market and the volumes on the market. This will also provide a way to gather information about their toxicity and ecotoxicity, which is important for research and also for the assessment of possible risks and the determination of preventive measures. This initiative is a first in Europe. Will it be followed by other governments?

2. While the second document, adopted by the relevant committee of the CoE on 19 November 2012 and officialised on 17 January 2013, contains similar elements to the American directives, it is nevertheless less directive, more timid, and less precise regarding the fabrication and use of nanometric objects. Articles 1–4 list the (very) general commitments of the committee to establish regulations

guaranteeing the safety of citizens and the need to introduce international regulation. Article 5 is clearer. The stipulation in Sect. 5.1 is to respect the precautionary principle while taking into account freedom of research and encouraging innovation. Restriction or freedom? Who decides?

The Position of Vivagora

In a press statement of 22 February 2010, the day before the end of the public debate on nanotechnology organised by the Commission particulière du débat public chaired by Jean Bergougnoux, the association Vivagora used the concept of an 'authorisation for release into society' (*autorisation de mise en société*) when the changes induced by emerging technologies have significant social consequences, or are perceived to do so.

The recommendations for (nano)bioethics in Sect. 6 of the text are also somewhat vague, but at least they have been formulated: "The Assembly recommends that the Council of Europe's Committee on Bioethics (DH-BIO) be entrusted with a feasibility study on the elaboration of possible standards in this area."

In conclusion, given the complexity and the long delays required to carry out research into the potentially harmful effects of nano entities, the priority today is clearly given over to economic considerations and the distribution of profit. But we may hope that the above regulations can and will have some effect in encouraging the acquisition of reliable and well documented knowledge about the real effects of nano substances and the products made from them.

This suggests that certain areas should still be treated with extreme vigilance: the risks to personal health and to the environment, and especially the risks involved in manipulations of the human species. Ethical rules must therefore be established to avoid abusive use for criminal purposes.

Regulating Derived Usage: Electronics

While it is clear that microelectronics, since the 1960s, and nanoelectronics, since the 1990s, have greatly extended our means of expression and communication, even while itinerant, broadening access to information and knowledge, they have also opened the way to uses that were not part of the original intention, in fact, uses that the law, morality, and ethics reprove in some of their applications. They have contributed to the traceability of individuals by cross-referencing between civil records and biometric data, or indeed the traceability of communications and movements by geolocation and by profiles established from traces left by Internet users in the cybersphere. The civil society movement known as Do Not Track (DNT), relayed by the bill 'Do not track online' put forward by the US Senate in May 2011, warns Internet users that, when the regulatory framework is out of phase with civil society, it will be hard to implement such regulations. The same warning was also expressed by the Court of Justice of the European Union in its judgement of 24 November 2011, when it clearly stated that the Internet Protocol (IP) address used to distribute electronic mail on Internet is protected personal data.[3]

It is also worth mentioning the work by the eminent economist and sociologist Vance Packard who, back in the 1950s, had drawn the attention of the American authorities to techniques of mental and psychological manipulations using subliminal messages on the television. We may cite two of his writings: *The Hidden Persuaders* in 1958 [9] and *The Naked Society* in 1965 [10].

17.2 The Scope of Social Disruption Induced by Nanotechnology

Although priority has been given to the trade aspect, the NNI and Observatory Nano have nevertheless fixed guiding principles for the 2020 horizon. These concern future production units, the contribution of nanotechnology to environmental questions, including energy, new frontiers in biosystems and medicine, information technologies, new materials and the associated systems and processes, teaching, research, and international regulation.

There are two main worries: the manipulation of the human genome (pushing for new frontiers?) and protection of the environment. In both cases, the question is whether nanotechnology might affect the evolutionary process or lead to mutations without prior assessment of the consequences?

17.2.1 Manipulating the Genome

From Nanotechnology to Synthetic Biology

In 1978, in an editorial of the journal *Gene*, Waclaw Szybalski wrote [11]:

> The work on restriction nucleases not only permits us easily to construct recombinant DNA molecules and to analyze individual genes, but also has led us into the new era of synthetic biology where not only existing genes are described and analyzed but also new gene arrangements can be constructed and evaluated.

---

[3]Decision of the European Court of Justice on 24 November 2011, prohibiting systematic filtering for Internet.

Extending this observation concerning the birth of synthetic biology some thirty years on, the *Journal officiel de la République française* (JORF) defined it in May 2013 as follows:

> […] the cross-disciplinary branch of molecular biology which uses physics, computing, and chemistry to invent genomes, obtain new enzyme reactions, and create cells capable of novel functional and metabolic activity.

We realise here that, on the one hand, synthetic biology came into being before nanoscience and nanotechnology had made their official appearance on the scene, and on the other, nanotechnology can provide it with much greater resources for its development.

This discipline and those associated with cognitive technologies are the ones that deserve the full attention of the powers that be, and this at a stage well upstream in the processes of research, development, and technological innovation, because they may bring about irreversible changes in society and the dividing line between their use and their misuse must necessarily be debated, established, and strictly controlled.

In theory, synthetic biology breaks with the Darwinian representation of a human species passively undergoing the consequences of the theory of evolution and those of its more or less hostile natural environment. It builds up gradually, starting from systems biology, and passing through molecular biology to epigenomics. It emerges from this as a genuine breakthrough technology. With bioinformatics, it benefits from the resources of digitisation and simulation, exploiting today's unprecedented technological capacities for the organisation of information, which combine with the ability to observe, interpret, understand, and manipulate matter on the nanometric scale.

The resulting professionals will be architects of living systems and engineers integrating fabrication processes to begin with, but then industrial producers, with important consideration of ethical, quality, hygiene, and safety issues. Future companies involved in synthetic biology should design and implement a precisely adjusted governance regarding their social and environmental responsibility, so that they may engage in responsible innovation. Without this, they may find that they have no sales prospects whatever in the eyes of the general public and disappear through lack of growth (and hence also, lack of funding for research and innovation).

Three Complementary Approaches to Synthetic Biology

These may be described as follows:

- *Simplification of living systems down to the smallest structure capable of duplicating itself*, and hence of efficiently integrating a genetic code that will be transferred during duplication. Starting from this elementary organism, one then adjoins and perfects more and more complex functionalities.

- *Recomposition of naturally existing living systems from basic building blocks*, which are simplified, standardised, reproducible, and can be controlled by humans. In view of the first results, the engineering of artificial genetic circuits with programmed behaviour presupposes such powerful software capacity and mastery of complexity that the transition from the laboratory to mass production will necessarily be very slow, unless we are able to standardise processes on a massive scale and opt for an open and cooperative approach.
- *Creation of new information codes for the species to be invented*, hybrid or otherwise, building upon the genetic basis provided by DNA, RNA, and the amino acids making up proteins. Here, the leap into the unknown may have unpredictable consequences, accelerating or producing bifurcations in otherwise slow processes or mutations of species.

Examples of Developments in Synthetic Biology

Here we give three examples:

- Synthetic artemisinin is an active principle against advanced stages of malaria, which went into industrial production by Sanofi in 2013 on the basis of a research partnership with the University of California at Berkeley (laboratory run by Jay Keasling). This application, which can produce the drug at a relatively low cost, is considered as a triumph of synthetic biology in the United States, with considerable social benefits, while the communication of this success in France does not employ the term 'synthetic biology'.
- The relatively cheap detection of arsenic in water using sensors before it reaches the toxicity threshold may concern some 10 million wells in a country like Bangladesh.
- The detection using bacteria of certain explosives is potentially useful for locating land mines.

From Nanotechnology to Epigenetics

The information codes used by living systems have gradually been revealed in their remarkable diversity by genome sequencing, but with increasing rapidity and much reduced and fast decreasing cost (see Fig. 17.3).

Research scientists have been investing human genetic mutations and the roles played by heredity and the environment in these changes, which are observed after the event and not deliberately chosen. Here, when we consider techniques of laboratory observation and analysis developed from progress in nanotechnology over the last twenty years, and the improved performance of other investigative techniques, combined with the power of bioinformatics, all these factors taken together have allowed researchers to validate the transmissibility of mutations generated by the environment. The publications of Eva Jablonska at the University of Tel Aviv already envisaged this type of mutation as early as 1992 and it was confirmed scientifically by recent work at the *Institut national de la recherche agronomique* (INRA) in France in 2012.

These reproducible in vitro experiments seem to demonstrate epigenetic heritability—the possibility of epimutation, or more specifically, a situation where not all features of living beings are transmitted by the genes—an observation with radical consequences. Evolutionary change can thus take place much faster than was generally accepted. While in the second half of the twentieth century, evolutionary biology was dominated by the synthetic (genetic) theory of evolution, in which the genes represent the exclusive means for transmission of information from one generation to the next, it turns out that non-genetic mechanisms are also able to code heritable information.

The synthetic biologist Drew Endy, who designed and set up the Registry of Standard Biological Parts at the Massachusetts Institute of Technology (MIT) to support the techniques of nanomanufacture of living systems developed by Christina Smolke at the University of Stanford, has reached the same conclusions about epigenomics. He adopts an engineering–architectural approach to living organisms, starting out with the basic building bricks of living systems (DNA, RNA, proteins, etc.), rather than a genetics approach.

Once the mechanisms of genetic evolution triggered by the environment have been fully identified, humans with their boundless imagination will want to put this knowledge to use, just as they have always played with what was accessible to them, sometimes at their cost. But this raises a question that would have been quite unthinkable until now: the question of a choice, and hence a responsibility, regarding the future pathways for the evolution of living species in general, and the human species in particular. A choice also regarding the speed with which we would like

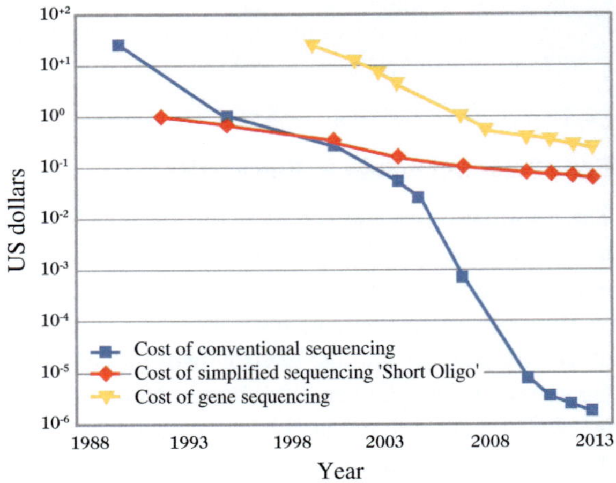

Fig. 17.3 Fall in the cost of sequencing and genetic synthesis over the past few decades

to follow these (technical) side roads and the conditions, notably ethical conditions, we should commit ourselves to in this in principle irreversible process.

With synthetic biology, the only constant would be change, and the only variable, the rate of this change, for which a determining parameter would be the regulatory framework we choose to apply.

It is precisely the implicit refusal to even discuss this prospect, i.e., deliberately engineered genetic mutations, that explains the violence with which certain stakeholders prevented several meetings of the national public debate on nanotechnology from taking place; nanotechnology whose aim in their view would be nothing other than choosing the future of the human species.

17.2.2 Toward the New Green Deal?

There can be no doubt that energy will be a major area of activity for nanotechnology. As for medicine and pharmacy, access to energy is a prerequisite, without which our safety and quality of life can no longer be guaranteed. As time goes by, the trend toward democracy and technological dependence have gone hand in hand. The transition from dependence on human energy, the typical situation in the nineteenth century, with the industrial revolution, to the use of new techniques of mechanisation or rationalisation of labour, even though they were implemented to raise productivity, has led to a reduction in the arduousness, monotony, and repetitiveness of the tasks accomplished by workers, and it has changed modes of governance within society.

To satisfy their energy requirements and other fundamental needs, societies have designed and adopted technological solutions whose negative consequences are accelerating anthropic climate change, as well as causing the extinction of many species and a drastic reduction in biodiversity. The sixth mass extinction is well under way and, for the General Secretary of the United Nations Ban Ki Moon, we are keeping our foot on the accelerator simply because we are unable to overcome our egoism and thereby avoid a situation where human industrial activity finally turns against us.

Technology is a powerful driving force in the relationship between humans and their environment, and increasingly, in the relationship that humans have with themselves, while the final goal remains for us to decide. Hence, a determined and concerted public action for a new green deal,[4] i.e., a new social contract for sustainable development, requires investment in technologies which will allow us, by substitution, to reduce greenhouse gas emissions produced mainly by burning fossil fuels (coal, petrol, and gas), to capture those that cannot be suppressed, and to increase energy efficiency and economy.

The new green deal is an approach first developed in the 1980s by the American economist Vance Packard. It associates the economy and the environment with

[4]This is in reference to the New Deal, which was the recovery plan launched by President Franklin Delano Roosevelt in 1933 to take America out of the 1929 depression.

the principal aim of controlling carbon dioxide emissions through the development of renewable energy forms, limiting the risks for ecosystems, and generating new professions across a whole range of different areas, and all this in the service of the general social well-being and through the resulting boost to the economy. These basic principles have already been implemented in many countries and their effectiveness is currently being monitored.

While they may not represent the only contribution, we may ask whether nanotechnology and synthetic biology offer prospects which could, either alone or combined with technologies like the information and communication technologies (ICT), make a real difference to the carbon balance of the planet through photovoltaic applications, the development of biofuels, smart meters and networks, but also by better sharing the access to energy resources. Indeed, theoretically, they should effectively lead to savings with regard to the extraction of primary raw materials and they release us from the constraint imposed by rarity. In this respect, the applications of nanotechnology to renewable energy would completely shake up the economic models arising from subscriptions paid by long term customers. Through their convergence, they would also rearrange the geographical disposition of the world's industry, because they would provide low cost solutions with local production and short distribution channels, based on open software architectures.

Contributions of Nanotechnology to Reducing Greenhouse Gas Emissions

Here we list four areas where techniques are already operational or under test:

- Energy: high efficiency photovoltaic cells, thin film nanomaterials, batteries for electric vehicles, biofuels.
- Transport: capture and storage of carbon dioxide, nanocatalysts.
- In industry: insulating glasses, heat pumps, lighting.
- In the home: insulating and energy recovery materials, photovoltaic materials incorporated into surfaces (without the solar panel frame).

Contribution to Energy Production from the Biomass and Synthetic Biology

The production of biofuels not competing with food resources is also under test. This is the case for production of isobutene obtained experimentally by setting up an artificial metabolic pathway exploiting novel enzyme activities in microorganisms. Isobutene is a gaz that can be converted into liquid fuel and various polymers, and the transition to industrialisation by the company Global Bioenergies, based at the Genopole site in Evry, is supported by Bpifrance (OSEO). An example of a biofuel produced by the biomass is the fibrous plant matter or bagasse which remains once the juice has been extracted from sugar cane. Using 90 % of this plant matter rather than the current value of 10 % in combustion techniques, many island or subcontinental populations could be self-sufficient in a sustainable, clean, and renewable energy (Haiti, Antilles, the Carribean, islands in the Indian Ocean, the China Sea, and the Pacific Ocean with tropical vegetation, but also Brazil and equatorial Africa, etc.).

Note that around 1.5 billion people do not have access to electricity. Such off-the-grid populations must depend permanently on solutions that are not connected to major transport and electricity distribution networks in order to reach the minimal standards of human dignity, such as access to drinking water by extraction, conveyance, and filtering techniques that do not depend on a centralised supply of electrical energy, but also fundamental rights including the possibility of remote communications.

17.2.3 Developing Responsible Innovation Upstream of Supply

Nanotechnology brings us new scientific knowledge about the singular behaviour of matter, and if this allows us to imagine novel forms and substances in nature, its applications must be fully understood and properly controlled, especially when interaction with living systems is one of the desired results. But the phenomena of affinity, metabolism, and translocation within the human body and dissemination in the environment are far from being perfectly understood and controlled. With the exception of therapeutic applications where the cost–benefit ratio must be interpreted in the light of a precise prognostic, a systemic and long term approach must be integrated as far upstream as possible in the design and formulation of the conditions of use at every stage in the life cycle of products incorporating manufactured nanomaterials. Concretely, this is how responsible innovation can play a positive role in the real economy.

17.2.4 Potential Risks and Gains of Nanotechnology Through Biogenetics

The risks induced by nanotechnology can be identified. They are biosafety, biosecurity, intellectual property in a field where patents that are too broad, dealing with critical production processes, would block any possibility of responsible innovation at a reasonable cost. The International Risk Governance Council (IRGC) also identifies the ethical questions with the difficulty of separating natural risks from risks induced by industrial synthesis [12].

Nanotechnology offers the prospect of a personalised offer, for better or for worse. Apart from nanomedicine as we conceive of it through applications like tissue regeneration, what is positive here lies in a range of bio-inspired or bio-enhanced materials, or complex hybrid systems, implanted or integrated like permanently incorporated prostheses, whose functionalities allow them to modulate their effects depending on how they interpret the environment. These systems, which have a certain similarity with robotics, can be autonomous or remotely controlled by dematerialised electronic communications, e.g., to modulate a dose or trigger a natural production (one may think of insulin here, but it is not the only possibility).

An international company which has made integration of nanoelectronics into a strong comparative advantage is considering the possibility of endowing mobile phones with systems interacting permanently with the human body, including the option of 24/7 remote data transmission to health systems and medical emergency services. Implanted or hybrid systems could then benefit from surveillance in the same way as vital biological parameters. The rapidity with which information could be acquired about inception of disease, combined with personalised medicine integrating physical, chemical, and biological properties, may make it possible to curb a pandemic in its early stages, or reduce the resulting morbidity, at the point where traditional pharmacy produced by synthetic chemistry encounters a conceptual difficulty and hence reaches a ceiling.

In any case, the discovery and elucidation of metabolic mechanisms, the transmission of information in the cell, and regulatory gene expression networks are fundamental for a synthetic biology whose ambition is to read more, and write more and better genomic sequences, i.e., artificial information code, using the basic building blocks supplied by life. At least, this is the opinion of the director of the Beijing Genomics Institute, Huanming Yang. Considering that life's blueprint, the system of DNA, genes, and genomes which governs every single living entity, was written down 4 billion years ago, he asserts that it is time to rewrite the program [13]. This is also the goal of the International Genetically Engineered Machine (iGEM) competition, at which French teams were awarded prizes in 2011.

The International Genetically Engineered Machine (iGEM) Competition

Set up in 2003, the aim of this competition is to have teams of students build and operate simple biological systems inside living cells, each time with a specific goal, and this using standard biological building blocks supplied by the Registry of Standard Biological Parts. Two French teams were awarded a gold medal at this competition in 2011, a year which saw the involvement of 160 teams from 30 countries, representing more than 2 000 people (Lyon-INSA-ENS was rewarded in France) (Fig. 17.4). The competition was started by three engineers, Tom Knight, Randy Rettberg, and Drew Endy, two working in computing.

Fig. 17.4 In 2011, the Hong Kong University of Science and Technology (HKUST) held the first Asian meeting for the International Genetically Engineered Machine (iGEM) competition. The meeting brought together more than 40 teams from different universities on the Asian continent. This competition was held for the first time in 2003 at the Massachusetts Institute of Technology (MIT)

As an example of the convergence of technologies at the nanometric scale, an arsenic biosensor was developed in the context of this competition.

Regarding the risks, we shall give just three examples of the potential impact of nanotechnology and synthetic biology.

First Example

Do they open the way to eugenics on a grand scale, with the prospect of achieving some perfect beauty (Apollo!) and even immortality? Naturally, there are many real and perfectly contemporary problems of differentiation by wealth and by values.

While this goes beyond the strict context of nanotechnology, but without excluding it, it should be borne in mind that disparities in the face of the genetic lottery are less and less well tolerated in countries where risk aversion grows with the economic cost of care for difference (handicap, dependence, etc.). What would society be like without physical imperfection, without weakness and pain, without handicap and without temporality? Such a society would probably be unbearable, governed by a requirement of absolute conformity, under the guise of free choice, appearances, and behaviour. Such a society would deny the very possibility of contemplating diversity as an asset, perceiving it rather as a threat. Paradoxically, the promise of a more freely chosen future (at the meeting point of libertarian and transhumanist thinking, which we shall discuss later, and according to which the right to individual self-determination should not be restricted by institutions) would lead precisely to voluntary selection practices, themselves based on coercion: standardised elimination or enhancement. Speaking of this, surrogacy, at the very heart of a topical societal debate at the present time, involves the major risk of making the transition, through individual choices and without an informed public debate regarding the long term systemic consequences, enlightened by ethical principles, from the biological child to the biotechnological child, some of whose characteristics could be selected from a catalogue, thereby extending a long established veterinary practice, these characteristics being verifiable during a prenatal 'quality control' as a basis for decision-making. Would the acceptance of chance-like differences be the same once such practices had spread through society? Technological convergence on the nanometric scale would then have the in principle uncontrolled effect of causing a societal divergence, bringing with it conflict and costs.

Synthetic Biology: Open or Closed?

There have been two proposals regarding access to information about the products of synthetic biology. The first advocates standardisation and free access to genetic materials without any form of code. The second would restrict access, advocating non-disclosure of genetic codes with arguments of intellectual property or else for reasons of safety and security, e.g., to block the reproduction of viruses or other pathogens.

The combination of synthetic biology and nanotechnology yields a highly complex situation for the analysis of deliberate or unintended toxicity. In return, however, it supplements current industrial technologies in three different ways:

- A systemic culture coming from biology, encouraging consideration of systemic risk from the moment a product is first conceived. For example, the late Henri Laborit, author of the visionary works *La Vie antérieure* in 1989 [14] and *L'Homme imaginant* in 1970 [15], was inspired by the endocrine system, a map of which occupied a whole wall of his study. This provides a didactic example of a complex dynamic system which will perhaps never be completely decoded.
- Simulation tools and language which, with bioinformatics, underpin this emerging field of synthetic biology. Taken together, these make it possible, without necessarily carrying out any laboratory work, to fully imagine the scenario in an informational context and to select, responsibly and in good conscience, the in vitro and even the in vivo stages, well before the legislator comes on the scene, compelled by some public health scandal, to determine the boundaries of the framework in which the products can find their way to professional or mass market applications.
- Intimate knowledge of the risks of dissemination and proliferation, inducing cautious behaviour, protective toward people and goods, and standardised (hence auditable) in laboratories and production units. Research laboratories working on the prion and neurovirology, set up by the CEA at Fontenay-aux-Roses with the backing of the Europrion project, have shown the way in this respect. For the record, this laboratory seeks to understand the mechanisms

and scenarios of translocation and differentiated activation of the various components of the prion in the brain. This research arose from scientific questions and public consternation over the Creutzfeldt–Jakob disease (also known as mad cow disease), which in the end only had a very small number of human victims, considering the amount of livestock that had to be destroyed, but whose long term effects and activation processes await more accurate determination. The prion, a generic term covering a wide variety of very small biological entities, is suspected in its various forms of playing a role in triggering degenerative brain disorders like Parkinson's and Alzheimer's diseases, although no formal scientific proof has yet been established.

Second Example

Concerning food, the debate about genetically modified organisms will certainly be reinvigorated by synthetic biology, since the objective here is not simply to modify genetic codes in some minor way; it may eventually be able to design and produce the building blocks needed for the ideal daily ration. Naturally, for starving populations, this is an interesting prospect in the short term. However, no one knows the medium and long term impacts of completely synthetic foods on the metabolism of living beings ingesting, either deliberately or otherwise, new generation foods produced by a technological process.

Third Example

Apart from the preoccupations of public health, the possible use of nanotechnology to develop chemical weapons or weapons of targeted mass destruction cannot be ignored. With a view to national security and defence, such possibilities really need to be envisaged and assessed as a technological and social risk, and in a continuous manner given their potential for proliferation.

These few examples show an urgent need to develop specific biosafety strategies. Throughout the history of nanotechnology and its applications, there have been 'sentinels' observing the evolution of the state of the art and the way it has been appropriated by extremely varied scientific, technical, and industrial communities. These sentinels have thus had the opportunity to develop a skill, namely the translation of very technical observations into an assimilable vocabulary, without at the same time misrepresenting the reality of those observations and measurements. But this must still be transformed in such a way as to confer upon nanoscience, nanotechnology, and the changes induced by their expansion, a genuine social utility without which supply and demand could never find common ground.

Here we may quote Henri Laborit [16]:

> "While there can be no doubt that genetic evolution occurred through genetic combinatorics brought about by sexuality, we may also assert that it is through conceptual combinatorics, interference, exchange, and the spreading of ideas that humanity, at the stage it has now reached, will be able to evolve still further. While it has no doubt reached a biological dead end, possibly temporary, it can still hope to constitute a great planetary body, by combination of knowledge, from its simplest to its most abstract forms."

17.3 The Systemic Nature of Risks Induced by Nanotechnology

In a democratic society whose development is based on science and technology, no ethics is possible without procedures agreed by all its citizens to determine a framework and, when necessary, the limits on what can be undertaken. There is a fundamental tension here between the imaginary phase which precedes technological innovation and the very conditions of ethics. Given the scope of the changes induced by nanotechnology, we shall have to remain perfectly lucid about this tension in order to avoid violent reaction and rejection by consumers and citizens of absolutely all the applications of nanotechnology, without distinction or discernment.

17.3.1 The Effects

We prefer the study of changes induced by nanotechnology, or again the effects of nanotechnology, since this avoids any attempt to qualify them a priori as positive or negative, leaving the way open to a rigorous and scientific approach.

We can identify five dimensions in which we may apprehend the main ethical questions raised by the rapid growth of nanotechnology. In order to adopt a balanced and neutral approach, refusing a priori normative judgements about the possible changes, beneficial or otherwise, they may induce, we shall speak about the *effects* of nanotechnology. The following is based on the report by Jean-Pierre Dupuy and Françoise Roure on the ethics and industrial prospects of nanotechnology, (CGM-CGTI 2005) [17].

According to Jean-Pierre Dupuy, a distinction must be made between the following:

- Effects on nature (ontological effects). It is clearly above all with respect to themselves and their own nature, if they accept that they have one, that humans will have to decide the thresholds for transformation beyond which they do not wish to progress.

- Effects on our relationship with knowledge (epistemic effects). With nanotechnology, the philosophy of science which has prevailed since the eighteenth century, according to which we can only know by appealing to experience, should be pushed to its ultimate limits. However, with nanotechnology and the possibility of creation by reverse engineering (bottom–up, self-assembly), it is no longer only by carrying out experiments on nature, and it is no longer only by making models, that humans will come to know this nature. It is literally by rebuilding it. But at the same time, it is not nature that they will come to know, but something they themselves will have made. For Jean-Pierre Dupuy, it is the very idea of nature, hence of something external to us, which will appear to have been superseded. The very distinction between knowing and doing will tend to lose its meaning, like the distinction which still exists between thinker and engineer.
- Ethical effects, or the very possibility of ethics. Nanotechnology opens up an enormous field in which humans will have to introduce standards if they are to attribute meaning and objectives to it. The human subject will have to have recourse to a surfeit of willpower and conscience to determine, not what he *can* do, but what he *must* do. A whole ethical system will be needed, infinitely more demanding than the one which is slowly settling into place today to contain the fast development and possible excesses of biotechnology. The proliferation of scientific publications, patents, and applications, identified in the scientometric studies presented to the *Agence nationale de la recherche* (ANR) and the Organisation for Economic Co-operation and Development (OECD), makes this consideration more relevant every day.
- Metaphysical effects. With the meta-convergence on the nanometric scale of transformational technologies, viz., nanotechnology, biotechnology, information technology, and cognitive science and technology (NBIC),[5] and bearing in mind the prospective time scales, the most worrying effect is undoubtedly the blurring of the category distinctions with which humanity, ever since it has been around, has always found its place in the world. The natural non-living, the living, and the artifact are well on the way to merging into one. One may well consider that the mechanistic and informational metaphors upon which cognitive science and molecular biology are based are scientifically and philosophically false, and at the same time concede that they give us a power to act and a radically new control over the natural and the living. *But if this is the case*, the very successes of the new technologies will render the mechanistic and informational representations undeniable, and nobody could persist in saying that they are illusory. There is no exaggeration here in speaking of metaphysical effects. And these are not neutral with respect to the representation and staging of the nanotechnology roadmap, and in particular the branch of it furthest from large scale applications, viz., synthetic biology.

---

[5]NBIC is used in the reports of the United States National Science Foundation, while the term Converging Technologies for the European Knowledge Society (CTEKS) is used in reports by the high level group convened by the European Commission to examine the prospects for the new technology wave on the 2020 horizon.

- Effects on relations of domination (power effects). Asymmetric conditions of access to knowledge, intellectual property, production, profits, and so on.

These effects may be understood by analysing the more singular long term goals, like those of the transhumanist movement, which advocates the enhancement of human performance with the help of cognitive science and the human–machine interface (HMI), and even the hybridisation or instrumentalisation of DNA, in the complete absence of regulatory control. (In contrast to this demand, organisations have come together to forbid doping in sport in the name of competitive equality, using regulatory frameworks that are regularly transgressed.) Other agendas would work toward the same goal, such as those that seek the supremacy of a system of forces in which the combatant (public or private, army or militia) benefits from a specific environment and/or treatment, generally reversible. As early as 1982, Isaac Asimov used the neologism 'mentalics' to name the ultimate weapon [18]. This refers to domination by the power of thought, but of course assisted by electronics, and it prefigures other neologisms introduced in the second half of the twentieth century, such as moletronics for molecular electronics.

Among the visual representations which have fashioned the way of thinking here features the identification of landmarks in the nanotechnology roadmap as presented by Mihail Roco of the National Science Foundation.

If the eminently ethical question of limits to the possibilities opened by nanotechnology were inadequately posed, or not posed at all, one immediate consequence would be the risk of a significant short term discrepancy between the perception of public opinion in the European Union and in the rest of the world, with highly uncertain effects on the continued public support for scientific and technological developments in this area, as has been illustrated by the philosopher Étienne Klein in his work *Allons-nous liquider la science? Galilée contre les Indiens* [19].

17.3.2 Consequences of the Systemic Nature of Risk on the Assessment of Changes Induced by Nanotechnology

Raised at the first international dialogue on responsible research and development of nanotechnology in Alexandria, Virginia, in 2004, the question of the emergence of a unified, dynamic, and normative method for assessing the changes induced by nanotechnology has become central to the issue of regulation. Indeed, without it, it seems likely that the potential divergence of national regulatory measures in a globalised market, which is a risk in itself, might become a reality. If that happened, we would collectively miss an opportunity to benefit from feedback from past and present crises of confidence and governance between civil society and public decision-makers (think of nuclear power, asbestos, genetically modified organisms (GMO), bedrock hydrocarbons such as shale gas in France) by working to design a suitable normative framework for nanotechnology and its applications. In this area, researchers in economics and the social sciences (the reader may refer to the article [20] by the Norwegian researcher Fern Wickson) have made a considerable contribution to the legislative and the executive, as have historians of science and the organised civil society.

The European Trade Union Conference (ETUC) clearly emphasises this point in its resolution of 25 June 2008 on nanoscience and nanotechnology:

> After the asbestos scandal which cost the lives of hundreds of thousands of workers, and when the EU has recently introduced new legislation on chemicals that puts the onus of proof onto manufacturers, the ETUC finds it unacceptable that products should now be manufactured without their potential effects on human health and the environment being known unless a precautionary approach has been applied and made transparent to the workers.

The declaration refers to the novel behaviour of matter on the nanometric scale to criticise, from the point of view of worker protection, the effectiveness of a regulatory framework and market surveillance which, in the view of the unions, remains fragmentary, vertical, and better suited to the regulation of finished products than to dealing with the specific characteristics of advanced manufactured nanomaterials and nanoparticles present as intermediate items in the value chain of products belonging to distinct sectors of application (cosmetics and personal hygiene products, paints and coatings, pharmaceutical products, food additives, and nanomaterials present in packaging, electronic components, etc.).

Now the risks which arise on the nanometric scale can be reduced by sharing knowledge, research, and test methods, and this would seem likely to reduce costs while at the same time making regulation, when it exists or comes into being, truly applicable.

Market surveillance and a fortiori early warning mechanisms will not work properly down to nanometric scales until reproducible, validated, and standardised tests become available. The processes of translocation of free nanoparticles in the human body, e.g., by inhalation or ingestion, remain largely unknown and must be determined by toxicokinetic studies. A network of nanoscience research centres would help to accelerate the pooling of resources. The necessary standardisation, which would come after characterisation and metrology of the nanomaterials as such and in their interaction with their environment as a result of their specific properties, provides a justification for the *Direction générale de la concurrence, de la consommation et de la répression des fraudes* (DGCCRF) and the *Direction générale de la santé* (DGS) for the support they bring to international institutional efforts whose mission is to hasten identification of priorities and ensure that such tests become rapidly available, in particular, the International Standards Organisation (ISO) and the OECD.

Logically, since this is an area of common ground, a certain consensus should arise, when the time is right and through extension of responsible dialogue at the relevant levels of subsidiarity, regarding the economic and social utility of a unified method for assessing the changes induced by nanotechnology and convergent technologies at the nanometric scale. Here, too, researchers have a key role to play in guiding prototype forms of governance by establishing evaluation models, providing feedback, identifying learning and appropriation processes, and decoding the appropriate value bases in order to choose the most relevant criteria. For example, in therapeutic applications, toxicity is assessed by a cost–benefit approach with reference to the prognosis, quite different from the approach of a consumer to a widely available product containing nanomaterials, e.g., mobile phones, smartphones, and portable computers and tablets with a wireless communication facility.

There are many other relevant criteria than those relating to toxicity and ecotoxicity when it comes to dynamic normative evaluation, including ethical, legal, and social criteria, which come to light in a science–society dialogue, provided that there is a commitment to invest in knowledge and integrate recommendations in the public decision; the criteria of education, training, and information; the criteria of access to the state of the art and its evolution, which raises the question of the existence of freely accessible cross-disciplinary databases concerning the fundamental properties of nano-objects and nanomaterials; the criteria of public security, and so on.

However, even if it is indispensable, it will not necessarily be an easy matter to set up, on the European and international level, multi-criteria evaluation methods allowing national governments, civil society, and consumers to determine what is desirable for, or what works against the common good, and this for three main reasons:

- Competition between nations often takes precedence over cooperation, except in very rare cases of ethical emergency (medicine). Fundamental research programs involved in innovation are getting under way and accelerating simultaneously in many parts of the world: the United States, Canada, Brazil, the European Union (Germany, France, the United Kingdom, Italy, the Netherlands, Belgium, Finland, Romania), the European programs EUREKA (Cluster for Application and Technology Research in Europe on NanoElectronics, CATRENE) and more recently the Graphene Flagship, but also Israel, Russia, Iran, Saudi Arabia, South Africa, India, China, South Korea, Taiwan, Singapore, Australia, etc., with public expenditure of the order of 15 billion dollars in 2010, of which 2 billion dollars for the US national initiative alone.
- Trade secrets often take precedence over appreciation of the advantages of pooling knowledge, e.g., by setting up a classification of nanoparticles based on criteria of identification, naming, and behaviour in natural or artificial media, inspired by Mendeleev's periodic table, but for sizes the next step up from the atom. An *open* software architecture could be used to build this inventory on a unified digital language describing the behaviour of matter on the nanometric scale. In 2008, the French observatory NanoTrend identified 11 000 companies around the world which have published or filed patents in the areas of nanoscience and nanotechnology, among which nanoelectronics is well represented, thus giving Asia the role of driving force. There is a growth of more than 10 % in the number of patents filed each year, and China now contributes more than half of this growth.

- Competition between researchers tends to take precedence, at the funding and development stage, over the question of public utility and even over the very possibility of responsible innovation. This is the case when patents are granted with too broad a scope (e.g., for some unavoidable process of generic nature) or when the number of patents required to develop an innovative application or product makes it economically inaccessible for an SME or start-up company.

17.4 Legal Problems Raised by Nanotechnology

These problems are symbolised by the request, formulated by the European Parliament at the European Commission, to provide a definition of nanomaterials that could be used for European regulatory purposes, then transposed in each of the member states. By the end of 2011, only one European recommendation supplies a complex and revisable definition of nanomaterials which member states are supposed to apply in their national laws.

Nanotechnology applications are being developed across a very broad range of areas that differ enormously in the kind of risks involved or indeed the risk–benefit balance: for example, nanoelectronics, nanomedicine, and nanomaterials are three areas with well distinguished goals from this point of view.

These applications must conform to existing laws, and in particular to a set of rules transposed from European to French law. This is the case, for example, for the European CLP regulations applying to the classification, labelling, and packaging of substances and mixtures representing some danger for health or the environment, in the perspective of sustainable development [21].

As a preamble to this section, it is useful to note that, taken in its generality, the annual world production of chemical products is some 400 million tonnes, with hundreds of thousands of different molecules being publicly commercialised. Within this volume, nanomaterials represent at the very most several tens of thousands of tonnes. However, specific regulation has been required for these in regard to the following issues:

- The adaptation of the law to different properties of matter at certain nanometric scales, typically but not exclusively in the range 1–100 nm, which supposes that the risks relating to exposure and sensitivity to the dangerous nature of a substance or mixture has been proven or that, in the absence of such proof, the precautionary principle should apply. As requested by France, the question of application of the REACH regulations (Restriction, Evaluation and Authorization of Chemicals) to nanoparticle substances is currently being examined by European member states and the European Commission. The legal framework for the European cosmetics market has been revised to regulate the use of insoluble nanomaterials. The EC regulation proposal regarding foodstuffs incorporating nanostructured additives was not validated in the European inter-institutional dialogue. Legal arrangements are not therefore frozen. Indeed, as mentioned above, legal arrangements arising from the French *Grenelle de l'environnement* led to an obligation to declare manufactured substances containing nanoparticles.

- The applicability of legal measures to the area of market surveillance, i.e., to products and systems commercialised, produced, or imported into the relevant territory (e.g., France, the European Union), which presupposes that adequate tests and measurement instruments are available and economically accessible, and that public authorities actually fulfill their general mission of market surveillance and consumer protection. At this stage, these conditions are not satisfied either everywhere or all the time in the common market of the European Union. The rules laid down by the World Trade Organisation (WTO), even though it is extremely vigilant with regard to technical barriers to international trade (TBT), and those laid down by the World Intellectual Property Organization (WIPO) regarding the patenting of inventions, allow exemptions for health matters.
- The scalability of sectoral and vertical legal mechanisms when the proposed nanotechnology products and systems fall outside the scope of the well established legal corpus, e.g., when a nanomedical system combines chemical and physical effects and the regulations in force only recognise and allow for the a priori evaluation of chemical characteristics.
- The time delay between a new scientific discovery that would require revision of dangerousness criteria and risk assessment methods on the one hand, and the introduction of restrictive standards for the conditions of use of nanotechnology products on the other. The standardisation process depends initially on voluntary actions and assumes that the scientific representation, definition, nomenclature, and taxonomy are perfectly understood, coherent, and transparent, which is not yet the case for nanotechnology.
- The most delicate question concerns the bio-accumulation of synthetic nanoparticles and many physical and chemical agents disseminated by release and which constitute a long term systemic risk for which there is no judicial solution in conventional law able to establish a demonstrable relation of cause and effect. In this situation, the law is unable to establish a direct, explicit, and indisputable causal link between a generating event and an impact. As a consequence, it can neither identify responsibility with certainty, nor a fortiori punish an action that has caused harm. It is the role of the health and environmental protection agencies to ensure that preventive measures and suitable controls are established and revised in close relation with the state of the art, and that research is oriented in such a way as to improve understanding of toxicity and ecotoxicity, starting with those products destined for the human body, i.e., foods, products and processes for personal hygiene and treatment of disease.

17.5 Ethical Questions

Not only does nanotechnology constitute a legal challenge to the regulatory framework in force, but it is also a challenge for the very design of a future judicial framework. Ethical choices will be essential in order to fix the possibilities and limits for implants, e.g., cochlear implants, and other applications of nanotechnology whose aim is to enhance human performance beyond simple maintenance or the repair of defective functions, given the individual (irreversibility and damage), social, safety, and defence risks that an absence of judicial standards would bring about.

Concerning the benefits, judicial consideration of ethical matters also means favouring legal authorisation for marketing innovative solutions whenever the expected reduction of morbidity justifies it, and this notably in the treatment of rare or orphan diseases by nanomedicine (or the treatment of rare forms of a disease, as in the case of certain cancers) where treatments never reach the threshold of economic viability.

17.5.1 Need for an International Normative Framework Applicable to Nanotechnology

For an overview of institutional dynamics, we may refer to the report emitted by the *Conseil national de la consommation* (the French national consumer council, CNC), and in particular the synoptic table. Here we note that the OECD is ahead of the field compared with the specialised UN bodies when it comes to identifying the relevant areas for governance of nanotechnology on the international and domestic levels. This organisation has set up a working group for nanomanufactured materials. It depends on its chemicals committee, and one of its aims is to arrive at guidelines and tests, in partnership with the International Standards Organisation (ISO), for fourteen priority nanoparticles. The nanotechnology working group is a subsidiary instance of its Committee for Scientific and Technological Policy (CSTP), and one of the actions of its strategic plan focuses on social issues with the production of a guide and recommendations for conducting a successful public debate on nanotechnology. However, these efforts by the OECD are not inclusive enough to be relevant to the markets and the ethical, legal, and societal issues at stake: China, India, Brazil, and developing countries are not members. Russia will doubtless be so in the short term.

The Food and Agriculture Organisation (FAO) and the World Health Organisation (WHO) have clearly identified the question of risk governance with regard to nanotechnology, and the WHO has set up a working group on the toxicity of synthetic nanoparticles in the context of its evaluation program PAVEL. However, we are still a long way from having any normative systems and the question of the availability of tests and independent expertise remains open.

The European Union has integrated important research programs into the calls for tender of the Seventh Framework Programme for Research and Technological Development (FP7), notably Nanosafe 2, Observatory Nano, Framing Nano, and ICPC NanoNet, whose results taken together brought progress in setting up expert networks, and in establishing the research priorities of the eighth framework programme H2020 for the period 2014–2020 using an impact assessment methodology in terms of responsible research and innovation. The question of coordinating these actions with those initiated at the request of the European Commission or by the specialised European agencies—the European Chemical Agency (ECHA), the European

Medicines Evaluation Agency (EMEA), the European Food Safety Agency (EFSA), the Joint Research Centre (JRC)—has not yet been resolved. The European Union does not yet speak with a single voice when it comes to nanotechnology policy.

17.5.2 Labelling: Helping the Consumer to Choose in a Complex and Uncertain Situation

Information about the composition of products for the general public, such as foods, remains hard to access, except in certain cases. Crucial for consumers, the innocuousness of products on the open market (apart from therapeutic products) is an essential issue. At the same time, the creation of a hasty and poorly thought out national label that did not really boost well informed trust would very probably have quite the opposite effect, leading to a general feeling of mistrust. A label of the kind 'Nano inside' would not contribute to an informed choice about the absence of risk when using, consuming, or processing the waste from a finished product at the end of its life cycle. This would be as absurd as a label proclaiming 'atoms inside'.

A label able to announce the information 'nano safe and sustainable by design' would bring consumers the information they require. At the plenary meeting of the ISO in Shanghai in November 2008, a technical working group on nanotechnology for sustainable development was set up, and this is certainly a step in the right direction, because it may provide a way to back up the label with a norm or standard. This said, the creation of this label and its subsequent use will take the time required for industrial production to evolve from the design stage right up to the processing of industrial and household waste. The emergence of international standards would lead to savings for producers by avoiding multiple standards, whence they would be quickly and generally adopted on a voluntary basis, and it would also lead to an opening up of the markets, avoiding the phenomena of environmental dumping or 'precaution dumping' with regard to consumer health depending on whether the consumer lives in a highly regulated region or a less advanced country.

However, in concrete terms, the step taken by the ISO has not led to the development of a consumer label, and it is now the European Committee for Standardisation (CEN) which is developing a standardisation project under the mandate of the European Commission. It will be important to mobilise suitable expertise at the CEN, and it is desirable that a broad range of stakeholders should contribute to this, e.g., industry, consumers, public metrology agencies. The issue of physical traceability of nanomaterials likely to exhibit toxicity or ecotoxicity under specific conditions of use is not part of the ISO program, even though its technical working group on the needs of consumers and citizens in matters of nanotechnological standards expressed the relevance of developing a physical traceability standard during its meeting in Queretaro, Mexico, in March 2013, and Belo Horizonte (Brazil) in November 2013.

At this stage, we must examine in more detail the ethical aspects of therapeutic products and processes involving nanotechnology. Medical applications relating to the enhancement of human performance—excluding those with therapeutic goals—are not part of the roadmap for public investment in R&D in Europe. Indeed, the European Parliament expressed itself to be very clearly in favour of a preliminary ethical review of the aims of this kind of research, with a guarantee to publish the results, before committing any funding to it.

The problem facing the public authorities has more to do with market forces which, because they deal in mature products and seek to prolong the profitability of old molecules, would try to block any change in regulations in favour of using physical nanotechnological processes, e.g., for cancer treatment, while chemotherapy has shown its limitations and prognostics are not always very good.

As an example, in 2006, Naomi J. Halas, director of the nanophotonics laboratory at Rice university in Texas, called for a radical change of thinking in this area, referring to cancer therapies exploiting physical phenomena as an alternative to chemotherapies developed on protocol. Indeed, for certain tumours, she obtained remission rates of the order of 100 % using functionalised nanoparticles, directed as markers to the regions to be treated by laser. In her view, the regulations in force at the time would not have authorised commercialisation of the proposed protocol on the US or European markets. This authorisation was only obtained several years later by a German company operating in these markets, viz., MagForce, whose nanotechnology products and processes were invented and made safe by Dr Andreas Jordan and his team at the Berlin Charity Hospital (see Chap. 10).

If it turned out that, through errors of appreciation on the part of public policy, through delays in companies adopting a responsible approach to innovation, or simply through conflicts of interest with ethical consequences, we were to reach an irreversible situation in which the freedom of fundamental research were hampered in any way, along with the authorisation to use radically innovative processes, we would be faced with a particularly serious ethical miscarriage.

Adaptation takes time and the rewards, beyond the unrealistic promises sometimes artificially amplified for financial reasons, have not discouraged the pioneers. But having said this, it remains to accelerate the possibilities for society to reap the rewards from the resources it devotes to nanoscience and nanotechnology.

17.6 The Key Role of Databases and a Better Understanding of Technology

Depending on whether the public comes from a scientific background (and even within this community), the definition of nanoscience and nanotechnology can vary considerably. Naturally, this is partly due to differences in scientific and technical culture between research scientists and the general public, but it is not the only factor. Language and codes play a crucial role here. In contrast to French-speaking Canada, France has not considered it a priority to develop an elementary vocabulary which would have made it possible to devise a body of terms with definitions within reach of those who disseminate information, viz., teachers and journalists. This in turn would bring about a faster rise in the general level of understanding and expression among the population. However, efforts of this kind have nevertheless been mooted by the *Commission nationale de terminologie et de néologie* (COGETERM) since the middle of the 2000s, but without any real mobilisation. The designated group at the French Academy of Science never actually met. English-speaking countries, and even a country like Iran, did not make the same mistake: they made suitable lexicons available to their populations, to raise the level of awareness and understanding in the new field of knowledge opened by nanotechnology. Even the ISO recognised in 2011 the usefulness of a vocabulary guide to key terms, standardised by international consensus, on the suggestion of its technical working group on the advantages of international standardisation of nanotechnology for consumers and citizens (consumers and societal dimension, CASD).

This disaffection on the part of the French academic establishment can be interpreted as a lack of interest for cross-disciplinary approaches. It is still reflected in matters of priority and their consequences for funding by most of the relevant ministerial departments (with the exception of the ministry dealing with public health). In a country like France where industry is poorly represented in education, this has negative consequences when it comes to stimulating curiosity in new areas, improving the attractiveness of new professions, and more generally preparing people for work in the modern world. It does not accompany the citizen in a public debate turning upon established scientific fact. For information alone is not enough. It must be preceded by an education which trains the individual to consider it in a critical way, through an effort of reasoning and discernment.

In its dialogue between the Commission, the Council, and the Parliament, the European Union was faced early on with the complex question of the terminology used by the relevant industrialists or scientific interlocutors. For example, when the directive on cosmetics was revised, the definition adopted for nanomaterials was restricted to a certain category of synthetic nanoparticles, namely insoluble ones, considered as potentially toxic in the present state of our understanding. Discussions about the applicability of French legal mechanisms tend to focus on the appropriate definition of nanoparticle substances, while the European terminology is having a hard time stabilising, given the problem of coexisting regulatory texts based upon different definitions.

The question of terminology thus precedes, in its resolution, the question of the necessary identification, declaration, and traceability of synthetic nano-objects or nanomaterials entering the composition of marketed products and whose effects would be such as to cause harm to the environment or health. In this context, the *Agence nationale de sécurité sanitaire pour l'évaluation des risques sanitaires* (ANSES), has played a determining role in France.

Actions Undertaken by ANSES with Regard to Nanotechnology

ANSES is a French government organisation set up to evaluate health risks. It is not responsible for managing these risks, a task that befalls other operators, and in particular the government. For this evaluation, it appeals to independent experts identified through open calls for tender. Its recommendations are made public.

It plays the role of a scientific watchdog and makes the results of this monitoring process publicly available, in particular by broadcasting the corresponding information in a bulletin. Through its extreme vigilance with respect to deontological rules, it benefits from an expertise bereft of conflicts of interest or relations, something which is liable to increase trust in the quality of that expertise. It takes part in European and international standardisation work in the field of nanotechnology when there are possible impacts on health and the environment.

In this rapidly expanding area, which includes DNA and technologies involving living systems on the nanometric scale, the amount of data is growing exponentially, with a great many distinct fields of application, but also a demand for international integrated harmonisation and interoperability of open source computer formats. Apart from principles of governance, it will be necessary to bring together and synchronise multiple sources of content and funding to accomplish the data sharing required for integration purposes in order to allow for responsible innovation on the nanometric scale. This is the whole issue of cross-disciplinarity in nanoscience. We have feedback on the design, organisation, and implementation of very large universally accessible scientific databases. The International Council for Scientific Unions (ICSU) has been a driving force in the design of successful internationally integrated data gathering, analysis, and management models, requiring structures able to handle very large volumes of data in a sustainable way.

These successful references for transposition to the nanoworld are as follows:

- The scientific databases needed for work by the Intergovernmental Panel on Climate Change (IPCC).
- The Worldwide Protein Data Bank (wwPDB) [22].
- The International Virtual Observatory on Astronomy (IVOA) containing astronomical data. The issues of conflict regarding intellectual property, standardised formats of digital data, and the principle of open access and scientific data sharing were discussed at ICSU.
- The OneGeology database containing geological data.

As far as nanoscience and nanotechnology are concerned, ICSU is now interested in the principles of governance that would preside over the implementation of an integrated terminology database on the international level, providing a universal way to describe nano-objects, as well as its subsystems organised by the relevant fields of application and properties. It considers it motivating to play an upstream role in the joint definition of this cross-disciplinary project and would like to put into practice its own experience and that of its dedicated entity the Committee on Data for Science and Technology (CODATA), in matters of data formats and interoperability, as required to ensure scalability and open access since there is question here of fundamental knowledge. In order to launch this program, it has taken the initiative of producing a white paper dealing with a universal description of nano-objects. Above all, this prospect requires scientists to express their desire for international cooperation, to be relayed by the relevant intergovernmental organisations (OECD, UNESCO).

Public Debate

The demand for transparency regarding the presence of synthetic nanoparticle substances in commercialised products was clearly expressed at the *Grenelle de l'environnement* (round table talks) during debates on emerging risks for the environment and sustainable development. Measures have since been introduced: there is an obligation to declare the production, importation, and commercialisation of nanoparticle substances and materials liable to release them into the environment, with the first public feedback session in November 2013. The environmental transition law extends this mechanism to products that are already regulated, such as pharmaceutical products and food additives. These measures serve as a reference in the context of free circulation of products at the European and international levels. A comparative study of European and American regulations considered it necessary to introduce a compulsory declaration for all commercialised products containing nanomaterials.

The question of counterpowers implicitly assumes that there is an asymmetry of information and decision-making capacity between researchers and industrialists on the one hand and the various components of civil society on the other. Concerning researchers, it should be recognised that research freedom must remain a respected principle, and that discoveries of unsuspected social utility are sometimes the result of a whim, i.e., ideas judged to be fanciful and unreasonable by evaluators. What is at stake here is rather the ability of civil society to divert research funding, and in particular public research funding, toward measures of significant social utility in the face of global challenges relating to the worthy survival of the species in general and individuals in particular (food, health, adaptation to climate change, energy, drinking water, etc.), and this by correctly anticipating rather than trying to remedy the potential negative impacts of scientific and technical choices.

The question of regulation of industrial activities is raised for applications of nanotechnology, as in other areas, given that there exist, in principle, regulatory and even self-correcting mechanisms: the governing bodies of a company watch over and ensure respect for the *affectio societatis*,[6] i.e., accomplishment of the corporate purpose, but taking into account the company's social and environmental responsibilities. From this point of view, and as a feature contributing to a balanced approach, in resonance with civil society, a certain diversity of origins, competence, and career

[6]A legal term in France referring to the *shared aim* bringing together several physical or legal entities to create a company and share the profits or losses.

paths among board members is desirable to ensure that strategic orientations, such as internal auditing, can prevent companies from deviating from their corporate purpose.

The Role Expected of Social Networks

Not all technological products are popular among consumers, in fact, far from it. Companies thus have every interest in marketing products that are in phase with the expectations of consumers and citizens, particularly regarding product life cycles. The social networks, including Internet, with the emergence of a blogosphere, constitute a possible counterpower for any information that might seem too biased. South Korea has developed social networks dealing with products and after-sales services, to which businesses pay more and more attention.

Finally, the lively exercise of representative democracy on the national and European levels is a guarantee for the expression of citizen's expectations, in the field of nanotechnology applications as in any other. Access to the information that underpins this exercise is guaranteed by the gradual construction of a system of open access archives containing any nanotechnology publications resulting from publicly funded research. The websites of the Observatory Nano and ICPC NanoNet, financed by the Seventh Framework Programme for Research and Technological Development (FP7), are based upon these principles.

At the moment when the United Nations Framework Convention on Climate Change (UNFCCC) is setting up an executive committee for technology whose aim is to accelerate technology transfer under economically acceptable conditions, it would be altogether unacceptable if it became impossible for researchers and innovators to exploit the specific nanoscale properties of nano-objects and nanomaterials. Indeed, we have by now established their potential in terms of their contribution to carbon-free energy forms, water treatment, depollution, and public health, all indispensable if we are to achieve the goals of the new millennium, and in particular sustainable development through growth characterised by a very low ecological footprint. A responsible development of these substances necessarily depends on organising and sharing knowledge about toxicity and ecotoxicity that may result when they are put to use [23]. It is by sharing fundamental knowledge that we may maintain a critical distance and exercise reason.

At the same time, there does indeed exist a spiritual dimension to the presumed or imagined impacts of nanotechnology. It is important to recognise this fact and take suitable account of it.

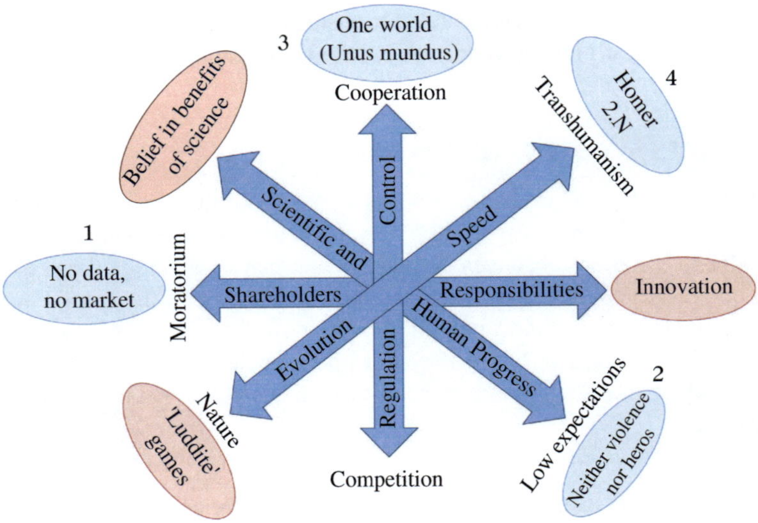

Fig. 17.5 Scenarios for acceleration of evolution and attitudes with regard to technical progress

17.7 Religion, French Transhumanism, and the Question of Ends

Transhumanism asserts that technology and nanotechnology can and should be used for the purposes of human development (see Fig. 17.5). Brought to attention in France by the stakeholder report presented to the *Commission particulière du débat public* (CPDP) by the *Association française de transhumanisme* (AFT) in 2010, this school of thought, still rather in the minority, is perhaps a weak sign of a transition occurring under our very eyes and calling for our attention, namely, a growing desire to overcome human finiteness through all kinds of material and virtual prostheses, or by irreversibly extending or modifying our way of interacting with our environment. The manifold promise of NBIC convergence, including elimination of suffering and even death, features on the agenda of transhumanists and post-humans. Transhumanism is an attempt to provide an outcome for modernism through a clearly stated desire for individual enhancement with the help of machines, even genetic machines if necessary.

Transhumanism is not a philosophy. It is a school of thought or opinion, oscillating between a materialist and a spiritualist identity. It believes in the regenerating power of machines. It does not see itself as normative, does not criticise the *status quo*, and does not envisage an ideal world which takes into account the imperfections of the real world. What it does do is merely to extrapolate the feasibility of its intentions on the basis of what is possible technologically, as expressed for example by Eric Drexler, author of *Engines of Creation* [24], or Ray Kurzweil in *The Singularity Is Near: When Humans Transcend Biology* [25]. Transhumanism focuses on possibilities for future evolution, e.g., Kevin Warwick, professor of cybernetics at Coventry university, presents an implanted prosthesis facilitating 'telepathy', and thereby making it unnecessary to use language.

Scientists also wonder about the scope of social change that may be induced by their research. In discussions with the CNRS ethical committee for science and information and communication technologies (ICT), the following questions were raised:

- In a system where direct causal relation becomes blurred, should the authorities seek to forbid potentially destructive initiatives?
- What is the collective responsibility?
- On what criteria should one judge the unethical nature of a research topic?
- How can one recognise the quality of a standard, a code of conduct, or a law in these areas?

As an illustration, these questions are relevant in the field of synthetic biology. The BioBricks Foundation [26] is a non-profit-making initiative, shared between MIT, Harvard, and the University of California in San Francisco (UCSF). It aims to facilitate the fabrication of DNA for nanoscale assembly of synthetic DNA sequences to be inserted in living organisms. Promoting software, archives, and open standards, it disseminates fundamental knowledge in digital form (see the work of the BioBricks Foundation, and in particular the development of software for structuring its open access database [27]). So we may ask whether this foundation bears any responsibility if its codes are misused in any way, and if so, to what extent. Perhaps, on the contrary, it contributes to strengthening the characterisation and management of risks. At this juncture, the European Research Network ERA SynBio cooperates with the Synthetic Biology Engineering Research Center (SynBERC) in California to ensure interoperability of digital data standards. This quest for common standards, based on a system for the universal description of nano-objects, has been presented to the ISO biotechnology group, set up in December 2013.

17.7.1 Hybrid Immersive Systems, the C Factor of NBIC Convergence, and Freedom of Thought

Apart from Promethean descriptions, the current convergence of NBIC technologies (nanotechnology, biology, information and communication technology, and cognitive science) invites us to consider spiritual and metaphysical issues in a particular light, namely, in terms of the influence of potential applications of nanoscale-convergent technologies on our way of seeing, thinking, and imagining ourselves three or four generations into the future. This influence affects the cognitive aspect of convergence. It has been perfecting itself since the invention of the printing press and the consequent availability of written knowledge, providing the layperson with access to the understanding of the material universe stored in books. Let us consider an example relating to current technology: nanoelectronics and nanostructured portable

Avatars in the world of cybernetics

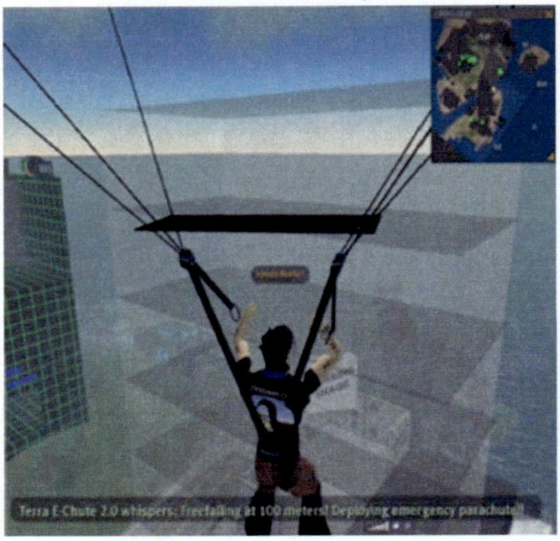

Fig. 17.6 Immersive technology, changing perceptions, and social engineering

batteries make it plausible to consider the immersion of vision and thought in virtual worlds that persist in memory and mobility. In 2010, images of the football world cup, a worldwide media event, were broadcast on 3D screens. Apart from sport, attention economics[7] finances 3D advertising, from which it expects increased revenues. Concerning fiction, the example of the 'hybrid man' in the film *Avatar*, specially 'modified' to communicate with extraterrestrials, provides a futuristic image of the extremes toward which transhumanism may tend (see Fig. 17.6).

Advertisers have long been investing in 'consent engineering', as described by Edward L. Bernays [28]. The idea is to create a desire and a purchasing act. With the persistent virtual worlds borne by information and communication technologies (ICT) and amplified by the behavioural engineering of social networks, the possibilities for influencing the way we think about and see things have never been greater. The Japanese school of technology and design has focused attention on the effectiveness of audiovisual technologies. Furthermore, it financed research on emotion technology, seeking to understand the mechanisms linking advertising content with the brain and emotions. If for Edward Bernays "the engineering of consent is the very essence of the democratic process, the freedom to persuade and suggest" [24], the duration of exposure to virtual worlds, combined with contents designed to persuade or suggest, generates a real risk, namely, the risk of manipulating our thoughts and our way of seeing things. This risk is one of the social changes wrought by (nano)technology and its intensive use by immersive technology.

Through the combination of nanoelectronics and portable photovoltaic cells, screens incorporate nanomaterials which make them flexible, very light, transparent, energy efficient, and hence portable and miniaturisable, whence the time an individual spends each day before a screen of some kind tends potentially to 100 % of his

[7] An approach to the management of information that treats human attention as a scarce commodity, and applies economic theory to solve various information management problems.

or her waking life. When interfaces become very ergonomic, e.g., by projection onto glasses, or even directly onto the retina, the individual becomes less and less aware of their presence and gradually becomes immersed in a world of virtual reality during his or her waking life. The technology platform of the TOTEM project[8] carried out by the *Institut Mines-Télécom*, was devoted to mobile mixed reality games, which offer an immersive experience: the games are not limited to the screen, but interact with the real world by incorporating surrounding objects and places, thereby benefiting from thoughts and emotions already present in the player, who can thus enjoy a much richer world and game experience.

In such a world, what about the notion of parental control, the ban on subliminal images, respect for the speaking time allowed to political figures during official periods of election campaigning, given the possible addiction to 'augmented' virtual reality? And what about reversibility, the freedom to choose other relationships and other lifestyles more respectful of the natural rhythms, as advocated in projects like the Italian *Citta Lente* (Slow City) movement?

The legal experts of the company commercialising the Nintendo 3DS game console had a health warning issued about the use of 3D vision technology without the need for special glasses. This company opted for a precautionary approach by recommending parents to stick with 2D portable game consoles for children less than 6 years old, owing to the potentially negative impacts on eyes and vision. Prior to this, Sony and Samsung had issued similar warnings for Playstation and TVs, respectively. But apart from the physical risk, there is also a potential risk of disorientation due to content. Here, ocular disorders 'hide the woods' as represented by the psycho-neuro-endocrine-immunology (PNEI) system, well known to physiologists dealing with stress, depression, and addiction.

Augmented virtual reality still depends on external prostheses, at least for a short time to come, but apart from ethical issues and consumer preferences, there is little reason why it should not become accessible through endogenous, incorporated, or hybrid interfaces, or even hybrid–native interfaces if the interactions between human neural systems and ICT systems interlock and coevolve. Then the whole range of sensations and emotions would be accessible to engineering techniques in real time in the computing sense of the term, that is, in a much shorter time than required for information processing by a non-equipped human central nervous system. Immersive virtual reality would then go beyond the fictional situations imagined by non-augmented human brains! Imagine the internalisation of nanostructured and biocompatible human–machine interfaces. Imagine also that it were possible to design cognitive capacities and stimulate them without limit, but in a completely controlled way, by very high accuracy brain imaging and synthetic biology. It is the imagination expressed by science fiction itself, resulting from a new type of cognitive (or 'cognifactured') travel, that would be modified. Proficiency in engineering such cognitive travel would then have a critical influence on choices: imagining, presenting, representing, and choosing for some, passive acceptance for others. These

[8]TOTEM is the acronym for theories and tools for distributed authoring of mobile mixed reality games.

possibilities must clearly be controlled, regulated, or prohibited, through a combination of ethics, context, and circumstances. Otherwise they will form the basis for a double-edged mental weapon.

17.7.2 Societal Questions Arising from Potential Uses of Nanoscale Technologies

The future developments of immersive technology based on nanoscience and nanotechnology remain unknown. What impacts will they have on future societies?

1. Following the desecration of the body, will our civilisation accept the same for consciousness, accept to make it available for design?
2. Will societies wish to promote an irreversible augmentation of the human species, going as far as a bifurcation in the odyssey of the human species?
3. Will our civilisation be forced to design and produce living humanoid species, over which it would have responsibility as 'products', to resist its own threat of extinction?

From an ethical standpoint, what company would have the resources to act proactively by regulating choices with a powerful and effective democratic control that is both dynamic and able to learn? Is there not a risk that, when confronted with natural or technological crises, fears, and cataclysms, it will merely react in haste, compelled to resort to radical and singular solutions with unknown impacts?

When it comes to considering future scenarios, freedom of thought is essential for imagining the world that we would like to live in. In the expression of these scenarios, who should have a say? How can we ensure a fruitful exchange between those who carry legislative, executive, and judiciary power and those who represent fragments of opinion in society? And how should we handle such problems in the context of globalisation? The very notion of society has itself evolved in these days of ubiquitous technology and social networks, and the globalisation of trade. This should be taken into account in the way the powers that be think about and exercise their responsibility to protect and plan for the future. The quality and efficacy of governance is a major factor here. The forward-looking approach is a help in representing the advantages and disadvantages of otherwise conventional options (cooperation, competition, wolves alliance, etc.), but whose beneficiaries will ultimately be groups of stakeholders with specific, different, even divergent interests. This is where governance, understood as the art of reconciling opposites by balancing powers, comes into its own.

17.8 The Fragile Beginnings of World Governance in the Field of Nanotechnology

Historically, technological developments with significant impacts on the environment, society, and individuals have always ended up being treated on a supranational scale, as illustrated by nuclear issues, or again substances arising from synthetic chemistry which affect biodiversity, particularly when the public authority, as a last resort, is the only body to guarantee citizen and consumer protection and it is unable to do so without intergovernmental or supranational agreement. For nuclear power, we may cite the International Atomic Energy Agency (IAEA), whose aim is to promote peaceful uses of nuclear energy and to restrict the development of military applications. For synthetic chemicals, the REACH program specifies a European regulatory framework to manage and control chemical products.

Nanoscience, nanotechnology, and the resulting innovations are already being deployed in a multipolar way with the involvement of private actors on globalised markets. In contrast, their public governance remains limited to well separated political and administrative territories.

17.8.1 The Long and Yet Unlikely Path Toward Diplomatic Action

Up to now, the traditional route toward the elaboration of a governance standard on the international level has followed a diplomatic process that may last over several decades, as witnessed in the area of climate change with the negotiation of the Kyoto protocol which seeks to reduce greenhouse gas emissions. It may stem from academic origins, with scientists assuming their share of responsibility by synchronising their research, sharing results with politicians, and disseminating the hypotheses and various scenarios arising from observation and simulation. The universal recognition of the utility of such an action came when the Nobel Prize was awarded jointly to the former vice-president of the United States Al Gore and the Intergovernmental Panel on Climate Change (IPCC) for their role in warning the world of the risks of climate change.

The case of nanotechnology is interesting when it comes to the question of governance because, while the different, even divergent interests were recognised right from the start at the beginning of the 2000s, the normative question was raised proactively at the European Union by the services of the US Trade Representative (USTR). Their intention was, and remains, to avoid at all costs the setting up of non-tariff trade barriers to international trade (NTBs), notably within the transatlantic zone, barriers that would have a social origin, thus reproducing the long-lasting quarrel over international trade of genetically modified organisms (GMO).

If the art of governance is to reconcile different, even opposing, interests, the question of governance for nanotechnology has been around for several years, not only for these technologies as such, but perhaps in a more forward-looking way, for all the applications arising from the combinations of nanotechnology, biotechnology, information and communication technology (ICT), and cognitive science, referred to in Europe as Converging Technologies for the European Knowledge Society (CTEKS).

17.8.2 When Different Interests Can Coexist on Harmonised Prerequisites

These interests may be industrial or commercial, but also strategic, given the intrinsically dual nature of research for high-tech applications. They may make highly contrasting prospective scenarios look more or less likely.

At the beginning of the 2000s, the first attempt to express the need for international governance in the area of convergent technologies and nanotechnology came from the academic world when, at the invitation of the US National Science Foundation (NSF), a first responsible international dialogue was organised in Alexandria (Virginia, United States). The aim was to establish responsible development of nanoscience and nanotechnology in a concerted way. This dialogue has since been organised on two other occasions, in Tokyo in June 2006, then at the invitation of the European Commission in March 2008, with the participation of 49 countries.

17.8.3 What Has Been Achieved by Intergovernmental Relations?

It should be noted that, while the international intergovernmental dialogue is informal, and will remain so—no text other than the conclusions and a report committing only the chairperson has been validated diplomatically—, it remains the only fully inclusive intergovernmental arena where countries who wish to do so, and even non-governmental organisations, can meet to express their point of view. But this dialogue, which has never been formalised by any text, was discontinued in 2010. The fact that no commitment was ever agreed, even about the very existence of the dialogue, and that preparations for the second dialogue fell through in 2005, can be put down to the position adopted by the US Department of State which refused the proposal of the European Commission to consent to a code of conduct that would be put forward for signature by the participants. Since then, the United States has sought and obtained, in close collaboration with the European Commission and members of the European Union—all aware of the fact that a partner like the United States could not be left out of any pre-normative process—the creation of an ad hoc intergovernmental dynamic, although unfortunately not inclusive, within the Organisation for Economic Co-operation and Development (OECD).

Priorities of the OECD Working Group on Nanotechnology from 2011 to 2014

These can be stated as follows:

- Statistical consideration of nanotechnology based on scientific and technical indicators, socioeconomic factors, production, and trade.
- Identification of issues specific to nanotechnology in the areas of innovation, industrial production, the life cycles of materials, and trade.
- Centres of cooperation for scientific research on nanoscience and nanotechnology (tools, specific themes, access to cooperative research facilities), taking into account nanobiotechnology, nanomedicine, and synthetic biology.
- Dialogue with the civil society, its goals, its modalities, and its results.
- Updated review of public policies implemented in countries taking part in the nanotechnology working group (NTWG).
- Contribution of nanotechnology to green growth, seeking to express the benefits of certain applications of nanotechnology for the greater good, focusing on the contributions to water treatment and depollution in 2008, carbon-free renewable energies in 2009, food and drugs in 2010, and nanomedicine in 2011.

It should be noted that the two working groups, NTWG mentioned above and the Working Party on Manufactured Nanomaterials (WPMN), reviewed available nanotechnologies and nanomaterials that could contribute to sustainable development under acceptable conditions of safety and security. The European Commission is associated with the work in these groups, having received a mandate from the European Parliament and Council to increase its participation in international, and in particular OECD bodies engaged in dialogue.

There is a certain convergence in the efforts being made within the OECD and in other forums, such as the US academies of science and engineering and the British Royal Society, viz., the convergence between nanotechnology, biotechnology, and information and communication technology for responsible development of synthetic biology. The changes induced by this convergence, based upon digital modelling, simulation, and storage of the behaviour of nano-objects and nanomaterials, are considered to be at least as important as those brought about by synthetic chemistry in the twentieth century. The nanotechnology and biotechnology groups of the OECD have a partially shared agenda.

Other specialised institutions of the United Nations have occasionally addressed the issue of manufactured nanomaterials, including the World Health Organisation (WHO) in the context of a group jointly financed by the European Union and not exclusively devoted to nanotechnology, and also UNESCO, the United Nations Industrial Development Organization (UNIDO), and the UN Food and Agriculture Organisation (FAO). The United Nations Economic Commission for Europe considers nanotechnology applications through its role in the transport of dangerous materials and the harmonised classification of chemical products.

Treaties dealing with the proliferation of biological and chemical weapons could usefully assess the idea of including risks introduced by synthetic biology (anti-error campaign, anti-terror campaign) in their mission and during revisions. However, ways of implementing controls, if only by peer reviewing according to the good practice of the International Atomic Energy Agency (IAEA), are not yet on the agenda. The responsibility for characterising and managing risks is thus shifted as far upstream as possible in partnership with those inclined to cooperate, and in particular industry. But then at the very least, joint evaluation of the risks would have to be placed at the heart of multilateral cooperation.

There is still no official recognition from country to country of the need for a harmonised and synchronised agenda for the various specialised institutions within the United Nations regarding changes brought about by nanotechnology. But an agenda of this kind would provide a way to address the key issues of governance, e.g., in support of the relevant specialised organisations, viz., the United Nations Environment Programme (UNEP) the World Health Organisation (WHO), and the United Nations Industrial Development Organization (UNIDO). For the record, it was indeed by entrusting to two specialised bodies, UNEP and the World Meteorological Organization (WMO), a precise mandate to work on an intergovernmental level in cooperation with the IPCC that international commitments were made to reduce greenhouse gas emissions, on the basis of status reports and simulations supplied by researchers. So given the similarities, would it not be a good idea to do the same thing for nanotechnology?

These questions of harmonisation and synchronisation in work contributing to the world governance of nanotechnology were raised at the meeting of the United Nations Commission on Sustainable Development (CSD) in New York in 2008 during a side event organised by the liaison office to the United Nations for non-governmental organisations. This was not followed up, however, except in requests formulated by the non-governmental organisations.

17.8.4 A More Mature and More Assertive European Impetus

The resolution concerning regulatory aspects of nanomaterials, adopted by the European Parliament on 24 April 2009 on the basis of the Schlyter report, considers that the notion of a safe, responsible, and integrated approach advocated by the European Commission on matters of nanotechnology is compromised by the lack of information about commercialised nanomaterials. In point 23, it invites the Commission and member states to launch a Europe-wide public debate on nanotechnology, nanomaterials, and associated regulatory matters, considering that the present debate is largely given over to experts, while synthetic nanomaterials seem likely to bring about profound societal changes requiring a much broader public consultation.

This resolution was adopted at a time when the outgoing Commission was preparing the main lines of its strategic action plan for nanotechnology for the period 2010–2014 in an international context where the many national initiatives had turned the nanotechnology agenda into a competitive quest to acquire sustainable comparative advantages.

The Commission, whose organisation and programs were supposed to recognise and include the strategic importance of nanotechnology and technology, nevertheless postponed the adoption and the announcement of this plan, even though research activities were still funded in the context of the Seventh Framework Programme for Research and Technological Development (FP7), which came to an end in 2013. The eighth framework programme (H2020) made use of ethical and social feedback regarding nanotechnology to propose, in a communication from the Commission to the European Council, to organise its funding with the help of an impact assessment methodology based on responsible research and innovation (RRI).

However, can we expect the European Union, aware of the criticisms and the limits encountered when piloting highly sensitive social dossiers such as commercialisation of GMOs, to take the lead in reviving the modes of governance for convergent technologies at the nanoscale? Against a background of accusations of third rate governance by some of the citizen's representation associations, will it take into account the theoretical considerations put forward by authors such as the sociologist Ulrich Beck in his book *Risk Society: Towards a New Modernity* [29], and echoed by the sociologist of innovation Arie Rip, regarding the part that could be played by a reflexive governance coevolving with technology and society? Or again, by the philosopher Jean-Pierre Dupuy when, with Alexeï Grinbaum, he pleads in favour of adopting a normative dynamic assessment of systemic risks brought about by convergent technologies on the nanoscale? By reflexive governance is understood the constructive dialogue between civil society and the authorities in such a way as to stimulate innovation exactly where it can be most helpful, while at the same time imposing strict limits on applications that do not conform to the general principles of the law with regard to protection of citizens, consumers, and the environment, at every relevant level of subsidiarity.

A public debate on nanotechnology and convergent technologies proposing a reflexive form of governance, appealing regularly to the various stakeholders involved in public actions and their impacts, might just be able to restore the social dimension to scientific and technical progress, particularly in the field of medicine and the new eco-friendly industries which will be promoted as a key driving force in reindustrialisation, a way out of systemic economic crises, and a way to create a net surplus of permanent skilled jobs.

In this context, the French public debate has contributed to identifying and promoting this form of reflexive governance, which is both inclusive and favourable for social cohesion, from the local to the global, through the European Union, right up to the level of intergovernmental dialogue. When he stressed the need to set up mechanisms favouring rapid appropriation of the relevant questions, a fair valorisation of the collective costs, and an acceleration of good choices and good types of behaviour—starting with consumers and citizens—, driving forces in the emergence of another form of innovation, the regional director for industry, research, and the environment in the Rhône-Alpes region was promoting the efficiency, even the pre-eminence, of this model when questions of responsibility are at stake in a context of long term systemic risks.

The European project Framing Nano, which came to an end in 2010, argued in favour of a change in the form of governance to move towards permanent action, the 'continuum'. Presented in the international cooperation and expert networks and put into practice in the scientific partnerships of the European Union, its vision of a continuum of governance for the changes induced by nanotechnology caused a great deal of interest in the capitals of Europe, but also among institutional correspondents in Asia (India, China, South Korea, Japan, Taiwan, Singapore, Malaysia, Thailand, etc.), Australia, Africa, the Near and Middle East, Latin America, the Caribbean countries, and Oceania. American academics interested in the interaction between technology and society have cited these studies and drawn the same conclusions.

But stressing sensitive areas of research and associated technology, the French presidency of the G20 and the G8 was hoping to change international values regarding the conception of new forms of governance, beyond the issues of systemic risk and financial regulation. Will the social responsibility of organisations, as specified by the guidelines laid out in the standard ISO 26000, be expressed identically or in a conflictual way between the different economic regions of the world when it comes to nanotechnology? What will be the terms and pathways of commercial exchange? The scientific arenas of open cooperation, the diplomatic counterparts?

But we still have to write history because state priorities advertised for research and private operators of any importance in the nano field (United States, European Union, South Korea, Brazil, China, Russia, and other emerging countries) reveal possible divergences between public and private, civilian and military roadmaps, against a diversity of values that is no less important regarding its impact on the freedom and the social utility of the sums invested. The OECD's policy of increased diversification will doubtless tend to slow this convergence down, but without containing it. Furthermore, could this be a goal in itself, and if so, to what advantage? Countries not belonging to the OECD clearly raise the question, to anyone who is prepared to listen, of the legitimacy of the 'pace setters', i.e., those who consider themselves the creators and guardians of standards in an area that is already globalised.

It looks as though an agreement is in the making between the United States, Asia, and the European Union to recognise that from now on there will be more drawbacks than advantages in any lasting interruption of the international dialogue for responsible development of nanoscience and nanotechnology.

However, this inclusive dialogue went beyond the criticism of the 'club' containing the OECD and its members. Originally called for by the United States along the lines of a consensus conference, then pursued in Tokyo and Brussels in 2006 and 2008, respectively, the Alexandria process was the sole multilateral opportunity, intergovernmental but informal, where all countries who wished to do so could take part in a dialogue on nanotechnology and its governance. It was also propitious for the participatory process which could give rise to the initiative of an intergovernmental panel for continuous assessment of changes brought about by nanotechnology. This shared assessment would have made it possible to prepare tools, methods, and processes of cooperation in such a way as to avoid divergence in the regulatory framework and dumping of all kinds, including ethical, environmental, and social.

While European determination seems to be intact here, it remains to implement it with the involvement of the specialised agencies of the United Nations whose missions are affected by nanotechnology in achieving the Millennium Development Goals (MDGs). The United Nations Environment Programme (UNEP) and also the United Nations Industrial Development Organization (UNIDO) should at least, with UNESCO, assume their role in this joint construction of harmonised, if not unified, governance of nanotechnology and emergent technologies.

The prospect of a transatlantic trade and investment partnership (TTIP) between the European Union and the United States by 2015 could be the moment to inaugurate, apart from dissension over the precautionary principle (a principle of action for some, a principle of inaction for others), a new form of international governance for emergent technologies, based on responsible research and innovation. Starting from a systemic approach to the changes brought about by nanotechnology, the European Commission has forged the concept of responsible research and innovation (RRI), extended to investment, as a principle of action bringing together all stakeholders to give the best chance of success to science and technology and bring about a better quality of life that is also more widely shared.

Nanotechnology has become part of the transatlantic dialogue through the Economic Forum and the recommendations of the Transatlantic Council. With the upcoming TTIP, a further step can be taken by introducing the RRI concept as a federating principle for negotiations opened up following this agreement, which has the potential to encourage sustainable codevelopment and progress. If this works, nanotechnology, which lies at the heart of the convergent technologies and new scientific discoveries, will be free to contribute, with the backing of and in harmony with civil society, to shaping our peaceful coexistence.

References

1. M. Crichton, *Prey* (HarperCollins, New York, 2002)
2. H. Jonas, *Pour une éthique du futur* (Rivages, Paris, 1998)
3. www.nano.gov
4. Nanotechnology Risk Governance. International Risk Governance Council (IRGC), June 2006
5. Observatory Nano ceased activity in 2012, but archives are still available at the website www.observatorynano.eu
6. www.smartregions.net
7. US-EPA Regulatory Notices Concerning Nanoscale Materials. Announcement of the publication in July 2013 of general regulations concerning nanomaterials and other nanoscale products
8. Council of Europe report, 17 January 2013. Committee for social affairs, health, and sustainable development. Nanotechnology: balancing benefits and risks to public health and the environment

9. V. Packard, *The Hidden Persuaders* (Ig Publishing, New York, 2007) (reissue)
10. V. Packard, *The Naked Society* (Ig Publishing, New York, 2014) (reissue)
11. W. Szybalski, A. Skalka, Nobel prizes and restriction enzymes. Gene **4**(3), 181–182 (1978)
12. Synthetic biology. Concept note. International Risk Governance Council (IRGC), 23 p
13. H. Yang, Synthetic biology and the future of man. Presented at the international symposium on opportunities and challenges in the emerging field of synthetic biology. Washington DC, USA, 9–10 July 2009, The National Academies' Keck Center, Washington DC
14. H. Laborit, *La Vie antérieure* (LGF, Paris, 1997)
15. H. Laborit, *L'Homme imaginant*, Union Générale d'Edition, Collection 10/18, 2nd edn (1993)
16. H. Laborit, Le rôle de la communication dans l'évolution des êtres vivants. Centre d'études expérimentales de physiobiologie, de pharmacologie et d'eutonologie de l'hôpital Boucicaut. Propos recueillis, Bulletin de l'IREPP, April 1990, p. 49
17. www.cgm.org/themes/deveco/develop/nanofinal.pdf
18. I. Asimov, *Foundations' Edge* (Doubleday, New York, 1982)
19. E. Klein, *Allons-nous liquider la science? Galilée contre les Indiens* (Flammarion, Collection Café Voltaire, Paris, 2008)
20. F. Wickson, Gobernanza nanotecnologica, por qué no podemos confiar en evaluaciones de riesgos cientificos. MundoNano, Revista interdiciplinaria en Nanociencia y Nanotecnologia, UNAM, Mexico, Vol. 4, no. 1, January–June 2011
21. Regulation (EC) no. 1272/2008 of the European Parliament and of the Council of 16 December 2008. JOCE of 31 December 2008
22. www.wwpdb.org
23. F. Roure, Introduction to Part III, in *Nanotoxicology and Nanoethics*, ed. by M. Lahmani, F. Marano, P. Houdy (Springer, Berlin, 2011), pp. 420–421
24. E. Drexler, *Engines of Creation: The Coming Era of Nanotechnology* (Anchor, New York, 1987)
25. R. Kurzweil, *The Singularity Is Near: When Humans Transcend Biology* (Penguin, New York, 2005)
26. http://openwetware.org/wiki/The_BioBricks_Foundation
27. www.biobricks.org
28. E.L. Bernays, The engineering of consent. Ann. Am. Acad. Polit. Soc. Sci. **250**, 113–120 (1947)
29. U. Beck, *Risk Society: Towards a New Modernity* (Sage Publications, London, 1992)

Glossary

Adatom Atom adsorbed on a surface. Adsorption is a phenomenon in which the atoms or molecules of a gas or liquid fix onto a solid surface through van der Waals interactions.

Adenosine triphosphate (ATP) A molecule which, in all living organisms, supplies by hydrolysis the energy required for the chemical reactions of the metabolism.

Aggregate Object resulting when a set of distinct entities, which may or may not be of the same kind, adhere to one another to form a single ensemble.

Alkane Saturated hydrocarbon with only single carbon–carbon bonds.

Alkene Unsaturated hydrocarbon with one carbon–carbon double bond.

Alkoxide Organic group bonded to a negatively charged oxygen atom ($R–O^-$).

Alkoxyamine group Chemical group initiating nitroxide-mediated polymerisation (NMP).

Alkyne Unsaturated hydrocarbon with a carbon–carbon triple bond.

Alkyl group These derive from alkanes by loss of one hydrogen atom. They may be radicals or substituents on complex organic molecules.

Allotrope Chemical element in one of its possible crystal structures, e.g., carbon in the form of graphite, diamond, or graphene.

Amide Organic compound derived from a carboxylic acid which has a nitrogen atom bonded to its carbonyl group.

Amine group Functional group derived from ammonia, e.g., $–NH_2$.

Amino acid Chemical compounds with two functional groups, viz., a carboxyl group and an amine group. These are the elementary building blocks of proteins.

Amphiphile A molecule with a hydrophilic polar head and a hydrophobic nonpolar tail.

Antibody Substance synthesised by a living being in response to stimulation by an antigen whose toxic effects are thereby neutralised.

Antigen An antigen is a natural or synthetic macromolecule, recognised by antibodies or cells of the immune system and able to generate an immune response.

Apoptosis Programmed cell death.

Aromatic compound In chemistry, a planar ring system with $4n + 2$ delocalised electrons, where n is the number of rings.

J.-M. Lourtioz et al. (eds.), *Nanosciences and Nanotechnology*,
DOI 10.1007/978-3-319-19360-1

Aryl group In chemistry, a functional group which derives from an aromatic hydrocarbon such as benzene.

Atheroma An accumulation of various materials (fats, fibrous tissues, calcium deposits, blood) on part of the inner wall of an artery.

Bacterium Single-celled organism or prokaryote present in all environments. Sizes range from 100 nm up to 2 microns.

Band gap In a solid, the energy or frequency spacing between the valence and conduction bands. In terms of energy, it is the energy that must be given to an electron before it can contribute to conduction in the material.

Biofilm A biofilm is a multicellular community of microorganisms such as bacteria, fungi, algae, or protozoa, sticking together and to a surface, where adhesion gives rise to the secretion of an adhesive and protective matrix. Biofilms generally form in an aqueous medium.

Biopsy Tissue sample on a living being, taken for microscopic examination.

Bipyridines Heterocyclic organic compounds with formula $(C_5H_4N)_2$, comprising two coupled pyridines. They are colourless solids, soluble in organic solvents, and slightly soluble in water.

Bit/byte The bit is the elementary unit of information, which can assume the values 0 or 1. A byte contains 8 bits and is the basic unit used to measure the amount of data in computing.

Cage molecule A molecule able to encapsulate another molecule.

Carbonyl Organic compound containing a double bond between a carbon atom and an oxygen atom, with the carbon atom bonded exclusively to hydrogen or carbon atoms.

Carboxyl group Functional group with general formula –COOH, comprising a carbon atom bound by a double bond to a first oxygen atom and by a single bond to a second oxygen atom.

Caspase assay Method for detecting cells and tissues undergoing apoptosis with the use of a molecular marker for apoptosis, viz., caspase 3.

Catalysis Process in which the rate of a chemical reaction is modified by the effects of a substance known as a catalyst.

Cement Powdery material which, when mixed with water or a saline solution, forms a plastic binding paste that can agglomerate various substances to produce a hard solid.

Chirality A chiral object is one that cannot be superposed on its mirror image.

Chromatography Physical technique for separating chemical species.

Clearance Ability to eliminate a substance or corpuscle from an organic fluid.

Click chemistry Set of chemical reactions able to create substances quickly, efficiently, and in a modulatable way by connecting up preexisting molecular fragments or ensembles. The starting materials are freely available and generate only harmless byproducts after the reaction.

Coating Treatment of a surface by applying a liquid coating to a substrate. Coating is generally followed by drying and possibly reticulation.

Colloid Substance made up of small particles (micelles) suspended in a medium. It reflects light rays when the medium is a liquid.

Complexing molecule In chemistry, a molecule that can capture and bond to ions (often metal ions) to form a complex.

Concrete Composite construction material made from natural granular substances (sand, gravel) or artificial granular substances (lightweight aggregates) held together by a binder which is usually cement.

Covalent bond A chemical bond between two atoms in which each atom shares one of the electrons in its outer shell to form an electron doublet which constitutes the bond.

Cracking A technique for breaking a complex organic molecule into smaller ones.

Cross-linking Transition of a polymer from a state in which the macromolecules are independent to one in which they are linked by chemical bonds.

Cryptography Discipline of cryptology dedicated to the protection of messages to ensure confidentiality, authenticity, and integrity, often using secret codes or keys. The information contained in the message can then only be decoded by the sender or the recipient.

Cyclisation Process whereby part of an organic molecule adopts a closed ring shape.

Cycloaddition Cyclisation reaction in which unsaturated molecules, or parts of the same molecule, combine to form a cyclic product with a reduction in bond multiplicity.

Cytotoxicity Toxicity for a species of cell. An example is genotoxicity, involving alteration of the genome.

Dendrimer Molecule that has a branched site with reactive endpoints and another site able to attach itself to a molecular or solid support.

Deoxycytidine deaminase Enzyme catalysing the metabolisation of gemcitabine.

Diamagnetic material A material which, when subjected to an external magnetic field, produces another magnetic field to oppose it (see the discussion of spin and magnetic domains in Sect. 2.5.2).

Diffusion length Length over which a physical quantity, a number of particles, etc., drops off by a factor of e, where $e \approx 2.7$ is the base of the natural logarithm.

Dihydroxylation Reaction fixing an OH group on each carbon atom of the double bond in an alkene.

Drug delivery Technique for modulating and controlling the distribution of an active principle with respect to a target by associating it with a carrier.

Electrolyte Electrically conducting liquid or solid substance, thanks to the presence of mobile ions.

Electronvolt Unit of energy equal to the energy acquired by an electron when it moves through a potential difference of one volt. Its value is $1\ \text{eV} = 1.6 \times 10^{-19}$ joule.

Electrophilic compound A compound attracted by negatively charged species, as opposed to a nucleophilic compound.

Emulsion Macroscopically homogeneous mixture of two immiscible liquids.

Endocytosis Mechanism of capture and transport of molecules and particles inside a cell.

Endogenous (exogenous) process One whose causes lie inside (outside) a given biological entity.

Energy See the entry for power/energy.

Equilibrative nucleoside transporter 1 (ENT1) Protein for intracellular transport of a nucleoside or nucleoside analogue.

Enzyme Protein playing the role of a biological catalyst.

Epigenetics Study of ways in which the environment and individual history can influence gene expression. This concerns all modifications of gene expression that are transmissible from one generation to the next, without alteration of nucleotide sequences, and reversible.

Epitaxy Technique for growing a monocrystalline film on a substrate that is itself monocrystalline. When the substrate and the film are made from the same material, this is called homoepitaxy, while heteroepitaxy refers to the opposite case.

Epoxidation Certain oxidation reactions of alkenes produce cyclic ethers in which the two carbon atoms connected by a double bond become bonded to the same oxygen atom. These products are known as epoxides.

Exciton A quasiparticle that can be represented as an electron–hole pair, bound together by the Coulomb force.

Exeresis Surgical removal of some harmful or useless element, such as an organ, a tumour, a foreign body, etc.

Exfoliation Removal of sheets one at a time from a layered structure.

Exogenous process See the entry for endogenous/exogenous process.

Extrusion Mechanical or thermomechanical fabrication process used to create objects of a fixed cross-sectional profile. A material is pushed or pulled through a die with the desired cross-section.

Ferroelectric material A material in which spontaneous electric polarisation can occur.

Ferromagnetic material A material that can be very strongly magnetised under the effect of an external magnetic field and retain a significant magnetisation even when the field is removed (see the discussion of spin and magnetic domains in Sect. 2.5.2).

Fibrosis Anomalous increase in the amount of fibrous conjunctive tissue in a tissue or organ.

Fluorescence Light emission caused by excitation of a molecule (generally by absorption of a photon), immediately followed by spontaneous emission.

Fluorescence imaging Imaging of cells or tissues through the presence of fluorescent entities which may either be present from the start or introduced into the organism artificially.

Fullerene Molecule made from carbon atoms and which can assume a geometric shape resembling a sphere, an ellipsoid, a tube, or a ring.

Fullerol A hydroxylated fullerene.

Galenic pharmacy Pharmaceutical discipline which studies drug formulation.

Gemcitabine Chemotherapy drug used in the treatment of non-small cell lung cancer, pancreatic cancer, breast cancer, and bladder cancer.

Gene therapy Therapeutic strategy in which genes are placed in the cells or tissues of a patient to treat a disease.

Genome Complete set of chromosomes and genes.

Genotoxicity See entry for cytoxicity.

Hall effect Discovered by Edwin Herbert Hall in 1879, this effect results from the force exerted by a magnetic field on electrons (the Lorentz force). When the field is normal to the direction of the electron current, the trajectory of the electrons bends and this leads to a potential difference in the perpendicular direction. This potential difference is proportional to the injected current and the magnetic field.

Helicity For a carbon nanotube, this is the angle between the nanotube axis and one of the carbon bonds. More precisely, it is a reference direction parallel to one of the sides of the hexagons in the unit cell.

Hematoxylin–eosin (HE) stain Classic histological staining method.

Heterostructure Term generally used to describe multilayers of different semiconductor materials.

Histology Branch of biology and medicine which studies biological tissues.

Hole Particle with a positive charge equal and opposite to that of the electron. A hole results from the absence of an electron.

Hydration In chemistry, hydration has three different meanings:

- When a product is soluble in water, the water molecules bind to the ion or the molecule to form a sort of layer.
- When a product can absorb water, hydration is the opposite of desiccation (dehydration, drying, lyophilisation).
- Reaction in which a water molecule is added to an organic molecule.

Hydrolysis Decomposition of a substance by H_3O^+ and HO^- ions produced by dissociation of water.

Hydroxylation Chemical treatment in which a hydroxyl group OH is fixed onto a surface.

ICT Information and communication technologies.

Immunohistochemistry Method for detecting proteins in the cells of a tissue section by marking with antibodies.

Indentation Technique for measuring the hardness of a material.

Intracellular enzyme Protein synthesised by living cells which catalyse one or more biochemical reactions within the cell.

Ionic–covalent bond An ionic–covalent bond between two atoms is a covalent bond in which there is a higher probability of finding the electron in the vicinity of one atom rather than the other.

Ionic strength Main factor influencing the activity of ions in an aqueous solution, expressed in $\mathrm{mol\,l^{-1}}$.

Junction oncogene Fused genes favouring the occurrence of cancers.

Ki67 staining Ki67 antigen is a marker for cell proliferation.

Lifetime Time after which a time-dependent quantity drops by a factor e, where $e \approx 2.7$ is the base of the natural logarithm.

Ligand Chemical species used by other molecules to bond together.

Lipophile Substance with an affinity for apolar solvents like lipids.

Liposome Vesicle made from lipids.

Lithography/nanolithography Process for imprinting motifs in a resist, used to reproduce large numbers of nanoscale and microscale devices.

Local hyperthermia Local rise in the temperature of the body above the normal value of 37 to 37.5 °C due to the accumulation of exogenous heat. This is also the name for the cancer treatment in which the local temperature of the body is raised by subjecting implanted nanoparticles to a magnetic field, if the particles are magnetic, or infrared radiation, if the particles are sensitive to those wavelengths. The nanoparticles then become a heat source which can destroy tumour cells, for example.

Luminescence Generic name for light emission by a source excited by light (photoluminescence), heat (thermoluminescence), chemical reaction (chemiluminescence), or electricity (electroluminescence).

Lymphoid organs Organs containing lymphocytes or other cells of the immune system.

Lysosome The digestive organs of the cell, these are intracellular vesicles characterised by an acidic pH and filled with enzymes.

Macrophage Cell carrying out phagocytosis within tissues. Arising from differentiation of blood leucocytes (monocytes), they contribute to innate immunity as a non-specific defence system.

Magnetic resonance imaging (MRI) A medical imaging technique producing 2D or 3D visualisations of the inside of the body in a non-invasive way and with relatively high contrast resolution.

Magnetoresistance Resistance whose value depends on the applied magnetic field. Classically, the magnetic field modifies the electric field lines and thus modulates the value of the resistance by a few percent (see Sect. 2.5.2 and Chap. 4).

Maleimide Contraction of maleic acid and imide, the –CNHC– functional group.

Mechanochemical polishing Process for smoothing surfaces involving the combined action of mechanical and chemical forces. It uses a polishing cloth and a chemical solution.

Mechanosynthesis Chemical synthesis in which reaction products result from the use of mechanical constraints to direct reactive molecules to specific molecular sites.

Memory There are two main types of semiconductor memory: read only memory (ROM) and random access memory (RAM), which is usually volatile. Magnetoresistive random-access memory (MRAM) is a non-volatile version of RAM.

Mesocosm Small controlled or semicontrolled part of the natural environment in which the experimenter can vary one or more parameters, such as the soil, hygrometry, temperature, CO_2 content of the air, concentration of other pollutants, etc.

Metabolisation Transformation of substances in a living organism under the effect of enzymes.

Metal Good electrical conductor thanks to the high density of quasi-free electrons. Unlike semiconductors, this density cannot be modulated.

Metamaterial In physics and electromagnetism, a metamaterial is an artifical composite with electromagnetic properties that cannot be found in natural materials.

A negative index metamaterial is one for which, at the interface between the metamaterial and a natural medium, the incident and refracted waves both lie at the same side of the normal to the interface.

Mobility Parameter expressing the ability of charge carriers to move under the effects of an electric field. It is defined as the ratio of the carrier speed and the electric field, expressed in metres per second divided by volts per metre ($m^2v^{-1}s^{-1}$).

Monocrystal Crystal in which the unit cell repeats in a strictly identical way throughout, and over distances that may include several billion crystal planes.

Mucus Viscous substance secreted by the mucous glands and serving as a protective coating for the mucous membrane. The secretion products are rich in glycoproteins and water.

Nanocarrier Molecular assembly for transporting drugs in the body. Also called nanodroplets, nanodrugs, etc.

NBIC technologies Nanotechnology, biotechnology, information technology, and cognitive science.

Nuclear magnetic resonance (NMR) Property of certain atomic nuclei which happen to have a magnetic moment, e.g., the proton, to resonate with radiofrequency radiation at a very precise frequency when they are placed in a magnetic field. The frequency depends on the magnetic field and other atomic or molecular factors, and in particular, the environment of the atomic nucleus. This phenomenon gave rise to magnetic resonance imaging (MRI).

Nucleophilic compound A compound attracted by positively charged species, as opposed to an electrophilic compound.

Nucleoside Constitutive element of RNA and DNA, comprising a base associated with a sugar.

Nucleoside analogue Synthetic compound resembling natural nucleosides, used in the treatment of serious diseases (mainly cancers and viral infections).

Opsonisation Immunological process used by organisms to mark exogenous particles such as bacteria, viruses, and nanoparticles, by adsorption of specific proteins called opsonins (fragments of immunoglobulins, fibronectin, complement, etc.). The particles are opsonised then recognised by certain macrophages in the organism which eliminate them.

Orthotopic mouse model Animal experimentation model using transplantation on mice.

Oxidative stress A form of aggression on cell constituents by reactive oxygen species.

Oxidosqualene cyclase Enzyme catalysing the cyclisation of squalene.

Paramagnetic material A material with no spontaneous magnetisation but which, when subjected to an external magnetic field, acquires a magnetisation directed in the same direction as that field (see the discussion of spin and magnetic domains in Sect. 2.5.2).

Pascal The pascal (Pa) is the SI unit of pressure (newton per square metre). Further, 100 kPa $=$ 1 bar $=$ 760 mm Hg $=$ 10 m of $H_2O \approx 1$ atm.

PEGylation PEGylated particles are coated with polyethylene glycol to reduce their elimination by macrophages in the organism.

Pellet Wood granules.

Phagocytosis Cell process in which microbes are destroyed by certain kinds of leucocytes, called phagocytes.

Phonon Quantum of vibration energy in a crystalline solid.

Phosphorylation Enzyme mechanism adding a phosphate group to an endogenous or exogenous molecule.

Photon Quantum of light energy.

Photonics Study of devices for generating, transmitting, processing (modulation, amplification), or converting optical signals. Photons are treated indifferently as wave or particle.

Piezoelectric effect Property of certain bodies to become electrically polarised under mechanical stress.

Plasma Fluid, often a gas, made up of charged particles, ions, and electrons.

Pneumocyte Cells lining the alveoli (air sacs) of the lungs and contributing to their function.

Power/energy Power is the ratio of energy to the time required to dissipate it. It is given in watts (W). Energy can be expressed in joules (J), electronvolts (eV), or watt hours (W h).

Protease Enzyme which hydrolyses proteins and polypeptides.

Protein Large molecules, made up of amino acids, and constituting the main part of organic matter and living beings.

Pyridines Simple heterocyclic organic compounds with similar structure to benzene, where one of the CH groups is replaced by a nitrogen atom.

Pyrolysis Decomposition of an organic compound by heat to produce products (gas and solids) that it did not contain previously.

Quality factor Quantity characterising the damping of an oscillator or resonator. The bigger it is, the greater the resonance of the oscillator and the narrower its bandwidth. A high quality factor is also a characteristic of a good frequency filter.

Quantum cascade Process in which, under an electric potential, an electron falls into a series of semiconductor quantum wells. In a quantum cascade laser, the energy lost by the electron within each quantum well is converted into emission of a photon.

Quantum dot A semiconductor nano-object which, thanks to its size, of the order of the electron wavelength, behaves as a potential well (quantum well) for electrons in the three space dimensions.

Quantum fluctuation In quantum physics, energy can be created at a point in space over a very short time, and this happens without violating energy conservation. This fluctuation obeys Heisenberg's uncertainty principle $\Delta E \times \Delta t \approx h$, where h is Planck's constant.

Quantum well A potential well with width less than the wavelength of the relevant particle (here, an electron or a hole). A semiconductor quantum well is obtained by stacking three semiconductor layers in such a way that the middle layer has lower band gap energy than the outer layers.

Rack Metal cabinet in which standard size boxes containing electronic or computer equipment can be slid on rails.

Raman effect Inelastic scattering of light by molecules or atoms.

Redox reactions Any reactions in which atoms are reduced or oxidised, i.e., electrons are transferred between chemical species. Reduction involves a gain of electrons and a decrease in the oxidation state of the molecule, atom, or ion. Oxidation involves a loss of electrons or an increase in the oxidation state of the molecule, atom, or ion. The redox potential is an empirical quantity measured in volts relative to the proton/hydrogen (H^+/H_2) pair, with zero potential. A positive redox potential corresponds to a strong oxidiser. A negative redox potential corresponds to a strong reducer.

Refraction Deflection of a wave when it crosses a surface separating two media in which it propagates at different speeds (quantified by the refractive index).

Registration, evaluation and authorization of chemicals (REACH) Regulations for registration, assessment, authorisation, and restrictions on chemical substances, adopted on 18 December 2006 by the European Parliament and the Council of Europe.

Ribosome A small structure in the cell cytoplasm which serves to assemble amino acids to form proteins.

Scale The macroscopic is our own scale. The microscopic is the scale of the microscope (the micrometre or above). The mesoscopic is the scale between quantum and classical physics (from the atom to the micrometre) and the nanoscopic is the scale of the nanometre, hence in the middle of the mesoscopic.

Semiconductor Material in which the density of mobile charge carriers can be varied, thereby modulating the conductivity. There are n-type and p-type semiconductors. In the former, the moving charges are negatively charged electrons, and in the latter, they are positively charged holes. A III–V semiconductor is a material made up of atoms from columns III and V of the periodic table, with as many from column III as from column V.

Speciation In chemistry, distribution of an element depending on its state in the medium. In biology, evolutionary process whereby new living species arise.

S phase Synthesis and replication of DNA during cell division.

Spin Intrinsic angular momentum of a particle (see the discussion of spin and magnetic domains in Sect. 2.5.2).

Spintronics Spintronics or spin electronics is an area of physics which studies and exploits the quantum properties of electron spin.

Squalene Lipid which can adopt a highly compact molecular conformation and which can be coupled to anticancer or anti-infection molecules, thus playing the role of a nanocarrier.

Stereochemistry A subdiscipline of chemistry, involving the study of the relative spatial arrangement of atoms within molecules. An important branch of stereochemistry is the study of chiral molecules.

Stereospecificity A chemical reaction is said to be stereospecific if reagents differing only in their stereochemistry are preferentially or exclusively transformed into products which only themselves differ by their stereochemistry.

Steric effect The steric effect arises from the fact that each atom contained in a molecule occupies a certain space. If some atoms are brought too close to others, there will be an energy cost associated with this.

Sublimation Direct transition from a solid to a gaseous state.

Semiconductor superlattice Periodic structure obtained from alternating multilayers of semiconductors with different band gaps. The thickness of a pair of different layers determines the period of the superlattice.

Superconductor Material with electrical resistance very close to zero at low temperatures.

Supramolecular chemistry Branch of chemistry, beyond molecular chemistry, which studies non-covalent or weak interactions between molecules within a molecular system.

Surface plasmon Collective oscillation of an electron plasma at the interface between two media. The surface plasmon is also the particular solution of the equations of propagation of electromagnetic waves or Maxwell's equations at the interface between these two media. The corresponding electromagnetic field is evanescent on either side of the interface.

Surfactant Compound modifying the surface tension between two surfaces.

Synapse Region of contact between two neurons or between a neuron and a cell.

Synthetic biology Cross-disciplinary branch of molecular biology using physics, computing, and chemistry to invent genomes, obtain new enzyme reactions, and create cells capable of novel functional and metabolic activities.

Theranostics Contraction of therapy and diagnostics.

Thermoelectricity Study of the relation between electrical and thermal phenomena.

Thesaurismosis Accumulation of drugs, excipients, or materials in cells, bringing about cell death (iatrogenic effect).

Thiol Organic compound carrying a sulfhydryl group –SH attached to a carbon atom.

Translocation Transfer of nanoparticles across cell or tissue barriers. More commonly, genetic translocation is a genetic mutation characterised by mutual exchange of chromosome material between non-homologous chromosomes, i.e., not belonging to the same pair.

Triazole Cyclic organic compound containing a five-atom ring with two double bonds and three nitrogen atoms, hence with formula $C_2H_3N_3$.

Tribology Science of friction, wear, and lubrication.

Trophic chain Network of food chains connected together within an ecosystem and through which energy and biomass circulate.

TUNEL assay Method for detecting cells undergoing apoptosis.

Ultrafine particle Defined traditionally as a particle with diameter less than 0.1 μm.

Van der Waals force Attractive force between closely spaced atoms and molecules (at nanometric distances or less).

Virus Agent of infection requiring a host, often a cell, whose constituents it uses to replicate. It has a size in the range 10–300 nm.

Vortex Rotating motion of fluid or particles.

Waveguide Device for guiding electromagnetic or acoustic waves, confining them in a particular medium over a certain distance. An optical waveguide is usually obtained from a material with high refractive index inserted in one or more materials of lower index, e.g., an optical fibre.

Xenograft Transplantation of a graft, e.g., a tumour, in which the donor belongs to a different biological species to the recipient.

Young's modulus Constant relating the stress and strain for an isotropic elastic material.

Index

A
Actin filament, 214
Activated carbon, 293
Activated sludge, 302
Active optical cable, 81
Active principle, 234, 240, 245, 254, 268, 269, 345, 369, 407
  coupled with squalene, 256
  encapsulation level, 254
Acute coronary syndrome, 284
Acyclovir, 258
Adatom, 56, 405
Adebiotech, 251
Adenosine, 258, 266
  with squalene, 266, 267
Adenosine triphosphate (ATP), 213, 301, 405
Adsorption, 405
Aeronautics, 345
Aerosol, 276, 279, 281, 284
Aerosol pyrolysis, 200
Affectio societatis, 390
AFT, 392
Ageing, 362
Agence nationale de la recherche (ANR), 319, 320, 352, 353, 379
Agence nationale de sécurité du médicament et des produits de santé (ANSM), 248
Aggregate, 405
Aggregation, 293–296
AIDS, 258, 264
Air–blood barrier, 280, 281
Airborne UFP, 276, 279, 284
Airbus, 345
Airways, 277–279
ALCATEL-THALES, 353
Algae, 302
Alkali reaction, 184
Alkane, 405
Alkene, 405, 407
Alkoxide, 405
Alkoxyamine group, 270, 405
Alkyl group, 405
Alkyne, 405
Allergic contact dermatitis (ACD), 282
Allotrope, 405
Aluminium–silicon nanotube, 289
Alveolar epithelium, 279
Alveolar interstitium, 279
Alveolar–capillary barrier, 280
Alveolus, 278, 279, 281
Alzheimer's disease, 243, 285, 377
AM1.5 spectrum, 99
Amide, 405
Amine group, 405
Amino acid, 369, 405, 412, 413
Amphiphile, 167, 405
Anaerobic media, 296
Anatase, *see* Titanium dioxide
ANSES, 363, 388, 389
Anti-infection drugs, 255
Anti-odour textiles, 147
Anti-reflective coating, 341
Anti-UV films, 201
Antibiotherapy, 265
Antibiotic, 257, 258, 265
Antibody, 405
Anticancer drug, 255–257, 264, 269, 270
Antigen, 405
Antioxidant, 301
Antioxidant enzyme, 285
Antiretroviral drug, 258
Antiviral drug, 256, 257, 264
ANVAR, 352

J.-M. Lourtioz et al. (eds.), *Nanosciences and Nanotechnology*,
DOI 10.1007/978-3-319-19360-1

Aorta, 284
Apollonian packing, 187
Apoptosis, 261–263, 406
Aqueous media, 294
Architecture
future, 61
neurally-inspired, 57, 70
von Neumann, 57
Arctic, 302
Aromatic compound, 405
Arsenic, 288, 307
biosensor, 375
detection, 369
Arsenic hydroxide, 293
Artemisinin, 369
Artificial genetic circuit, 369
Artificial retina, 231
Aryl group, 406
Asbestos, 285, 288, 357, 380, 381
fibre, 276
Asimov, I., 380
Aspdin, J., 185
Asthma, 278
Atherogenesis, 284
Atheroma, 406
Atheromatous plaque, 230, 236
Atherosclerosis, 284
Atom, 1
Atomic force microscope (AFM), 3, 24, 52, 191, 214, 336
Atopic eczema, 282
Attention economics, 394
Augmented virtual reality, 395
Autologous re-implantation, 243
Automobile industry, 103
Avatar, 394

B
Bachelor's degrees, 328
Bachelor–master–doctorate (BMD) system, 328
Bacteria, 258, 265, 301, 302, 369, 406
intracellular, 265, 266
marine, 302
marine sediment, 302
soil, 302
Bactericide, 275, 302, 340
Bagasse, 372
Bamboo nanofibre, 125
Band gap, 406, 414
Bangladesh, 369
Bar des sciences, 358, 359
Basic Technological Research (BTR) network, 316
Bayes, T., 307
Bayesian model, 307
Beck, U., 401
Behavioural engineering, 394
Beijing Genomics Institute, 374
Benzene, 306, 347, 406
Bernays, E.L., 394
Binnig, G., 2
Bio-accumulation, 384
of silver, 302
Bio-inspired materials, 171, 374
Bioactive material, 244
BioAlliance, 271
Bioavailability, 297
BioBricks Foundation, 393
Biochip, 86
BioChipLab, 225
Biocompatibility, 232, 245, 255, 269, 345
of gold nanoparticles, 158
Biodegradability, 263
Biodistribution, 284
Biodiversity, 302, 371, 397
Bioengineering, 242
Bioethics, 366
Biofilm, 302, 406
benthic, 302
desquamation, 302
soil bacteria, 302
Biofuels, 372
Biogenetics, 373
Biograph mMR scanner, 230
Bioinformatics, 368, 369, 376
Biokinetics, 277
Biological membrane, 300
Biology, 113
molecular, 379
synthetic, 367–369, 371, 375–377, 379, 393, 395, 399, 400, 414
systemic culture, 376
systems, 368
Biomagnetism, 107
Biomarker, 225, 231, 232
surface, 236
Biomass, 372
Biomaterials, 243, 244
intelligent, 246, 254
requirements for, 245
toxicity, 245
Biopersistence, 285
Biopsy, 225, 232, 262, 281, 406
Biosafety, 373, 377

Biosecurity, 373
Biosensor, 232
  arsenic, 375
Biotechnological child, 376
Biotechnology, 340, 346
  excesses, 379
Bipyridine, 144, 406
Bit, 406
Blogosphere, 391
Blood, 232, 260, 263, 265, 280, 283, 284
  capillary, 278, 279
  circulation, 281
  clearance, 266
  sugar level, 345
Blood–air barrier, 280, 281
Blood–brain barrier (BBB), 266, 267, 284, 345
Blood–spinal cord barrier (BSCB), 266
Bone, 245
Bone tissue induction, 244
Boron nitride, 175
Bottom–up approach, 9, 28, 114, 344, 379
Bpifrance, 352–354, 372
Bragg multilayer, 75
Brain, 233, 237, 243, 266, 278, 281, 284, 322, 395
  cancer, 240
  disease, 377
  oxidative stress damage, 285
  oxygen consumption, 285
  pathologies, 285
BRAIN project, 217
Breakthrough curve, 298, 299
Breakthrough technology, 368
Bronchi, 279
Bronchial epithelium, 278
Bronchioli, 279
Brust–Schriffrin synthesis, 156, 157
Building materials, 340
Butterfly, 326
Byte, 406

C

Cadmium telluride, 95
Café scientifique, 358, 359
Cage molecule, 406
Canadian Centre for Occupational Health and Safety (CCOHS), 306
Cancer, 123, 249, 281, 385
  bladder, 257, 408
  brain, 240
  breast, 257, 408
  bronchial, 257
  colon, 263
  drugs, 258
  early diagnosis, 345
  liver, 271
  local hyperthermia, 239
  lung, 263, 279, 285, 345, 408
  nanodrugs, 254
  pancreatic, 257, 261, 263, 269, 408
  papillary thyroid, 264
  pleura, 285
  therapy, 387
  treatment, 257, 264, 410
Carbon, 12, 116
  activated, 132, 293
  aerogel, 132
  agglomerate, 117
  aggregate, 117
  balance, 372
  carbide-derived, 133
  porous, 132
  template, 132, 133
Carbon black, 117, 275
  functionalisation, 152
Carbon dioxide capture, 372
Carbon meta-nanotubes, 124
Carbon nanotube, 37, 123, 175, 209, 216, 276, 299, 306, 409
  functionalisation, 151, 299
  helicity, 124
  in friction test, 177
  in fuel cells, 162
  in industry, 341
  multi-walled, 123, 124, 176
  nanotexture, 126
  price, 344
  purification, 125
  single-walled, 123, 124, 177, 306
  synthesis, 124
  texture, 126
  thermal plasma synthesis, 125
  toxicity, 281
  transistor, 69, 71
Carbon onion, 175, 177
  in friction test, 177
Carbonatation, 184
Carbonyl, 406
Carboxyl group, 405, 406
Cardiac tissue, 245
Cardiorespiratory disease, 276
Cardiotoxicity, 263
Cardiovascular
  disease, 284, 345

system, 284
Carnot excellence label, 319, 353
Carnot institutes, 317, 319, 352, 353
Carnot–TIC–MNT alliance, 353
Cartilage, 245
Caspase assay, 262, 406
Catalysis, 340, 406
Cathepsins, 260
Cathodoluminescence, 50
CATRENE, 352, 354, 382
CDC, 133
CEA-LETI, 316, 319, 343, 354
Cell
biology, 214
CIGS, 95
ciliated, 279
death, 301
differentiation, 243, 245, 246
epithelial, 279
membrane, 260, 264
signalling, 214
therapy, 243
Cell-on-a-chip, 228
Cellular automata, 192, 196
Cement, 183–197, 275, 288, 289, 406, 407
alkali reaction, 184
binding phase, 183
carbonatation, 184
compressive strength, 185
degree of hydration, 190
hydration, 184
intelligent, 340
Portland, 185
pozzolanic reaction, 184
Roman, 184
setting, 184, 189, 190
sulfation, 184
theory of hydraulicity, 185
water-repellent, 341
Central nervous system (CNS), 266
Central processing unit (CPU), 66
Centre national de la recherche scientifique (CNRS), 316
ethical committee, 393
Cerebral ischemia, 266
Cerium oxide, 290
Charge effect, 295
Chemical production, 383
Chemical vapour deposition, 154
Chemical weapons, 377
Chemiluminescence, 410
Chemistry, 113
Chemotherapy, 236, 387, 408
China, 382
Chirality, 406, 413
Chloroauric acid, 156
Chloroform, 290
Chlorophyll, 202
Chromatography, 406
Chromosome, 408, 414
Chronic obstructive pulmonary disease (COPD), 278
CIFRE, 320, 326, 352, 353
CIGS cell, 95
Cilia, 279
Ciliated cell, 279
Cisplatin, 268
Citizens' conference, 358
Città Lente, 395
Civil society, 358, 380, 382, 390
Clay, 185
Clean energy, 159
Clean room, 334, 335
Clearance, 406
mechanisms, 278
Click chemistry, 138, 254, 406
applications, 139
Climate change, 371, 390, 397, 400
UNFCCC, 391
Clinical trial, 249, 271
Club NanoMétrologie, 318
Cluster, 40, 277
CMOS inverter, 58, 59
C'Nano, 316, 318, 336, 337
CNDP, 363
CNET, 343
CNFM, 331
Coating, 381, 406
Cochlear implant, 231
COGETERM, 388
Cognitive
science, 379, 380, 398
technology, 368
travel, 395
Collateral damage, 305
Colloid, 3, 154, 307, 406
Colloidal solution, 38
Combustion products, 276
Commissariat à l'énergie atomique et aux énergies alternatives (CEA), 316, 320, 363
Committee for Scientific and Technological Policy (CSTP), 385
Committee on Data for Science and Technology (CODATA), 390
Common good, 382

Competition, 396
  between nations, 382
  between scientists, 383
Competitiveness clusters, 319, 353
Complexing molecule, 407
Compressive strength, 185
Concentrated photovoltaics (CPV), 95
Concrete, 183, 202, 407
  cracks, 190
  Ductal, 187
  durability, 188
  for radioactive waste, 188
  grain loosening, 187
  grain packing, 187
  high performance (HPC), 186
  macro defect free (MDF), 186
  mix water, 186
  porosimetry, 188
  poured, 185
  reactive powder, 187
  ready-to-use, 196
  self-cleaning, 194, 202
  self-compacting, 186, 196
  self-consolidating, 186
  shrinkage, 184, 187, 192
  ultra-high performance (UHPC), 186, 188, 190
  very high performance (VHPC), 186
Conduction in 1D system, 45
Conductivity, 46
Conseil national de la consommation (CNC), 385
Consensus conference, 402
Consent engineering, 394
Consumer, 360, 364, 378, 382, 384, 386, 388, 391, 401, 402
  confidence, 364
  preference, 395
  protection, 275, 397
Consumer protection associations, 359, 363, 364
Consumers and societal dimension (CASD), 388
Continuum, 402
Contrast agent, 216, 230, 240, 241, 254, 268, 345
Convergence, 376
Converging Technologies for the European Knowledge Society (CTEKS), 379, 398
Cooperation, 396, 400
Copper, 289
Corporate purpose, 390
Cosmetics, 275, 277, 282, 288, 340, 381
  directive, 388
  market, 383
Cost–benefit ratio, 247, 249
Coulomb blockade, 68
Counterpower, 390, 391
Covalent bond, 5, 407
CPDP, 363, 392
Cracking, 407
Creutzfeldt–Jakob disease, 377
Cross-disciplinarity, 389
Cross-disciplinary training, 331
Cross-linking, 407
Cryofracture, 259
Cryptography, 407
  quantum, 73
CTAB, 155
Cyclisation reaction, 407, 411
Cycloaddition, 138, 407
Cyclodextrins, 146, 156
Cyclotoxicity, 407
Cytarabine, 258
Cytokine, 284
Cytoplasm, 212, 213, 216, 254, 260
Cytotoxicity, 204, 260, 268, 301
  of TiO_2 nanoparticles, 204

D

Dangerousness, 290, 306
  criteria, 384
Dangling bond, 176
Daphnia, 302
Data
  sharing, 389
  storage, 140
Database, 387–391, 393
  cross-disciplinary, 382
DdA-TP, 265
Defect tolerance, 90
Democracy, 371
Democritus, 2
Demultiplexer, 76
Dendrimer, 139, 155, 156, 407
Deoxycytidine deaminase, 257
Depollution, 336, 391, 399
Deposition, 32
  atomic layer, 32
  chemical vapour (CVD), 32
  sputtering, 32
  vapour–liquid–solid (VLS) method, 32, 41
Dermis, 282

DGCCRF, 364, 381
DGS, 381
Diabetes, 231
Diagnosis, 216, 220, 226, 227, 230, 246, 320
 difficulties, 226
 in vitro, 226, 228
 in vivo, 229
 point-of-care, 228
Dialogue, 358, 359, 363, 380, 381, 388, 398, 399, 401, 402
 inter-institutional, 383
 science–society, 382
 transatlantic, 403
Diamagnetism, 63, 407
Diazonium, 151
Didanosine, 256, 258
Didanosine (ddI), 264
Diesel emissions, 276, 279
Diesel engine, 276
Differentiated activation, 377
Diffusion, 278, 297
 length, 407
Digestive system, 277
Digital optofluidics, 87
Dihydroxylation, 138, 407
Dip coating, 164
Diplomatic process, 397
DNA, 214, 260, 363, 367, 369, 370, 374, 389, 411
 alteration, 300, 301
 amplification, 86
 chip, 228
 elasticity, 214
 instrumentalisation, 380
 replication, 413
 synthetic, 393
Do Not Track (DNT), 366
Doctoral degrees, 328
Double layer, 136, 137, 154
Doxorubicin, 123, 258, 263, 268
Drexler, E., 392
Drexler, K.E., 11
DRT, 316
Drug delivery, 147, 236, 237, 253–271, 345, 407
 for central nervous system, 266
Drug encapsulation, 234, 235
Drug pump, 237
Drug release, 254
Drug targeting, 254
DSM, 316
Dumping, 403
Dupuy, J.-P., 401

E
Echography, 229
ECJS teachers, 327, 335
École Centrale Paris, 328
École normale supérieure (ENS), 328, 335
École Polytechnique, 328
Economic Forum, 403
Ecosystem, 204
Ecotoxicity, 347, 365, 384, 386, 391
Education, 361, 382
Egoism, 371
Ehrlich, P., 253
Eigler, D., 3
Elastomers, 147
Electric vehicle, 372
Electrografting, 151
Electroluminescence, 410
Electrolyte, 407
Electromagnetic invisibility, 72
Electron, 6
Electron energy loss spectroscopy (EELS), 49
Electron microscope, 2
Electron–hole pair, 201, 202
Electron–hole recombination, 75
Electronic components, 343
 passivation, 341
Electronic mail, 367
Electronics, 366
 industry, 340
 molecular, 380
Electrophilic compound, 407
Electrospinning, 244
Electrostatic interaction, 154, 298
Embolism, 254
Emulsion, 407
Enabling technology, 358
Enantiomer separation, 162
Encapsulation, 254
Endocytosis, 266, 407
Endogenous process, 407
Endogenous stimulus, 254
Endolysosomal membrane, 260
Endoplasmic reticulum
 membrane, 260
Endoscopy, 232
Endosome, 265
Endothelial cell, 284
Endothelial dysfunction, 284
Endothelium, 284
Endy, D., 370, 374
Energy, 371, 390
 carbon-free, 391, 399

consumption, 306
efficiency, 371
production, 372
renewable, 372, 399
saving, 362
storage, 134, 317
Engineering courses, 328
Enterocyte, 283, 284
Environment, 195, 287–309, 361, 363, 389
effect on genetic mutation, 369
freshwater, 302
marine, 302
protection, 367
UN program, 400
Environmental dumping, 386, 403
Environmental Protection Agency (EPA), 364
Environmental transition law, 390
Enzyme, 408, 410
activity, 372
intracellular, 409
lysosomal, 265
Epidemiological studies, 275, 276
Epidermis, 282
Epigenetics, 369
Epigenomics, 368
Epimutation, 370
Epistemic effects, 379
Epitaxial film, 343
Epitaxy, 408
Epithelial cell, 279
Epithelium, 283
single-layered, 284
Epoxidation, 138, 408
EPR effect, 236
Equilibrative nucleoside transporter 1 (ENT1), 408
Equipex, 320
ERA SynBio, 393
Erasmus Mundus program, 330, 331
Etching, 31
electron beam, 31
ion beam, 31
plasma, 31
RIE, 31
Ethera, 347
Ethics, 247, 327, 366, 371, 378, 384–387
dumping, 403
possibility of, 379
Ethylenediamine, 306
ETI, 352
Eugenics, 375
EUREKA, 354, 382
EuroNanoMed, 250
European Chemical Agency (ECHA), 385
European Committee for Standardisation (CEN), 386
European Court of Justice, 367
European Food Safety Agency (EFSA), 386
European Group on Ethics in Science and New Technology (EGE), 247, 248
European Institute of Innovation and Technology (EIT), 351
European Medicines Agency (EMA), 248, 386
European Patent Office (EPO), 355
European Trade Union Conference (ETUC), 381
Europrion project, 376
Evanescent field, 414
Evanescent wave, 54
Evolution, 367, 368
interfering with, 370
Exciton, 408
Exeresis, 238, 408
Exfoliation, 408
Exogenous process, 407
Exogenous stimulus, 254
Expert network, 402
Expert system, 307, 308
Extinction, 371
Extracorporeal magnetic field, 254, 267
Extrusion, 164, 408

F

Face-centered cubic structure, 295
Farnesyl, 269
FEMTO-Innovation, 319
Ferroelectric material, 408
Ferromagnetism, 63, 408
Fert, A., 324
Fertiliser, 289
Fibre optics, 80
Fibrosis, 281, 408
Filter organism, 302
Financing, 350
Fire resistance, 340
Flat lens, 77
Flex circuit, 46, 70
Flexible electronics, 70
Fluobeam probe, 241
Fluorescence, 408
Fluorescence imaging, 408
Fluorescent nano-object, 85
Focused ion beam (FIB), 31

Food, 377, 386, 390, 399
additives, 381, 383, 390
chain, 297
industry, 146, 275, 283
packaging, 283
security, 362
Food and Agriculture Organisation (FAO), 385, 399
Food and Drug Administration (FDA), 248
Footbridge of Peace, 185, 187
Force-feedback nanomanipulator, 337
Formaldehyde, 347
Fossil fuels, 371
FP7, 385, 391, 401
Fractal, 295
Framework Programme for Research and Technological Development, 385, 391, 401
Framing Nano, 385, 402
Fraunhofer institutes, 331, 353
Free radical, 284, 285, 290, 301
Freedom of thought, 393
Freescale, 353
French education system, 328–331
FRET, 85, 267
Friction, 414
Friction curve, 177
Fuel cell, 107, 159
electrode material, 159
proton exchange membrane, 159
Fullerene-like molybdenum disulfide, 176
Fullerenes, 37, 119, 293, 306, 408
aggregation, 294
hydrophobic/hydrophilic, 292
hydroxylated, 295, 308
inorganic, 175
interaction with water, 292, 299
non-hydroxylated, 295
Fullerol, 295, 299, 308, 408
aggregates, 308
Fulvic acid, 299, 302
Functional probe, 210
Functionalisation, 148–152, 210, 232, 254, 263
chemical, 150
electrochemical, 151
gas phase, 149
of nanodiamond, 149
Fungi, 302
Fusion oncogene, 264

G

G20, 402
G8, 402
Gadolinium, 268
Galenic pharmacy, 254, 408
Game console, 395
Gate, 57
length, 57
oxide, 59
voltage, 58
Geisha effect, 201
Gemcitabine, 256, 257, 260, 270, 407, 408
cytotoxicity, 260
free, 261
intracellular penetration, 257
magnetically guided, 267
metabolisation, 261
phosphorylation, 260
release kinetics, 269
resistance to, 260
targeted release, 260
with terpenes, 269
Gemzar, 257
Gene, 11, 408
therapy, 408
GeneChip, 228
Genepac project, 160
General public, 336, 358
Genetic mutation, 369
Genetically modified organism (GMO), 377, 380, 397, 401
Genome, 227, 367–369, 374, 407, 408
sequencing, 369
Genotoxicity, 204, 301, 363, 407
Geolocation, 366
GIANT, 317
Giant magnetoresistance (GMR), 64, 89, 103, 324, 342
Glass, 288
coloured, 290
insulating, 372
Global Bioenergies, 372
Globalisation, 396
Glucometer, 232
Glycemia, 345
Glycoprotein, 411
Goblet cell, 283
Gold, 139, 140, 290
biocompatibility, 158
functionalisation, 148
nanoparticle, 154–156, 213
Gore, A., 397
Governance, 363, 364, 368, 371, 380, 382, 385, 389, 390, 396–398, 400–403
continuum, 402

harmonised, 403
reflexive, 363, 401
risk, 385
standard, 397
Grafting, 139
Grand emprunt, 352
Granuloma, 281
Graphene, 38, 43, 126
chemical exfoliation, 127
chemically converted, 128
conductivity, 46
EU Flagship project, 321–322, 382
FET flagship, 127
synthesis, 127
transistor, 69
Graphite, 40, 116
exfoliation, 43
intercalation compound, 128
Grätzel cell, 202
Green growth, 399
Greenhouse gas, 195, 371, 397, 400
Grenelle de l'environnement, 347, 363, 383
Grinbaum, A., 401
Growth, 32
Guest molecule, 146
Gypsum, 194
Gyrolite, 188

H

H2020, 385, 401
Hafnium oxide, 347
Hair follicles, 282
Halas, N.J., 387
Hall effect, 409
Hall, E.H, 409
Hard disk, 62, 103, 341
Hardness, 409
Harmonisation, 389, 400
Health, 273–285, 357, 384, 389–391
and safety, 361, 365
crises, 285
industry, 340
warning, 395
Heart, 281, 284
Heat pump, 372
Helicity, 409
Hematopoietic toxicity, 261
Hematoxylin–eosin (HE) stain, 409
Hemolysis, 264
Hepatocarcinoma, 271
Herbicide, 294
Heredity, 369
Herpes, 258
Heteroepitaxy, 408
Heterostructure, 41, 409
High performance liquid chromatography, 162
High-K oxide, 60
High-throughput analysis, 226
High-throughput screening, 228
Higher education, 328
Histology, 409
Homeostasis, 282
Homoepitaxy, 408
Hong Kong University of Science and Technology (HKUST), 375
HORIBA, 353
Horizon 2020, 322, 350
Host molecule, 146
Human Brain Project, 217
Human–machine interface (HMI), 380, 393–396
Humic acid, 299
Humic substance, 293
Huntington's disease, 285
Hybrid materials, 165, 210
Hybrid nanomaterials, 163
classification, 164
legochemistry approach, 166
synthesis, 165
synthesis using amphiphilic molecules, 167
Hydrated calcium silicates, 183
Hydrates, 183
Hydration, 409
Hydrogen bond, 143
Hydrogen fuel cell, 317
Hydrolysis, 409
Hydrophilic–hydrophobic interaction, 143
Hydrophilicity, 295
Hydrophobic pocket, 255
Hydrophobicity, 341
Hydroxylation, 409
Hydroxypropyl methylcellulose (HPMC), 186
Hyperthermia, 239, 254
Hypodermis, 282

I

Iatrogenic effect, 414
IBM, 320, 354
ICPC NanoNet, 385
IEMN, 319
Ilmenite, 200

Imaging, 267
Immersive technology, 394–396
Immersive virtual reality, 395
Immortality, 375
Immune system, 245, 283, 405, 410
Immunoassay, 86
Immunohistochemistry, 262, 409
Imogolite, 288, 289
Impact assessment methodology, 385, 401
Impermeable cloth, 341
Implant, 231, 237, 384
  nanostructured, 243
Inac, 316
INC, 316
InCheck, 228
Inclusion complex, 146, 147
Incoming mobility, 334
Indentation, 409
Indium tin oxide (ITO), 130
Industrial property rights, 363
Industrial revolution, 371
Industrialisation, 361
Industry, 356
  in France, 340–347
  key sectors, 343–345
  worldwide, 348–354
INERIS, 363
Inertial impaction, 277
Infectious diseases, 264–266
Inflammation, 230, 236, 279, 281, 284, 301
Information and communication technologies (ICT), 321, 322, 349, 353, 372, 393, 394, 398
Information society, 353
Innovation, responsible, 373
Inorganic fullerene-like metal disulfide, 175, 177
INP, 316
INP Phelma, 328
INRA, 370
INSA, 328
INSB, 316
INSHS, 316
INSIS, 316
Institut Mines-Télécom, 395
Institut national de la santé et de la recherche médicale (INSERM), 316, 317
Instrumentation, 361
Insulin, 237, 238, 374
Integrated circuit, 3, 26
Intellectual property, 373
Interception, 277
Interconnects, 60
  intra-chip, 81
  optical, 73
Interface energy, 291
Interferometer, 76
Intergovernmental Panel on Climate Change (IPCC), 389, 397, 400
International Agency for Research on Cancer (IARC), 281
International Atomic Energy Agency (IAEA), 397, 400
International consensus meeting (ICM), 358
International cooperation network, 402
International Council for Scientific Unions (ICSU), 389
International dialogue on responsible research and development of nanotechnology, 380
International Genetically Engineered Machine (iGEM) competition, 374
International relations, 362
International Risk Governance Council (IRGC), 373
International Standards Organisation (ISO), 381, 385, 386, 393
International Technology Roadmap for Semiconductors (ITRS), 28
International Virtual Observatory on Astronomy (IVOA), 389
Internet, 391
Internet Protocol (IP), 367
Intestinal barrier, 284
Intestine, small, 284
Intracellular
  enzymes, 260
  infection, 265
  membranes, 260
Intravenous administration, 253, 255, 261, 263, 266, 268
Invertebrate, 302
Investissements d'avenir, 320, 352
  nanobiotechnologies, 250
Invisibility cloak, 77, 140
IOGS, 319, 328
Ionic strength, 299, 307, 409
Ionic–covalent bond, 409
Iramis, 316
Iridium, 281
IRT, 320, 352
  M2P, 320, 352
  NanoElec, 320, 352
ISO 26000, 402
Iso-Combretastatin, 263
Isobutene, 372

Isoprene, 269
Isothermal calorimetry, 190

J
Jablonska, E., 370
Joint Research Centre (JRC), 386
Jordan, A., 387
JORF, 368
Junction oncogene, 409

K
Keasling, J., 369
Ki67 staining, 409
Kilby, J., 3
Kinesin, 213, 214
Klein, E., 380
Knight, T., 374
Knoll, M., 2
Knowledge society, 353
Kurzweil, R., 392
Kyoto protocol, 397

L
LAAS, 319, 335
Lab-on-a-chip, 86, 225–228
Labelling, 386
Labex, 320
Laboratory animals, 277
Laborit, H., 376, 377
Land mine, 369
Laser
  quantum cascade, 82, 412
  semiconductor, 74, 341
  thresholdless, 73
  vertical cavity, 76
  waveguide, 76
Le Châtelier, H., 185
Legal issues, 383–384
Legislation, 327
Legochemistry, 166
Lehn, J.-M., 324
LETI, 317
Leucippus, 2
Leucocyte, 410, 412
Leukemia, 258
  lymphocytic, 261
  mouse, 261
  rat, 261
Libertarianism, 375
Lifetime, 409
Ligand, 409
Light emission, 73
Light-emitting diode, 140
Limestone, 185
Lipid, 255, 409, 413
  alteration, 300
  cationic, 264
Lipid nano-delivery system, 223
Lipidot, 223, 235
Lipophile, 409
Liposome, 263, 409
Liquid crystal, 140, 141
  hybrid, 167
LIST, 317
LITEN, 317
Lithography, 29, 410
Liver, 263, 265, 281
LNE, 318
Local hyperthermia, 410
Lorentz force, 409
Lotus effect, 341
Lubrication, 414
Lubrication additives, 175, 177
Luminescence, 73, 410
Lung, 263, 277, 279, 285, 412
  cancer, 345
  infection, 279
  particle retention, 279, 281
Lycurgus cup, 38, 46, 158
Lymphocyte, 264, 265, 410
Lymphoid organs, 261, 410
Lysosome, 254, 265, 266, 410

M
M2P, 320, 352
Macrophage, 256, 265, 266, 283, 410, 411
  alveolar, 279, 280
  clearance by, 278
  liver, 263
Macropore, 133
Mad cow disease, 377
MagForce, 387
Maghemite, 254, 292
Magic bullet, 253
Magnetic
  guidance, 267
  imaging, 293
  mapping, 106
  tunnel junction, 65
  tweezers, 216
Magnetic resonance imaging (MRI), 106, 107, 216, 229, 230, 268, 410, 411
Magnetic sensor, 103, 104

angle, 105
GMR, 104
TMR, 104
Magnetisation, 62, 63
Magnetite, 267
Magneto-encephalography (MEG), 108
Magnetocardiography (MCG), 106, 108
Magnetoresistance, 410
Magnetoresistive random access memory (MRAM), 410
Malaria, 369
Maleimide, 410
Manipulation
of atoms, 51, 56
of magnetisation, 62, 67
of single spins, 62
Marine sediment, 302
Market surveillance, 381, 384
Mask, 31
Massachusetts Institute of Technology (MIT), 370, 375
Master's degrees, 328
in Europe, 331–334
in France, 329
Materials chemistry, 254
M cell, 283, 284
Mechanical transduction mechanisms, 214
Mechanomechanical polishing, 410
Mechanosynthesis, 410
Medical imaging, 229–231, 320
combined modes, 230
intraoperative, 241
nanoparticles used for, 284
resolution, 345
Medical research, 317
Membrane integrity, 301
Membrane receptor, 214
Memory, 57, 410
cache, 66
flash, 67
hard disk, 62
magnetic, 62, 65
race track, 66
random access, 65
STT, 65, 66
Memristor, 70
Mentalics, 380
Mesenchymal cell, 245
Mesocosm, 302, 410
Mesopore, 133
Meta-convergence, 379
Metabolisation, 266, 410
Metabolism, 374, 377, 405
Metal, 410
disulfides, 175
oxides, 296
Metallic–organic framework (MOF), 166
Metalloid, 296
Metamaterial, 72, 410
Metaphysical effects, 379
Metrology, 285, 318
Micelle, 168, 406
Michelin, 345
Micro X-ray analysis, 205
Micro- and nanotechnology (MNT), 353
Microelectromechanical systems (MEMS), 83, 237
Microelectronics, 26
market, 348
Microemulsion, 164
Microfluidics, 226, 228
Micromanipulation, 214
Micromixer, 228
Micropore, 133
Microprocessor, 2, 30, 57
frequency, 57
Micropump, 224, 228
Microscope
atomic force, 3, 24, 52
electron, 2
optical, 53
scanning electron, 48
scanning near-field optical, 24, 54
scanning tunnelling, 2, 7, 24, 50, 251
transmission electron, 49
Microtubule, 213, 214
Microvalve, 228
Microvillus, 284
Mie theory, 157, 201
Millennium Development Goals (MDGs), 403
Minalogic, 353
MINATEC, 317, 336
Miniaturisation, 340
of microprocessors, 341
Mobile phone, 343, 374, 382
Mobility, 411
Molecular
biology, 368, 379
dynamics, 196
electronics, 57, 69
imaging, 229, 230
imprinting, 162–163
motor, 213, 214
scaffolds, 245
sorting, 225

Molecular beam epitaxy (MBE), 32, 342, 343
Molecularly imprinted polymer (MIP), 162
Molecule, 1
Moletronics, 380
Molybdenum dithiocarbamate, 177
Monocrystal, 411
Monocyte, 410
Moore's law, 26, 27
Moore, G., 26, 27
Morbidity, 374
Morpho butterfly, 326
Motor protein, 213
Mouse, 261, 263, 267, 275, 281, 284, 411
  stroke model, 266
MRAM, 66
Mucociliary epithelium, 279
Mucous cell, 284
Mucus, 279, 411
Multidrug nanoparticle, 263
Multilayer, 341
Multiple sclerosis, 266
Multisector applications, 340
Mutation, 370
Myocardial infarction, 284
Myosin, 214

N

Nano publications, 315
Nano strain gauge, 345
Nano-delivery system, 222
Nano-École, 327, 335, 337
Nano-enabled products, 348
Nano-harvesting, 232
Nano-INNOV, 251
Nano-inside, 222
Nano-onion, 119
Nano2, 361
Nano2012, 320
Nano2017, 320, 352, 354
Nanoantenna, 55, 78
Nano@school program, 327
Nanobioplasmonics, 79
Nanobiotix, 347
Nanocapsule, 235
Nanocarrier, 222, 230, 231, 234–236, 241, 253, 411, 413
  advantages, 236
  antibiotic, 265
  core, 267, 268
  doxorubicin-loaded, 263
  for theranostics, 268
  magnetic, 254
  market, 349
  nucleic acid, 264
Nanocatalyst, 372
Nanocathode, 341
Nanochemistry, 113
  market, 349
Nanoclay, 167, 283
Nanocomposite, 118
Nanocompression test, 179
Nanoconcrete, 194, 195
  environmental advantages, 195
Nanocrystallites in cement, 183–197
Nanodiamond, 120
  functionalisation, 149
Nanodroplet, 235–236, 411
  lipid, 235
Nanodrug, 13, 19, 222, 235, 284, 317, 411
  advantages, 254
  burst release, 253, 254
  intelligent, 254
  metabolisation, 255, 257, 260, 261
  regulation, 248
  squalenoylation, 253–271
Nanodrugs, 222
NanoElec, 320
Nanoelectronics, 317, 320, 324, 343, 361, 366, 374, 382, 393, 394
Nanofabrication, 340, 361
Nanofibre, 123, 243
Nanofiller, 118
Nanofilm, 283
Nanofluidics, 226
NanoGenotox, 363
Nanoimprinting, roll-to-roll, 34
Nanoindentation, 53
NanoInnov, 317, 319, 335, 336, 352, 353, 363
Nanolithography, 410
Nanolubricants, 175
Nanomagnetism, 62
Nanomanipulator, 337
Nanomaterial, 18, 209, 289
  biodegradable, 255
  contaminant adsorption, 307
  database, 306, 307
  definition, 383, 388
  environmental risks, 287–309
  exposure, 275
  health impacts, 273–285
  integration, 344
  life cycle, 304
  production, 343–344

redox reactivity, 307
registers, 358
risks, 303–309
second order effects, 307
structural properties, 307
Nanomedicine, 216, 219–252, 320, 374, 385, 399
certification, 248
clinical trials, 249, 251
definition, 221
French firms, 251
in France, 250
industry, 249
market, 249, 349
typical development period, 251
Nanometer, 1
Nanomonde platform, 336
Nanoparticle
accumulation, 277, 285
agglomeration, 154
aggregation, 293–296, 308
aluminium oxide, 299
alumoxane, 306
as fluidifier, 283
as lubricant, 175
as probe, 210
as trap for contaminants, 293
attachment efficiency, 298, 299
based on squalene, 256
bimetal, 161
bioaccumulation, 284
bioavailability, 297
biodistribution, 284
biopersistence, 285
boron nitride, 175
carbon, 12, 281
carbon black, 275
carcinogenic effects, 281
cardiovascular effects, 284
catalytic properties, 290
cerium oxide, 290
coating, 202
colloidal silica, 275
core–shell, 167
ddC–squalene (SQgem), 257
effect on heart, 284
electrostatic stabilisation, 154
elimination, 278
exposure routes, 277–284
cutaneous, 282–283
oral, 283–284
respiratory, 277–281
for fuel cell, 158
for local hyperthermia, 239
for radiotherapy, 238, 239
for targeting, 230
for theranostics, 240
fullerene, 293
functionalised, 211, 214–216, 344, 387
gemcitabine–farnesyl, 269
gemcitabine–polyisoprene, 270
gemcitabine–squalene, 269
gold, 213, 281, 290, 302
hafnium oxide, 347
health impacts, 317
in alveolus, 279
in food, 283
in medicine, 220
in nervous system, 284
in porous medium, 297–300
in water, 287
indium, 290
industrial production, 344
inorganic, 299
insoluble, 279
interaction with water, 291–296
iridium, 281
iron, 284, 301
iron oxide, 239, 267, 293
labelled, 277
lipid, 235
lipid solubility, 284
magnetic, 215, 216, 239, 268, 410
manufactured, 289
metal, 158
metal oxide, 295, 296
metallic silver, 302
migration, 297
mobility, 299
morphology, 291
mucociliary clearance, 278–280
multidrug, 263
NanoXray, 238, 239
naturally occurring, 288
of squalene conjugates, 257
optical properties, 290
paclitaxel–squalene (SQgem), 257
paramagnetic, 214
penicillin–squalene (SQgem), 257, 259, 265
persistence, 277
phagocytosis, 280
photocatalytic properties, 290
platinum, 161
polyacrylate, 281
produced by human activity, 289

professional exposure, 277, 281
sedimentation, 302
shape, 158
silica, 12, 283, 289, 299
silver, 12, 275, 291, 302
size, 158
size distribution, 294
solubility, 291, 300
soluble, 278
SQddC, 264
SQddl, 264, 265
squalene–gemcitabine (SQgem), 257–263
stabilisation, 154, 302
steric stabilisation, 154, 155
surface charge, 297
thermal properties, 290
tin, 290
titanium dioxide, 12, 200, 281, 290, 296, 302, 306, 336
titanium oxide, 275, 281, 282
toxicity, 204, 276, 284, 296, 300–303
transition metal, 154
translocation, 277, 281, 285, 381
Trojan horse, 292
zinc oxide, 282, 302
Nanophotonics, 72
Nanopillar, 233
Nanoprobe, 212, 213
Nanopump, 237, 238
Nanoribbon, 44, 209
NANORIS, 363
Nanosafe 2, 385
Nanosand, 12
Nanoscience, 4
at school, 334
Nanosensor, 224, 345
Nanosheet, 167
clay, 167
Nanosource, 73
Nanospaghetti, 258
Nanosphere, 235
Nanostructured media, 65
Nanostructured surface, 245
Nanotechnology, 4
Nanotechnology Center for Learning and Teaching (NCLT), 335
Nanotechnology Consumer Product Inventory, 273
Nanotechnology working group (NTWG), 399
Nanotechnology, biotechnology, information technology, and cognitive science (NBIC), 379, 392, 393, 411
Nanotheranostics, 268
NanoTherm particles, 239
Nanotoxicity, 273–285, 287–309
Nanotoxicology, 276
Nanotracer, 205
NanoTrend, 382
Nanowire, 37, 209
InP, 42
NanoXray platform, 238, 239, 347
Nasopharynx, 278, 279
National Aeronautics and Space Administration (NASA), 323
National Institute of Standards and Technology (NIST), 323
National Nanotechnology Initiative (NNI), 323, 349, 350, 361
National Science Foundation (NSF), 335, 379, 380, 398
NBIC technologies, 379, 392, 393, 411
Negative refractive index, 72
Nervous system, 249, 284
Neural tissue, 245
Neurodegenerative disorders, 285, 345
Neurological pathologies, 266
Neuron repair, 243
Neuronal cell, 212
Neurovirology, 376
New deal, 371
New energy forms, 317
New green deal, 371
Nickel, 289
Nintendo, 395
NIOSH, 204
Nitrogen, 302
Nitroxide mediated polymerization (NMP), 270
NMP, 320
Non-destructive testing, 106
Non-tariff trade barrier, 363, 397
Normative evaluation, 382
Nuclear imaging, 229
Nuclear magnetic resonance (NMR), 191
Nuclear power, 380, 397
Nucleic acid, 275
Nucleophilic compound, 411
Nucleoside, 408, 411
Nucleoside analogue, 255–257, 408, 411
antiretroviral, 264–265
Nucleoside transporter, 257, 260
Nucleotide, 264

Nucleus, 254

O

Observatory Nano, 362, 385, 391
Ocular disorders, 395
Olfactory nerve, 278
OMNT, 318
Oncogene, 264, 409
OneGeology, 389
Ontological effects, 378
ONXEO, 271
Open access, 389–391, 393
Opsonin, 411
Opsonisation, 265, 411
Optical
  disk, 341
  fibre, 415
  microscope, 53
  tweezers, 56, 214
Opto-microfluidics, 86
Optoelectronics market, 348
Oral administration, 255
Organisation for Economic Co-operation and Development (OECD), 379, 381, 385, 390, 398–399, 402
Orphan disease, 385
Orthotopic mouse model, 411
Osteopontin, 244
Outgoing mobility, 334
Outreach, 334–336, 350
Oxidative stress, 281, 284, 285, 300, 301, 411
Oxidosqualene cyclase, 255, 411

P

P2N program, 319, 353
Pace setter, 402
Packaging, 381
Packard, V., 367, 371
Paclitaxel, 257, 258, 268
Paint, 275, 288, 340, 381
Pancreas, 261, 263, 269
Pandemic, 374
Paramagnetism, 63, 411
Parental control, 395
Parkinson's disease, 243, 285, 377
Passive targeting, 236
Patent, 355, 373, 383, 384
PAVEL, 385
PCPC NanoNet, 391
Pearl necklace nanofibre, 125
Peer reviewing, 400
PEGylation, 236, 256, 263, 411
Pellet, 412
PEMFC, *see* Fuel cell
Penicillin, 257, 258, 265, 266
Percolation, 190
Personal hygiene products, 277
Personalised medicine, 216, 233, 268, 374
Pesticide, 288, 294
Peyer patch, 283, 284
PH metre, 296
Phagocytosis, 279, 410, 412
  frustrated, 301
Pharmaceutical formulation, 254
Pharmaceutical products, 381, 390
Pharmaceutics, 146
Pharmacognosy, 234
Pharmacy, 234
Phonon, 75, 412
Phosphorylation, 265, 412
Photocatalysis, 202, 296
Photodynamic therapy (PDT), 240
Photoluminescence, 410
Photon, 412
Photonic crystal, 72, 336
Photonics, 412
Photosynthesis, 202
Photovoltaic cell, 341, 372, 394
  organic, 129
Photovoltaic conversion, 73
Photovoltaics, 95, 202, 317, 372
  concentrated, 95
  organic, 129
Physics, 113
PIA, 334
PICOGIGA, 343
Picophysics, 89
Piezoelectric effect, 412
Piezoresistance, 84
Pig, 282
PillCam COLON capsule, 232
Placental barrier, 285
Planar technologies, 29
Plankton, 302
Plant nutrition, 302
Plasma, 412
Plasma membrane, 257
Plasmon, 47, 55
  band, 157
  surface, 414
  waveguide, 140
Plaster, 194
Platinum, 160
Pleura, 285

Pleural effusion, 281
Pneumocyte, 279, 412
Pollution, 175, 284, 288, 291
 atmospheric, 279
 indoor air, 347
 organic, 290, 293
 risks, 306
Poly alpha olefin oil, 177
Polycaprolactone (PCL) membrane, 244
Polycation, 264
Polycyclic aromatic hydrocarbons (PAH), 288, 306
Polyelectrolyte, adsorption of, 299
Polyethylene glycol (PEG), 236, 256, 411
Polyisoprene, 269
Polymer, 407
 π-conjugated, 130
 biocompatible, 269
 cationic, 264
 methacrylic, 187
Polymerisation
 inorganic, 164
 nitroxide mediated, 270
 radical, 270
Poly(methyl methacrylate) (PMMA), 245
poly(*N*-vinyl-2-pyrrolidone) (PVP), 156
Polysaccharide, 299
Poly(vinylalcohol) (PVA), 155, 156
Poly(vinylether) (PVE), 156
Pont du Gard, 185
Population inversion, 75
Pore, 132, 188
 IUPAC classification, 133
 size, 298
Porosimetry, 188
Porosity, 297
Porous media, 297–300
Portland cement, 185
Positron emission tomography (PET), 230
Potassium tetrachloroaurate, 156
Potential barrier, 25
Powder, 277
Power effects, 380
Power electronics, 353
Pozzolanic reaction, 184
Precaution dumping, 386
Precautionary principle, 366, 383, 403
Preconcentration, 225
Prion, 376, 377
Private capital, 358
Probability network, 307
Processor, multicore, 61
Professional opportunities, 334
Profil'air kit, 347
Prokaryote, 406
Prosthesis, 242, 374
Protease, 412
Protein, 275, 370, 405, 412, 413
 alteration, 300
 conformation, 301
Proteome, 227
ProTool technology, 233
Protozoa, 302
Psoriasis, 282
Psycho-neuro-endocrine-immunology (PNEI) system, 395
Public debate, 390
Public health, 317, 377
Pulmonary fibrosis, 281, 285
Pulmonary surfactant, 280
Pyridine, 412
Pyrolysis, 412

Q

Quality factor, 412
Quantitative structure–activity relationship (QSAR), 307
Quantum cascade, 412
Quantum components, 57
Quantum computer, 89
Quantum cryptography, 73
Quantum dot, 7, 37, 209, 211, 216, 290, 412
 CdSe, 38
 GaAs, 39
 in nanowire, 42
 InAs, 39
 toxicity, 220
 ZnSe, 306
Quantum electrodynamics, 46
Quantum fluctuation, 412
Quantum size effects, 44
Quantum teleportation, 73
Quantum well, 412

R

Radioactive tracer, 277
Radiotherapy, 238, 239, 347
 with iron oxide nanoparticles, 239
Ragone plot, 134
Raman effect, 413
Random access memory (RAM), 65, 410
Rare earth, 105
Rat, 245, 261, 275, 281, 284
 spinal cord injury model, 266
Rayleigh's law, 202

REACH regulations, 383, 397, 413
Reactive oxygen species (ROS), 290, 295, 300–302, 411
Read head, 341, 342
Read only memory (ROM), 410
Redox reaction, 413
Refraction, 413
Refractive index, 415
  negative, 76, 411
Regenerative medicine, 242
Registry of Standard Biological Parts, 370, 374
Regulations, 347, 359, 381
  CLP, 383
  in Europe, 365
  REACH, 383, 397, 413
Regulatory gene expression network, 374
Reindustrialisation, 401
Religion, 392–396
RENATECH, 316, 319
Renewable energy, 159, 372
Reprecipitation, 287
Research
  Chinese programs, 324
  European programs, 320–322
  international programs, 320–324
  Japanese programs, 323
  Taiwanese programs, 324
  US programs, 323
Research and development
  in France, 318–320
Research and Development Framework Programme (RDFP), 320, 350
Research tax credit, 318, 352, 354
Resist, 29
Resources, 362
Respiration enzyme, 301
Respiratory
  barrier, 276
  pathologies, 281
  system, 277
Responsible research and innovation (RRI), 401, 403
Restriction nuclease, 367
Rettberg, R., 374
Reverse engineering, 379
Reverse iontophoresis, 232
Rhodamine B, 205
Ribosome, 413
Rip, A., 401
Risk
  assessment, 303–309, 384
  capital, 350
  forecasting models, 306
  governance, 385
  management, 303
  normative assessment, 401
  synthetic biology, 400
  systemic, 378–384, 401, 402
Risk–benefit analysis, 246, 383
RNA, 258, 369, 370, 411
  polymerase, 214
Roadmap, 379, 380
Roco, M., 380
Rohrer, H., 2
Roman cement, 184
Roosevelt, F.D., 371
RTRA, 319
Ruska, E., 2
Rutile, *see* Titanium dioxide

S

Samsung, 395
Sand, 299
Sanofi, 369
Scalability, 384, 390
Scale, 413
Scanning electron microscope (SEM), 48
Scanning near-field optical microscope (SNOM), 24, 54
Scanning tunnelling microscope (STM), 2, 7, 24, 50, 251
School, 334
Science and society, 327
Science café, 358
Sediment, 294
Sedimentation, 277
Selective extraction, 162
Self-assembled monolayer (SAM), 150
Self-assembly, 9, 114, 141, 254, 269, 379
  of amphiphilic molecules, 167
  of squalene conjugates, 257
Self-cleaning windows, 202
Self-healing rubbers, 147
Semi-metal, 296
Semiconductor
  band gap, 97
  III–V, 81, 95, 100
  laser, 341
  n-type, 413
  p-type, 413
  superlattice, 414
  thin film, 343
Sensor, 152, 162, 210, 231, 347
  acceleration, 341

chemical, 153
for diagnosis, 345
GMR, 342
magnetic, 341
position, 341
Shale gas, 380
Shear wave ultrasonic testing, 190
Sherbrooke Footbridge, 187
Signalling protein, 216
Silica, 12, 169, 283
colloidal, 187, 275
mesoporous, 133, 141
microbeads, 298
microporous, 133
nanoparticle, 141, 289
toxicity, 285
Silicon, 30, 95, 226, 252
wafer, 227
Silicon carbide, 252
Silicon-on-insulator (SOI) technology, 60, 343
Silver, 12, 275
bio-accumulation, 302
nanofilm, 283
Single-electron components, 57
Single-photon source, 42
SiRNA, 257, 258, 264
coupled to squalene, 264
Skin, 282
disease, 282
reconstituted, 282
Slow City, 395
Small and medium-sized enterprises (SME), 318, 352
Smart meter, 372
Smart network, 372
Smartphone, 382
Smolke, C., 370
SNRI, 320
Social debate, 357
Social dumping, 403
Social networks, 391, 394, 396
Social sciences, 318
Socio-scientific debate, 327
Soft chemistry, 163
Soft matter, 164
SOITEC, 343
Solar cell, 95, 202
conversion efficiency, 202
dye sensitised, 202
energy efficiency, 98
multijunction, 97–99
Solar farm, 95
Sol–gel process, 163–165, 201, 347
Solubility, 277, 291
Solubility product, 291
Solvent, 294, 306
Sonodrug, 230
Sony, 395
South Korea, 391
Speciation, 302, 413
Specific energy, 134
Specific power, 134
Specific surface area, 276
of activated carbons, 132
of CDCs, 134
Spectroscopic imaging, 229
S phase, 413
Spin, 63, 413
magnetic moment, 63
Spin caloritronics, 67
Spin coating, 164
Spin current, 62
Spin electronics, 61
Spin-lattice relaxation, 192
Spin-off companies, 346
Spin valve, 63, 103, 104
fixed layer, 104
free layer, 104
pinned layer, 104
supramolecular, 70
Spinal cord, 243
injury, 266
Spintronics, 62, 103, 413
Spirituality, 391
Spleen, 265, 281
Spontaneous emission, 73, 74, 408
Sputtering, 164
SQ-Dox, 263
SQAd nanoassembly, 266, 267
SQddC, 264
SQddl, 264, 265
SQgem, 257–263
in vitro trials, 260
in vivo trials, 261–263
magnetic, 268
on mouse leukemia, 261
on pancreatic cancer, 261
SQsiRNA, 264
Squalene, 255, 269, 411, 413
coupled to siRNA, 264
coupling with drugs, 256
cyclisation reaction, 255
isoprene units, 269
maleimide, 258, 264
with gadolinium, 268

Squalene–penicillin nanoparticle, 265
Squalenic acid, 256
Squalenoylation, 253–271
Stabilisation of nanoparticles, 154
Stakeholder report, 392
Standard development organisations (SDO), 363
Staphylococcus aureus, 266
Start-up companies, 318, 345–347
Starvation, 377
Stealth particle, 236
Steel, 195
Stem cell, 243–245
Stereochemistry, 413
Steric effect, 414
Stimulated emission, 74
STMicroelectronics, 317, 320, 324, 343, 354
Strategic action plan, 361
Stratum corneum, 204, 282
Stroke, 266
STT-RAM, 65, 66
Sublimation, 414
Subsidiarity principle, 364, 382, 401
Sulfation, 184
Sulfhydryl group, 414
Sun cream, 201, 204, 275, 282
SUPELEC, 328
Supercapacitor, 134
  description, 135
Superconductor, 414
Superhydrophobicity, 341
Superlattice, semiconductor, 414
Superlens, 77, 140
Superparamagnetic oxide, 292
Superprism, 76
Supramolecular chemistry, 141, 165, 324, 414
Surface energy, 291
Surface forces, 191
Surface treatments, 340
Surfactant, 414
  in cement, 186
  sulfonated, 186
Surgical ablation, 238
Surrogacy, 375
Suspension, 294
Sustainable development, 362, 371, 383, 390, 391, 399
Sweat glands, 282
Symbiosis, 302
Synapse, 212, 414
SynBERC, 393
Synchronisation, 400
Synchrotron X-rays, 191
Synthetic biology, 11, 367–369, 371, 372, 374–377, 379, 393, 395, 399, 400, 414
  definition, 368
Systemic injection, 234
Systemic risk, 402
Systems biology, 216, 368
Szybalski, W., 367

T

Taniguchi, N., 4
Tannic acid, 299, 308
Targeted drug delivery, 122
Targeting ligand, 230, 236, 256
Teaching package, 335, 336
Technical barriers to trade (TBT), 384
Technological node, 59
Technology platform, 251
Technology transfer, 391
Telepathy, 393
Teleportation, quantum, 73
Terahertz wave, 82
Terpene, 255, 269
Terpenoylation, 269
Tetrahydrofuran, 295
Textiles, 340
Textured substrate, 341
TGIR, 316
Theranostics, 216, 230, 233, 240–241, 254, 267, 414
Therapeutic index, 254
Therapy, 216, 220, 227, 230, 243, 246, 254, 340, 345, 382, 386
  cancer, 387
  gene, 408
  hyperthermy, 217
Thermal agitation, 6
Thermal therapy, 217
Thermoelectricity, 414
Thermoluminescence, 410
Thesaurismosis, 255, 414
Thin film, 83, 341, 372
  III–V semiconductor, 343
Thiol, 414
Thiophene, 306
Thomson CSF, 343
Thymidine, 258
Tissue engineering, 243, 244
Tissue regeneration, 374
Titanium dioxide, 12, 199, 275, 290
  anatase, 199, 201

as pigment, 199, 200
as semiconductor, 201
band gap energy, 201
biocompatibility, 200
chemical inertness, 200
doped with nitrogen, 202
effects on ecosystems, 204
in anti-UV films, 201
in ceramics, 200
in cosmetics, 200
in foodstuffs, 200
in paints, 200, 202
in photocatalysis, 202
in photovoltaics, 202
in prosthesis, 200
in self-cleaning windows, 202
in sun creams, 201
melting temperature, 200
nanoparticles, 200
particles, 200
production, 199
refractive index, 199
rutile, 199, 201
toxicity, 200, 204
TOAB, 157
Tobacco mosaic virus, 144
Toluene, 295, 306
Top–down approach, 10, 28, 114
TOTEM, 395
Touch screen, 46
Toxicity, 217, 255, 264, 347, 365, 376, 382, 384–386, 391
cardiac, 263
hematopoietic, 261
mechanisms, 300–303
of active molecules, 254
of polycations, 264
of siRNA, 264
of TiO_2 nanoparticles, 204
Toxicology, 276, 287
Traceability, 358, 365, 386, 388
of communications, 366
of individuals, 366
Tracer, 230, 234
Tracheobronchial tree, 278
Trade secrets, 382
Training, 382
cross-disciplinary, 331
programs, 325–337
Transatlantic Council, 403
Transatlantic trade and investment partnership (TTIP), 403
Transhumanism, 375, 380, 392–396
Transistor, 16
carbon nanotube, 69, 71
graphene, 69
heterojunction bipolar, 343
MOS, 26, 58
semiconductor channel, 58
single-electron, 68
ultra-miniaturised, 341
Transition metal, 154
Translational science, 322
Translocation, 373, 377, 381, 414
Transmission electron cryomicroscopy (cryoTEM), 259
Transmission electron microscopy (TEM), 49, 191
Transparency, 390
Transplantation, 242
Transport industry, 340
Triazole, 139, 414
Tribofilm, 178, 179
Tribology, 175, 181, 414
Trichloroethylene, 290
Trojan horse, 292
Trophic cascade, 287
Trophic chain, 302, 414
Trust, 386, 389
Tumour, 217, 230, 231, 236, 269, 347, 415
acidic environment, 254
active and passive targeting, 237
brain, 233
destruction, 410
early detection, 345
endocrine, 264
local hyperthermia, 239
mouse, 267
nodule, 268
photodynamic therapy, 240
physical destruction, 238
pulmonary, 281
solid, 257, 261
targeted treatment, 241
targeting, 236
TUNEL assay, 262, 414
Tunnel effect, 7, 24, 25, 104
optical, 54
Turkevich–Frens method, 156
Twin photons, 73
Tyndall effect, 336
Tyre, 118, 288, 289, 340
green, 345

U

UHVCVD, 154

Ultra-high performance concrete (UHPC), 194
Ultrafine particle (UFP), 276, 279, 284, 287, 414
  airborne, 288
Ultrahigh frequency, 343
Ultrasound spectroscopy, 191
Uncertainty principle, 412
UNESCO, 390, 399
United Nations, 371
United Nations Commission on Sustainable Development (CSD), 400
United Nations Environment Programme (UNEP), 400, 403
United Nations Framework Convention on Climate Change (UNFCCC), 391
United Nations Industrial Development Organization (UNIDO), 399, 400, 403
United States Patent and Trademark Office (USPTO), 355
Unsaturated fatty acid, 285
Upstream research, 319
  in France, 315–318
US Department of Defence (DoD), 323
US Environmental Protection Agency (EPA), 323
US Food and Drug Administration (FDA), 323
US National Institute of Health (NIH), 323
User guide, 358
USPIO, 268
UTC, 328

V

Value chain, 304, 345, 381
Van der Waals interaction, 6, 52, 154, 191, 405, 414
Vascular remanence, 258
Very small enterprises (VSE), 318
Vicat, L., 185
Virtual world, 394, 395
  persistent, 394
Virus, 2, 414
Vivagora, 366
Volcanic activity, 289
Vortex, 414

W

Wafer bonding, 36
Warwick, K., 393
Waste, 386
Water
  catchment area, 300
  contamination, 369
  drinking, 287, 390
  treatment, 293, 294, 302, 391, 399
  treatment station, 304
Wave function, 44
Waveguide, 55, 415
Weapons, 400
  of mass destruction, 377
Wear, 414
  debris, 179
Wetting, 186
Wickson, F., 381
Woodrow Wilson International Center for Scholars, 199, 274, 287
Work station, 358
Worker protection, 275
Working Party on Manufactured Nanomaterials (WPMN), 399
World Health Organisation (WHO), 385, 399, 400
World Intellectual Property Organization (WIPO), 384
World Meteorological Organization (WMO), 400
World Trade Organisation (WTO), 349, 384
Worldwide Protein Data Bank (wwPDB), 389

X

Xenograft, 415
Xonotlite, 188
X-ray imaging, 229
Xylene, 306

Y

Yang, H., 374
Young's modulus, 415

Z

Zalcitabine, 256
Zalcitabine (ddC), 264
Zebra fish, 302
Zeolite, 133
Zero point charge (ZPC), 296
Zidovudine, 256, 258
Zinc oxide, 202
Zsigmondy, R., 3

MIX
Papier aus verantwortungsvollen Quellen
Paper from responsible sources
FSC® C105338
FSC
www.fsc.org

If you have any concerns about our products,
you can contact us on
ProductSafety@springernature.com

In case Publisher is established outside the EU,
the EU authorized representative is:
Springer Nature Customer Service Center GmbH
Europaplatz 3, 69115 Heidelberg, Germany

Printed by Libri Plureos GmbH
in Hamburg, Germany